T0202544

Editorial Policy

§ 1. Lecture Notes aim to report new developments - quickly, informally, and at a high level. The texts should be reasonably self-contained and rounded off. Thus they may, and often will, present not only results of the author but also related work by other people. Furthermore, the manuscripts should provide sufficient motivation, examples and applications. This clearly distinguishes Lecture Notes manuscripts from journal articles which normally are very concise. Articles intended for a journal but too long to be accepted by most journals, usually do not have this "lecture notes" character. For similar reasons it is unusual for Ph. D. theses to be accepted for the Lecture Notes series.

§ 2. Manuscripts or plans for Lecture Notes volumes should be submitted (preferably in duplicate) either to one of the series editors or to Springer- Verlag, Heidelberg . These proposals are then refereed. A final decision concerning publication can only be made on the basis of the complete manuscript, but a preliminary decision can often be based on partial information: a fairly detailed outline describing the planned contents of each chapter, and an indication of the estimated length, a bibliography, and one or two sample chapters - or a first draft of the manuscript. The editors will try to make the preliminary decision as definite as they can on the basis of the available information.

§ 3. Final manuscripts should preferably be in English. They should contain at least 100 pages of scientific text and should include
- a table of contents;
- an informative introduction, perhaps with some historical remarks: it should be accessible to a reader not particularly familiar with the topic treated;
- a subject index: as a rule this is genuinely helpful for the reader.

Further remarks and relevant addresses at the back of this book.

Lecture Notes in Mathematics 1538

Editors:
A. Dold, Heidelberg
F. Takens, Groningen

Springer
Berlin
Heidelberg
New York
Barcelona
Budapest
Hong Kong
London
Milan
Paris
Santa Clara
Singapore
Tokyo

Paul-André Meyer

Quantum Probability
for Probabilists

Second Edition

 Springer

Author

Paul-André Meyer
Institute de Recherche Mathématique Avancée
Université Louis Pasteur
7, rue René Descartes
67048 Strasbourg-Cedex, France

Library of Congress Cataloging-in-Publication Data

Meyer, Paul André.
   Quantum probability for probabilists / Paul André Meyer. -- 2nd
ed.
      p.   cm. -- (Lecture notes in mathematics ; 1538)
   Includes bibliographical references and index.
   ISBN 3-540-60270-4 (pbk.)
   1. Probabilities. 2. Quantum theory.   I. Title. II. Series:
Lecture notes in mathematics (Springer-Verlag) ; 1538.
QA3.L28 no. 1538 1995
[QC174.17.P68]
510 s--dc20
[530.1'2'015192]   95-37123
                                                                CIP

The present volume is a corrected second edition of the previously published first edition (ISBN 0-540-56476-4).

Mathematics Subject Classification (1991):
Primary: 81S25
Secondary: 60H05, 60H10, 60J55, 60J60, 46LXX

ISBN 3-540-60270-4 Springer-Verlag Berlin Heidelberg New York

© Springer-Verlag Berlin Heidelberg 1995
Printed in Germany

Typesetting: Camera-ready TeX output by the author
SPIN: 10479489   46/3142-543210 - Printed on acid-free paper

Introduction

These notes contain all the material accumulated over six years in Strasbourg to teach "Quantum Probability" to myself and to an audience of commutative probabilists. The text, a first version of which appeared in successive volumes of the *Séminaire de Probabilités*, has been augmented and carefully rewritten, and translated into international English. Still, it remains true "Lecture Notes" material, and I have resisted suggestions to publish it as a monograph. Being a non-specialist, it is important for me to keep the moderate right to error one has in lectures. The origin of the text also explains the addition "for probabilists" in the title : though much of the material is accessible to the general public, I did not care to redefine Brownian motion or the Ito integral.

More precisely than "Quantum Probability", the main topic is "Quantum Stochastic Calculus", a field which has recently got official recognition as 81S25 in the Math. Reviews classification. I find it attractive for two reasons. First, it uses the same language as quantum physics, and *does* have some relations with it, though possibly not with the kind of fundamental physics mathematicians are fond of. Secondly, it is a domain where one's experience with classical stochastic calculus is really profitable. I use as much of the classical theory as I can in the motivations and the proofs. I have tried to prepare the reader to make his way into a literature which is often hermetic, as it uses physics language and a variety of notations. I have therefore devoted much care to comparing the different notation systems (adding possibly to the general confusion with my personal habits).

It is often enlightening to interpret standard probability in a non–commutative language, and the interaction has already produced some interesting commutative benefits, among which a better understanding of classical stochastic flows, of some parts of Wiener space analysis, and above all of Wiener chaos expansions, a difficult and puzzling topic which has been renewed through Emery's discovery of the chaotic representation property of the Azéma martingales, and Biane's similar proof for finite Markov chains.

Anyone wishing to work in this field should consult the excellent book on Quantum Probability by K.R. Parthasarathy, *An Introduction to Quantum Stochastic Calculus*, as well as the seven volumes of seminars on QP edited by L. Accardi and W. von Waldenfels. It has been impossible to avoid a large overlap with Parthasarathy's monograph, all the more so, since I have myself learnt the subject from Hudson and Parthasarathy. However, I have stressed different topics, for example multiplication formulas and Maassen's kernel approach.

Our main concern being stochastic calculus on Fock space, we could not include the independent fermion approach of Barnett, Streater and Wilde, or the abstract theory of stochastic integration with respect to general "quantum martingales" (Barnett and Wilde ; Accardi, Fagnola and Quaegebeur). This is unfair for historical reasons and unfortunate, since much of this parallel material is very attractive, and in need of systematic exposition. Other notable omissions are stopping times (Barnett and Wilde, Parthasarathy and Sinha), and the recent developments on "free noise" (Speicher). But also entire fields are absent from these notes : the functional analytic aspects of the dilation problem, non-commutative ergodic theory, and the discussion of concrete Langevin equations from quantum physics.

These notes also appear at a crucial time : in less than one year, there has been an impressive progress in the understanding of the analytic background of QP, and the non-commutative methods for the construction of stochastic processes are no longer pale copies of their probabilistic analogues. This progress has been taken into account, but only in part, as it would have been unreasonable to include a large quantity of still undigested (and unpublished) results.

A good part of my pleasure with QP I owe to the openmindedness of my colleagues, which behaved with patience and kindness towards a beginner. Let me mention with special gratitude, among many others, the names of L. Accardi, R.L. Hudson, J. Lewis, R. Streater, K.R. Parthasarathy, W. von Waldenfels. I owe also special thanks to S. Attal, P.D.F. Ion, Y.Z. Hu, R.L. Hudson, S. Paycha for their comments on the manuscript, which led to the correction of a large number of errors and obscurities.

P.A. Meyer, October 1992

Note on the second printing. A few misprints have been corrected, as well as an important omission in the statement of Dynkin's formula (Appendix 5). Above all, I have made some additions to Chapter 1 (developing more completely the point of view that a random variable is a homomorphism between algebras), to Appendix 2 (including the recent probabilistic interpretation, due to Bhat and Parthasarathy, of the Stinespring construction as the quantum analogue of Markov dependence), and I have added a new Appendix containing recent results on stochastic integration.

December 1994

Table of Contents

CHAPTER IV : FOCK SPACE (1)

Symmetric and antisymmetric tensor products, norms, bases (1). Boson and fermion Fock spaces (2). Exponential or coherent vectors (3). Creation and annihilation operators (4). Canonical commutation (CCR) and anticommutation (CAR) relations (5). Second quantization of an operator (6). Some useful formulas (7). Creation/annihilation operators on the full Fock space (8). Weyl operators (9). A computation of generators (10). Exponentials of simple operators (11).

Wiener–Ito multiple integrals (1). Extensions : square integrable martingales and the chaotic representation property (2 a)), the symmetric description (2 b)) and the "shorthand notation" (2 c)). Interpretation of exponential vectors (3) and of the operators P_t, Q_t. The number operator and the Ornstein–Uhlenbeck semigroup (4). Compensated Poisson processes as processes of operators on Fock space (5). The Ito integral in Fock space language (6).

The Wiener multiplication formula : classical form (1) and shorthand form (2). Maassen's sum/integral lemma, associativity, Wick and other products (3). Gaussian computations (4–5). The Poisson multiplication formula (6). Relation with toy Fock space (7). Grassmann and Clifford multiplications (8). Fermion brownian motion (9).

Kernels with two and three arguments (1). Examples of kernels (2). Composition of kernels (3). Kernel computation on Gaussian spaces (4). The algebra of regular kernels : Maassen's theorem (5). Fermion kernels (6). The Jordan–Wigner transformation (7). Other estimates on kernels (8).

CHAPTER V. MULTIPLE FOCK SPACES

Fock space over $L^2(\mathbb{R}_+, \mathcal{K})$, multiple integrals, shorthand notation (1). Some product formulas (2). Some non–Fock representations of the CCR (3), their Weyl operators (4), Ito table and multiplication formula (5). A remark on complex Brownian motion (6).

Exchange operators, Evans' notation (1). Kernels on multiple Fock space (2). Composition of kernels (3). Uniqueness of kernel representations (4). Operator valued kernels (5). Parthasarathy's construction of Lévy processes (6).

Imitating simple Fock space (1). Belavkin's notation (2). Another approach to kernels (2).

CHAPTER VI. STOCHASTIC CALCULUS ON FOCK SPACE

Initial space, adapted operators (1). Integration of an adapted process, informal description, simple Fock space (2–4). Computation on exponential vectors (5). Basic inequalities (6–7). Extension to multiple Fock spaces (8). The general formulas (9). Uniqueness of the representation as stochastic integrals (10). Shift operators on Fock space (11). Cocycles (12). Some examples (13).

Algebraic stochastic integration : the Skorohod integral (1). Extension to kernels : regular and smooth processes (2). Another definition of the stochastic integral (3).

Table   IX

Elementary geometric properties (1). The main operators (2–3). Interpretation of the adjoint (4). The modular property (5). Using the linear RN theorem (6). The main computation (7). The three main theorems (8). Additional results (9). Examples (10).

APPENDIX 5 : LOCAL TIMES AND FOCK SPACE

Symmetric Markov semigroups and processes (1). Dynkin's formula (2). Sketch of the Marcus–Rosen approach to the continuity of local times (3).

Notations of complex Brownian motion (1). Computing the Wiener product (2). Extension to other bilinear forms (ξ-products) (3). Stratonovich integral and trace (4). Expectation of the exponential of an element of the second chaos (5). Return to Clifford algebra : antisymmetric ξ-products (6). Exponential formula in the antisymmetric case (7). Supersymmetric Fock space : the Wick and Wiener products (8). Properties of the Wiener product (9). Applications to local times (sketch) (10). Proof of associativity of the ξ-product (11).

APPENDIX 6 : MORE ON STOCHASTIC INTEGRATION

Semimartingales of vectors and operators (1). An algebra of operator semimartingales (2).

Non-commutative regular semimartingales (1). A lifting theorem (2). Proof of the representation theorem (3–5).

Probabilistic interpretations of Fock space (1–2).

Chapter I

Non-Commutative Probability

These notes are a revised version of notes in French published in successive volumes (XX to XXII) of the *Séminaire de Probabilités*, and we will not offer again a slow introduction to this topic, with detailed justifications of the basic definitions. Non-commutative probability is the kind of probability that goes with the non-commutative world of quantum physics, and as such it is a natural domain of interest for the mathematician. We are eager to get to the heart of the subject as soon as possible. The reader that thinks we are too quick may be referred to the original French version, or (better) to the book of Parthasarathy [Par1].

§1. BASIC DEFINITIONS

1 The shortest way into quantum probability is to use *algebras*. By this word we always understand *complex algebras $\mathcal{A}$ with involution* * *and unit* I. Again we offer no apology for the role of complex numbers in a probabilistic setup. We recall that $(ab)^* = b^*a^*$, $I^* = I$, and set a few elementary definitions.

An element $a \in \mathcal{A}$ being given, we call a^* its *adjoint*, and we say that a is *hermitian* or *real* if $a = a^*$ (we avoid at the present time the word "self-adjoint"). For any $b \in \mathcal{A}$, b^*b and bb^* are real (as well as $b + b^*$ and $i(b - b^*)$).

A *probability law* or a *state* on $\mathcal{A}$ is a linear mapping ρ from $\mathcal{A}$ to $\mathbb{C}$, which is *real* in the sense that

$$\rho(a^*) = \overline{\rho(a)} \quad \text{for all } a \in \mathcal{A},$$

is *positive* in the sense that $\rho(b^*b) \geq 0$ for all $b \in \mathcal{A}$, and has *mass* 1 ($\rho(I) = 1$).

> One can easily see that positivity implies reality in *-algebras with unit : we do not try to minimize the number of axioms.

Real elements a will represent real random variables, and we call $\rho(a)$ the *expectation of a in the state ρ* — but arbitrary non-real elements do not represent complex random variables : these correspond to elements a which commute with their adjoint (*normal* elements).

Given a real element a, a^k is also real, and we put $r_k = \rho(a^k)$. We have the following easy lemma.

LEMMA. *The sequence* (r_k) *is a moment sequence.*

PROOF. Let $b = \sum_n \lambda_n a^n$, where the λ_n are arbitrary complex numbers, only a finite number of which are different from 0. Writing that $\rho(b^* b) \geq 0$, we find that

$$\sum_{m,n} \overline{\lambda_m} \lambda_n \, r_{m+n} \geq 0 \, .$$

According to Hamburger's answer to the classical moment problem on the line (see [BCR] p. 186), there exists at least one probability measure π on the line such that

$$r_k = \rho(a^k) = \int_{-\infty}^{\infty} x^k \, \pi(dx) \, .$$

If the moment problem is determined (the classical sufficient condition implying determinacy is the analyticity near 0 of $\sum_k r_k z^k / k!$), we call π *the law of the random variable* a. At the present time, this is merely a name.

2 Let us give at once an interesting example. We consider the Hilbert space $\mathcal{H} = \ell^2$ of all square summable sequences $(a_0, a_1 \ldots)$, the shift operator S on this space, and its adjoint S^*

$$S(a_0, a_1 \ldots) = (0, a_0, a_1 \ldots) \, , \quad S^*(a_0, a_1 \ldots) = (a_1, a_2 \ldots) \, ,$$

and take for $\mathcal{A}$ the complex algebra generated by S and S^*. Since $S^* S = I$ the identity need not be added. This relation implies

$$S^{*p} S^q = S^{*(p-q)} \quad \text{if } p \geq q, \quad = S^{q-p} \quad \text{if } p \leq q \, .$$

Then it is easily seen that $\mathcal{A}$ has a basis, consisting of all operators $S^p S^{*q}$. The multiplication table is left to the reader, and the involution is given by $(S^p S^{*q})^* = S^q S^{*p}$. Thus we have described $\mathcal{A}$ as an abstract *–algebra.

Let $\mathbf{1}$, the *vacuum vector*, be the sequence $(1, 0, 0 \ldots) \in \mathcal{H}$. In the language that will be ours from the following chapter on, we may call $S = A^+$ the *creation operator,* and $S^* = A^-$ the *annihilation operator.* The basis elements are then "normally ordered", *i.e.* they have the creators to the left of annihilators.

It is well known, and easy to see, that the mapping $\rho(A) = \langle \mathbf{1}, A\mathbf{1} \rangle$ is a state on the algebra of all bounded operators on ℓ^2. Its restriction to $\mathcal{A}$ is given by

$$\rho(S^p S^{*q}) = 1 \quad \text{if } p = q = 0, \quad 0 \text{ otherwise.}$$

We call ρ the *vacuum state* on $\mathcal{A}$. We ask for the law under ρ of the hermitian element $Q = (S + S^*)/2$.

Let us compute the moments $\rho((S + S^*)^n)$. Expanding this product we find a sum of monomials, which may be coded by words of length n of symbols $+$ (for S) and $-$ (for S^*). The n-th moment of $S + S^*$ is the number of such words that code the identity. Consider the effect of such a monomial on a sequence $(a_i) \in \ell^2$. Each time a $+$ occurs the sequence is moved to the right and a zero is added to its left, and for a $-$ the sequence is moved to the left, its first element being discarded — and if it was one of the original a_i's, its value is lost forever. Since at the end we must get back the original sequence, the balance of pluses and minuses must remain ≥ 0 all the time, and return to 0 at the last step. It now appears that we are just counting Bernoulli paths which return to 0 at time n after spending all their time on the positive side. The number of such paths is well known to be 0 for odd n, and $\frac{1}{k+1}\binom{2k}{k}$ for even $n = 2k$. To get the moments of Q we must divide it by 2^n

The *Arcsine law* on $[-1, 1]$, *i.e.* the projection of the uniform distribution on the unit half-circle, has similar moments, namely $2^{-2k}\binom{2k}{k}$ for $n = 2k$. To insert the factor $\frac{1}{k+1}$ we must project the uniform distribution on the half-disk instead of the half-circle. This law is called (somewhat misleadingly) "Wigner's semicircle law", and has been found (Voiculescu [Voi], Speicher [Spe]) to play the role of the Gauss law in "maximally non-commutative" situations.

3 This approach using moments is important, but it has considerable shortcomings when one needs to consider more than one "random variable" at a time.

It is a general principle in non-commutative probability that *only commuting random variables have joint distributions in arbitrary states*. For simplicity we restrict our discussion to the case of two commuting elements a, b of $\mathcal{A}$. We copy the preceding definitions and put $r_{k\ell} = \rho(a^k b^\ell)$. Then setting $c = \sum_{k\ell} \lambda_k a^k + \mu_\ell b^\ell$ the condition $\rho(c^* c) \geq 0$ again translates into a positive definiteness property of the sequence $r_{k\ell}$. However, our hope that it will lead us to a measure π (possibly not unique) on the plane such that $r_{k\ell} = \int x^k y^\ell \pi(dx, dy)$ is frustrated : *in more than one dimension, moment sequences are not characterized by the natural (or any other known) positive definiteness condition*. For details on this subject, see Berg-Christensen-Ressel [BCR] and Nussbaum [Nus]. Another interesting reference is Woronowicz [Wor].

> This does not prevent the moment method from being widely used in non-commutative situations. Whenever the sequence $(r_{k\ell})$ happens to be the moment sequence of a unique law on $\mathbb{R}^2$, it can safely be called the joint law of (a, b). The moment method has been used efficiently to investigate central limit problems.

The way out of this general difficulty consists in replacing moments by Fourier analysis. Then Bochner's theorem provides a positive definiteness condition that implies existence and uniqueness of a probability law, and extends to several dimensions. Thus we try to associate with a, or with a, b (or any number of commuting real elements of $\mathcal{A}$) functions of real variables $s, t \ldots$

$$\varphi(s) = \rho(e^{isa}) , \qquad \varphi(s, t) = e^{i(sa+tb)} \qquad \ldots$$

and to check they are positive definite in the sense of Fourier analysis. However, defining these exponentials requires summing the exponential series, while in the moment approach no limits in the algebra were involved. Assuming $\mathcal{A}$ is a Banach algebra would make things easy — but, alas, far too easy, and too restrictive.

For simplicity let us deal with one single element a. If we can define in any way $U_s = e^{isa}$, we expect these elements of $\mathcal{A}$ to satisfy the properties

$$U_s U_t = U_{s+t} , \qquad U_0 = I , \qquad (U_s)^* = U_{-s} ,$$

implying that for every s we have $U_s^* U_s = I = U_s U_s^*$. Elements of $\mathcal{A}$ satisfying this property are called *unitary*, and a family (U_s) of unitaries satisfying the first two properties is called a (one parameter) *unitary group*.

Putting $\varphi(t) = \rho(U_t)$, $c = \sum_i \lambda_i U_{t_i}$ and writing that $\rho(c^* c) \geq 0$ yields the positive type condition $\sum_{ij} \overline{\lambda}_i \lambda_j \varphi(t_j - t_i) \geq 0$. If φ is continuous, Bochner's theorem shows there is a unique probability law μ on $\mathbb{R}$ such that $\varphi(t) = \int e^{itx} \mu(dx)$. If the semigroup U_t is given as e^{ita}, we may call μ the *law of* a, but in most cases the generator a of the semigroup does not belong to the algebra itself, but to a kind of closure of $\mathcal{A}$. This is the place where the concrete interpretation of $\mathcal{A}$ as an algebra of operators becomes essential.

Von Neumann's axiomatization

4   We are going now to present the essentials of the standard language of quantum probability, as it has been used since the times of von Neumann. We refer for proofs to Parthasarathy [Par1] or to standard books on operator theory, the most convenient being possibly Reed and Simon [ReS].

> We use the physicists' convention that the hermitian scalar product $<, >$ of a complex Hilbert space is linear with respect to its *second* variable.

We start with the algebra $\mathcal{L}$ of all bounded operators on a complex (separable) Hilbert space $\mathcal{H}$. The real elements of $\mathcal{L}$ are called *bounded hermitian, symmetric or selfadjoint operators,* or we may call them (real) *bounded random variables,* and physicists call them *bounded observables.* The algebra $\mathcal{A}$ of the preceding sections may be a subalgebra of $\mathcal{L}$, but for the moment we do not insist on restricting the algebra. Complex (bounded) random variables correspond to *normal* operators in $\mathcal{L}$, *i.e.* to operators a which commute with their adjoint a^*.

Intuitively speaking, the range of values of a random variable a, whether real or complex, is the *spectrum* of a, *i.e.* the set of all $\lambda \in \mathbb{C}$ such that $a - \lambda I$ is not a 1-1 mapping of $\mathcal{H}$ onto itself. For instance, hermitian elements assume real values, unitaries take their values in the unit circle, and *projections, i.e.* selfadjoint elements p such that $p^2 = p$, assume only the values 0 and 1. Thus the projection on a closed subspace E corresponds to the indicator function of E in classical probability (the *events* of classical probability may thus be identified with the *closed subspaces* of $\mathcal{H}$). The classical boolean operations on events have their analogues here : the intersection of a family E_i of "events" is their usual intersection as subspaces, their "union" is the closed subspace generated by $\cup_i E_i$, and the "complement" of an event E is its orthogonal subspace $E^\perp$. However, these operations are of limited interest unless they are performed on events whose associated projections all commute.

Most of classical probability deals with completely additive measures. Here too, the interesting states (called *normal states*) are defined by a continuity property along monotone sequences of events. However, it is not necessary to go into abstract discussions, because all normal states ρ can be explicitly described as

$$(4.1) \qquad\qquad \rho(A) = \mathrm{Tr}(AW) = \mathrm{Tr}(WA) \qquad (A \in \mathcal{L})$$

where Tr is the trace and W is a positive operator such that $\mathrm{Tr}(W) = 1$, sometimes called the *density operator* of the state (see Appendix 1). We shall have little use for such general states, which are most important in quantum statistical mechanics. Our concrete examples of states are mostly *pure states* of the form

$$(4.2) \qquad\qquad\qquad \rho(A) = <\varphi, A\varphi>$$

where φ is a given unit vector, which is determined by ρ up to a factor of modulus 1 ("phase factor"). This corresponds in (4.1) to the case where W is the orthogonal projection on the one-dimensional subspace generated by φ. Non-pure states are often called *mixed states*.

The pure state associated with the unit vector φ is simply called "the state φ" unless some confusion can arise. The set of all states is convex, compact in a suitable topology, and the pure states are its extreme points. Hence they correspond to the point masses ε_x in classical probability theory, and are the most "deterministic" of all states. However, and contrary to the classical situation, an observable which does not admit φ as eigenvector has a non-degenerate law in the pure state φ.

Selfadjoint operators

5 The algebra $\mathcal{L}$ is a (non-separable) Banach algebra in the operator norm, but the introduction of the Hilbert space $\mathcal{H}$ allows us to use also the more useful *strong* and *weak* operator topologies on $\mathcal{L}$. In particular, the most interesting unitary groups (U_t) are the *strongly continuous* ones, which generally have unbounded generators (in fact, the generator is bounded if and only if the group is norm-continuous). Let us recall the precise definition of the generator

$$A = \frac{1}{i} \frac{d}{dt} U_t \Big|_{t=0} \ .$$

The *domain* of A consists exactly of those $f \in \mathcal{H}$ such that $\lim_{t\to 0}(U_t f - f)/t = iAf$ exists in the (norm) topology of $\mathcal{H}$. This turns out to be a *dense* subspace of $\mathcal{H}$, and A on this domain is a *closed* operator, which uniquely determines the one-parameter group (U_t). These results are rather easy, and can be deduced from the general Hille-Yosida theory for semigroups. One formally writes $U_t = e^{itA}$.

A densely defined operator A, closed or simply closable, has a densely defined and closed *adjoint* A^*, whose domain consists exactly of all $f \in \mathcal{H}$ such that $g \longmapsto \, < f, Ag >$ on $\mathcal{D}(A)$ extends to a continuous linear functional on $\mathcal{H}$, and whose value is then defined uniquely by the relation

(5.1) $$< A^* f, g > \, = \, < f, Ag > \quad \text{for } g \in \mathcal{D}(A).$$

The following result is *Stone's theorem*. It describes completely the "random variables" in the unitary group approach of subsection 3.

THEOREM 1. *An operator A is the generator of a strongly continuous unitary group if and only if it is densely defined, closed and selfadjoint, i.e. $A = A^*$, including equality of domains.*

The classical examples on $\mathcal{H} = L^2(\mathbb{R})$ are the unitary group of translations, $U_t f(x) = f(x-t)$, whose generator is id/dt, and the unitary group $U_t f(x) = e^{itx} f(x)$, whose generator is multiplication by x.

It is usually rather difficult to prove that a given operator is selfadjoint, except in the case of an everywhere defined, bounded operator A, in which case selfadjointness reduces to the *symmetry* condition $< f, Ag > \, = \, < Af, g >$. The following happy situation sometimes occurs in probability, in relation with Markov processes :

THEOREM 2. *Let* (P_t) *be a strongly continuous semigroup of positive, symmetric contractions of* $\mathcal{H}$, *and let* A *be its generator (Hille–Yosida theory). Then* $-A$ *is selfadjoint and positive.*

The proof of this result is rather easy. The classical example is $-A = -\Delta$, also called the "free Hamiltonian", which corresponds to a Brownian motion semigroup.

Criteria for selfadjointness

6 The domain of a selfadjoint operator generally cannot be given in an explicit way. The operator is usually defined as an extension of some symmetric, non-closed operator A on a dense domain $\mathcal{D}$ consisting of "nice" vectors. Symmetry easily implies that A is closable. Then two problems arise:

— Is the closure $\overline{A}$ of A selfadjoint? (If the answer is *yes*, A is said to be *essentially selfadjoint on* $\mathcal{D}$, and $\mathcal{D}$ is called a *core* for $\overline{A}$).

— If not, has $\overline{A}$ any selfadjoint extensions?

We shall deal here with the first problem only. The most important practical criterion is Nelson's theorem on *analytic vectors* ([ReS], Theorem X.39). It is implicit that $\mathcal{D}$ is dense and A is defined on $\mathcal{D}$ and symmetric.

THEOREM 3. *Assume the domain* $\mathcal{D}$ *is stable under* A, *and there exists a dense set of vectors* $f \in \mathcal{D}$ *such that the exponential series*

$$(6.1) \qquad\qquad \sum_n \frac{z^n}{n!} A^n f$$

has a non-zero radius of convergence. Then A *is essentially selfadjoint on* $\mathcal{D}$. *Besides that, whenever the series (6.1) converges for* $z = it$, *its sum gives the correct value for the unitary operator* e^{itA} *acting on* f.

Conversely, it can be shown that any strongly continuous unitary group has a dense set of *entire* vectors, *i.e.* of vectors for which the exponential series is convergent for all values of z.

The second result is the spectral criterion for selfadjointness. It is very important for quantum mechanics, but we will not use it. See [ReS], Theorem X.1.

THEOREM 4. *The operator* A *is essentially self-adjoint on* $\mathcal{D}$ *if and only if there exist two numbers whose imaginary parts are of opposite signs (in the loose sense : one real number is enough), and which are not eigenvalues of* A^*.

Spectral measures and integrals

7 Let $(E, \mathcal{E})$ be a measurable space. By a *spectral measure* on E we mean a mapping J from $\mathcal{E}$ to the set of projections on the Hilbert space $\mathcal{H}$, such that $J(\varnothing) = 0$, $J(E) = I$, and which is *completely additive* — as usual, this property splits into additivity and continuity. The first means that $J(A \cup B) = J(A) + J(B)$ if $A \cap B = \varnothing$, with the important remark that the left side is a projection by definition, and the sum of two projections is a projection if and only if they are orthogonal, and in particular commute. As for continuity, it can be expressed by the requirement that $A_n \downarrow \varnothing$ should imply $J(A_n) \to 0$ in the strong topology.

Given a real (finite valued) measurable function f on E, one can define a spectral integral $\int_E f(x) J(dx)$, which is a selfadjoint operator on $\mathcal{H}$, bounded if f is bounded. Instead of trying to develop an abstract theory, let us assume that $(E, \mathcal{E})$ is a very nice (= Lusin) measurable space[1]. It is known that all such spaces are isomorphic either to a finite set, or a countable one, or to the real line $\mathbb{R}$. Thus we may assume that E is the real line, with its Borel σ–field $\mathcal{E}$. Then to describe the spectral measure we need only know the operators $E_t = J(]-\infty, t])$, a right-continuous and increasing family of projections, such that $E_\infty = I$, $E_{-\infty} = 0$. Such a family is called a *resolution of identity*.

Denote by $\mathcal{H}_t$ the range of E_t ; then $\mathcal{H}_t$ is an increasing, right continuous family of closed spaces in $\mathcal{H}$, which plays here the role of the filtration $\mathcal{F}_t$ in martingale theory. For any $x \in \mathcal{H}$, the curve $x(t) = E_t x$ plays the role of the martingale $\mathbb{E}[x \mid \mathcal{F}_t]$: we have $x(t) \in \mathcal{H}_t$ for all t, and for $s < t$ the increment $x(t) - x(s)$ is orthogonal to $\mathcal{H}_s$. The bounded increasing function $< x >_t = \| x_t \|^2$ plays the role of the angle bracket in martingale theory, in the sense that

$$< x >_t - < x >_s = \| x(t) - x(s) \|^2 \quad \text{for} \ s < t.$$

And then we may define a "stochastic integral" $\int f(s) \, dx(s)$ under the hypothesis $\int |f(s)|^2 \, d< x >_s < \infty$. The construction is so close to (and simpler than) the Ito construction of stochastic integrals that no details are necessary for probabilists. We note the useful formula

$$< \int f(s) \, dx(s), \int g(s) \, dy(s) > = \int f(s) \, g(s) \, d< x(s), y(s) >$$

The right side has the following meaning : the bracket (ordinary scalar product in $\mathcal{H}$) is a function of s of bounded variation, w.r.t. which $f(s) \, g(s)$ is integrable. This formula corresponds in stochastic integration to the isometry property of the Ito integral.

These were *vector* integrals. To define $F = \int f(s) \, dE_s$ as an *operator* stochastic integral, we take the space of all $x \in \mathcal{H}$ such that $\int |f(s)|^2 \, d< x >_s < \infty$ as domain of F, and on this domain $F x = \int f(s) \, dx(s)$. Then we have :

THEOREM 5. *For any real (finite valued) function f on the line, the spectral integral $F = \int f(s) \, dE_s$ is a densely defined selfadjoint operator. If $|f|$ is bounded by M, so is the operator norm of F.*

Returning now to the original space E, it is not difficult to show that the spectral integral does not depend on the choice of the isomorphism between E and $\mathbb{R}$, and then the theory is extended to (almost) arbitrary state spaces. Spectral measures on E are the non-commutative analogues of classical random variables taking values in E.

For instance, let Z be a classical r.v. on a probability space $(\Omega, \mathcal{F}, \mathbb{P})$, taking values in E. We may associate with Z a spectral measure J on $\mathcal{H} = L^2(\Omega)$, such that for $A \in \mathcal{E}$

$$J(A) \quad \text{is the operator of multiplication by} \ I_{Z^{-1}(A)},$$

obviously a projection. Given a real valued r.v. f on E, the spectral integral $\int f(x) J(dx)$ is multiplication by $f \circ Z$.

---

[1]   All complete separable metric spaces are Lusin spaces. See Chapter III of [DeM1].

Given a state ρ on $\mathcal{H}$ and a spectral measure J, the mapping $A \longmapsto \rho(J(A))$ is an ordinary probability measure π on E, called the *law of J in the state ρ*. Spectral measures behave much like ordinary random variables. For instance given a measurable mapping $h : E \to \Gamma'$, we define a spectral measure $h \circ J$ on F as $B \longmapsto J(h^{-1}(B))$, whose law in the state ρ is the direct image $h(\pi)$ in the classical sense.

In order to construct joint laws of random variables in this sense, the following result is essential : given two (Lusin) measurable spaces E_1, E_2 and two *commuting* spectral measures J_1, J_2, there is a unique spectral measure J on $E_1 \times E_2$ such that $J(A \times E_2) = J_1(A)$ for $A \in \mathcal{E}_1$ and $J(E_1 \times A) = J_2(A)$ for $A \in \mathcal{E}_2$. This is closely related to the *bimeasures theorem* of classical probability (see [DeM1], III.74). The theorem also applies to an infinite number of spectral measures, as a substitute of Kolmogorov's theorem on the construction of stochastic processes.

Spectral representation of selfadjoint operators

8   We have mentioned two interpretations of real valued random variables : as self adjoint operators and as spectral measures on $\mathbb{R}$, the first one being more natural in physics, and the second closer to probabilistic intuition. Here is von Neumann's theorem, which draws a bridge between them.

THEOREM 6. *For every selfadjoint operator A, there exists a unique resolution of identity (E_t) on the line, such that*

$$A = \int s \, dE_s \ .$$

Then, given any (real or complex) Borel function f on the line, the spectral integral $\int f(s) \, dE_s$ is denoted by $f(A)$.

An example of such integrals in the complex case is the unitary group

$$U_t = e^{itA} = \int e^{its} \, dE_s \ .$$

Then the formula

$$<\varphi, U_t\varphi> = \int e^{its} d<\varphi, E_s\varphi>$$

shows there is no contradiction between the two definitions of the law of A in the pure state φ, by means of unitary groups and spectral measures.

Given any two real valued Borel functions f, g, $g(A)$ is a selfadjoint operator, and we have $(f(g(A)) = (fg)(A)$. For a complex function f, the adjoint of $f(A)$ is $\bar{f}(A)$.

WARNING. The issue of defining the joint law of several commuting operators, which has plagued the moment approach, has not become a trivial one. Except in the bounded case, it is not easy to decide whether the spectral measures associated with several selfadjoint operators do commute. In particular, *it is not sufficient to check that they commute on a dense common stable domain*. The best general result seems to be an extension of the analytic vectors theorem to the Lie algebras setup, also due to Nelson. However, the commutation of the associated unitary groups is a (necessary and) sufficient condition, and in all practical cases in these notes the unitary groups will be explicitly computable.

Elementary quantum mechanics

9 Let us show how the above language applies to the elementary interpretation of quantum mechanics, in the case of a selfadjoint operator with discrete spectrum.

Physicists like the *Dirac notation*, in which a vector $x \in \Omega$ is denoted by $|x>$ (*ket* vector), the linear functional $<y, \cdot>$ on Ω is denoted by $<y|$ (*bra* vector), and the value of the linear functional (bra) on the vector (ket) is the *bracket* $<y|x>$ or $<y, x>$. However, this explanation does not convey the peculiar kind of intuition carried by Dirac's notation, which uses bras and kets "labelled" in a concrete way, and interprets the scalar product $<y|x>$ between labelled vectors as a "transition amplitude" from the "state" labelled x, to the state labelled y, whose squared modulus is a "transition probability". We shall see later examples of such a "labelling" (see the first pages of Chapter II). Taking an orthonormal basis (e_k), and expanding the vector f as $\sum_k f^k e_k$, we may also consider the ket vector $|f>$ as a column matrix (f^k), the bra vector $<g|$ as a row matrix $g^* = (g_k)$ with $g_k = \sum_j \bar{g}^i \delta_{ik}$, and the bracket $<g, f>$ is the usual matrix product $g^* f$. In this context the linearity of the scalar product w.r.t. its second variable is very natural. A slightly unusual feature of this notation is its tendency to write scalars on the right, as in the following formula for the representation of a vector f in the o.n. basis (e_k)

$$|f> = \sum_k |e_k> <e_k|f> \quad \text{or} \quad I = \sum_k |e_k> <e_k|.$$

Let A be a selfadjoint operator. If the basis vectors e_k belong to the domain of A, the brackets $<e_k, Ae_j>$ (Dirac : $<e_k|A|e_j>$) are called the *matrix elements* of A. Assume now that A has discrete spectrum, every e_k being an eigenvector of A with eigenvalue λ_k. Then the unitary group associated with A is

$$U_t = \sum_k e^{it\lambda_k} |e_k> <e_k|$$

Let ρ be the pure state associated with φ, a unit vector. Then we have, putting $<e_k, \varphi> = \varphi_k$

$$<\varphi, U_t \varphi> = \sum_k e^{it\lambda_k} |\varphi_k|^2 .$$

Inverting the Fourier transform, we identify the law of A in the state ρ as the discrete law $\sum_k |\varphi_k|^2 \varepsilon_{\lambda_k}$, in conformity with the rules of elementary quantum mechanics. On the other hand, the spectral subspace E_t for A is the closed subspace generated by all e_k such that $\lambda_k \leq t$.

About quantization

10 The following comments are not of a mathematical nature, but provide some hints on what will be done in these notes.

Each time we want to provide a quantum analogue to some given classical probabilistic situation $(\Omega, \mathcal{F}, \mathbb{P})$, be it simple random walk, Brownian motion, or Poisson processes, we take as Hilbert space $\mathcal{H}$ the space $L^2(\Omega)$, and interpret the classical random variables as multiplication operators on this space. The classical probability law also provides a natural state on $\mathcal{H}$, the pure state associated with the function 1. Then the interest of the quantum probabilistic situation depends on finding natural sets of random variables that do not commute with the classical ones — they often occur as generators of groups of symmetries operating on the space. These two non-commuting sets then are "mixed together" to produce new observables or sets of observables, whose probabilistic structure is investigated. Non-commutative probability has the same purpose as classical probability, namely to compute laws of random variables and stochastic processes. The difference lies in a richer space of random variables, and the fact that seemingly innocuous operations, like addition of non-commuting r.v.'s, sometimes produce unexpected results.

We are going to investigate successively the analogues of Bernoulli space and the finite random walk (Chapter II), of finite dimensional Gaussian random variables (Chapter III), of Brownian motion and Poisson processes (Chapter IV). Then we shall study the simplest non-commutative extension of the classical Ito Calculus.

Back to algebras [1]

11 Let us now return to the general problem of defining a random variable in quantum probability, trying to get a broader view. We started from the simple idea that a r.v. is a hermitian element of some algebra $\mathcal{A}$ (with unit and involution). Then we described the traditional approach of von Neumann, which specifies $\mathcal{A}$ to be $\mathcal{L}(\mathcal{H})$, the algebra of all bounded operators on some separable Hilbert space, and defines a r.v. to be a spectral measure on $\mathcal{H}$ over $\mathbb{R}$, and more generally a r.v. taking values in a measurable space $(E, \mathcal{E})$ to be a spectral measure on $\mathcal{H}$ over E. Neither of these two point of views is the modern definition. Roughly speaking, the modern idea of a r.v. contains both an arbitrary algebra $\mathcal{A}$ and a Hilbert space $\mathcal{H}$, as follows :

A r.v. X is a homomorphism from some algebra $\mathcal{A}$ into $\mathcal{L}(\mathcal{H})$ (preserving the involution and the identity [2]), and possibly satisfying some additional continuity property.

This new point of view on random variables and stochastic processes was introduced in quantum probability by an influential paper of Accardi, Frigerio and Lewis. In these notes it is used only in the section on Evans–Hudson flows.

First of all, let us consider commutative situations. Let X be a classical random variable defined on a probability space $(\Omega, \mathcal{F}, \mathbb{P})$ and taking values in some measurable space $(E, \mathcal{E})$. Then we make take for $\mathcal{A}$ the algebra of bounded measurable functions f on E, for $\mathcal{H}$ the Hilbert space $L^2(\Omega)$, and then mapping $f \in \mathcal{A}$ to the multiplication operator by $f \circ X$ defines a homomorphism from $\mathcal{A}$ to $\mathcal{L}(\mathcal{H})$. Our new definition amounts to saying that this mapping — which uniquely determines X — is the random variable. Note also that this mapping is continuous w.r.t. monotone convergence (if a bounded sequence f_n increases to f, then $f_n \circ X$ converges to $f \circ X$ in the strong convergence of operators.)

Next, consider a spectral measure X on a Hilbert space $\mathcal{H}$ over some measurable space $(E, \mathcal{E})$. Then the spectral integral

$$f \longmapsto X(f) = \int_E f(s)\, dX(s)$$

---

[1] This subsection is an addition, and may be omitted.

[2] Sometimes identity preservation is omitted, but never without a warning.

defines a homomorphism from the algebra $\mathcal{A}$ of bounded measurable functions into $\mathcal{L}(\mathcal{H})$. If E is a compact metric space, we may restrict $\mathcal{A}$ to be the smaller algebra $\mathcal{C}(E)$ — knowing this restriction uniquely determines X, since we recover all measurable functions by monotone convergence. But restricting the algebra in this way leads to an interesting converse :

THEOREM. *If E is compact metric[1], any homomorphism X from $\mathcal{C}(E)$ to $\mathcal{L}(\mathcal{H})$ is associated as above with a spectral measure.*

PROOF (sketch). For $u \in \mathcal{H}$, the mapping

$$f \longmapsto\ <u,\, X(f)\, u>$$

is a linear functional μ_u on $\mathcal{C}(E)$ such that $\mu_u(f^* f) \geq 0$, hence a positive measure on E. It follows by polarization that for $u, v \in \mathcal{H}$, $f \longmapsto\ <v,\, X(f)\, u>$ is a complex measure $\mu_{u,v}$. Then it is easy to extend the mapping $X(f)$ to Borel bounded functions on E in such a way that $<u,\, X(f)\, u> = \int f(x)\, \mu_{u,v}(dx)$. Then taking for f the indicator function I_A of a Borel set, we define a spectral measure, and it easy to check that X is the corresponding spectral integral.

Let us turn now to the non-commutative case : there are two interesting situations, that of a C^* algebra $\mathcal{A}$, which is the non-commutative analogue of $\mathcal{C}(E)$ for a "non-commutative compact space E" (which however has no independent existence : it is entirely subsumed in its algebra), and that of a von Neumann algebra $\mathcal{A}$, which is the non-commutative analogue of a "non-commutative space $L^\infty(\lambda)$". In the latter case, the homomorphism X should also satisfy a continuity property called *normality* (we will not discuss it here).

> The case of an arbitrary hermitian element a in an algebra, which was our starting point in this chapter, can also be translated using the language of homomorphisms, but we leave it aside.

COMMENT. As the preceding theorem concerns only *compact* spaces, it doesn't apply to the most common example, that of real valued random variables! The obvious way out of this difficulty consists in using the compact space $\overline{\mathbb{R}}$ instead of $\mathbb{R}$, that is, in allowing the values $\pm\infty$ in our definition of random variables. This requires only trivial modifications, except that only *finite* random variables correspond to self-adjoint operators.

The additional introductory material has been given in appendices at the end of the volume, so that our reader is not tempted to read it at once, but goes directly to the heart of the matter. Chapter II in particular contains non-trivial finite dimensional examples.

---

[1] We assume metrizability only for definiteness.

Chapter II : Spin

This chapter has little to do with the geometric property of elementary particles called *spin*. Historically spin 1/2 provided the first basic example of a "two-level quantum system", so that till now those simplest of all quantum systems are introduced to students of physics with the help of "fictitious spins". The truth is, that two-level systems (or spins) play in quantum probability the role of *Bernoulli random variables*. As in the classical case, they may be considered the cornerstone of probability theory.

The chapter is organized as follows : we describe the elementary (Bernoulli) case, and study finite systems of such "quantum Bernoulli" random variables, commutative in section 2, and anticommutative in section 5. We prove in §3 a quantum version of the most elementary central limit theorem of probability theory, namely the de Moivre-Laplace theorem; it slightly anticipates Chapter III. Section 4 studies "spin n" (angular momentum), one of the possible quantum generalizations of a random variable assuming $n > 2$ values, with an application to Biane's example of a commutative random walk arising naturally in a non-commutative setup.

§1. TWO–LEVEL QUANTUM SYSTEMS

Two-dimensional Hilbert space

1   The simplest non-trivial Hilbert space $\mathcal{H}$ has dimension 2. Let us denote by (e_0, e_1) an orthonormal basis. As usual we identify these basis elements with column matrices

$$e_0 = \begin{pmatrix} 1 \\ 0 \end{pmatrix} \quad ; \quad e_1 = \begin{pmatrix} 0 \\ 1 \end{pmatrix}.$$

These two vectors are usually "labelled" as follows (the general idea being that the first vector is somehow "below" the second one)

$$e_0 = |\,\text{down}\!> \quad ; \quad e_1 = |\,\text{up}\!>$$

(for a spin), or

$$e_0 = |\,\text{empty}\!> \quad ; \quad e_1 = |\,\text{full}\!>$$

(for an electronic layer), or

$$e_0 = |\,\text{fundamental}\!> \quad ; \quad e_1 = |\,\text{excited}\!>$$

(for a two-level atom), etc.

Of course there is nothing mathematical about "up" and "down" and the basis vectors can be labelled as well $|slit1>$ and $|slit2>$ in the well known two-slit experiment. Another useful interpretation of the Hilbert space $\mathcal{H}$ is the following. Consider a (classical) symmetric Bernoulli random variable x on some probability space (for instance the line with the symmetric Bernoulli measure carried by $\{-1,1\}$, x being the identity mapping). Then we may identify e_0 with the r.v. 1, and e_1 with x.

2 In the two-dimensional case, it is possible to visualize all states and all observables (random variables). Let us spend a little time doing this, though we do not need it for the sequel.

A normalized vector $\omega = u e_0 + v e_1$ can be considered as an element of the 3–sphere $S_3 = \{|u|^2 + |v|^2 = 1\}$. To describe the pure states, *i.e.* the normalized vectors up to a phase factor, we put

$$|u| = \cos\frac{\theta}{2} \ , \ |v| = \sin\frac{\theta}{2} \qquad (0 \le \theta \le \pi) \ .$$

Then $u = |u|e^{ia}$, $v = |v|e^{ib}$ and we may add the same real number to a and b without changing the state. We fix this number by taking $a + b = 0$. Thus we don't lose pure states if we take

$$(2.1) \qquad \omega = \omega_{\theta,\varphi} = \cos\frac{\theta}{2} e^{-i\varphi/2} e_1 + \sin\frac{\theta}{2} e^{i\varphi/2} e_2$$

with $0 \le \theta \le \pi$, $0 \le \varphi < 2\pi$. It is easy to check that in this way, taking θ to be the latitude and φ to be the longitude, we define a 1–1 continuous mapping of the 2–sphere S_2 on the set of all pure states (then it is natural to expect the set of all states to be described by the convex hull of the sphere, *i.e.* the ball; we shall see this later).

Let us return to the probabilistic interpretation using the Bernoulli r.v. x. As explained in Chapter I, it may be interpreted in the operator sense as multiplication by x. Since $x \cdot 1 = 1 \cdot x = x$, $x^2 = 1$, the matrix representing this observable in the basis $(1, x)$ is $\begin{pmatrix} 0 & 1 \\ 1 & 0 \end{pmatrix}$, commonly denoted by σ_x or σ_1.

The most general s.a. operator on $\mathcal{H}$ is given by a matrix

$$(2.2) \qquad A = \begin{pmatrix} t+z & x-iy \\ x+iy & t-z \end{pmatrix}$$

with real coefficients x, y, z, t. It can be written as $tI + x\sigma_x + y\sigma_y + z\sigma_z$, a linear combination of I and the three *Pauli matrices* given by

$$(2.3) \qquad \sigma_x = \begin{pmatrix} 0 & 1 \\ 1 & 0 \end{pmatrix} \ ; \ \sigma_y = \begin{pmatrix} 0 & -i \\ i & 0 \end{pmatrix} \ ; \ \sigma_z = \begin{pmatrix} 1 & 0 \\ 0 & -1 \end{pmatrix}$$

(also denoted $\sigma_1, \sigma_2, \sigma_3$). Among the s.a. operators, the density matrices of states D are characterized by the property of being positive and having unit trace. This implies $t = 1/2$, so it is more convenient to set (z being real and $r = x + iy$ complex, contrary to all good manners)

$$(2.4) \qquad D = \frac{1}{2}\begin{pmatrix} 1+z & \bar{r} \\ r & 1-z \end{pmatrix} \ .$$

The positivity property $< \binom{u}{v}, D \binom{u}{v} > \geq 0$ amounts to $\bar{r}r \leq 1 - z^2$ or $x^2 + y^2 + z^2 \leq 1$. This parametrizes the set of all states, in an affine way, by the points of the unit ball of $\mathbb{R}^3$, and of course the extreme (= pure) states by points of the unit sphere. We leave it to the reader to check that indeed, if we take in (2.4)

$$(2.5) \qquad x = \sin\theta\cos\varphi \quad ; \quad y = \sin\theta\sin\varphi \quad ; \quad z = \cos\theta$$

then the density matrix

$$(2.6) \qquad D_{\theta,\varphi} = \frac{1}{2}(I + \sigma_{\theta,\varphi}) \quad ; \quad \sigma_{\theta,\varphi} = \begin{pmatrix} \cos\theta & \sin\theta e^{-i\varphi} \\ \sin\theta e^{i\varphi} & -\cos\theta \end{pmatrix}$$

represents the projection on the one dimensional subspace generated by the vector given by (2.1). The matrices $\sigma_{\theta,\varphi}$ represent all possible "Bernoulli r.v.'s", *i.e.* all observables admitting the eigenvalues ± 1.

In the concrete interpretation of a spin-$1/2$ particle, $\sigma_{\theta,\varphi}$ describes the outcome of a spin measurement in $\mathbb{R}^3$ (in units of $\hbar$) along the unit vector (2.5). Assume the system is found in the pure state $w = e_1$ (in the representation (1.1) this corresponds to $\theta = 0$ and φ arbitrary). Then a spin measurement along Oz is described by σ_z and gives the probability 1 for "up" and 0 for "down", an apparently deterministic situation, physically realizable by means of a filter along Oz which lets through only objects with spin up. In classical probability, this would be the end of the matter. Here, a measurement along (2.5) will give a "probability amplitude" $< w, \sigma_{\theta,\varphi} w > = \cos\theta$ of finding the spin along the axis, corresponding to a probability $\cos^2\theta$, and there is a probability $\sin^2\theta$ to find it in the reverse direction. Thus we see at work the new kind of indeterminism due to non-commutativity.

Creation and annihilation

3 We are going to introduce new sets of matrices, which will play a fundamental role in the sequel.

We define the *vacuum vector* as the first basis vector e_0. It corresponds to the function 1 in the standard probabilistic interpretation using the Bernoulli measure, and we therefore denote it by **1** (a notation general among physicists is $|0>$). The second basis vector e_1 was identified, in the probabilistic interpretation, with the Bernoulli r.v. x. Next we define the three matrices

$$(3.1) \qquad b^- = \begin{pmatrix} 0 & 1 \\ 0 & 0 \end{pmatrix} \quad ; \quad b^+ = \begin{pmatrix} 0 & 0 \\ 1 & 0 \end{pmatrix} \quad ; \quad b^\circ = \begin{pmatrix} 0 & 0 \\ 0 & 1 \end{pmatrix} .$$

called respectively the *annihilation, creation and number operators* (in the beautiful Hindu interpretation of Parthasarathy, the last one is the *conservation operator*). These are not the standard notations (physicist use rather b^*, b, n). The (number and) annihilation operator kill the vacuum vector **1**, while the creation operator transforms it into the occupied state (it "creates a particle"). On the other hand, the creation operator transforms into 0 the occupied state (a state cannot be doubly occupied : this model obeys *Pauli's exclusion principle*). These unsociable particles are toy models for real particles of the *fermion* family. Note that b^+ and b^- are mutually adjoint, and $b^0 = b^+b^-$ is selfadjoint. Note also the relation

(3.2) $\{b^-, b^+\} = b^-b^+ + b^+b^- = I$

(3.3) $[b^-, b^+] = b^-b^+ - b^+b^- = I - 2b^0$.

Relation (3.2) is the simplest example of the so called *canonical anticommutation relations* (CAR); (3.3) has no special name.

The matrices I and b^ε ($\varepsilon = -, \circ, +$) form a basis for all (2,2) matrices. In particular, the Pauli matrices are represented as follows

(3.4) $b^+ + b^- = \sigma_x$; $i(b^+ - b^-) = \sigma_y$; $[b^-, b^+] = \sigma_z = I - 2b^0$.

4 It is easy to give examples of unitary groups in this situation. In fact, for complex $z = x + iy = re^{i\varphi}$, the operator $zb^+ - \bar{z}b^-$ is $-i$ times the s.a. operator $x\sigma_y - y\sigma_x$, hence its exponential is unitary. Let us call it a *discrete Weyl operator*

(4.1) $W(z) = \exp(zb^+ - \bar{z}b^-) = \begin{pmatrix} \cos r & i\sin re^{-i\varphi} \\ i\sin re^{i\varphi} & \cos r \end{pmatrix}$.

For every z, the family $W(tz)$, $(t \in \mathbb{R})$ is a unitary group. On the other hand, let us call *discrete coherent vectors* the vectors

$$\psi(z) = W(z)\mathbf{1} = \begin{pmatrix} \cos r \\ \sin re^{-i\varphi} \end{pmatrix} .$$

Taking r in the interval $[0, \pi]$ and φ in the interval $[0, 2\pi[$ this realizes a parametrization of unit vectors modulo a phase factor, slightly different from that in (2.1).

§2. COMMUTING SPIN SYSTEMS

In this section, the L^2 space over classical Bernoulli space becomes a finite dimensional model ("toy Fock space") which approximates boson Fock space (the L^2 space over Wiener space). We underline the analogies between the role played by the Bernoulli process for vectors, and that played for operators by the non-commutative triple of the creation/number/annihilation processes. Later, this analogy will be carried over to continuous time.

Commuting spins and toy Fock space

1 Let ν_i, $i = 1, \dots, N$ be a finite family of independent, identically distributed Bernoulli random variables, each one taking the value 1 with probability p, the value 0 with probability $q = 1 - p$. As usual, we define

(1.1) $x_i = \dfrac{\nu_i - p}{\sqrt{pq}}$

to get r.v.'s with zero expectation and unit variance. It is convenient to consider the r.v.'s x_i as the coordinate mappings on the product space $E = \{-1, 1\}^N$, on which the measure is denoted by μ_p (by μ if $p = \frac{1}{2}$, the most usual situation).

We denote by $\mathcal{P}_N$ or simply $\mathcal{P}$ the set of all subsets of $\{1, \ldots, N\}$, and for $A \in \mathcal{P}$ we put

$$(1.2) \qquad x_A = \prod_{i \in A} x_i \qquad (x_\emptyset = 1)$$

These r.v.'s constitute an orthonormal basis of the Hilbert space $\Psi = L^2(\mu_p)$. When $p = \frac{1}{2}$, it is interesting to give an interpretation of this basis as follows : E is a finite group under multiplication, the Haar measure and the characters of which are exactly the measure μ and the mappings x_A for $A \in \mathcal{P}$ (the dual group of E is isomorphic with $\mathcal{P}$ equipped with the symmetric difference operation Δ). Then the expansion in the basis (x_A) is simply a finite Fourier–Plancherel representation.

Let Ψ_n be the subspace of $L^2(\mu_p)$ generated by all x_A such that $|A| = n$, $|A|$ denoting here the number of elements of A. The space Ψ_0 is one dimensional, generated by $x_\emptyset = 1$. It is convenient to give names to these objects : we call Ψ_n the n-th *(discrete) chaos*, and $x_\emptyset$ the *vacuum vector*, also denoted by **1**. Note that all these Hilbert spaces are naturally isomorphic, no matter the value of p. Given the chaotic expansion of a random variable $Y = \sum_A c_A x_A$, we may compute the scalar product of Y with any other r.v. defined in the same way, and in particular the expectation of Y, equal to $<1, Y>$. What makes the distinction between the different values of p is not the Hilbert space structure, but the *multiplication of random variables*. Taking it for granted that the product is associative, commutative, and that **1** acts as unit element, it is entirely determined by the formula giving x_i^2

$$(1.3) \qquad x_i^2 = 1 + c x_i \quad \text{with} \quad c = \frac{1 - 2p}{\sqrt{pq}}$$

In particular if $p = \frac{1}{2}$, this reduces to $x_i^2 = 1$ and we have the explicit formula

$$(1.4) \qquad x_A x_B = x_{A \Delta B}$$

where Δ is the symmetric difference operation. From the above almost trivial example, one may get the feeling that different probabilistic interpretations can be described on the same Hilbert space by different multiplicative structures. This idea will not be given a precise form, but still it pervades most of these notes.

Operators on toy Fock space

2   We are going to define for each $k \in \{1, \ldots, N\}$ *creation and annihilation operators* a_k^+ and a_k^-. We put

$$a_k^+(x_A) = x_{A \cup \{k\}} \quad \text{if} \quad k \notin A \,, \; = 0 \;\; \text{otherwise} \qquad \text{(creation)}$$

$$a_k^-(x_A) = x_{A \setminus \{k\}} \quad \text{if} \quad k \in A \,, \; = 0 \;\; \text{otherwise} \qquad \text{(annihilation)}$$

It is easy to check that $<x_B, a_k^- x_A> = <a_k^+ x_B, x_A>$, *i.e.* these operators are mutually adjoint. From a physical point of view, they are creation and annihilation operators "at the site k" for a distribution of spins located at the "sites" $1, \ldots, N$. Let us introduce also the corresponding number operators $a_k^0 = n_k = a_k^+ a_k^-$, so that

$$a_k^0 x_A = x_A \quad \text{if} \quad k \in A, \quad = 0 \quad \text{otherwise}.$$

Denoting by greek letters $\varepsilon, \eta, \ldots$ symbols taken from the set $\{-, 0, +\}$, we see that operators a_j^ε and a_k^η relative to different sites always commute, while at one given site k we have

(2.1) $$[a_k^-, a_k^+] = I - 2a_k^0 .$$

These relations may be considered in a very precise way (which will be described later on) as discrete analogues of the CCR. This is why we shall call *(symmetric) toy Fock space* the finite model we have been describing. The idea of considering seriously toy Fock space as a discrete approximation of symmetric Fock space (to be described in Chapter IV) is due to J.-L. Journé, under the French name of *bébé Fock*. The formal analogies between the discrete and continuous situations have been noticed much earlier (see for instance the paper by Arecchi and *al.* [ACGT], and other papers mentioned in the book [KlS]).

At each site, we consider the self-adjoint operators

(2.2) $$q_k = a_k^+ + a_k^- , \quad p_k = i(a_k^+ - a_k^-) .$$

The way q_k operates is easily made explicit :

$$q_k x_A = x_{A \Delta \{k\}} \quad ;$$

comparing this with (1.4), we see that q_k *is the operator of multiplication by x_k in the symmetric Bernoulli case*. The operator of multiplication by x_k in the non-symmetric cases turns out to be equal to $q_k + c n_k$, the constant c being the same as in (1.3).

What about p_k ? Let us denote by $\mathcal{F}$ the unitary operator on toy Fock space Ψ defined b $\mathcal{F} x_A = i^{|A|} x_A$. This operator preserves $\mathbf{1}$, and we have

(2.3) $$\mathcal{F}^{-1} a_k^+ \mathcal{F} = -i a_k^+ \quad ; \quad \mathcal{F}^{-1} a_k^- \mathcal{F} = i a_k^- \quad ; \quad \mathcal{F}^{-1} a_k^0 \mathcal{F} = a_k^0$$

(2.4) $$\mathcal{F}^{-1} q_k \mathcal{F} = -p_k \quad ; \quad \mathcal{F}^{-1} p_k \mathcal{F} = q_k .$$

The letter $\mathcal{F}$ is meant to recall a Fourier transform, though it should not be confused with the Fourier transform for the group structure mentioned above.

3   As we have expanded all vectors in E in the basis (x_A), we look for a similar expansion of all *operators*. To this end, we put $a_A^\varepsilon = \prod_{i \in A} a_i^\varepsilon$. for $\varepsilon = -, 0, +$ and $A \in \mathcal{P}$. Then we have

(3.1) $$a_A^+ x_B = x_{A+B} \quad , \quad a_A^- x_B = x_{B-A}$$

where x_{A+B} (x_{B-A}) means $x_{A \cup B}$ ($x_{B \setminus A}$) if A, B are disjoint (if $A \subset B$) and 0 otherwise. Similarly, $a_A^0 x_B = x_B$ if $A \subset B$ and 0 otherwise.

Let $\mathcal{A}$ be the algebra generated by the creation and annihilation operators (note that it contains the number operators too). Every non-zero vector $x \in \Psi$ has some non-zero coefficient c_A in its chaotic expansion; choosing $|A|$ maximal, x is transformed into $c_A \mathbf{1}$ by the pure annihilator a_A^-. On the other hand, $\mathbf{1}$ can be transformed into x_B by the pure creator a_B^+, and hence can be transformed into any non-zero vector by a linear combination of creators. Consequently, there is no subspace of E invariant under the algebra $\mathcal{A}$ except the trivial ones $\{0\}$ and E, and a simple theorem in algebra implies that $\mathcal{A}$ contains all operators on the finite dimensional space E[1]. Using the commutation properties of the operators $a_k^{\pm}$, it is easy to show that $\mathcal{A}$ consists of all linear combinations of products $a_A^+ a_B^-$, *i.e.* products for which creators stand to the left of annihilators (such products are said to be *normally ordered*). On the other hand, the number of all such products is $(2^N)^2$, which is the dimension of the space of all $(2^N, 2^N)$ matrices, *i e.* of all operators on E. Otherwise stated, we have constructed a basis for this last space.

There is another interesting basis : in a product $a_A^+ a_B^-$, we may use the commutation at different sites to bring successively into contact all pairs of operators a_k^+ and a_k^-, where k belongs to the intersection $A \cap B$. Having done this, we see that our algebra is also generated by products $a_U^+ a_V^\circ a_W^-$, where now U, V, W constitute a *disjoint triple*. Counting these operators, we see again that their number is 2^{2N}, hence they form a basis too, though it is less familiar to physicists than the preceding one.

Adapted operators

4 The preceding description of all operators did not depend on the order structure of the set $\{1, \ldots, N\}$. We are going to use this structure to describe operators as "stochastic integrals", as random variables are in classical probability.

Let us return to the probabilistic interpretation. The Bernoulli random variables $x_1, \ldots, x_N$ are independent. They generate σ–fields, for which we introduce some notation : $\mathcal{F}_{\leq k}$ is the *past* σ–field, generated by $x_1, \ldots, x_k$, and one may consider also the *strict past* σ–field $\mathcal{F}_{k-1} = \mathcal{F}_{<k}$ (trivial for $k = 1$). There are also *future* and *strict future* σ–fields, denoted respectively by $\mathcal{F}_{\geq k}$ and $\mathcal{F}_{>k}$. Since the past σ–fields occur more often than the future ones, and heavy notation is not convenient, we write $\mathcal{F}_k$ and $\mathcal{F}_{k-}$ for the past and strict past, whenever the context excludes any ambiguity. For the corresponding L^2 spaces we use similarly the notations $\Phi, \Phi_k, \Phi_{>k}$

We denote by E_k the orthogonal projection on the past Φ_k (from the probabilistic point of view, this is a conditional expectation).

Independence implies that Φ is isomorphic to the tensor product $\Phi_{<k} \otimes \Phi_{>k}$, the operation $\otimes$ being read as the ordinary product of random variables, restricted to pairs of r.v.'s which depend only on pre–k and post–k co-ordinates. Why restrict it at all? Some attention should be given to this point, which will be important in continuous time. Toy Fock space is a quantum probabilistic object, which admits several classical

---

[1] This may be considered as an elementary case of von Neumann's bicommutant theorem.

probabilistic interpretations, and the full product of random variables *depends on the interpretation* (in a way, it even *is* the interpretation itself!). However, for pairs of r.v.'s depending on different sets of coordinates as above, it reflects the algebraic operation $\otimes$ and has an intrinsic meaning.

Traditionally, a r.v. measurable w.r. to $\mathcal{F}_k$ ($\mathcal{F}_{k-}$) is said to be *k-adapted* (*k-previsible*) and families of r.v.'s indexed by time which are adapted (previsible) at each time are called *adapted (previsible) processes* — the word "predictable" is commonly used too. We are going to extend these notions to operators on Φ.

An operator A on Φ is said to be *k-adapted* if, in the tensor product decomposition $\Phi_{\leq k} \otimes \Phi_{>k}$, it is read as the tensor product $B \otimes I$ of an operator on the first factor and the identity on the second one. Otherwise stated, if f and g are functions depending respectively on the first k and the last $N - k$ co-ordinates, then 1) Af still depends only on the first k coordinates, 2) $A(fg) = (Af)g$. For example, the multiplication operator by a random variable h satisfies 2), and it satisfies 1) iff h is $\mathcal{F}_k$-measurable. The conditional expectation operator E_k satisfies 1), but not 2). The creation, counting (number) and annihilation operators a_A^ε are k-adapted iff $A \subset \{1, \ldots, k\}$.

How can we transform an operator A into a k-adapted operator? It is very natural to define (with the same notations as above)

$$A_k(fg) = (E_k(Af))g$$

or in more algebraic notation $A_k = (E_k A E_k) \otimes I_{>k}$. Then it is easy to see that A_k plays the role of a conditional expectation of the operator A *in the vacuum state*. This means that given any two k-adapted operators G, H we have

(4.1) $<1, G^* A H 1> = <1, G^* A_k H 1>$.

This is a nice example of a case where quantum conditional expectations do exist — in general, they do not.

The three basic "martingales"

5 It is well known that any martingale M_k on Bernoulli space has a *predictable representation*

(5.1) $$M_k = m + \sum_{j=1,\ldots,k} u_k x_k$$

where m is a constant (the expectation of the martingale) and u_k is a k-previsible vector. It is interesting to remark that the product which appears here is precisely of the kind considered in the preceding section, *i.e.* has an intrinsic meaning, independent of the probabilistic interpretation of toy Fock space.

The predictable representation theorem for operators can be stated as follows : *Every operator A on toy Fock space has a unique representation as*

(5.2) $$A = mI + \sum_{\substack{j=1,\ldots,N \\ \varepsilon=+,0,-}} u_j^\varepsilon a_j^\varepsilon$$

where the operators u_j^ε *are* j-*predictable.* This theorem amounts to the choice of a third basis for the space of operators : introduce, besides the three symbols $+, \circ, -$ a fourth symbol $\bullet$ and make the convention that for every k, $a_k^\bullet$ is the identity operator. For every sequence $\xi = \varepsilon_1, \ldots, \varepsilon_N$ of these four symbols, set $a^\xi = a_1^{\varepsilon_1} \ldots a_N^{\varepsilon_N}$. Then these 4^N operators form a basis, and the predictable representation theorem amounts to arranging this basis according to the last index k such that $\varepsilon_k \neq \bullet$.

Discrete coherent vectors

6 Given a martingale M of expectation 0, with its predictable representation (5.1), probabilists often use the *(discrete) stochastic exponential of* M, a martingale of expectation 1 whose value at time N is the r.v.

$$(6.1) \qquad Z = \prod_{k=1,\ldots,N} (1 + u_k x_k) .$$

Most important is the case where the r.v.'s u_k are constants, *i.e.* $M_k = \mathbb{E}\,[\,u\,|\,\mathcal{F}_k\,]$ where $u = M_N = \sum_k u_k x_k$ belongs to the first chaos. Then we put $Z = \mathcal{E}(u)$; vectors of this form are often called *(discrete) exponential vectors* (or *discrete coherent vectors* after normalization). We have

$$(6.2) \qquad <\mathcal{E}(u), \mathcal{E}(v)> = \prod_k (1 + \overline{u}_k v_k) .$$

The word "coherent vectors" has many uses, and the whole book [KlS] has been devoted to its different meanings. We have already met it in §1, for the elementary situation of the two-dimensional spin space $\mathcal{H}$. These two meanings are closely related : the toy Fock space Φ is isomorphic to the tensor product of N copies of $\mathcal{H}$, and the new coherent (or exponential vectors) are tensor products $\psi(z_1) \otimes \ldots \otimes \psi(z_n)$ of old coherent (or exponential) vectors.

> We find it convenient to use the word *exponential vectors* to denote vectors of expectation 1 in the vacuum state, and *coherent vectors* for vectors of norm 1. This use is not standard in the literature.

§3. A DE MOIVRE–LAPLACE THEOREM

In this section we prove, by elementary methods, a theorem which corresponds to the classical de Moivre–Laplace theorem on the convergence of sums of independent Bernoulli r.v.'s to a Gaussian r.v.. The Bernoulli r.v.'s are replaced by spin creation, annihilation and number operators, (commuting at different sites), and the analogue of Gaussian variables are the *harmonic oscillator* creation, annihilation and number operators, to be defined in the next chapter. The reader may prefer to become acquainted first with the harmonic oscillator, and postpone this section. Elementary results in this direction have been known since Louisell's book and the paper [ACGT] of Arecchi and *al.* (1972). We follow the modern version of Accardi–Bach [AcB1],

In classical probability, the de Moivre–Laplace theorem is the first step into a vast category of general problems : weak convergence of measures, central limit theorems. Only special cases of these problems have been considered yet in quantum probability. The method generally used is convergence of moments (see Giri and von Waldenfels [GvW], von Waldenfels [vW1] for the fermion case, and Accardi–Bach [AcB3]).

Preliminary computations

1 We first recall, for the reader's convenience, the notations for the elementary spin operators. First of all, the three basic creation, number and annihilation operators on the two dimensional Hilbert space $\mathcal{H} = \mathbb{C}^2$ (Bernoulli space, with its basis consisting of **1** and **x**). To avoid confusion with the corresponding harmonic oscillator notation, we put (as in subsection 5)

$$b^- = \begin{pmatrix} 0 & 1 \\ 0 & 0 \end{pmatrix} \quad ; \quad b^\circ = \begin{pmatrix} 0 & 0 \\ 0 & 1 \end{pmatrix} \quad ; \quad b^+ = \begin{pmatrix} 0 & 0 \\ 1 & 0 \end{pmatrix} .$$

The most general *skewadjoint* operator g such that $<\mathbf{1}, g\mathbf{1}> = 0$ can be written as

(1.1) $$g = \begin{pmatrix} 0 & -\zeta \\ z & 2i\lambda \end{pmatrix} \text{ with } \zeta = \bar{z} \text{ and } \lambda \text{ real.}$$

We begin with the computation of the unitary operator e^g : when $\lambda = 0$, this operator is the *elementary Weyl operator* $w(z) = \exp(zb^+ - \bar{z}b^-)$. For $\lambda \neq 0$ we denote it by $w(z, \lambda)$. The vectors $\psi(z) = w(z)\mathbf{1}$ are the *elementary coherent vectors*.

We take n independent copies of the preceding space, *i.e.* build the tensor product $\mathcal{H}_n = \mathcal{H}^{\otimes n}$, with its vacuum vector $\mathbf{1}^{\otimes n}$ denoted $\mathbf{1}_n$ or simply $\mathbf{1}$. Given an operator h on $\mathcal{H}$, we may copy it on $\mathcal{H}_n$ in the 1-st,... n-th position : thus h_1 denotes $h \otimes I \otimes ... \otimes I$ and h_n $I \otimes ... \otimes I \otimes h$. Classical probability suggests that a $\sqrt{n}$ normalization is convenient for a central limit theorem, and we therefore put

(1.2) $$B_n^\pm = \frac{b_1^\pm + ... + b_n^\pm}{\sqrt{n}}$$

(1.3) $$W_n(z) = w(z/\sqrt{n})_1 \otimes ... \otimes w(z/\sqrt{n})_n = \exp(zB_n^+ - \bar{z}B_n^-)$$

(1.4) $$\psi_n(z) = W_n(z)\mathbf{1} = (\psi(z))^{\otimes n}$$

These last operators and vectors are called respectively the *discrete Weyl operators* and *discrete coherent vectors*. The set of coherent vectors is the same physicists use, but the mapping $z \to \psi_n(z)$ is not the same, since we have included the normalization in our definition. Nothing has been said yet about B_n°.

We also introduce the notations for the harmonic oscillator. It suffices to say that we are given a Hilbert space generated (through linear combinations and closure) by a family of coherent vectors $\psi(u) = W(u)\mathbf{1}$ satisfying the relation

(1.5) $$< \psi(u), \psi(v) > = e^{-(|u|^2 + |v|^2)/2} e^{\bar{u}v} ,$$

(these vectors differ by a normalization constant from the exponential vectors $\mathcal{E}(u)$ which are used in the harmonic oscillator section). The vacuum vector is $\mathbf{1} = \psi(0)$.

The annihilation operator is defined by $a^-\psi(u) = u\psi(u)$, the creation operator a^+ is its adjoint, and $a^\circ = a^+a^-$. The Weyl operators $W(z)$ are given formally by $W(z) = \exp(za^+ - \bar{z}a^-)$. We only need to know how they operate on the coherent vectors, namely

$$(1.6) \qquad\qquad W(z)\psi(u) = e^{i\Im(\bar{z}u)}\psi(u+z) .$$

The basic computation we use is that of the matrix elements

$$(1.7) \qquad\qquad <\psi(u), W(z)\psi(v)> = e^{i\Im(\bar{z}v)}<\psi(\dot{u}), \psi(z+v)> ,$$

a somewhat lengthy expression if (1.5) is taken into account. Indeed, the result we want to prove is : *matrix elements of the discrete Weyl operators $W_n(z)$ between discrete coherent states $\psi_n(u)$ and $\psi_n(v)$ tend to the matrix elements (1.7) with the same values of z, u, v.* In the last part we shall include the number operator too.

2   We start from the following relation, where g is given by (1.1)

$$g^2 = \begin{pmatrix} -z\zeta & -2i\lambda\zeta \\ 2i\lambda z & -z\zeta - 4\lambda^2 \end{pmatrix} = -z\zeta I + 2i\lambda g$$

(this is of course a trivial case of the Hamilton–Cayley theorem). We change variables, assuming at the start that the denominator does not vanish

$$j = \frac{g - i\lambda}{\sqrt{\lambda^2 + z\zeta}} \quad ; \quad g = i\lambda + j\sqrt{\lambda^2 + z\zeta} .$$

Since $j^2 = -I$, Euler's formula gives

$$e^g = e^{i\lambda}(\cos\sqrt{\lambda^2 + z\zeta} + j\sin\sqrt{\lambda^2 + z\zeta})$$

and we have, replacing j by its value

$$(2.1) \qquad e^g = e^{i\lambda}\left(\cos\sqrt{\lambda^2 + z\zeta} + \frac{\sin\sqrt{\lambda^2 + z\zeta}}{\sqrt{\lambda^2 + z\zeta}}(g - i\lambda)\right)$$

which is correct even in the case $\lambda^2 = -z\zeta$ if $\sin 0/0 = 1$. The interest of taking a general ζ will be clear later on. There is no harm in using complex square roots : the functions $\cos\sqrt{z}$ and $\sin\sqrt{z}/\sqrt{z}$ are single valued and entire.

We first apply this result to Weyl operators, taking $\lambda = 0$, $\zeta = \bar{z}$ and $|z| = r$. We get

$$(2.2) \qquad\qquad w(z) = \cos r + \frac{\sin r}{r}Z \quad \text{with} \quad Z = \begin{pmatrix} 0 & -\bar{z} \\ z & 0 \end{pmatrix} .$$

Therefore

$$\psi(z) = w(z)\mathbf{1} = \cos r\mathbf{1} + \frac{\sin r}{r}z\mathbf{x} ,$$

which gives the matrix element

$$< \psi(u), \psi(v) > = \cos|u| \cos|v| + \frac{\sin|u|}{|u|} \frac{\sin|v|}{|v|} \bar{u}v$$

and if we replace z by $z/\sqrt{n}$ we get (the symbol $\sim$ meaning that only the leading term is kept)

$$(2.3) \qquad \psi(z/\sqrt{n}) \sim (1 - \frac{|z|^2}{2n})\mathbf{1} + \frac{z}{\sqrt{n}} \mathbf{x} .$$

with the trivial consequence

$$\lim_n < \psi_n(u), \psi_n(v) > = < \psi(u), \psi(v) > ,$$

which is already an indication of convergence to the harmonic oscillator.

Let us now state (and sketch a proof of) an important result of Accardi–Bach : *for every finite family of complex numbers* $u, z_1, \ldots, z_k, v$, *the matrix element*

$$< \psi_n(u), W_n(z_1) \ldots W_n(z_k) \psi_n(v) >$$

tends to the corresponding matrix element for the harmonic oscillator,

$$< \psi(u), W(z_1) \ldots W(z_k) \psi(v) > .$$

In this general form, no generality is lost by computing the matrix elements relative to the vacuum vector $\mathbf{1}_n$. Indeed, it is sufficient to add $W_n(-u) = W_n^*(u)$ and $W_n(v)$ to the set of our Weyl operators to get the general case.

On the other hand, it is interesting to consider more general matrix elements, and to replace the Weyl operators $w(z) = \exp(zb^+ - \bar{z}b^-)$ by the operators $\omega(z, \zeta) = \exp(zb^+ - \zeta b^-)$, and similarly $W_n(z)$, $W(z)$ by the corresponding (non–unitary) operators $\Omega_n(z, \zeta) = \exp(zB_n^+ - \zeta B_n^-)$, $\Omega(z, \zeta) = \exp(za^+ - \zeta a^-)$. The interest of doing so comes from the fact that such matrix elements

$$<\mathbf{1}, \Omega_n(z_1, \zeta_1) \ldots \Omega_n(z_k, \zeta_k) \mathbf{1}> =$$
$$<\mathbf{1}, \omega(z_1/\sqrt{n}, \zeta_1/\sqrt{n}) \ldots \omega(z_k/\sqrt{n}, \zeta_k/\sqrt{n}) \mathbf{1}>^n .$$

are *entire functions* of the complex variables $z_1, \zeta_1, \ldots, z_k, \zeta_k$ and (if we are careful to find locally uniform bounds for these entire functions) convergence of the functions will imply convergence of all their partial derivatives or arbitrary orders. This means that matrix elements (relative to $\mathbf{1}$ or to arbitrary coherent vectors) of arbitrary polynomials in the operators $B_n^{\pm}$ converge to the matrix elements of the corresponding polynomials in the operators $a^{\pm}$.

The principle of Accardi–Bach's proof is simple : we expand the exponentials and get

$$\sum_p n^{-p/2} \sum_{m_1+\ldots m_k=p} \frac{1}{m_1! \ldots m_k!} <\mathbf{1}, (z_1b^+ - \zeta_1 b^-)^{m_1} \ldots (z_k b^+ - \zeta_k b^-)^{m_k} \mathbf{1}>$$

to be raised to the power n. The term corresponding to $p=0$ is 1, and the following term is 0 since $<1, b^{\pm}1>=0$. For $p=2$ we get

$$(2.4) \qquad -\frac{1}{n}\left(\frac{1}{2}\sum_i z_i\zeta_i + \sum_{i<j}\zeta_i z_j\right),$$

which we denote by S/n. Finally, the sum of the remaining terms is the product of $n^{-3/2}$ by a factor which remains uniformly bounded as the complex variables z_i, ζ_i range over a compact set. The limit of $(1+(-S+o(1))/n)^n$ is e^{-S}, and it remains to check that this limit is equal to

$$(2.5) \qquad <1, e^{z_1a^+-\zeta_1a^-}\ldots e^{z_ka^+-\zeta_ka^-}1>,$$

to which a meaning should be given first. In order to do this, we anticipate Chapter III and remark that $e^{za^+-\zeta a^-}$ may be given a definition in analogy with (and in analytic continuation of) that of the Weyl operators, on the stable domain $\mathcal{E}$ generated by the exponential vectors :

$$(2.6) \qquad e^{za^+-\zeta a^-}\,\mathcal{E}(u) = e^{-\zeta u - z\zeta/2}\mathcal{E}(u+z)$$

and that on this domain it satisfies an extension of the Weyl commutation relation allowing the computation of the matrix element (2.5), which is found indeed to be e^{-S} (or just check it for the Weyl operators and deduce the general equality by analytic continuation).

Including the number operator

3   In this section we are going to discuss the same topic, but including the number operator. The first problem is the following : how to choose a constant c_n such that the matrix elements

$$<\psi_n(u), e^{i\lambda B_n^o}\psi_n(v)>$$

relative to the operator $B_n^o = (b_1^o + \ldots + b_n^o)/c_n$ converge to a finite and non-trivial limit. The surprising answer is that we may take $c_n = 1$. From a classical point of view, this is paradoxical : should not a r.v. with zero expectation and finite variance σ^2 be normalized by $\sigma\sqrt{n}$? Precisely, this is the point : in the vacuum state, the number operator has variance *zero*, and therefore does not need to be divided by something which tends to infinity.

To see that this choice is convenient, let us compute the matrix element

$$<\psi_n(u), e^{i\lambda B_n^o}\psi_n(v)> = \mu^n$$

where $\mu = <\psi(u/\sqrt{n}), e^{i\lambda b^o}\psi(v/\sqrt{n})>$. Using the relation $(b^o)^2 = b^o$, we find that $e^{i\lambda b^o} = I + (e^{i\lambda} - 1)b^o$, hence

$$\mu \sim 1 - \frac{|u|^2}{2n} - \frac{|v|^2}{2n} + e^{i\lambda}\frac{\bar{u}v}{n}\ .$$

and the limit of the matrix element is seen to be $<\psi(u), \psi(e^{i\lambda}v)>$, while $a^o\psi(v) = \psi(e^{i\lambda}v)$. Thus our "normalization" works.

It is therefore natural to use this normalization to define the operators B_n^o and the discrete "Weyl like" operators[1] $\Omega_n(z, 2i\lambda) = \exp(z B_n^+ - \bar{z} B_n^- + 2i\lambda B_n^o)$ involving the number operator, and to discuss the convergence of their matrix elements between discrete coherent vectors. More generally, it would be interesting 1) to study the corresponding operators $\Omega_n(z, \zeta, 2i\lambda)$ which are entire functions of three complex variables, and reduce to the preceding ones for $\zeta = \bar{z}$ and λ real; this would justify as above the convergence of all partial derivatives of the matrix elements; 2) to discuss the convergence of matrix elements for finite products of such operators. This program, however, will be "left to the interested reader", and consider only the simplest case (see Accardi–Bach [AcB2] for a detailed discussion).

The limit we have to compute is

(3.1) $\lim_n \left(< \psi(u/\sqrt{n}), \exp(2i\lambda b^o + k/\sqrt{n})\psi(v/\sqrt{n}) > \right)^n$

where we have put

$$k = \begin{pmatrix} 0 & -\zeta \\ z & 0 \end{pmatrix} .$$

We shall deduce this limit from formula (2.1). To abbreviate the writing, we also put

$$\alpha = \sqrt{\lambda^2 + \frac{z\zeta}{n}} \sim \lambda + \frac{z\zeta}{2\lambda n} .$$

The matrix $\exp(2i\lambda b^o + k/\sqrt{n})$ is equal to

$$e^{i\lambda} \begin{pmatrix} \cos\alpha - i\lambda\frac{\sin\alpha}{\alpha} & -\frac{\sin\alpha}{\alpha}\frac{\zeta}{\sqrt{n}} \\ \frac{\sin\alpha}{\alpha}\frac{z}{\sqrt{n}} & \cos\alpha + i\lambda\frac{\sin\alpha}{\alpha} \end{pmatrix} .$$

Next, $\cos\alpha \sim \cos\lambda - \sin\lambda\frac{z\zeta}{2\lambda n}$ and

$$\lambda\frac{\sin\alpha}{\alpha} \sim \sin\lambda + \cos\lambda\frac{z\zeta}{2\lambda n} - \sin\lambda\frac{z\zeta}{2n\lambda^2} .$$

The leading term is therefore

(3.2) $\Omega_n(z, \zeta, 2i\lambda) \sim e^{i\lambda} \begin{pmatrix} e^{-i\lambda} - i\frac{z\zeta}{2\lambda n}(e^{-i\lambda} - \frac{\sin\lambda}{\lambda}) & -\frac{\sin\lambda}{\lambda}\frac{\zeta}{\sqrt{n}} \\ \frac{\sin\lambda}{\lambda}\frac{z}{\sqrt{n}} & e^{i\lambda} + i\frac{z\zeta}{2\lambda n}(e^{i\lambda} - \frac{\sin\lambda}{\lambda}) \end{pmatrix} .$

It remains to compute $< \psi(u), \Omega_n(z, \zeta, 2i\lambda)\psi(v) >$, to raise it to the power n and take a limit. Neglecting second order terms in $1/n$ we have for the matrix element (before raising to the power n)

$$1 - \frac{|v|^2}{2n} - \frac{|u|^2}{2n} - \frac{z\zeta}{2n}\frac{i}{\lambda}(1 - e^{i\lambda}\frac{\sin\lambda}{\lambda}) - \frac{\zeta v}{n}e^{i\lambda}\frac{\sin\lambda}{\lambda} + \frac{\bar{u}z}{n}e^{i\lambda}\frac{\sin\lambda}{\lambda} + e^{2i\lambda}\frac{\bar{u}v}{n} .$$

---

[1]   They are not exactly Weyl operators, and for this reason we avoid the notation W_n.

After raising to the power n and passing to the limit, the matrix element is the exponential of

$$-\frac{|v|^2}{2} - \frac{|u|^2}{2} - \frac{z\zeta}{2}\frac{i}{\lambda}(1 - e^{i\lambda}\frac{\sin\lambda}{\lambda}) + (\bar{u}z - \zeta v)e^{i\lambda}\frac{\sin\lambda}{\lambda} + \bar{u}ve^{2i\lambda} \ .$$

This looks complicated, and needs some comments. First of all, what we are supposed to have computed is

$$< \psi(u) \, , \, \Omega(z, \zeta, 2\lambda)\,\psi(v) > \quad \text{where} \quad \Omega(z, \zeta, 2\lambda) = e^{za^+ - \zeta a^- + 2i\lambda a^\circ} \ .$$

The language of future chapters will replace coherent vectors $\psi(u)\,\psi(v)$ by *exponential vectors* $\mathcal{E}(u)\,\mathcal{E}(v)$ proportional to the preceding ones, but with a different normalization so that $< \mathcal{E}(u)\,, \mathcal{E}(v) > = e^{< u, v >}$. This replacement eliminates $-(|v|^2 + |u|^2)/2$ to the left. This being done, let us put $e_1(z) = (e^z - 1)/z$ and $e_2(z) = (e^z - 1 - z)/z^2$. The preceding expression can be written

$$< \mathcal{E}(u)\,, \, \Omega(z, \zeta, 2i\lambda)\,\mathcal{E}(v) >$$

where the right side is equal to

$$C\mathcal{E}(e^{2i\lambda}v + e_1(2i\lambda z)) \quad \text{and} \quad C = \exp(-\zeta(ve_1(2i\lambda) + ze_2(2i\lambda))) \ .$$

That is, the operators we have constructed act on a rather simple way on exponential (or coherent) vectors : the argument undergoes an affine transformation, and a scalar factor is applied. They are simply related to the general Weyl operators to be introduced in Chapter IV, and we shall also see in Chapter V that the preceding formula can be extended to infinite dimensional situations.

This concludes the discussion of the de Moivre–Laplace theorem for commuting spins.

§4. ANGULAR MOMENTUM

As every reader of Feller's famous book knows, the most interesting object of study in elementary probability theory is *Bernoulli random walk*, its zeros, maxima and fluctuations. The subject has not yet been studied in a systematic way in the framework of von Neumann probability theory, and the following discussion is a first step in this direction. I am grateful to W. von Waldenfels for discussions which led to considerable improvements.

Bernoulli random walk

1   As we have seen in §1, a non-commutative "Bernoulli random variable" is a triple of elementary creation/annihilation/number operators a^+, a^-, a°, or equivalently of Pauli matrices $x = a^+ + a^-$, $y = i(a^+ - a^-)$, $z = I - 2a^\circ$. Taking independent commuting copies of these operators as in the preceding section, we define

$$(1.1) \qquad X_\nu = x_1 + \ldots + x_\nu \ , \ Y_\nu = y_1 + \ldots + y_\nu \ , \ Z_\nu = z_1 + \ldots + z_\nu \ .$$

These three processes of selfadjoint operators satisfy at each time ν the following *angular momentum commutation relations*

(1.2) $[X_\nu, Y_\nu] = 2iZ_\nu \ , \ [Y_\nu, Z_\nu] = 2iX_\nu \ , \ [Z_\nu, X_\nu] = 2iY_\nu$

(the last two relations will usually be replaced by the mention "circ. perm."). The factor 2 appears here, because Bernoulli random variables assume the values ± 1 and not $\pm 1/2$; it may be considered as another case of the probabilists' normalization $\hbar = 2$. More generally

(1.3) $[X_m, Y_n] = 2iZ_{m \wedge n}$ (circ. perm.)

The relation between these three processes is symmetric. In this section, we are going to study these commutation relations for their own sake, and thus get new examples of quantum probability spaces.

REMARK.   We discussed in §3 the central limit behaviour of (1.1) in the vacuum state (or in states close to it, due to our normalization of the state) : the symmetry between the three processes disappears in the limit. In the vacuum state X and Y are Bernoulli random walks, but Z_ν is a.s. equal to ν. Thus if we divide X_ν and Y_ν by $\sqrt{\nu}$ and pass to the limit formally in (1.2) we get (see the preceding section for a rigorous discussion) two selfadjoint operators X and Y which satisfy a classical Heisenberg commutation relation, while Z degenerates. Privileging other pure states amounts to the same up to a rotation, and the symmetry again breaks. If we try to take a perfectly symmetric state, namely on each $\mathbb{C}^2$ factor the state with density matrix $I/2$, each one of the random variables x_ν, y_ν, z_ν then is a symmetric Bernoulli r.v. and the symmetric normalization by $\sqrt{\nu}$ works for all three, but then the commutation relation (1.2) degenerates and the limit r.v.'s commute. On the other hand, physicists do consider, under the name of "Heisenberg field", a continuous one dimensional system of spins in which the symmetry is preserved.

Ladder spaces

2   We are going to present (with notations appropriate to our probabilistic applications) the elementary theory of angular momentum. The subject can be found in all books on quantum mechanics, with slightly different notations.

Consider three selfadjoint operators on some Hilbert space $\mathcal{H}$ satisfying the relations

(2.1) $[X, Y] = 2iZ$ (circ. perm.)

(the Lie algebra commutation relations for the group $SU(2)$). For simplicity, we assume that $\mathcal{H}$ is finite dimensional. Hence all operators are bounded, and no domain problems arise.

The basic operator which classifies the representations is the *spin operator* J. Its standard definition becomes in our notation

$$J(J+I) = (X^2 + Y^2 + Z^2)/4 \ ,$$

the factor 4 coming from the fact that our operators are twice those of the physicists. Taking this factor to the left, and adding I to get a perfect square, it is clear that in our situation the natural operator to consider is $S = 2J + I$ (which has integer eigenvalues, as we shall see). It is defined by

$$(2.2) \qquad\qquad S^2 = X^2 + Y^2 + Z^2 + I \quad ;$$

S will then be the positive square root of S^2. For instance, in the elementary case of §1, we had $X^2 = Y^2 = Z^2 = I$, hence $S = 2I$. Expressing $[Y^2, X]$ as $Y[Y, X] + [Y, X]Y$, one sees that $Y^2 + Z^2$ commutes with X, and by symmetry S *commutes with all three operators*. The same is then true of all projections in the spectral resolution of S. Therefore, in an irreducible situation, *i.e.* one in which every subspace invariant under X, Y, Z is trivial, a condition of the form $S = \sigma I$ must be fulfilled; S being a positive operator, σ is a positive number.

In the next step we imitate the elementary spin case, introducing creation, annihilation and number operators. Our definition is not exactly that of the physicists, our apology for this (hateful) change of notation being coherence with the elementary spin case, and with the harmonic oscillator in the limit. We define $A^\pm$ and N by

$$(2.3) \qquad\qquad X = A^+ + A^-\ ,\ Y = i(A^+ - A^-)\ ,\ Z = (S - I) - 2N\ .$$

Our creation/annihilation pair $A^\pm$ corresponds to the physicists' $J_\mp$, and our Z is $2J_z$ in their notation. We call N the "number operator" for reasons to appear later (we avoid here the notation A^0 for the number operator). Using (2.3) a direct computation gives

$$X^2 + Y^2 = S^2 - Z^2 - I = 2(A^- A^+ + A^+ A^-)\ ,$$

and combining this with $2Z = 2(A^- A^+ - A^+ A^-)$, we get formulas which simplify nicely

$$(2.4) \qquad\qquad A^+ A^- = N(S - N)\ ,\qquad A^- A^+ = (N + I)(S - N - I)\ .$$

On the other hand, the basic commutation relations become (since S commutes with the other operators)

$$(2.5) \qquad [N, A^-] = -A^-\ ,\ [N, A^+] = A^+\ ,\ [A^-, A^+] = Z = S - I - 2N\ .$$

A fundamental consequence of (2.5) is the following : if x is an *eigenvector of N with eigenvalue λ*, $A^\pm x$ *is either 0, or an eigenvector of N with eigenvalue $\lambda \pm 1$*. Therefore, the sequence of vectors $A^{+m} x$ consists of mutually orthogonal vectors, and since the space is finite dimensional it gets to 0 after finitely many steps. The same is true for the downward sequence $A^{-m} x$.

Assume now that x also is an eigenvector of S with eigenvalue σ (such joint eigenvectors exist since S and Z commute). Starting from x and applying the powers $(A^+)^m$ upwards and the powers $(A^-)^m$ downwards we get the finite *spin ladder* of x, which generates linearly a *ladder space*. Since $A^\pm$ and S commute it remains in the σ-eigenspace of S. It follows from (2.4) that applying $A^\pm$ to ladder vectors yields linear combination of ladder vectors. More precisely, the vectors x and $A^\pm x$ generate the same ladder space, and we may therefore take as generator the bottom vector of the ladder. We normalize it and call it x_0 : it plays the role of the vacuum state $\mathbf{1}$. Then we put $x_m = (A^+)^m x_0$. Calling x_k the last non-zero vector, the dimension of the ladder space is $k+1$.

Here is the information we already possess on the structure of the ladder space

$$N x_0 = \lambda x_0 , \qquad A^- x_0 = 0 , \qquad A^+ x_0 = x_1 ,$$
$$N x_1 = (\lambda + 1) x_1 , \qquad A^- x_1 = ? , \qquad A^+ x_1 = x_2 ,$$
$$\cdots$$
$$N x_k = (\lambda + k) x_k , \qquad A^- x_k = ? , \qquad A^+ x_k = 0 .$$

We add to this the relations $S x_i = \sigma x_i$. Applying (2.4) one may answer the question marks

$$A^- x_i = A^- A^+ x_{i-1} = (N + I)(S - N - I) x_{i-1} = i(\sigma + i - 1) x_{i-1} .$$

Also, applying (2.4) to the relations $A^+ A^- x_0 = 0$ and $A^- A^+ x_k = 0$ we get

$$\lambda(\sigma - \lambda) x_0 = 0 = (\lambda + k + 1)(\sigma - \lambda - k - 1) x_k .$$

The choice $\lambda = \sigma$ in the first relation cannot satisfy the second one since $\sigma \geq 0$. Therefore we must choose $\lambda = 0$ and $\sigma = k + 1$. Therefore σ is a strictly positive integer (the total number of ladder rungs), and the spin parameter $j = (\sigma - 1)/2$ can take integer and half-integer values starting with $0, 1/2, \ldots$. Since $\lambda = 0$ we have $N x_j = j x_j$, and N is a genuine number operator.

Note that $Z = S - I - 2N$ takes integer values, odd for S even (half-integer spin), even for S odd.

To determine the structure of the ladder space we must also know the scalar product : the basis vectors are mutually orthogonal, and to compute their norms we use for the first time the fact that A^+ and A^- are mutually adjoint. Then we have using (2.4)

$$< x_{i+1}, x_{i+1} > = < A^+ x_i, A^+ x_i > = < x_i, A^- A^+ x_i > = (i + 1)(k - i) < x_i, x_i >$$

from which the relation $< x_i, x_i > = \frac{k! \, i!}{(k-i)!}$ follows. The simplest formulas occur if we take as non-normalized basis elements the vectors $y_i = x_i / i!$. Then we have $y_i = 0$ for $i > k$, and for $i \leq k$

(2.6) $N y_i = i y_i , \quad A^- y_i = (k + 1 - i) y_{i-1} , \quad A^+ y_i = (i + 1) y_{i+1} , \quad < y_i, y_i > = \binom{k}{i} .$

Formulas (2.6) describe completely the structure of the ladder space, and it is a matter of routine to check that a representation of the angular momentum relations is indeed defined in this way. It is called the *spin $j = k/2$ representation* and commonly denoted by $\mathcal{D}_j$. It is easily shown to be irreducible. Indeed, any non-zero vector can be transformed into a multiple of x_0 by a convenient annihilator, and then transformed into any other non-zero vector by a convenient linear combination of creators. Hence there is no proper invariant subspace.

Conversely, given any (finite-dimensional) representation of the angular momentum commutation relations by selfadjoint operators X, Y, Z, one may choose a common eigenvector of the commuting selfadjoint operators S and N, and construct the corresponding stable ladder space. It is easy to see that its orthogonal subspace is stable too, so that we extract from it a new ladder space, etc. In this way the representation is decomposed into a direct sum of ladder representations.

Each ladder space of dimension $2j$ carries a spin label j (integer or half-integer) and a label $\alpha = 1, \ldots$ telling how many spaces of that dimension occurs in the representation. Within the ladder space itself it is traditional to label the basis by the $2j + 1$ values of $J_z = Z/2$ (magnetic quantum number m), which increase by 1 between $-j, -j+1, \ldots, j-1, j$. This leads to an orthonormal basis of the Hilbert space $\mathcal{H}$ labelled $|\alpha, j, m>$ in Dirac's notation.

It is true on any ladder space of spin j that the greatest eigenvalue of Z is $k = 2j$. Thus it is true for the whole representation that the greatest eigenvalue of Z and the greatest eigenvalue of J are related by

$$(2.7) \qquad\qquad 2J_{\max} = Z_{\max},$$

and in fact the common eigenvectors for Z, J with eigenvalue $Z_{\max}$ for Z must have the eigenvalue $J_{\max}$ for J.

Some comments

3 We point out here some analogies between the angular momentum space and the harmonic oscillator (anticipating on chapter III) and a few additional remarks. This subsection may be skipped without harm.

a) The central limit theorem in §3 suggests to study operators of the form

$$e^{p^+ A^+ + p^- A^- + qZ},$$

which for $p^+ = z$, $p^- = -\bar{z}$ and q purely imaginary will be unitary "Weyl operators". These exponentials can be explicitly computed as a product

$$e^{r^+ A^+} e^{sZ} e^{r^- A^-}.$$

This is called the *disentangling formula* for angular momentum operators. It is essentially the same as in the elementary case since operators at different sites commute, and one may find it in Appendix A to the paper [ACGT] of Arecchi and *al.*, p. 703 of the book of Klauder and Skagerstam [KlS] (remembering that $J_\pm = A^\mp$, $2J_z = Z$).

These Weyl operators (with $q = 0$) also appear on p. 690 of [KlS], formula (3.7a). The same trend of ideas leads to a family of coherent vectors (called *Bloch coherent states*). The corresponding unnormalized *exponential vectors* are given by the formula (*cf.* [KlS] p. 691, formula (3.12))

$$\mathcal{E}(z) = e^{zA^+}\mathbf{1} = \sum_{j=0}^{k} \frac{z^j}{j!}\, x_j.$$

Using (2.6) it is easy to find that

$$<\mathcal{E}(z), \mathcal{E}(z'), > = (1 + \bar{z}z')^k.$$

These vectors are also studied in a paper by Radcliffe in the same book. There are other sets of coherent and exponential vectors of physical interest.

b)   For large values of S the ladder representation is close to the quantum harmonic oscillator. To see this let us perform a change of operators and basis vectors : if we put $S = k + 1$ and

$$\alpha^+ = A^+/\sqrt{k}\ ,\ \alpha^- = A^-/\sqrt{k}\ ,$$

and replace x_j by $h_j = \alpha^{+j}x_0$, so that $[\alpha^-, \alpha^+] = I - \frac{2N}{k}$. The operators and the Hilbert space structure are described as follows in the new representation

$$\alpha^+ h_j = h_{j+1}\ ,\ N h_j = j h_j\ ,\ \alpha^- h_j = j(1 - \tfrac{j+1}{k}) h_{j-1}$$
$$\|h_j\|^2 = j!\,(1 - \tfrac{1}{k})(1 - \tfrac{2}{k})\dots(1 - \tfrac{j-1}{k})\ .$$

For fixed j and large k this tends to the description of the harmonic oscillator.

c) The case $k = 1$ was the analogue of simple Bernoulli space in quantum probability. One can consider the angular momentum space with spin j as the quantum probabilistic analogue of the measurable space with $\nu = 2j+1$ points, but this is not the most natural identification. The natural one consists in taking as Hilbert space the L^2 space of the uniform distribution on ν points, *i.e.* the Haar measure on the group $G = \mathbb{Z}/\nu\mathbb{Z}$. This L^2 space carries two simple unitary operators

$$U f(x) = f(y - 1)\ ,\ V f(x) = \xi^x f(x)$$

where ξ is a primitive ν–th root of unity. They satisfy a kind of Weyl commutation relation

$$V^p U^q = \xi^{pq} U^q V^p\ ,$$

and one may develop a discrete theory which closely parallels that of the canonical pair (Chapter III). See the paper [CCS] by Cohendet et *al.* in the references).

d) We are careful to work on a finite tensor product $\mathcal{H}_\nu \approx (\mathbb{C}^2)^{\otimes \nu}$, while the natural tendency in probability theory is to jump at once to the L^2 space of an infinite random walk. There is a difficulty here in von Neumann probability, of which our reader should be aware, since it is the first place where the elementary Hilbert space probability becomes insufficient, and the need for C^*–algebraic probability is felt. We can take an infinite product of measures, but not of Hilbert spaces — or rather, there exist infinite tensor products depending on the choice of a state in each Hilbert space, playing the role of a measure. For details, see the basic article [vN2] of von Neumann.

Addition of angular momenta

4   Consider independent copies of two representations of the angular momentum commutation relations, X_i, Y_i, Z_i ($i = 1, 2$), on Hilbert spaces $\mathcal{H}_i$. This means, as in classical probability, that we form the tensor product $\mathcal{H} = \mathcal{H}_1 \otimes \mathcal{H}_2$, and identify X_1 with $X_1 \otimes I_2$, X_2 with $I_1 \otimes X_2$, etc. Then taking $X = X_1 + X_2$, etc. we get a new representation X, Y, Z, whose structure we want to describe, assuming that the factor representations are irreducible, with spins j_1, j_2. Note that in this case the values of $Z = Z_1 + Z_2$ either are all even or all odd : hence all the spin ladders occurring in the decomposition are of the same kind, all with half-integer or integer spin.

The addition theorem for angular momenta asserts that *the possible values j for the spin of the sum are given by the inequalities*

$$|j_1 - j_2| \leq j \leq j_1 + j_2 \,,$$

and that each one of these values occurs exactly once in the new representation. We shall prove it in detail in the case $j_2 = 1/2$ only. We may consider Z_1 and Z_2 (two commuting selfadjoint operators) as two ordinary functions, and thus the multiplicity of an eigenvalue z of Z is the number of decompositions $z = z_1 + z_2$ of z into admissible values z_1, z_2 ($-2j_i \leq z_i \leq 2j_i$). Since $z_2 = \pm 1$, this number is at most 2. Looking at the multiplicity of the eigenvalue 0 or 1, which appears in all spin ladders, it is easy to see that there are at most 2 spin ladders, and their sizes are deduced from the relation

$$2 \times 2j_1 = (2j_1 - 1) + (2j_1 + 1) \,.$$

The trivial case $j_1 = 0$ was implicitly excluded. The general addition theorem is proved essentially in the same way, assuming $j_1 \geq j_2$ for instance : there are at most $j_1 - j_2 + 1$ spin ladders, and their sizes are deduced from the relation

$$(2j_1 + 1)(2j_2 + 1) = \sum_{j=|j_1-j_2|}^{j_1+j_2} (2j + 1) \,.$$

The consequence of interest for us is that, if we add a spin $1/2$ representation to *any* representation $\mathcal{H}$, every spin ladder in $\mathcal{H}$ of spin j gives birth to two spin ladders of spins $j \pm 1/2$, except for $j = 0$ which generates one single ladder of spin $1/2$.

Another topic which may be mentioned here (though it is not necessary for the sequel) is that of *Clebsch–Gordan coefficients* for the addition of two spins j_1 and j_2. First of all let us rewrite the structure of the spin j ladder space using the standard physicists' orthonormal basis $|j, m>$

$$A^+|j,m> = \rho(j,m)|j,m+1> \,, \quad A^-|j,m> = \rho(j,m-1)|j,m-1> \,,$$
$$Z|j,m> = 2m|j,m> \,.$$

The value of $\rho(j,m)$ is $\sqrt{(j-m+1)(j+m)}$. For each representation $\mathcal{H}_i$ we have a basis for $\mathcal{H}_i$ consisting of the vectors $|j_i, m_i>$, with m_i ranging from $-j_i$ to j_i. Then we have for the tensor product $\mathcal{H}$ a basis $|j_1, m_1> \otimes |j_2, m_2>$, which is simply denoted by $|j_1, m_1, j_2, m_2>$ On the other hand, $\mathcal{H}$ has its own spin basis $|j, m>$ with j ranging from $|j_1 - j_2|$ to $j_1 + j_2$. The Clebsch–Gordan coefficients $C(j_1 m_1, j_2 m_2 ; jm)$ give the decomposition of this spin basis over the tensor product basis.

They can be computed as follows. We express $A^+ |j, m>$ in two ways, first as

$$\rho(j,m)|j,m+1> = \rho(j,m) \sum_{k_1+k_2=m+1} C(j_1 k_1, j_2 k_2 ; j\, m+1) |j_1 k_1, j_2 k_2> \,,$$

and on the other hand as

$$\sum_{m_1+m_2=m} C(j_1 m_1, j_2 m_2 ; jm) A^+ |j_1 m_1, j_2 m_2> \,.$$

Then the last vector is replaced by its value

$$\rho(j_1, m_1) \mid j_1\, m_1+1, j_2\, m_2 > + \rho(j_2, m_2) \mid j_1\, m_1, j_2\, m_2+1 > \quad ;$$

from which we easily deduce the following induction formula : if $k_1 + k_2 = m+1$

(4.1)
$$C(j_1\, k_1, j_2\, k_2; j\, m+1) = \frac{\rho(j_1, k_1-1)}{\rho(j, m)} C(j_1\, k_1-1, j_2\, k_2; j\, m)$$
$$+ \frac{\rho(j_2, k_2-1)}{\rho(j, m)} C(j_1\, k_1, j_2\, k_2-1; j\, m) \ .$$

If one gets down to $k_1 = -j_1$ or $k_2 = -j_2$, the right hand side contains only one term, and the downward induction continues in the same way to the last step $C(j_1 - j_1, j_2 - j_2, j - j) = 1$. Thus all the coefficients can be computed (more efficient methods have been devised for practical applications). Using A^- instead of A^+ would give a similar upward induction formula.

Structure of the Bernoulli case

5 We are going to apply the preceding general results to the quantum Bernoulli random walk. Let $\mathcal{H}$ be the L^2 space of the elementary symmetric Bernoulli measure, with its basis $\mathbf{1}, x$. The representation space is $\mathcal{H}_\nu = \mathcal{H}^{\otimes \nu}$ identified with the L^2 space of the classical Bernoulli random walk up to time ν (toy Fock space). At each time ν the angular momentum representation is provided by the three random variables X_ν, Y_ν, Z_ν,

(5.1) $$X_\nu = x_1 + \ldots + x_\nu, \quad \text{etc.}$$

and the corresponding creation and annihilation operators A_ν^+, A_ν^- are just the sums of the elementary creation and annihilation operators at times $i \le \nu$. The Pauli matrix σ_z keeps $\mathbf{1}$ fixed and maps x into $-x$. Hence on the usual o.n. basis x_A (z_A would be better here!) of toy Fock space, Z_ν acts as follows

(5.2) $$Z_\nu\, x_A = (|A^c| - |A|)\, x_A = (\nu - 2|A|)\, x_A \ .$$

So the eigenspaces of Z_ν are the chaos of toy Fock space, the eigenvalue being $\nu - 2p$ on the p-th chaos C_p. Since S_ν commutes with Z_ν it keeps C_p stable, and decomposing the representation amounts to finding the eigenvectors of S in each C_p. Recalling that $Z = S - I - 2N = 2J - 2N$, we see that for ν even (odd) all components have integer (half-integer) spin. We have

$$A_\nu^+\, x_A = x_A \cdot \sum_{i \notin A} x_i \, , \quad A_\nu^-\, x_A = x_A \cdot \sum_{i \in A} x_i \ .$$

By (2.7) the greatest allowed value of spin is exactly $\nu/2$, and corresponds to the chaos C_0 of order 0. The vacuum vector $\mathbf{1} \in C_0$ generates a ladder of spin $\nu/2$ (*i.e.* of size $\nu + 1$) : the vector $A^{+i}\mathbf{1}$ of this ladder is the symmetric polynomial of degree i in the Bernoulli r.v.'s, conveniently normalized. On the first chaos C_1, the eigenvalue of Z is $\nu/2 - 1$, and therefore any ladder built on an eigenvector of J in C_1 has at least ν rungs. Thus the ladder size can only be ν or $\nu + 1$, and to have $\nu + 1$ steps it is necessary to push the foot of the ladder down to C_0. This corresponds to the ladder of the vacuum vector, and therefore we have $\binom{\nu}{1} - 1$ ladders of size ν. Similarly in the second chaos we find $\binom{\nu}{2} - \binom{\nu}{1}$ ladders of size $\nu - 1$, etc. Finally, the decomposition of the representation can be described by the following drawing

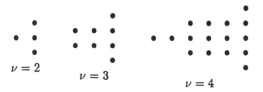

$$\nu = 2 \qquad \nu = 3 \qquad \nu = 4$$

A commutative process of spins

6 At each time ν let us define in the standard way

$$(5.3) \qquad S_\nu^2 = X_\nu^2 + Y_\nu^2 + Z_\nu^2 + I \, .$$

S_ν commutes with X_ν, Y_ν, Z_ν; on the other hand it commutes with all x_k, y_k, z_k for $k > \nu$, hence also with X_k, Y_k, Z_k, and finally with S_k. Then it is very easy to see that (S_ν) *is a commutative process*. However, the two processes (Z_ν) and (S_ν) commute at each fixed time, but not between different times.

The representation at time $\nu + 1$ is the tensor product of the representations at time ν and a spin $1/2$ representation, and by the addition theorem each spin ladder of size $s = 2j + 1$ at time ν generates two ladders of sizes $s \pm 1$ at time $\nu + 1$, or one exceptionally if $s = 1$. Thus we may describe the time evolution of spin by discrete paths ω from the integers $0, \ldots, \nu$ to the integers ≥ 1, which start from 1 at time 0 and increase by ± 1 between successive integers; each path at time ν with $\omega(\nu) > 0$ can be extended in two ways to time $\nu + 1$, and in just one way if $\omega(\nu) = 1$. Thus *the irreducible components in the Bernoulli representation at time ν are in 1–1 correspondence with the paths ω*, the value of $S_\nu = 2J_\nu + 1$ being $\omega(\nu)$.

Each path corresponding to a spin ladder of the representation, knowing the path provides us with the two labels α, j in the standard description. Thus to completely describe the basis we only need the label m, which corresponds to knowing Z_ν. Therefore the path $(S_k)_{k \leq \nu}$ and the operator Z_ν form a complete observable.

If we choose a state, we may talk about the probability law of the process (S_k). It is trivial in the vacuum state, since then $Z_k = k$, the maximum allowed value, and therefore $S_k = k + 1$. Ph. Biane computed this law in the tracial state, under which the probability distribution of a self-adjoint operator A on $\mathbb{C}^{2\nu}$ with eigenvalues a_i and eigenspaces E_i is given by

$$\mathbb{P}\{A = a_i\} = 2^{-\nu} \dim(E_i) \, .$$

Since the dimension of the ladder space corresponding to a path $\{S_1 = s_1, \ldots, S_\nu = s_\nu\}$ is equal to s_ν, the corresponding probability is $2^{-\nu} s_\nu$. Then it is very easy to see that the process (S_k) is a Markov chain on the integers, with $S_1 = 2$ and for $k > 1$

$$\mathbb{P}\{S_{\nu+1} = k + 1 \mid S_\nu = k\} = \frac{k+1}{2k} \, , \quad \mathbb{P}\{S_{\nu+1} = k - 1 \mid S_\nu = k\} = \frac{k-1}{2k}$$

while if $S_\nu = 1$ one a.s. has $S_{\nu+1} = 2$. This Markov chain is the discrete analogue of a Bessel$_3$ process.

For a computation concerning general product states, see the papers of Biane [Bia] and von Waldenfels [vW2]. It would be very interesting to know more about non commutative conditioning in the Bernoulli setup.

§5. ANTICOMMUTING SPIN SYSTEMS

This section provides an example of a truly non-commutative situation, of great physical interest. It also constitutes an elementary introduction to some aspects of the theory of Clifford algebras. Its starting point was the appendix in the book by D. Kastler [Kas].

Fermion creation and annihilation operators

1 Let us recall the basic notations. We start with a family of N independent, identically distributed, symmetric Bernoulli random variables x_k. The canonical sample space to realize them is $\{-1, 1\}^N$ equipped with the measure μ which ascribes the weight 2^{-N} to every point, and the corresponding L^2 space is the toy Fock space Ψ. We denote by $\mathcal{T}$ the "time set" $\{1, \ldots, N\}$, and by $\mathcal{P}$ the set of all subsets of $\mathcal{T}$. The space Ψ has an o.n. basis consisting of all products $x_A = \prod_{k \in A} x_k$, A ranging over $\mathcal{P}$.

We introduce some combinatorial definitions : given a subset A and an index $k \in \mathcal{T}$, $n(k, A)$ is the number of elements of A which are strictly smaller than k (that is, the number of strictly decreasing pairs one gets by writing first k, then the elements of A in increasing order). Similarly, $n'(k, A)$ is the number of elements of A strictly larger than k, so that $n(k, A) + n'(k, A) = |A| - I_A(k)$. Given now two subsets A and B, we define $n(A, B) = \sum_{k \in A} n(k, B)$ (the total number of inversions when one writes first A, then B), and similarly $n'(A, B) = n(B, A)$. It is clear that

$$(1.1) \qquad n(A, B) + n(B, A) = |A|\,|B| - |A \cap B| \ .$$

In particular, $n(A, A) = |A|(|A| - 1)/2$.

We define new creation and annihilation operators, which differ from those of the §2 subsection 2 by the presence of alternants

$$b_k^+(x_A) = (-1)^{n(k,A)} x_{A \cup \{k\}} \quad \text{if} \quad k \notin A, \ = 0 \quad \text{otherwise} \qquad \text{(creation)}$$

$$b_k^-(x_A) = (-1)^{n(k,A)} x_{A \setminus \{k\}} \quad \text{if} \quad k \in A, \ = 0 \quad \text{otherwise} \qquad \text{(annihilation)}$$

These operators are mutually adjoint for every k. It is easy to verify that (denoting anticommutators by $\{\,,\,\}$)

$$(1.2) \qquad \{b_j^-, b_k^-\} = 0 = \{b_j^+, b_k^+\} \quad ; \quad \{b_j^-, b_k^+\} = \delta_{jk} I \quad ;$$

the meaning of these formulas being that each pair (b_k^-, b_k^+) is a spin, and *all these spins anticommute*. These relations are the CAR (canonical anticommutation relations) in their exact form, unlike the CCR in §2. For this reason, there is no "toy Fock space" in the antisymmetric case, only a finite dimensional Fock space.

The operator $b_k^+ b_k^-$ is the same as the operator a_k^o of toy Fock space, but we denote it by b_k^o for consistency; it is called as before the *number operator*.

For each k we introduce, as in the case of toy Fock space, self-adjoint linear combinations of creation and annihilation operators

$$(1.3) \qquad r_k = b_k^+ + b_k^- \quad ; \quad s_k = i\,(b_k^+ - b_k^-) \; .$$

These operators may be considered as a pair of Pauli matrices σ_x, σ_y, the third matrix being $\sigma_z = I - 2b_k^o$. Spins relative to different sites all *anticommute*. Given a subset $A = \{i_1 < i_2 \ldots < i_n\} \in \mathcal{P}$ we define $r_A = r_{i_1} \ldots r_{i_n}$ ($r_\emptyset = I$ by convention), and we define similarly s_A and b_A^ε where $\varepsilon = -, o, +$. The ordering of the time set is important because of anticommutativity : for instance, the adjoint of b_A^+ is not b_A^-, but $(-1)^{n(A,A)} b_A^-$. The algebra of operators generated by all the $b_k^\pm$ and I consists of all operators on Ψ, the proof being the same as in §2. Using the anticommutation relations one can show that the operators $b_A^+ b_B^-$ define a basis for all operators. Another useful basis consists of all operators $r_A s_B$ (the proof is easy, and left to the reader).

2 The fermion operators $b_k^\pm$ can be constructed from the toy boson operators $a_k^\pm$ through the *Jordan–Wigner transformation*. The measure μ is a product of N Bernoulli measures, one for each coordinate, and thus $L^2(\mu)$ is a tensor product of N elementary Bernoulli spaces, each site i carrying its three Pauli matrices $\sigma_x(i), \sigma_y(i), \sigma_z(i)$ as in §1. In particular, the operators $\sigma_x(i)$, $\sigma_y(i)$ are the same as the toy boson selfadjoint operators q_i, p_i (commuting between different sites) from §2. Then we have

$$r_1 = \sigma_x \otimes I \otimes I \ldots \,, \qquad s_1 = \sigma_y \otimes I \otimes I \ldots \,,$$
$$r_2 = \sigma_z \otimes \sigma_x \otimes I \otimes I \ldots \,, \qquad s_2 = \sigma_z \otimes \sigma_y \otimes I \otimes I \ldots \,,$$
$$r_3 = \sigma_z \otimes \sigma_z \otimes \sigma_x \otimes I \otimes I \ldots \,, \qquad s_3 = \sigma_z \otimes \sigma_z \otimes \sigma_y \otimes I \otimes I \ldots \,,$$

and so on. For $N = 2$ we get in this way four 4×4 matrices, which are (one of the possible representations of) the famous Dirac matrices. This explicit construction of anticommuting matrices from tensor products, *i.e.* from matrices which commute, can be interpreted as a *stochastic integral representation* of the "fermion martingales"

$$R_k = r_1 + \cdots + r_k \,, \qquad S_k = s_1 + \cdots + s_k \,,$$

by means of the "toy boson" martingales

$$Q_k = q_1 + \cdots + q_k \,, \qquad P_k = p_1 + \cdots + p_k \; .$$

This representation can be inverted, and leads to a stochastic integral construction of "toy bosons" from fermions.

This construction has been used by Onsager to compute the partition function of the two dimensional Ising model by means of spin matrices. Thus anticommutative algebra may be useful in commutative problems.

3 Let us write down a few formulas concerning the fermion operators.

$$(3.1) \qquad r_k x_B = (-1)^{n(k,B)} x_{B\Delta\{k\}} \quad \text{hence} \quad r_A x_B = (-1)^{n(A,B)} x_{A\Delta B}$$

One may also compute $b_A^{\pm} x_B$ as in §2 (3.1). We have

$$(3.2) \qquad r_A \, r_B = (-1)^{n(A,B)} r_{A\Delta B} = (-1)^{n(A,B)+n'(A,B)} r_B \, r_A \; .$$

In particular, we have $r_A^2 = (-1)^{n(A,A)} I$. On the other hand, direct computation from (1.3) gives

$$s_k x_B = i(-1)^{n(k,B)} \big(I_{B^c}(k) - I_B(k) \big) x_{B\Delta\{k\}}$$

The middle difference has the value 1 for $k \in B$, -1 for $k \notin B$, and can be written $(-1)^{n(k,B)+n'(k,B)+|B|}$. It follows that

$$(3.3) \qquad s_k \, x_B = i(-1)^{n'(k,B)}(-1)^{|B|} x_{B\Delta\{k\}} \; .$$

It does not seem necessary to write explicitly the formula for $s_A \, x_B$, but we have the same relation as (3.2) with s instead of r. On the other hand, we have

$$(3.4) \qquad s_B r_A = (-1)^{|A||B|} r_A \, s_B \; .$$

We shall summarize these results in the next section, as follows : the algebra of all operators on Bernoulli space can be considered as a Clifford algebra, generated by all r_i, s_i together, and the r_i and s_i taken separately also generate two (isomorphic) Clifford algebras.

All these results have parallels in continuous time.

Clifford algebras

4 In toy Fock space theory, q_k was the multiplication operator by the Bernoulli r.v. x_k. We are going to interpret similarly the operators r_k as multiplication by x_k in a non commutative algebra structure.

The space of all operators on toy Fock space has a basis consisting of the operators $r_A s_B$. A look at (3.2) shows that linear combinations of the operators r_A alone (including I) form a subalgebra. The composition of operators being associative, and the operators r_A being linearly independent, (3.2) is the complete multiplication table for an associative algebra. Since the vectors x_A form a basis of Ψ, we may carry this multiplication back to Ψ itself : x_A is read as a product $x_{i_1} \ldots x_{i_n}$ indexed by the elements of A in increasing order, and then we have (in close analogy with formula (1.4) of §2)

$$(4.1) \qquad x_A \, x_B = (-1)^{n(A,B)} x_{A\Delta B} \; .$$

The formula

$$x_k x_B = (-1)^{n(k,B)} x_{B\Delta\{k\}}$$

then shows that r_k is the left multiplication operator by x_k. We computed above the adjoint of the operator r_A, which belongs to the same subalgebra. Thus we may also provide Ψ with an involution

$$(4.2) \qquad (x_A)^* = (-1)^{n(A,A)} x_A .$$

If we also carry back to Ψ the usual norm for operators, Ψ becomes a normed algebra (a C^*–algebra).

The concrete *–algebra we have thus constructed will be called *the* complex Clifford algebra $C(N)$ with N generators. We are going to widen this definition, characterizing this algebra up to isomorphism, first using a system of generators, and then in an intrinsic way.

5 Let ν be an integer, and let $\mathcal{A}$ be an algebra over $\mathbb{C}$, generated (as an algebra) by its unit $\mathbf{1}$ and by generators γ_k $(k = 1, \ldots, \nu)$ satisfying the following relations

$$(5.1) \qquad \gamma_k^2 = \mathbf{1} \quad ; \quad \gamma_j \gamma_k + \gamma_k \gamma_j = 0 \quad (j \neq k) .$$

Given a subset $A = \{j_1 < \ldots < j_n\} \subset \{1, \ldots, \nu\}$ we define as above $\gamma_A = \gamma_{j_1} \cdots \gamma_{j_n}$, and it is easy to see that

$$(5.2) \qquad \gamma_A \gamma_B = (-1)^{n(A,B)} \gamma_{A \triangle B}$$

as in (4.1). Hence l'near combinations of the elements γ_A form a subalgebra, which obviously contains the γ_k and $\mathbf{1}$, and therefore is the whole algebra. *If the elements γ_A are linearly free,* a situation which arises in the concrete case considered above, (5.2) is the complete multiplication table of the algebra, which is then completely described up to isomorphism, and called a *Clifford algebra* $C(\nu)$ [1], of dimension 2^ν. The last result in subsection 3 can now be stated as follows,

The operators r_k and s_k taken separately generate an algebra $C(N)$; taken together they generate an algebra $C(2N)$.

One can prove that the italicized condition about the elements γ_j being linearly free is *unnecessary for even* ν. Indeed, assume $\nu = 2N$ and consider the mapping which assigns to the operators $r_1, \ldots, r_N, s_1, \ldots, s_N$ the elements $\gamma_1, \ldots, \gamma_\nu$. Since the two algebras have the same multiplication rules, it can be extended to a homomorphism of the algebra generated by all r_j, s_k, *which is the algebra of all linear operators on* Ψ, to the algebra $\mathcal{A}$. The kernel of a homomorphism is a two-sided ideal, and it is a classical (and not difficult) result that a full matrix algebra is *simple,* i.e. has no two-sided ideals except zero and the algebra itself. So the homomorphism is an isomorphism, and we see that for even ν, $C(\nu)$ is isomorphic to the full matrix algebra over a space of dimension $2^{\nu/2}$, and is a simple algebra.

Let us anticipate a little to describe the case of odd ν. We shall construct in 9 a linear operator σ on Ψ such that $\sigma^2 = I$, and which anticommutes with all r_j, s_k. Since the $r_A s_B$ are sufficient to generate the full matrix algebra, the algebra generated

---

[1] We have learnt much about Clifford algebras from excellent mimeographed notes by J. Helmstetter, *Algèbres de Clifford et Algèbres de Weyl*, Cahiers Math. Université de Montpellier n° 25, 1982. We are far from exhausting their contents.

by the r_i, s_j and σ cannot be the Clifford algebra $\mathcal{C}(2N+1)$, which algebra thus admits a non-trivial homomorphic image, and therefore is not simple.

EXAMPLES. Let us describe the low dimensional Clifford algebras. We begin with a remark : let $\gamma_1, \ldots, \gamma_\nu$ be the generators of $\mathcal{C}(\nu)$. For $j < k$ we have $\gamma_{1j}\gamma_{1k} = -\gamma_{jk}$, and therefore the $\nu - 1$ elements $i\gamma_{12}, \ldots, i\gamma_{1n}$ of the second chaos satisfy the anticommutation relations (5.1). Since the linear independence hypothesis is trivially satisfied, they generate a copy of $\mathcal{C}(\nu-1)$, which is called the *even subalgebra* since as a linear space it is the sum of the even numbered "chaos".

The Clifford algebra $\mathcal{C}(1)$ is commutative. The case $\nu = 2$ may be identified with the algebra of all 2×2 matrices (of all operators on the elementary Bernoulli space, with its basis consisting of $\mathbf{1}$ and one single Bernoulli variable x). The (linear) basis consisting of the vectors $r_A s_B$ has four elements $\mathbf{1}, r, s, rs$ such that $r^2 = \mathbf{1} = s^2$, $rs + sr = 0$. If we take as basis $\mathbf{1}, r, s, irs$ the corresponding 2×2 matrices are $I, \sigma_x, \sigma_y, \sigma_z$. If we choose instead the basis $\mathbf{1}, ir, is, -rs$, the multiplication table is that of quaternions (with complex coefficients). The even subalgebra, generated by $\mathbf{1}$ and $irs = \sigma_z$, is isomorphic to the algebra of diagonal matrices, *i.e.* to $\mathbb{C}^2$.

The case of $\mathcal{C}(4)$ is fundamental in physics : it is generated by the four 4×4 Dirac matrices, and has dimension 16. Its even subalgebra is $\mathcal{C}(3)$ of dimension 8 (instead of 4 for the algebra generated by the three Pauli matrices).

Basis-free definition of Clifford algebras

6 We now give a definition of Clifford algebras which does not depend on an explicit choice of generators. Consider a complex linear space $\mathcal{H}$ of finite dimension ν, and a non degenerate symmetric *bilinear* form B on $\mathcal{H}$ — in our case, $\mathcal{H}$ is the first chaos over Bernoulli space, *i.e.* the space of complex linear combinations of the Bernoulli random variables x_i, and $B(u,v) = \mathbb{E}[uv]$. We identify Ψ, as a vector space, with the exterior (Grassmann) algebra $\Lambda(\mathcal{H})$ over $\mathcal{H}$. Indeed, the k-th exterior power of $\mathcal{H}$ has a basis consisting of the vectors $x_{i_1} \wedge \ldots \wedge x_{i_k}$ with $i_1 < \ldots < i_k$, hence it can be identified with the k-th chaos of toy Fock space (with $\mathbb{C}$ for $k = 0$). Then the Clifford product is characterized as the unique associative multiplication on Ψ with $\mathbf{1}$ as unit, such that

$$(6.1) \qquad\qquad uv + vu = B(u,v)\mathbf{1} \quad \text{for } u,v \in \mathcal{H}$$

Indeed this says the same thing as (5.1) for the basis (x_i), and the linear independence statement in subsection 5 has been hidden in the assumption that the underlying space of the algebra is $\Lambda(\mathcal{H})$.

> This description can be extended to possibly degenerate bilinear forms (when the bilinear form is 0, one gets the exterior multiplication). We give additional details in Appendix 5.

To develop non-commutative probability, an involution is necessary. It requires an additional element of structure : $\mathcal{H}$ must be a complex Hilbert space with a *conjugation* $u \longmapsto \bar{u}$. In the Bernoulli case it is the standard conjugation in the space of complex linear combinations of the real r.v.'s x_i. This conjugation is used to construct the bilinear form $B(u,v)$ from the complex scalar product, as $< \bar{u}, v >$. It is then easy to show there is a unique involution of the Clifford algebra such that $x^* = \bar{x}$ on the first chaos $\mathcal{H}$.

For algebraists, exterior multiplication is more elementary than Clifford multiplication, but for us the "elementary" thing was Bernoulli space, and we did not mention at all exterior algebra. The advantage of our apparently crooked path is its easy extension to infinitely many dimensions, replacing Bernoulli space by Wiener space.

7 Looking back on this discussion, exterior multiplication appears as another interesting multiplication on Bernoulli space, with multiplication table

$$(7.1) \qquad x_A \, x_B = (-1)^{n(A,B)} \, x_{A \cup B} \quad \text{if} \quad A \cap B = \emptyset \, , \quad = 0 \quad \text{otherwise.}$$

This illustrates the vague philosophy that many interesting algebra structures, commutative or not, can grow on the same space. The exterior and Clifford multiplications are related on $\mathcal{H}$ by the equality $u \wedge v = uv - B(u,v)\mathbf{1}$, and there are interesting formulas (some of which will appear in the course of these notes) relating the two products for higher order monomials.

Returning now to Bernoulli, we are led by analogy with (7.1) to a new *commutative* multiplication on toy Fock space, whose table is given by

$$(7.2) \qquad x_A \, x_B = x_{A \cup B} \quad \text{if} \quad A \cap B = \emptyset \, , \quad = 0 \quad \text{otherwise} \quad .$$

This multiplication is called the *Wick product* on Bernoulli space.

Linear spaces of Bernoulli random variables

8 Probabilists are familiar with *Gaussian spaces, i.e.* linear spaces of random variables, every one of which has a Gaussian law. If $\mathcal{H}$ is a real Gaussian space, we have for $h \in \mathcal{H}$

$$\mathbb{E}\left[e^{iuh}\right] = e^{-q(h)/2} \quad \text{with } q(h) = \mathbb{E}\left[h^2\right]$$

and therefore q is a positive quadratic form, which determines the probabilistic structure of the space. Then the Wiener chaos theory (chapter IV) describes the full *algebra* generated by the space $\mathcal{H}$.

The probabilistic idea behind the preceding description of Clifford algebras is that of a *linear space $\mathcal{H}$ of Bernoulli random variables*, "Bernoulli" meaning that the support of the r.v. has at most two points. This cannot be realized in commutative probability, but here every real linear combination $y = \sum_i a_i x_i$ of the basic Bernoulli random variables x_i has the property that y^2 (Clifford square) is a constant, and therefore y assumes only two values. In the vacuum state y has a symmetric Bernoulli law, but in other states it may not be symmetric. An additional term $a_0 \mathbf{1}$ may be allowed in the sum, thus allowing degeneracy at points other than 0. Conversely, Parthasarathy has shown in [Par2] that linear spaces of two-valued random variables lead necessarily to Clifford algebras. The analogy between Gauss in the commutative case and symmetric Bernoulli in the anticommutative case is important, and should be kept in mind.

There is a commutative analogue of Clifford algebras, namely the *Gaussian algebras* defined by Dynkin [Dyn], but it is far less important from the algebraic point of view.

Automorphisms of the Clifford structure

9 The Hilbert space Φ (anticommutative Fock space) carries three distinguished operators, which behave in a definite way with respect to the Clifford multiplication on Ψ.

Given an integer n, let us define

$$\rho(n) = (-1)^{n(n+1)/2}\ , \quad \sigma(n) = (-1)^n\ , \quad \tau(n) = (-1)^{n(n-1)/2}\ .$$

Given an element x of the n-th chaos in Ψ, define $\sigma(x) = \sigma(n)\,x$ (and similarly for ρ, τ) and extend to Ψ by linearity. Then it turns out that $\tau = \rho\sigma = \sigma\rho$, that σ is an automorphism of the exterior and Clifford algebra structures, and ρ, τ are antiautomorphisms (*i.e.* reverse the order of products). We have $\rho^2 = \sigma^2 = \tau^2 = I$. Sometimes σ is called the *main automorphism* and τ the *main antiautomorphism* of the structure. The involution on the Clifford algebra is the composition of ρ and the conjugation.

> One may add to this list the involution $*$, and the "Fourier transform" $\mathcal{F}$ we introduced in the commutative case, which exchanges the operators r_i and s_i.

The operator σ is also called the *parity* : it has the even and odd subalgebras as eigenspaces with eigenvalues ± 1 respectively. It follows easily that it anticommutes with every operator r_k or s_k, since these exchange the two subalgebras.

In the tensor product representation of subsection 2, σ is written as $\sigma_z^{\otimes N}$. Looking at the computation of the operators r_i, s_i in the same tensor product representation, and using the relation $\sigma_x\sigma_y = i\sigma_z$, we see that the product $r_1 s_1 \ldots r_N s_N$ is equal to $i^N \sigma_z^{\otimes N}$. However, σ has square I and anticommutes with the $2N$ operators r_i, s_j.

> In particular, for $N = 2$ r_1, s_1, r_2, s_2 are the four Dirac matrices $\gamma_1, \ldots, \gamma_4$, and their product is called $-i\gamma_5$ and anticommutes with all of them.

Integrals

10 The concrete Clifford algebra carries a natural probability law, which is the state associated with the vacuum vector. It is given by

$$\rho(x_A) = <\mathbf{1}, r_A \mathbf{1}> = 1 \quad \text{if } A = \emptyset,\ 0 \text{ otherwise.}$$

It is important to note that $\rho(x_A x_B) = \rho(x_B x_A)$: the vacuum vector defines a tracial state. On the other hand, the vacuum vector defines also a state on the full matrix algebra (which is also the Clifford algebra generated by the r_A and s_B taken together), on which it is clearly not a trace.

The value of the vacuum state on an operator expanded as $\sum_A c_A x_A$ is the coefficient $c_\emptyset$ in the lowest chaos. If we take the coefficient in the highest chaos, we get a linear functional (which is not a state for the ordinary involution) called the *Berezin integral*. It has many interesting properties, and no obvious extension to the infinite dimensional case.

REMARK. Let $\mathcal{H}$ be a real or complex linear space with a bilinear scalar product. The elements of the associated Clifford algebra (and modules over this algebra) are called *spinors*. They appear in physics in two ways : the "quantum probabilistic" way described in this chapter, and also as a way of handling the orthogonal group of $\mathcal{H}$ (the case of the Lorentz group being particularly important). This is a topic we leave aside.

Chapter III

The Harmonic Oscillator

This chapter is the closest in these notes to what is usually called "Quantum Mechanics". The present version is considerably shorter than the original French. It thus becomes more obvious that its main topic is not really elementary quantum mechanics, but rather elementary Fock space, and the quantum analogue of finite dimensional Gaussian random variables.

§1. SCHRÖDINGER'S MODEL

The canonical pair

1 Everyone has heard about the "quantization procedure" which leads from a classical Hamiltonian

$$(1.1) \qquad H\left(p_1, q_1, \ldots, p_n, q_n\right),$$

on the $2n$–dimensional phase space $\mathbb{R}^{2n}$, to the quantum mechanical Hamiltonian

$$(1.2) \qquad H\left(P_1, Q_1, \ldots, P_n, Q_n\right),$$

Q_j denoting the multiplication operator on $L^2(\mathbb{R}^n)$ corresponding to the coordinate mapping x_j, while $P_j = -i\hbar D_j$ is a partial derivative operator. For instance, if (1.1) is the Hamiltonian for a system of n particles of mass m interacting through a potential V

$$\sum_j \frac{p_j^2}{2m} + V(q_1, \ldots, q_n),$$

then the quantum mechanical Hamiltonian is the well known Schrödinger operator $-\frac{\hbar^2}{2m}\Delta + V$. Except in special cases like this one, the non-commutativity of the operators P_j and Q_j makes it a difficult problem to give a definite meaning to the operator (1.2), and much work has been devoted to this "quantization problem", which does not concern us here.

Let us deal with the case $n = 1$ for simplicity. A *canonical pair* is informally described as a pair of self-adjoint operators P, Q on a Hilbert space $\mathcal{H}$, satisfying (Heisenberg's) *canonical commutation relation* (CCR)

$$(1.3) \qquad [P, Q] = -i\hbar I .$$

However, this definition is not satisfactory : even if one postulates that (1.3) holds on a dense stable domain $\mathcal{D}$ which is a core for both operators, pathological examples can be given (see Reed and Simon [ReS], end of VIII.5). The standard way out of this difficulty consists in translating the CCR into a statement concerning bounded operators, the *Weyl form* of the CCR, described in the next subsection.

2   Schrödinger's model of the canonical pair is given by

$$(2.1) \qquad \mathcal{H} = L^2(\mathbb{R}) \ , \quad Qf(x) = xf(x) \ , \quad Pf(x) = -i\hbar Df(x) \ .$$

In the commutation relation (1.3) position Q and momentum P play nearly symmetric roles, but the explicit realization (2.1) privileges (diagonalizes) position. The corresponding "momentum representation" is deduced from this one using a conveniently normalized Fourier–Plancherel transform,

$$(2.2) \qquad \widehat{g}(u) = \int e^{-iux/\hbar} g(x) dx \ , \quad dx = dx/\sqrt{2\pi\hbar} \ .$$

which maps $Q_x g$ into $P_u \widehat{g}$ and $P_x g$ into $-Q_u \widehat{g}$, so that P gets diagonalized.

Domains can be exactly described : the domain of Q consists of those "wave functions" f such that $xf \in L^2$, and the domain of P of those f whose derivative in the distribution sense belongs to L^2. The commutation relation (1.3) holds on the Schwartz space $\mathcal{S}$, which is a core for both operators. The unitary groups $U_r = e^{-irP}$ and $V_s = e^{-isQ}$ can be made explicit

$$(2.3) \qquad U_r f(x) = f(x - \hbar r) \ , \quad V_s f(x) = e^{-isx} f(x) \quad ;$$

and it is easy to verify that

$$(2.4) \qquad U_r V_s = e^{i\hbar rs} V_s U_r \ .$$

This is the most elementary version of the Weyl form of the CCR, and what we precisely call a (one dimensional) canonical pair in this chapter will be *a Hilbert space $\mathcal{H}$ provided with two unitary groups $(U_r), (V_s)$ satisfying* (2.4). The canonical pair is said to be *irreducible* if the only closed subspaces invariant under both groups are the trivial ones $\{0\}$ and $\mathcal{H}$. The interest of this definition lies in the fact that *every irreducible canonical pair is isomorphic to Schrödinger's model* (the Stone–von Neumann theorem, subs. 6).

One can also express (2.4) as a relation between one of the unitary groups (say, U_r) and the spectral family (E_t) of the other one. In the explicit Schrödinger representation, we have

$$E_t f(x) = I_{]-\infty, t]}(x) f(x)$$

and (2.4) is equivalent to

$$(2.5) \qquad U_r E_t = E_{t + \hbar r} U_r \ .$$

This can be interpreted as a *covariance property* of the spectral measure under the group G of translations of the line : let $g \in G$ be a translation $x \longmapsto x + u$, acting on functions by $\tau_g f(x) = f(g^{-1}x) = f(x - u)$. Then we have, for $A =]-\infty, t]$ and then for every Borel set of the line

$$(2.6) \qquad \tau_g E_A = E_{g(A)} \ .$$

This is a simple example of an *imprimitivity system* in the sense of Mackey, the general form of which concerns the covariance of a spectral measure on a space E under a locally compact group G of symmetries of E.

Weyl operators

3 Weyl suggested the following approach to the quantization problem of associating a quantum observable $F(P, Q)$ with a function $F(p, q)$ on classical phase space. First define the unitary operators $e^{-i(rP+sQ)}$ for real r, s. Then take the Fourier transform of the classical observable

$$\widehat{F}(r, s) = \int e^{-i(rp+sq)} F(p, q) \, dp \, dq \ ,$$

and transform it backwards into the operator,

$$F(P, Q) = \int e^{i(rP+sQ)} \widehat{F}(r, s) \, dr \, ds \ .$$

For a discussion of this method, which is extremely interesting as a version of pseudo-differential calculus, see Folland [Fol]. Here we shall be concerned only with the definition of the Weyl operators

(3.1) $$W_{r,s} = e^{-i(rP-sQ)}$$

and, (by differentiation) of the selfadjoint unbounded operators $rP - sQ$. To conform with "symplectic" tradition, we have inserted a minus sign in (3.1).

A reasonable meaning for (3.1) is suggested by the elementary Campbell–Hausdorff formula for a pair of finite dimensional matrices A and B which commute with their commutator $[A, B]$

(3.2) $$e^{A+B} = e^A \, e^B \, e^{-[A,B]/2} = e^B \, e^A \, e^{[A,B]/2} \ .$$

We try this on $A = -irP$, $B = isQ$, $[A, B] = rs[P, Q] = -i\hbar rs\, I$, and we have on the model

(3.3) $$W_{r,s}f(x) = e^{-i\hbar rs/2} \, e^{isx} \, f(x - \hbar r) \ ,$$

which is obviously unitary. It is interesting to define (just for a few lines) $W_{r,s,t} = e^{-it}W_{r,s}$, in order to check on (3.3) that

(3.4) $$W_{r,s,t} \, W_{r',s',t'} = W_{r+r',\, s+s',\, t+t'-\hbar(rs'-sr')/2} \ .$$

The composition law of the parameters defines the so called *Heisenberg group* multiplication on $\mathbb{R}^3$. In particular, taking $t = 0$ we get the *Weyl form of the CCR*

(3.5) $$W_{r,s} \, W_{r',s'} = e^{-i\hbar(rs'-sr')/2} \, W_{r+r',\, s+s'} \ ,$$

which shows in particular that the operators $W_{r,s}$ constitute a *projective unitary representation* (*i.e.* a unitary representation up to a phase factor) of the additive group $\mathbb{R}^2$.

If we put $W_z = W_{r,s}$ for $z = r + is$, we get the *complex form of the CCR*, which is particularly useful

(3.6) $$W_z W_{z'} = e^{-i\hbar \Im(\bar{z}z'/2)} W_{z+z'}$$

It becomes meaningful for any complex Hilbert space, if $\bar{z}z'$ is understood as $<z, z'>$. We shall return to this general form of the CCR in chapter IV.

As for the operator $rP - sQ$ itself, we may now *define* it as the selfadjoint operator $i \frac{d}{dt} W_{tr,ts}|_{t=0}$. One can prove that on the Schwartz space $\mathcal{S}$ this operator coincides with the linear combination $rP - sQ$, and that it admits $\mathcal{S}$ as a core.

REMARKS. a) In classical mechanics, much importance is given to *canonical transformations*, which preserve the symplectic structure. Here, a linear transformation

$$P' = aP + bQ, \quad Q' = cP + dQ$$

preserves the Heisenberg commutation relation, *i.e.* satisfies $[P', Q'] = -i\hbar I$, if and only if $ad - bc = 1$. It is safer not to deal with unbounded operators, and to remark that the operators $W'_{r,s} = W_{ar+cs, br+ds}$ satisfy the Weyl CCR.

b) Given any probability law ρ for the canonical pair (*i.e.* any positive operator ρ of trace 1 on $L^2(\mathbb{R})$), we may define its *(quantum) characteristic function* by the formula

(3.7) $$F(r,s) = \mathbb{E}[e^{-i(rP-sQ)}] = \mathrm{Tr}(\rho W_{r,s}) .$$

Note that, for $r = 0$ and $s = 0$, the quantum c.f. gives the ordinary characteristic functions of the observables Q, P. For instance, if ρ is the pure state associated with a normalized "wave function" f, we have according to (3.4)

$$F(r,s) = <f, W_{r,s}f> = e^{-i\hbar rs/2} \int \bar{f}(y) f(y - \hbar r) e^{isy} dy$$

(3.8) $$= \int \bar{f}(x + \hbar\tfrac{r}{2}) f(x - \hbar\tfrac{r}{2}) e^{isx} dx$$

(we find it convenient here to use the Hilbert space $L^2(dx)$ instead of dx). The properties of $F(r,s)$ are similar to those of a characteristic function in classical probability : it is continuous and bounded by 1 in modulus, and satisfies a suitable "positive type" property. See for instance [Moy], [Fan], [Poo], [CuH].

Gaussian states

4   As in classical probability theory, a state ρ for the canonical pair is called (centered) *Gaussian* if its characteristic function is the exponential of a quadratic form. This form must be negative since F is bounded, hence the quantum c.f. is of positive type in the usual sense. For example, let us take $\hbar = 1$ and consider the pure state corresponding to the square root of the Gaussian density w.r.t. the measure dx : $f = c^{-1/4} e^{-x^2/4c}$. Then the characteristic function is easily seen to be the exponential of the quadratic form $-(cs^2/2 + r^2/8c)$, which at first sight seems not to be symmetric. However, we have been working in the "position representation" which privileges the operator Q. To reestablish the balance between P and Q let us set $c = c_q$ and define c_p by the (uncertainty) relation $c_p c_q = 1/4$. Then the quadratic form can be written symmetrically as $-\frac{1}{2}(c_p r^2 + c_q s^2)$. We shall see later other examples of quantum Gaussian laws.

Minimal uncertainty

5 The minimal uncertainty states of the canonical pair are Gaussian. In order to find them, we first reproduce the classical proof (by H. Weyl) of the Heisenberg uncertainty relation. In this section we assume $\hbar = 1$.

> This is not the best normalization for probability theory. The product of variances $\hbar^2/4$ in the Heisenberg uncertainty relation should be made equal to 1, and thus $\hbar = 2$, the standard choice in these notes. On the other hand, having the momentum operator equal to $-2iD$ is hardly acceptable for physics.

A normalized element ω of $\mathcal{H}$ such that $\mathbb{E}_\omega[Q^2] < \infty$ and $\mathbb{E}_\omega[P^2] < \infty$ is a function $\omega(x)$, with weak derivative $\omega'(x)$ (defined a.e.) such that

$$\int |\omega(x)|^2\, dx = 1\,, \quad \int |\omega'(x)|^2\, dx < \infty\,, \quad \int |x|^2 |\omega(x)|^2\, dx < \infty\,.$$

The existence of a derivative in the L^2 sense implies that ω has a continuous representative. We are going to prove

(5.1) $$\mathbb{E}_\omega[Q^2]\,\mathbb{E}_\omega[P^2] = \Big(\int |x|^2 |\omega(x)|^2\, dx\Big)\Big(\int |\omega'(x)|^2\, dx\Big) \geq 1/4$$

This will follow from $\int |\, x\omega(x)\,\omega'(x)|\, dx \geq 1/2$, and the Schwarz inequality. On every finite interval (a,b) we have, using integration by parts

$$2\int_a^b x\omega\omega'\, dx = -\int_a^b \omega^2\, dx + x\omega^2(x)\,\Big|_a^b\,.$$

Since the continuous function $x\,\omega(x)$ belongs to L^2, it cannot remain bounded away from 0 at $\pm\infty$, and we can let $b\uparrow +\infty$ and $a\downarrow -\infty$ in such a way that the last term on the right tends to 0, thus getting

$$2\int x\omega\omega'\, dx = -\int \omega^2\, dx\,.$$

If ω is real, the right side is equal to -1 and we are finished. If ω is complex, we remark that $|\omega|$ still is derivable in L^2 sense, with a derivative whose modulus is smaller than $|\omega'|$; applying the preceding inequality to $|\omega|$ then gives the required result.

This proof allows us to find the normalized functions ω which achieve the minimum in the preceding inequality. First of all, if ω achieves the minimum so does $|\omega|$, therefore we may assume that ω is real. Then we have equality in the Cauchy–Schwarz inequality $(\int x\omega\omega'\, dx)^2 \leq (\int \omega^2\, dx)(\int x\omega'^2\, dx)$, hence $\omega'(x) = c\,x\omega(x)$ for some constant c (which cannot be ≤ 0 if ω has to be in L^2). Finally we get a Gaussian wave function (normalized here with respect to dx)

(5.2) $$\omega(x) = c^{-1/4} e^{-x^2/4c}$$

There might exist complex functions with the same modulus, also achieving the minimum, so let us set $\omega = \theta|\omega|$ with $|\theta| = 1$; since $|\omega|$ is C^∞ and does not vanish, θ is locally derivable in L^2. We may write $\omega' = \theta|\omega|' + \theta'|\omega|$, from which follows easily that $|\omega'| > |\omega|'$ unless $\theta' = 0$ and equality in (5.1) is achieved only if θ is constant.

From these states we get all wave functions of minimal *uncertainty* as not necessarily centered Gaussian wave functions, *i.e.* wave functions (5.2) shifted in position and momentum space by some constant.

> We shall see later the relation of these minimal uncertainty wave functions with the *coherent states* of the harmonic oscillator.

The Stone–von Neumann theorem

6 The *Stone-von Neumann theorem* shows that any irreducible realization of the Weyl commutation relations is isomorphic to Schrödinger's model. This result is particularly attractive for probabilists, since it has an interpretation in the theory of flows : in discrete time it has been thus rediscovered by O. Hanner (1950), and in continuous time by Kallianpur and Mandrekar [KaM]. See also J. de Sam Lazaro and P.A. Meyer [LaM]. We are going to sketch the probabilistic proof, translating it into pure Hilbert space language.

> On the other hand, the theorem has a more general interpretation in Mackey's theory of imprimitivity systems.

We are given a Hilbert space $\mathcal{H}$, provided with an increasing and right continuous family of subspaces $(\mathcal{H}_t)_{t\in\mathbb{R}}$. It is convenient to assume that $\mathcal{H}_\infty = \mathcal{H}$ and $\mathcal{H}_{-\infty} = \{0\}$, but this last hypothesis will be used only at the end of the proof. We denote by E_t the orthogonal projection on $\mathcal{H}_t$. In the probabilistic interpretation, $\mathcal{H}$ and $\mathcal{H}_t$ are spaces $L^2(\Omega,\mathcal{F},\mathbb{P})$ and $L^2(\Omega,\mathcal{F}_t,\mathbb{P})$ relative to an increasing and right continuous family of σ–fields $\mathcal{F}_t$, and E_t is the conditional expectation $\mathbb{E}[\,\cdot\,|\,\mathcal{F}_t]$.

We are also given a unitary group U_t, which in the probabilistic situation is a group of automorphisms Θ_t of the measure space Ω. The covariance property (2.5) means that for all s,t

$$(6.1) \qquad\qquad f \in \mathcal{H}_s \implies U_t f \in \mathcal{H}_{s+t} .$$

A *screw line* (or *helix*) is a curve $x(t)$ in $\mathcal{H}$ with the following properties. First, $x(0) = 0$, the curve is adapted (*i.e.* $x(t) \in \mathcal{H}_t$) for $t > 0$ and has orthogonal increments, *i.e.* for $0 \le s \le t$ $x(t) - x(s)$ is orthogonal to $\mathcal{H}_s$ — in the probabilistic context, this is the martingale property. Next, the curve has stationary increments : $U_h(x(t) - x(s)) = x(t + h) - x(s + h)$ for all s,t,h (in probabilistic language, it is an "additive functional"). The heart of the Stone-von Neumann theorem is the following result :

THEOREM.   *If the filtration is not deterministic (i.e. if $\mathcal{H}_{-\infty} \neq \mathcal{H}$) there exists a (non-zero) screw line.*

PROOF.   Due to the stationarity, the hypothesis that the filtration is not deterministic means that the Hilbert space $\mathcal{E} = \mathcal{H}_0$ is not equal to $\mathcal{H}$. Given now $f \in \mathcal{E}$, put for $t \ge 0$

$$P_t f = E_0 U_t f \in \mathcal{E} .$$

We leave it to the reader to check that $(P_t)_{t\ge0}$ is a strongly continuous semigroup of contractions of $\mathcal{E}$. We denote by A its generator, with domain $\mathcal{D}(A)$. We also put $f \circ X_t = U_t f$, a notation which looks strange, but has an intuitive content since, in the probabilistic case, X_t may be interpreted as a mapping from $(\Omega, \mathcal{F})$ to the "state space" $\mathcal{E} = (\Omega, \mathcal{F}_0)$, and these mappings constitute a *Markov process* with semigroup (P_t).

Our next point (almost trivial) is to check that for $f \in \mathcal{D}(A)$ the following curve, defined for $t \geq 0$

(6.2)
$$x(t) = f \circ X_t - f \circ X_0 - \int_0^t Af \circ X_s \, ds$$

is a screw line, or rather can be extended to the whole of $\mathbb{R}$ so as to become a screw line. This is also left to the reader. So finally we must only show that if all these screw lines are equal to zero, then the filtration is deterministic, *i.e.* $\mathcal{E} = \mathcal{H}$.

The assumption that all the screw lines (6.2) are equal to zero can be stated as follows : if $f \in \mathcal{D}(A)$, then we have (recalling that $f \circ X_t = U_t f$ for $f \in \mathcal{E}$)

$$U_t f = U_0 f + \int_0^t U_s Af \, ds \ .$$

Otherwise stated, if we denote by $(B, \mathcal{D}(B))$ the generator of the semigroup $(U_t)_{t \geq 0}$, $f \in \mathcal{D}(A) \Longrightarrow f \in \mathcal{D}(B)$ and $Af = Bf$. Now let $R_p = \int_0^\infty e^{-ps} P_s \, ds$ be (for $p > 0$) the resolvent of (P_t) and similarly S_p be the resolvent of (U_t). For $u \in \mathcal{E}$ the vector $f = R_p u$ belongs to $\mathcal{D}(A)$ and $Af = pf - u$, therefore it also belongs to $\mathcal{D}(B)$ and $Bf = pf - u$, which implies $S_p u = f$. Since this is true for all $p > 0$ inverting the Laplace transform gives $U_t u = P_t u$ on $\mathcal{E}$, implying that for $t \geq 0$ U_t maps $\mathcal{E}$ into itself. Then it follows from (6.1) that $\mathcal{H}_0 = \mathcal{H}$, which implies easily that the filtration is deterministic.

It is worthwhile to describe the screw line $x(t)$ which generates the model itself : for $t > 0$ $x(t)$ is the indicator function $I_{]0,t]}$, and for $t < 0$ it is equal to $-I_{]t,0]}$.

The non-trivial part of the proof is finished. We consider a non-zero screw line, and remark that $\| x(t) - x(s) \|^2 = c|t-s|$ by the stationarity of increments. We may assume the screw line is normalized so that $c = 1$. Then we may define for $f \in L^2(\mathbb{R})$ the "stochastic integral" $I(f) = \int f(s) \, dx(s)$, which carries isometrically $L^2(\mathbb{R})$ onto a closed subspace $M(x)$ of $\mathcal{H}$. It is trivial to check that $M(x)$ is invariant under the operators E_t and U_t, the first one corresponding to the multiplication of f by $I_{]-\infty,t]}$ and the second one to a shift of f by t. *Otherwise stated, the situation induced on* $M(x)$ *is isomorphic to Schrödinger's model*. Now we restrict ourselves to the stable space $M(x)^\perp$; if the induced filtration is not deterministic we can extract again a screw line and a copy of the model, and so on until we reach a residual stable space on which the induced filtation is deterministic (the "and so on" can be replaced, as usual in mathematics, by a rigorous argument beginning with the sentence "consider a maximal family of mutually orthogonal normalized screw lines..."). If we have $\mathcal{H}_{-\infty} = \{0\}$, this residual space is also equal to $\{0\}$, and we have decomposed $\mathcal{H}$ into a direct sum of copies of the model (at most countably infinite since $\mathcal{H}$ is always assumed to be separable).

7 If an irreducible canonical pair exists at all (*i.e.* one for which the only closed stable subspaces are the trivial ones) then by the above proof it must consist of one single copy of the model. Let us prove that the model itself is irreducible. According to the Stone–von Neumann theorem, if the model were reducible into two orthogonal stable subspaces, it would contain two orthogonal screw-lines; thus it is sufficient to prove there is only one normalized screw-line $\xi(t)$. Let $f(x)$ be the function $\xi(1) \in L^2$; according to adaptation, we have $f(x) = 0$ (a.e.) for $x \geq 1$. The orthogonality of increments implies that $\xi(1) = \xi(1) - \xi(0)$ is orthogonal to $E_0 \mathcal{H}$, i.e. $f(x) = 0$ a.e. for $x < 0$. Writing that $f(x) - f(x-t) = 0$ a.e. on $] - \infty, t]$ for $0 < t < 1$, we find that f is a.e. constant on $[0,1]$, and by normalization $\xi(t)$ is the fundamental screw line $I_{]0,t]}$ $(t \geq 0)$, whence the conclusion.

8 The Stone–von Neumann theorem is a powerful result. For instance, set $U'_t = V_{-t}$, $V'_t = U_t$; this is a second pair of unitary groups which satisfies the Weyl relation, and which is evidently irreducible; since the model is unique there must exist a unitary operator on $L^2(\mathbb{R})$ which transforms the first pair into the second one. We have proved by "abstract nonsense" that the Fourier–Plancherel transform is unitary ! Also, consider a minimal uncertainty Gaussian wave function in the position representation. Minimal uncertainty is a property of the state itself, and does not depend on the representation of the canonical pair we use; hence the Fourier transform of a minimal uncertainty wave function is a minimal uncertainty wave function in the momentum representation, and we get the well known result that the Fourier transform of a Gaussian is Gaussian[1].

9 Here is another interesting consequence of the Stone–von Neumann theorem. Let us consider on some (separable) Hilbert space $\mathcal{K}$ a family of operators $\widetilde{W}_{r,s}$ which satisfy the Weyl commutation relations. According to the Stone–von Neumann theorem, $\mathcal{K}$ is isomorphic to a countable sum of copies $\mathcal{H}_n$ of the model. Let f be a normalized vector, and let $f = \sum_n c_n f_n$ its decomposition along the spaces $\mathcal{H}_n$, the vectors f_n being normalized. Then we have

$$< f, \widetilde{W}_{r,s} f > = \sum_n |c_n|^2 < f_n, W_{r,s} f_n >$$

On the right side, we have done as if f_n were a vector in the model $\mathcal{H}$, not in its copy $\mathcal{H}_n$, and what we get is the characteristic function of the *mixture* $\sum_n |c_n|^2 \varepsilon_{f_n}$, denoting by ε_h the pure state corresponding to the vector h. This is an equality between two quantum characteristic functions, computed on the left side on a pure state of a reducible canonical pair, and on the right side on a mixture in the irreducible model. Thus, every law that can be realized on any canonical pair can also be realized on the model itself. The idea is somewhat the same as that of "canonical processes" in classical probability theory : one starts realizing the law of the process on some huge probability space, using possibly auxiliary processes, and then the law may be mapped to a reduced space (consisting generally of a minimal set of sample functions) from which all the auxiliary information has been erased.

We apply this idea as follows : we take a tensor product of two copies (P_1, Q_1) and (P_2, Q_2) of the Schrödinger model. More concretely, the basic Hilbert space is $L^2(\mathbb{R}^2)$, the first pair consists of the multiplication operator by the coordinate x_1 and the partial derivative operator $-i\hbar D_1$, and similarly for the second pair using the coordinate x_2. Then we set

$$\widetilde{Q} = aQ_1 + bQ_2 \quad ; \quad \widetilde{P} = aP_1 - bP_2$$

---

[1]   Another wonderful application concerns the structure of the endomorphisms of a *-algebra $\mathcal{L}(\mathcal{H})$. See Parthasarathy's book, Prop. 29.4, p. 253.

where a, b are real and such that $a^2 - b^2 = 1$: a formal computation shows that $[\widetilde{P}, \widetilde{Q}] = -i\hbar I$, so we have built a new (reducible) canonical pair. It is more rigorous to define the Weyl operators

$$\widetilde{W}_{r,s} f(x_1, x_2) = e^{-i\hbar rs/2} e^{-is(ax_1 + bx_2)} f(x_1 - a\hbar r, x_2 + b\hbar s)$$

and check that they satisfy the CCR. Let now f_1 and f_2 be two normalized vectors in the model, and f be their tensor product. Then the quantum characteristic function of the state f is

$$< f, \widetilde{W}_{r,s} f > \, = \, < f_1, W_{ar, as} f_1 > < f_2, W_{-br, bs} f_2 > .$$

In particular, if we take for f_1, f_2 two identical real Gaussian wave functions, we get a characteristic function of the form

$$\exp\left(-\frac{1}{2}(\lambda r^2 + \mu s^2)\right) \quad \text{with} \quad \lambda = (a^2 + b^2)c_p , \ \mu = (a^2 + b^2)c_q .$$

In this way we have constructed quantum Gaussian laws of arbitrary uncertainty product $> \hbar^2/4$. We shall use the same construction in the infinite dimensional situation, using the tensor product of two copies of the simple Fock model. But there one cannot get back to the model (the Stone–von Neumann theorem holds only for a finite number of canonical pairs), and some "non-Fock" representations of the CCR appear.

Up to now, we have seen only Gaussian quantum characteristic functions corresponding to diagonal quadratic forms. It is easy to deduce from them Gaussian characteristic functions which possess an off-diagonal term. Keep the same state (pure or mixed), but replace the Weyl family by a new one, using a linear canonical transformation (§1, subsection 2). Then the quadratic form gets "rotated" and is no longer diagonal, its discriminant remaining invariant in this transformation because of the relation $ad - bc = 1$. These general quantum Gaussian laws constitute the class of all possible limit laws in the *quantum central limit theorem* of Cushen–Hudson [CuH].

§2. CREATION AND ANNIHILATION OPERATORS

In this section, attention shifts from the canonical pair Q, P towards a pair of non-selfadjoint operators, the *creation and annihilation pair*, soon to be completed by the *number operator*, which is closely related to the harmonic oscillator Hamiltonian.

From now on, we adopt systematically the convention that $\hbar = 2$. This will eliminate many unwanted factors of 2 in probabilistic formulas. However, there exists no system of units such that $2 = 1$, and unwanted factors of 2 will appear elsewhere.

The harmonic oscillator, or elementary Fock space

1 We first review quickly Dirac's approach to the theory of the *quantum harmonic oscillator*. We start with the canonical pair in Schrödinger's form

$$(1.1) \qquad\qquad Q = x \ , \ P = -2iD \ ,$$

and define new operators $a^{\pm}$ on the common stable domain $\mathcal{S}$ for P and Q as follows

$$(1.2) \qquad\qquad Q = a^+ + a^- \ , \ P = i\,(a^+ - a^-) \ .$$

They are called respectively the *creation* and *annihilation* operator. The standard notation in physics books is $a = a^-$, $a^* = a^+$. They satisfy on $\mathcal{S}$ the commutation relation $[a^-, a^+] = I$, as well as the relation $<a^+ f, g> = <f, a^- g>$. Therefore each of them has a densely defined adjoint, and hence is closable. Their closures are denoted by the same notation, and one can prove they are exactly adjoint to each other (we shall never use this fact).

The operator $a^{\circ} = a^+ a^-$ is well defined on $\mathcal{S}$, and called the *number operator*. The notation a° is not standard, the usual one being N. On $\mathcal{S}$ the number operator is positive symmetric, and one can prove that it is essentially selfadjoint.

The explicit expression of the three operators on $\mathcal{S}$ is

$$(1.3) \qquad a^+ = \frac{1}{2}\,(x - 2D) \ , \ a^- = \frac{1}{2}\,(x + 2D) \ , \ a^{\circ} = \frac{1}{4}\,(x^2 - 4D^2 - 2I) \ .$$

All unpleasant factors of 2 will be eliminated in (2.2).

The general form of the Hamiltonians describing quantum harmonic oscillators (with one degree of freedom) is $H = aQ^2 + bP^2$ with $a, b > 0$. All elementary textbooks on quantum mechanics explain how they can be reduced to a standard form, for instance $(P^2 + Q^2)/4 = (x^2 - 4D^2)/4$, which by (1.3) can be expressed as $a^{\circ} + I/2$. We are going to compute all eigenvalues and eigenvectors of H (or equivalently of N).

Knowing these eigenvalues and eigenvectors would allow us to compute the unitary group e^{-itH} which describes the quantum evolution of a harmonic oscillator. However, we are not really interested in this *dynamical* problem. What we are doing will remain at a *kinematical* level, namely, will amount to another description of the canonical pair.

2 The wave function

$$(2.1) \qquad\qquad \varphi(x) = C = (2\pi)^{-1/4} e^{-x^2/4} \ ,$$

is normalized in $\mathcal{H} = L^2(\mathbb{R}, dx)$ and belongs to $\mathcal{S}$. It is killed by a^-, and therefore by $a^{\circ} = a^+ a^-$. Its square is the density $\gamma(x)$ of the standard Gaussian measure $\gamma(dx)$. Since φ is strictly positive, the mapping $f \to f/\varphi$ is an isomorphism of $L^2(\mathbb{R})$ onto $L^2(\gamma)$, and we may carry the whole structure on this last Hilbert space. We thus get a new model for the canonical pair, called the *real Gaussian model* :

$$(2.2) \qquad \mathcal{H} = L^2(\gamma) \ , \ a^+ = (x - D) \ , \ a^- = D \ , \ a^{\circ} = N = -D^2 + xD$$

The operator $Q = a^+ + a^-$ is still equal to x, while the operator $D^2 - xD = -N$ is well known to probabilists as the generator L of a Markov semigroup (P_t), symmetric with respect to the measure γ, which is called the *Ornstein–Uhlenbeck* semigroup. Therefore L is negative selfadjoint and N itself has a natural positive selfadjoint extension. It is true (we shall neither prove nor need it) that this extension is simply the closure of N on $\mathcal{S}$, i.e. N is essentially s.a. on $\mathcal{S}$. Finally, the vector φ has become the constant function 1. We call it the *vacuum vector* and denote it by $\mathbf{1}$.

In the Gaussian model it is clear that the space $\mathcal{P}$ of all polynomials is stable under a^+, a^-. On the other hand, it is dense in $\mathcal{H}$: this is a classical result, which can be deduced from the fact that trigonometric polynomials are dense, and the exponential series for e^{iux} is norm convergent. One can show that P and Q are essentially selfadjoint on $\mathcal{P}$.

3 We are going to find all the eigenvalues and eigenfunctions of $N = a^\circ$. The computation rests on the simple formulas, valid on the stable domain $\mathcal{P}$

$$Na^+ = a^+(N+I) \,, \qquad Na^- = a^-(N-I) \,.$$

The first one implies that, given an eigenvector $f \in \mathcal{P}$ of N with eigenvalue λ, $a^+ f$ is either 0, or an eigenvector of N with eigenvalue $\lambda + 1$. Starting with $f = \mathbf{1}$, $\lambda = 0$ we construct vectors $h_n = (a^+)^n \mathbf{1}$, such that $Nh_n = n\,h_n$, and

$$< h_{n+1}, h_{n+1} > \; = \; < a^+ h_n , a^+ h_n > \; = \; < h_n , a^- a^+ h_n > \; =$$
$$= \; < h_n , (N+I)\, h_n > \; = \; (n+1) < h_n , h_n > \,,$$

from which one deduces that none of them is 0, and

(3.1) $$< h_n , h_n > \; = n!$$

The vectors h_n are read in the Gaussian model as the polynomials $h_n(x) = (x - D)^n \mathbf{1}$: they are the Hermite polynomials with the normalization convenient for probabilists (not for analysts). The first ones are

$$1 \,, \; x \,, \; x^2 - 1 \,, \; x^3 - 3x \dots \,.$$

Since there is one of them for each degree, they form a basis of the space of all polynomials. Since polynomials are dense in $\mathcal{H}$, the vectors $\varphi_n = (n!)^{-1/2} h_n$ constitute an orthonormal basis of $\mathcal{H}$ which completely diagonalizes the number operator. Here are the expressions of $a^\pm$ in this basis :

(3.2) $$a^+ \varphi_n = \sqrt{n+1}\, \varphi_{n+1} \,, \; a^\circ \varphi_n = n\varphi_n \,, \; a^- \varphi_n = \sqrt{n-1}\, \varphi_{n-1} \,.$$

From this we can deduce the matrices of P and Q, which were at the root of Heisenberg's (1925) "matrix mechanics".

In fact this orthonormal basis is less useful than the *orthogonal* basis (h_n). Here is the formula to be memorized : every element of $\mathcal{H}$ has an expansion

(3.3) $$f = \sum_n \frac{c_n}{n!} h_n \quad \text{with} \quad \| f \|^2 = \sum_n \frac{|c_n|^2}{n!} \,.$$

The Hilbert space consisting of all such expansions, and provided with the three operators a^ε, is the most elementary example of a *Fock space*. More precisely, it will be interpreted in Chapter IV as the (boson) Fock space over the one dimensional Hilbert space $\mathbb{C}$.

Weyl operators and coherent vectors

4   The Weyl operator W_z for $z = r + is$ (§1, subs. 3) was written formally as $e^{-i(rP-sQ)}$; replacing P, Q by their expressions above we can write

$$(4.1) \qquad W_z = e^{za^+ - \bar{z}a^-} = e^{za^+} e^{-\bar{z}a^-} e^{z\bar{z}[a^+,a^-]/2} = e^{-z\bar{z}/2} e^{za^+} e^{-\bar{z}a^-} .$$

According to (4.1) and the fact that a^- kills the vacuum vector, we have

$$(4.2) \qquad W_z \mathbf{1} = e^{-z\bar{z}/2} e^{za^+} \mathbf{1} = e^{-|z|^2/2} \sum_k \frac{z^k}{k!} a^{+k} \mathbf{1} = e^{-|z|^2/2} \sum_k \frac{z^k}{k!} h_k .$$

Since W_z is unitary, these vectors have norm 1. They are called *coherent vectors* and denoted by $\psi(z)$. Let us show that their linear span is dense in $\mathcal{H}$. To this end, it is more convenient to normalize them by the condition of having expectation 1 in the vacuum state instead of having norm 1. The vectors one gets in this way are called *exponential vectors*, and denoted by

$$(4.3) \qquad \mathcal{E}(z) = \sum_k \frac{z^k}{k!} h_k .$$

Given (3.1), we have

$$(4.4) \qquad \mathcal{E}(0) = \mathbf{1} , \qquad <\mathcal{E}(u), \mathcal{E}(v)> = e^{<u,v>} .$$

To see that the space $\mathcal{E}$ generated by all exponential (or coherent) vectors is dense in $\mathcal{H}$, it suffices to note that the n–th derivative at 0 of $\mathcal{E}(z)$ with respect to z (which belongs clearly to the closure of $\mathcal{E}$) is equal to h_n. The space $\mathcal{E}$ is used as a convenient domain for many operators.

Let us return for a short while to the Schrödinger representation. The vacuum vector then is read as $Ce^{-|x|^2/4}$ (2.1), and if $z = r + is$, formula (3.3) in §1 gives us for the coherent vector $\psi(z)$ the formula

$$(4.5) \qquad \begin{aligned} \psi(z) &= W_{r,s}\mathbf{1} = Ce^{-irs} e^{isx} e^{-(x-2r)^2/4} = Ce^{-r^2-irs} e^{-|x|^2/4} e^{zx} \\ &= Ce^{-(|z|^2+z^2)/2} e^{-|x|^2/4} e^{zx} . \end{aligned}$$

These are Gaussian states, closely related to the minimal uncertainty states of §1 subsection 4. The exponential vectors have the slightly simpler form $Ce^{zx-|z|^2/4}$.

Returning to the Gaussian representation, we see that the exponential vector $\mathcal{E}(z)$ is read as the function

$$(4.6) \qquad \mathcal{E}(z) = \sum_k \frac{z^k}{k!} h_k = e^{zx-z^2/2} .$$

We get in this way the well known generating function for Hermite polynomials. Also, we have an explicit formula for the action of Weyl operators on exponential vectors

$$(4.7) \qquad W_z \mathcal{E}(u) = e^{-\bar{z}u - |z|^2/2} \mathcal{E}(z+u) ,$$

which will become the basis for the extension of the theory to several (possibly infinitely many) dimensions, since the product $\bar{z}u$ can be interpreted as a complex scalar product, and formula (4.7) thus depends only on the complex Hilbert space structure.

We are going to describe in Chapter IV, in a very general situation, a more complete family of unitary "Weyl operators", which in our case act as follows (λ being a real parameter)

$$(4.8) \qquad W(z,\lambda)\,\mathcal{E}(u) = \exp\left(-\bar{z}ue^{i\lambda} - \frac{|z|^2}{2}\right)\mathcal{E}(ze^{i\lambda}+u)\,.$$

These operators are related to the limits of discrete Weyl operators in the de Moivre–Laplace theory of Chapter II, §3. We shall return to this subject in Chapter V.

The complex Gaussian model

5   Let us associate with every $h \in \mathcal{H}$ the function $\varphi_h : z \longmapsto\ <h,\mathcal{E}(z)>$ on $\mathbb{C}$. The mapping φ is antilinear and injective, since the exponential vectors are total in $\mathcal{H}$. The image of the vector $\sum_n \frac{c_n}{n!}h_n$ is the entire function $\sum_n \frac{c_n}{n!}z^n$, and in fact $\mathcal{H}$ is mapped onto a *Hilbert space of entire functions*, the exponential vector $\mathcal{E}(u)$ being transformed into the exponential function $e^{\bar{u}z}$, and the creation and annihilation operators being read respectively on this space as the operators of multiplication by z and of derivation. The scalar product is given by the formula

$$\text{if} \quad F(z) = \sum_n a_n \frac{z^n}{n!} \quad \text{and} \quad G(z) = \sum_n b_n \frac{z^n}{n!}\,, \quad \text{then} \quad <F,G> = \sum_n \frac{\bar{a}_n\,b_n}{n!}\,.$$

This means that $<z^m,z^n> = \delta_{mn}n!$, and this property is realized by the ordinary L^2 scalar product with respect to the Gaussian measure ϖ with density $(1/\pi)\,e^{-|z|^2}$ relative to the Lebesgue measure $dx\,dy$ on $\mathbb{C}$. Therefore instead of mapping $\mathcal{H}$ onto all of $L^2(\gamma)$ for a *real* Gaussian measure, we have mapped it on a subspace of $L^2(\varpi)$ for a *complex* Gaussian measure. This interpretation is due to Bargmann in finite dimensions, to Segal in infinitely many dimensions (Segal prefers to deal with a linear mapping on a space of antiholomorphic functions, but the idea is the same).

Thus we end up with four interpretations of the canonical pair Hilbert space : Schrödinger's $L^2(\mathbb{R})$, the real Gaussian $L^2(\gamma)$, the discrete Hilbert space ℓ^2 with the Heisenberg matrices, and the subspace of entire functions in the complex Gaussian space $L^2(\varpi)$.

Classical states

6   This subsection is borrowed mostly from Davies' book [Dav]. Our aim is to compute some explicit laws, and to give still another construction of quantum Gaussian laws.

First of all, let us denote by $\mathbb{P}_w$ the probability law (pure state) corresponding to the coherent vector $\psi(w)$, with $w = u + iv$. Setting $z = r + is$ we compute the characteristic function

$$\mathbb{E}_w\left[e^{-i\,(rP-sQ)}\right] = <\psi(w), W_{r,s}\psi(w)> = e^{-|w|^2}<\mathcal{E}(w), W_{r,s}\mathcal{E}(w)>$$

$$(6.1) \qquad = e^{-|w|^2}e^{-\bar{z}w-|z|^2/2}<\mathcal{E}(w),\mathcal{E}(z+w)> = e^{-\bar{r}w+\bar{w}z-|z|^2/2}\,.$$

This is a Gaussian ch.f. of minimal uncertainty, and we can deduce from it the classical ch.f.'s of the individual variables P, Q

$$\mathbb{E}[e^{isQ}] = e^{-2isu - s^2/2} \ , \quad \mathbb{E}[e^{-irP}] = e^{-2irv - r^2/2} \ ,$$

which means that replacing the vacuum by a coherent state has shifted the means of P and Q by $2r$ and $-2s$ respectively. Let us compute also the distribution of the number operator N, using the expansion of $\psi(w)$ in the orthonormal basis which diagonalizes N

$$(3.2) \qquad \mathbb{P}_w\{N = k\} = <\psi(w), I_{\{N=k\}}\psi(w)> = e^{-|z|^2}\frac{z^k}{k!}$$

which is a Poisson law with mean $e^{-|z|^2}$.

   Classical states are mixtures of coherent states, *i.e.* are integrals $\int \mathbb{P}_z \, \mu(dz)$ for some probability measure μ on $\mathbb{C}$. The corresponding density operator ρ is given by

$$<f, \rho g> = \int <f, \psi(z)> <\psi(z), g> \mu(dz)$$

and the matrix elements of ρ in the Hermite polynomial basis are easily computed. Taking for μ a Gaussian measure, we may construct in this way all possible Gaussian laws for the canonical pair.

   To describe the law of the observable N in a classical state, we use the generating function $\mathbb{E}[\lambda^N]$. Under the law $\mathbb{P}_z$ this function is $e^{(\lambda-1)|z|^2}$, which we integrate w.r.t. $\mu(dz)$. In particular, if μ is complex Gaussian law with density $\frac{a}{\pi}e^{-a|z|^2}$, ρ is diagonal in the Hermite polynomial basis (*i.e.* is a function of N) and we have

$$\mathbb{E}[\lambda^N] = \frac{a}{a + 1 - \lambda} = \frac{1 - b}{1 - \lambda b} \quad \text{with} \quad b = 1/1 + a \ .$$

Then $\mathbb{P}\{N = n\} = (1 - b)\, b^n$, a geometric law, and therefore

$$(6.3) \qquad \rho = e^{-cN}/\text{Tr}\,(e^{-cN}) \quad \text{with} \quad b = e^{-2c} \ .$$

The Ornstein–Uhlenbeck semigroup (P_t) has $-N$ as its infinitesimal generator, *i.e.* we have $P_t = e^{-tN}$. The density operator ρ thus appears to be an operator of this semigroup, normalized to have unit trace. This has a physical interpretation as follows. Consider a quantum system with a hamiltonian H bounded from below; then the density operator $e^{-H/kT}$ normalized (if possible) to have unit trace represents a statistical equilibrium of the system at absolute temperature T, k denoting the Boltzmann constant. Since the harmonic oscillator hamiltonian differs from N by a constant multiple of the identity, which disappears by normalization, the law (6.3) appears as an equilibrium state of the harmonic oscillator at a strictly positive temperature, while minimal uncertainty corresponds to zero temperature. These features appear again in the infinite dimensional case, however the Ornstein–Uhlenbeck semigroup operators no longer have a finite trace and the temperature states cannot be represented as mixtures of coherent states : they must be described by different representations of the CCR. This is similar to the fact that dilations of a Gaussian measure give rise to mutually singular measures if the space is infinite dimensional.

Chapter IV

Fock Space (1)

The preceding chapters dealt with the non-commutative analogues of discrete r.v.'s, then of real valued r.v.'s, and we now begin to discuss stochastic processes. We start with the description of Fock space (symmetric and antisymmetric) as it is usually given in physics books. Then we show that boson Fock space is isomorphic to the L^2 space of Wiener measure, and interpret on Wiener space the creation, annihilation and number operators. We proceed with the Poisson interpretation of Fock space, and the operator interpretation of the Poisson multiplication. We conclude with multiplication formulas, and the useful analogy with "toy Fock space" in chapter II, which leads to the antisymmetric (Clifford) multiplications. All these operations are special cases of Maassen's kernel calculus (§4).

§1. BASIC DEFINITIONS

Tensor product spaces

1   Let $\mathcal{H}$ be a complex Hilbert space. We consider its n–fold Hilbert space tensor power $\mathcal{H}^{\otimes n}$, and define

$$(1.1) \qquad u_1 \circ \ldots \circ u_n = \frac{1}{n!} \sum_{\sigma \in G_n} u_{\sigma(1)} \otimes \ldots \otimes u_{\sigma(n)} \, ,$$

$$(1.2) \qquad u_1 \wedge \ldots \wedge u_n = \frac{1}{n!} \sum_{\sigma \in G_n} \varepsilon_\sigma \, u_{\sigma(1)} \otimes \ldots \otimes u_{\sigma(n)} \, ,$$

G_n denoting the group of permutations σ of $\{1, \ldots, n\}$, with signature ε_σ. The symbol $\wedge$ is the *exterior product*, and $\circ$ the *symmetric product*. The closed subspace of $\mathcal{H}^{\otimes n}$ generated by all vectors (1.1) (*resp.* (1.2)) is called the *n–th symmetric (antisymmetric) power* of $\mathcal{H}$, and denoted by $\mathcal{H}^{\circ n}$ ($\mathcal{H}^{\wedge n}$). Usually, we denote it simply by $\mathcal{H}_n$, adding $\circ$ or $\wedge$ only in case of necessity. We sometimes borrow Wiener's probabilistic terminology and call it the *n–th chaos* over $\mathcal{H}$. We make the convention that $\mathcal{H}_0 = \mathbb{C}$, and the element $1 \in \mathbb{C}$ is called the *vacuum vector* and denoted by **1**.

Given some subspace U of $\mathcal{H}$ (possibly $\mathcal{H}$ itself) we also define the *incomplete n–th chaos* over U to be the subspace of $\mathcal{H}_n$ consisting of linear combinations of products (1.1) or (1.2) with $u_i \in U$. It does not seem necessary to have a general notation for it.

In the physicists' use, if $\mathcal{H}$ is the Hilbert space describing the state of one single particle, $\mathcal{H}_n$ describes a system of n particles of the same kind and is naturally called the *n–particle space*. The fact that these objects are identical is expressed in quantum mechanics by symmetry properties of their joint wave function w.r.t. permutations, symmetry and antisymmetry being the simplest possibilities. Mathematics allow other "statistics", but they have not (yet?) been observed in nature. We tend to favour bosons over fermions in these notes, since probabilistic interpretations are easier for bosons.

One can find in the literature two useful norms or scalar products on $\mathcal{H}_n$, differing by a scalar factor. The first one is that induced by $\mathcal{H}^{\otimes n}$,

$$(1.3) \qquad < u_1 \otimes \ldots \otimes u_n, v_1 \otimes \ldots \otimes v_n > \; = \; < u_1, v_1 > \ldots < u_n, v_n > \; .$$

According to (1.2) we then have, in the antisymmetric case for instance (with σ and τ ranging over the permutation group G_n)

$$< u_1 \wedge \ldots \wedge u_n, v_1 \wedge \ldots \wedge v_n > \; = \; (\frac{1}{n!})^2 \sum_{\sigma,\tau} \varepsilon_\sigma \, \varepsilon_\tau \; < u_{\sigma(1)}, v_{\tau(1)} > \ldots < u_{\sigma(n)}, v_{\tau(n)} > \; .$$

Summing over τ cancels one of the factors $1/n!$, but it is more convenient to cancel both of them, and to define

$$(1.4) \qquad < u_1 \wedge \ldots \wedge u_n, v_1 \wedge \ldots \wedge v_n >_\wedge \; = \; \det < u_i, v_j > \; .$$

Similarly we have, replacing the determinant by a *permanent* (which has the same definition as a determinant, except that the alternating factor ε_σ is omitted)

$$(1.5) \qquad < u_1 \circ \ldots \circ u_n, v_1 \circ \ldots \circ v_n >_\circ \; = \; \operatorname{per} < u_i, v_j > \; .$$

In particular, the norm^2 for symmetric or antisymmetric tensors of order n is $n!$ times their norm^2 as ordinary tensors — differential geometers often make this convention too. In cases of ambiguity the two norms will be called explicitly the *tensor norm* and the *(anti)symmetric norm*, and a convenient symbol like $\| \; \|_\otimes$ will be added.

As an illustration, let us start with an o.n. basis $(e_i)_{i \in I}$ for $\mathcal{H}$ and construct bases for $\mathcal{H}_n$. First of all, in the antisymmetric case, vectors of the form $e_{i_1} \wedge \ldots \wedge e_{i_n}$ *with* $i_1 < \ldots < i_n$ *in some arbitrary ordering of the index set I* constitute an o.n. basis of the n–th chaos space in the antisymmetric norm. Once the ordering has been chosen, such a vector can be uniquely denoted by e_A where A is the finite subset $\{i_1, \ldots, i_n\}$ of I, a notation we used in Chapter II, and which is now extended to infinitely many dimensions. In the Dirac notation, vectors e_i are states labelled $|i>$, e_A is labelled $|i_1, \ldots, i_n>$, and the indicator function of A appears as a set of "occupation numbers" equal to 0 or 1 ("Pauli's exclusion principle"). The vacuum (no-particle state) is $|0>$ in this notation.

In the symmetric case, we get an orthogonal, but not orthonormal, basis consisting of the vectors $e_{i_1}^{\circ \alpha_1} \ldots e_{i_n}^{\circ \alpha_n}$, with $i_1 < \ldots < i_n$ as above, $\alpha_1, \ldots, \alpha_n$ being strictly positive integers. Ascribing the occupation number 0 to the remaining indices, we define an "occupation function" $\alpha : i \longmapsto \alpha(i)$ on the whole of I, and the basis vectors can be unambiguously denoted by e_α. We denote by $|\alpha|$ the sum of the occupation numbers and set $\alpha! = \prod_i \alpha(i)!$ (with the usual convention $0! = 1$). Then the squared tensor norm of e_α is $\alpha!/|\alpha|!$, and the squared symmetric norm is simply $\alpha!$. It is sometimes convenient to consider the *ket* $|e_i>$ as a coordinate mapping X^i (an *antilinear* coordinate mapping : if we had used the linear *bra* mapping, the multiplication of a mapping by a complex scalar would not be the usual one). Then e_α is also interpreted as a mapping, the (anti)monomial X^α, and the elements of the n-th chaos appear as homogeneous *anti*polynomials of total degree n. The symmetric norm on the space of polynomials, given by $\| X^\alpha \|^2 = \alpha!$, and the corresponding scalar product, appear in classical harmonic analysis on $\mathbb{R}^n$. They are usually given the form $< P, Q > = \overline{P}(D)\, Q(X) \big|_{X=0}$, $P(D)$ being understood as a differential operator with constant coefficients.

Fock spaces

2 To describe systems of an arbitrary (maybe variable, maybe random) number of particles, we take the direct sum of the spaces $\mathcal{H}_n$. More precisely, the *symmetric (antisymmetric) Fock space* $\Gamma(\mathcal{H})$ *over* $\mathcal{H}$ is the Hilbert space direct sum of all the symmetric (antisymmetric) chaos, with the corresponding sym.(ant.) scalar product. However, we prefer the following representation, which we already used in the harmonic oscillator case : an element of Fock space is a series

$$(2.1) \qquad\qquad F = \sum_n \frac{f_n}{n!} \quad \text{with} \quad f_n \in \mathcal{H}_n \quad \text{for all} \quad n$$

of elements $f_n \in \mathcal{H}_n$ such that

$$(2.2) \qquad\qquad \|F\|^2 = \sum_n \frac{\|f_n\|^2}{n!} < \infty .$$

In this formula the tensor norm is used, the factorials taking care of the change in norm : the n-th chaos subspace with the induced norm then is isometric to the n-th symmetric (antisymmetric) tensor product *with its symmetric (antisymmetric) norm.*

REMARK. In the probabilistic interpretations of Fock space, F is interpreted as a *random variable*, and (f_n) as its *representing sequence* in the chaos expansion (comparable to a sequence of Fourier coefficients). The "=" sign then has a non-trivial meaning, namely the Wiener–Ito multiple integral representation. Many results can thus be expressed in two slightly different ways, considering either the r.v. F or its representing sequence (f_n).

As we defined in subsection 1 the incomplete chaos spaces over a prehilbert space U, we may define the *incomplete Fock space*, consisting of finite sums (2.1) where f_n belongs to the incomplete n-th chaos. A convenient notation for it is $\Gamma_0(\mathcal{H})$ (Slowikowski [Slo1]). It is useful as a common domain for many unbounded operators on Fock space.

In this chapter, we deal essentially with the case of the Fock space Φ over $\mathcal{H} = L^2(\mathbb{R}_+)$. An element of $\mathcal{H}_n$ then is a function (class) $f(s_1, \ldots, s_n)$ in n variables, and since it is either symmetric or antisymmetric, it is determined by its restriction to the *increasing simplex* of $\mathbb{R}_+^n$, i.e. the set $\Sigma_n = \{s_1 < \ldots < s_n\}$. Therefore, *the symmetric and antisymmetric Fock spaces over $\mathcal{H}$ are naturally isomorphic.* In strong contrast with this, the antisymmetric Fock space over a finite dimensional Hilbert space is finite dimensional, while a symmetric Fock space is always infinite dimensional.

Since all separable infinite dimensional Hilbert spaces $\mathcal{H}$ are isomorphic, our restriction to $L^2(\mathbb{R}_+)$ is inessential in many respects. One may object that using such an isomorphism is artificial, but there is a ready answer : in the case of a finite dimensional space E, no one finds it artificial to order a basis of E to construct a basis of the exterior algebra $\bigwedge E$: what we are doing is the same thing for a continuous basis instead of a discrete one. Though the level of "artificiality" is the same, it requires a deeper theorem to perform this operation : every reasonable non-atomic measure space $(E, \mathcal{E}, \mu)$ is isomorphic with some interval of the line provided with Lebesgue measure, and this is what we call "ordering the basis".

REMARK. The set $\mathcal{F}$ of all sums $\sum_n f_n/n!$ with $f_n \in \mathcal{H}^{\otimes n}$ and $\sum_n \| f_n \|^2/n! < \infty$ is called the *full Fock space over* $\mathcal{H}$. As it has recently played an interesting role in non-commutative probability, we will devote to it occasional comments.

We begin with a remark from Parthasarathy–Sinha [PaS6]. If $\mathcal{H} = L^2(\mathbb{R}_+)$, the mapping $(s_1, s_2, \ldots, s_n) \to (s_1, s_1 + s_2, \ldots s_1 + \ldots + s_n)$ is a measure preserving isomorphism between $\mathbb{R}_+^n$ and the increasing simplex Σ_n, and leads to an isomorphism between the Hilbert space direct sums $\bigoplus_n L^2(\mathbb{R}_+^n)$ (the full Fock space) and $\bigoplus_n L^2(\Sigma_n)$ (the symmetric/antisymmetric Fock space). However, this isomorphism is less simple (uses more of the structure of $\mathbb{R}_+$) than the isomorphism between the symmetric and antisymmetric spaces.

Exponential vectors

3 This subsection concerns symmetric Fock space only. Given $h \in \mathcal{H}$, we define the *exponential vector* $\mathcal{E}(h)$

$$(3.1) \qquad \mathcal{E}(h) = \sum_n \frac{h^{\otimes n}}{n!} \qquad (h^{\circ n} = h^{\otimes n}) .$$

The representing sequence of $\mathcal{E}(h)$ is thus $(h^{\otimes n})$. In particular, the vacuum vector $\mathbf{1}$ is the exponential vector $\mathcal{E}(0)$.

Let us compute the scalar product of two exponential vectors $\mathcal{E}(g)$ and $\mathcal{E}(h)$. Since $g^{\circ n} = g^{\otimes n}$, we have $<g^{\circ n}, h^{\circ n}> = <g, h>^n$, hence using (2.2)

$$(3.2) \qquad <\mathcal{E}(g), \mathcal{E}(h)> = e^{<g,h>} .$$

As in the case of the harmonic oscillator, we call *coherent vectors* the normalized exponential vectors, which also define *coherent states*.

Let us indicate some useful properties of exponential vectors. First of all, differentiating n times $\mathcal{E}(th)$ at $t = 0$ we get $h^{\circ n}$. Linear combinations of such symmetric powers generate $\mathcal{H}_n$, by a polarization formula like

$$(3.3) \qquad u_1 \circ \ldots \circ u_n = \frac{1}{n!} \sum_\varepsilon (\varepsilon_1 u_1 + \ldots + \varepsilon_n u_n)^{\circ n} ,$$

over all choices $\varepsilon_i = \pm 1$. Thus *the subspace generated by exponential vectors is dense in* $\Gamma(\mathcal{H})$. We call it the *exponential domain* and denote it by $\mathcal{E}$.

On the other hand, *any finite system* $\mathcal{E}(f_i)$ *of (different) exponential vectors is linearly independent.* Let us reproduce the proof of this result from Guichardet's notes

[Gui]. We assume a linear relation $\sum_i \overline{\lambda}_i \mathcal{E}(f_i) = 0$ and prove that its coefficients λ_i are all equal to 0. For every $g \in \Gamma$ the relation $\sum_i < \overline{\lambda}_i \mathcal{E}(f_i), \mathcal{E}(g) > = 0$ gives $\sum_i \lambda_i e^{< f_i, g >} = 0$. Replacing g by $g + th$ and differentiating k times at $t = 0$, we get

$$\sum_i \lambda_i < f_i, \cdot >^p = 0 \quad (p = 0, 1, \ldots, k - 1) \, .$$

On the other hand, $\det (< f_i, \cdot >^p) = \prod_{i<j} <f_i - f_j, \cdot>$ (Vandermonde determinant). For every pair (i,j) the set $<f_i - f_j, \cdot> \neq 0$ is the complement of a hyperplane, hence the intersection of these sets is not empty, and the determinant cannot vanish identically. Choosing a point where it does not vanish, we see that all λ_i must be equal to zero.

Symmetric Fock space over a Hilbert space can be described entirely in terms of exponential vectors. Consider two Hilbert spaces $\mathcal{H}$ and Φ and a mapping $\varphi : \mathcal{H} \to \Phi$ such that $< \varphi(f), \varphi(g) > = e^{< f, g >}$ for $f, g \in \mathcal{H}$, and the image $\varphi(\mathcal{H})$ generates Φ. Then it is easy to construct an isomorphism between Φ and the symmetric Fock space over $\mathcal{H}$ which transforms φ into the exponential mapping.

Creation and annihilation operators

4 The *creation operators* a_h^+ (b_h^+) are indexed by a vector $h \in \mathcal{H}$: they climb one step on the ladder of chaos spaces, mapping $\mathcal{H}_n$ into $\mathcal{H}_{n+1}$ as follows

$$(4.1) \qquad a_h^+(u_1 \circ \ldots \circ u_n) = h \circ u_1 \ldots \circ u_n \quad ; \qquad b_h^+(u_1 \wedge \ldots \wedge u_n) = h \wedge u_1 \ldots \wedge u_n \, .$$

In particular, $a_h^+ \mathbf{1} = h$, $b_h^+ \mathbf{1} = h$. In the symmetric case, and using the symmetric norm on both spaces, a_h^+ has norm $\sqrt{n + 1} \, \|h\|$. Indeed, we may assume that $\|h\| = 1$, then take h as the vector e_1 in some o.n. basis (e_n) of $\mathcal{H}$, and the result follows from the computation of the squared norm of e_α at the end of subsection 1. The same reasoning in the antisymmetric case gives $\| b_h^+ \| = \|h\|$.

The *annihilation operator* $a_{h'}^-$ or $b_{h'}^-$ is indexed by an element h' of the *dual space* $\mathcal{H}'$ of $\mathcal{H}$, and goes one step down the ladder : a hat on a vector meaning that it is omitted, we have

$$(4.2) \qquad\qquad a_{h'}^-(u_0 \circ \ldots \circ u_n) = \sum_i h'(u_i) \, u_0 \circ \ldots \widehat{u}_i \ldots \circ u_n$$

$$(4.3) \qquad\qquad b_{h'}^-(u_0 \wedge \ldots \wedge u_n) = \sum_i (-1)^i h'(u_i) \, u_0 \wedge \ldots \widehat{u}_i \ldots \wedge u_n \, .$$

In particular, $a_h^- \mathbf{1} = 0$, $b_h^- \mathbf{1} = 0$. The adjoint of $a_{|h>}^+$ then is $a_{<h|}^-$, a *bra* vector belonging to $\mathcal{H}'$ (same property for $b^\pm$). This notation, due to Parthasarathy, is by far the best, but it is more usual to omit the Dirac brackets and to write a_h^+, a_h^- with $h \in \mathcal{H}$, annihilation operators being then *antilinear* in h.

Creation and annihilation operators are extended by linearity to the incomplete Fock space, and then by closure to the largest possible domain. In the antisymmetric case, the extension consists of bounded operators. In the symmetric case, the domain of $a^\pm$ consists of those elements $\sum_n f_n/n!$ of Fock space such that $\sum_n \| a^\pm f_n \|^2/n! < \infty$, the value of the operator on this domain being the obvious one. In particular, creation and annihilation operators are defined on the exponential domain, and we have the very useful properties

$$(4.4) \qquad a_h^- \mathcal{E}(f) = <h, f> \mathcal{E}(f) \quad ; \quad a_h^+ \mathcal{E}(f) = \frac{d}{d\varepsilon} \mathcal{E}(f + \varepsilon h) \Big|_{\varepsilon=0} .$$

REMARKS. 1) The exponential domain $\mathcal{E}$ is stable by annihilation operators, but not by creation operators. One sometimes uses the *enlarged exponential domain* $\Gamma_1(\mathcal{H})$, generated by products $u \circ \mathcal{E}(v)$ where u belongs to the incomplete Fock space $\Gamma_0(\mathcal{H})$ and v to $\mathcal{H}$. For details, see for instance Slowikowski [Slo1]. This domain is clearly stable by all creation operators, since a_h^+ is symmetric multiplication by h. It is stable by annihilation operators, because a_h^- acts as a derivation w.r.t. the product $\circ$.

2) If instead of representing an element of Fock space as a sum $\sum_n f_n/n!$ one uses the representing sequence (f_n), then the description of $a^\pm$ becomes slightly simpler. The representing sequence of $g^\pm = a_h^\pm f$ is (in the symmetric case)

$$(4.5) \qquad g_n^- = <h, g_{n+1}> , \qquad g_n^+ = h \circ f_{n-1} ,$$

where on the left side we have a contraction with h.

3) A useful subspace included in the domain of all creation and annihilation operators consists of the sums $\sum_n f_n/n!$ such that $\sum_n n \|f_n\|^2/n! < \infty$ (the domain of the square root $\sqrt{N}$ of the number operator, as we shall see later).

A slightly more delicate point consists in checking that each operator a_h^ε as we have defined it is exactly the adjoint of its mate. We reproduce the proof from Bratteli–Robinson [BrR2], but for the sake of completeness only : the result will never be applied. The relation $<x, a_h^- y> = <a_h^+ x, y>$ is obvious for $x, y \in \Gamma_0$; keeping fixed $y \in \Gamma_0$ we extend it to $x \in \mathcal{D}(a_h^+)$. Then it is clear that the mapping $<x, a_h^- \cdot>$ is continuous on Γ_0, meaning that $x \in \mathcal{D}(a_h^-)^*$ and $a_h^{-*} x = a_h^+ x$. Conversely, assume that $x \in \mathcal{D}(a_h^{-*})$ and put $a_h^{-*}(x) = z$; expand along the chaos spaces $x = \sum_n x_n$ and $z = \sum_n z_n$. Then the relation $<x, a_h^- y> = <z, y>$ for $y \in \Gamma_0$ gives $z_n = a_h^+ x_n$, hence $\sum_n \|a_h^+ x_n\|^2 < \infty$, and finally $x \in \mathcal{D}(a_h^+)$, $a_h^+ x = z$.

Commutation and anticommutation relations

5 An obvious computation on the explicit form (4.1) of the operators $a^\pm$, $b^\pm$ shows that on the algebraic sum of the (uncompleted) chaos spaces we have the commutation/anticommutation relations

$$(5.1) \qquad [a_h^-, a_k^-] = 0 = [a_h^+, a_k^+] \quad ; \quad [a_h^-, a_k^+] = <h, k> I \ ;$$

$$(5.2) \qquad \{b_h^-, b_k^-\} = 0 = \{b_h^+, b_k^+\} \quad ; \quad \{b_h^-, b_k^+\} = <h, k> I .$$

It is thus tempting to define in the symmetric case the following *field operators* with domain Γ_0, which are good candidates for essential selfadjointness (there are similar definitions in the antisymmetric case)

$$(5.3) \qquad Q_h = a_h^+ + a_h^- \quad ; \quad P_h = i(a_h^+ - a_h^-) .$$

If h is allowed to range over all of $\mathcal{H}$, there is no essential difference between the two families of operators : we have $P_h = Q_{ih}$. On the other hand, if $\mathcal{H} = \mathcal{K} \oplus i\mathcal{K}$ is the complexification of some real Hilbert space $\mathcal{K}$, and h, k range over $\mathcal{K}$, the commutation relations then take an (infinite dimensional) Heisenberg-like form with $\hbar = 2$ (probabilistic normalization!)

$$(5.4) \qquad [P_h, P_k] = 0 = [Q_h, Q_k] \quad ; \quad [P_h, Q_k] = -2i < h, k > .$$

We shall not discuss essential selfadjointness of the field operators on the uncompleted sum Γ_0 (this can be done using analytic vectors). As in the case of the harmonic oscillator, we shall directly construct unitary Weyl operators, and deduce from them (as generators of unitary groups) selfadjoint operators extending the above operators on Γ_0.

> We shall describe later on a probabilistic interpretation of the operators Q_h for h real as multiplication operators, which makes much easier (if necessary at all!) the discussion of their essential selfadjointness. The antisymmetric case does not require such a discussion, the creation and annihilation operators being then bounded.

Second quantization

6 Let $A : \mathcal{H} \to \mathcal{K}$ be a bounded linear mapping between two Hilbert spaces; then $A^{\otimes n}$ is a bounded linear mapping from $\mathcal{H}^{\otimes n}$ to $\mathcal{K}^{\otimes n}$; if A is contractive or is an isomorphism, the same is true for $A^{\otimes n}$. Since this mapping obviously preserves the symmetric and antisymmetric subspaces, we see by taking direct sums that *a contraction A of $\mathcal{H}$ induces a contraction between the Fock spaces (symmetric and antisymmetric) over $\mathcal{H}$ and $\mathcal{K}$*, which we shall in both cases denote by $\Gamma(A)$. It is usually called the *second quantization of A*, and operates as follows

$$(6.1) \qquad \Gamma(A)(u_1 \circ \ldots \circ u_n) = A u_1 \circ \ldots \circ A u_n$$

(in the antisymmetric case replace $\circ$ by $\wedge$). For operators A of norm greater than 1, $\Gamma(A)$ cannot be extended as a bounded operator between Fock spaces. Given three Hilbert spaces connected by two contractions A and B, it is trivial to check that $\Gamma(AB) = \Gamma(A)\Gamma(B)$. If A is an isomorphism, the same is true for $\Gamma(A)$. Note that $\Gamma(A)$ preserves the symmetric (antisymmetric) product.

The most important case is that of $\mathcal{H} = \mathcal{K}$ and unitary A. Since we obviously have $\Gamma(A^*) = \Gamma(A)^*$, and $\Gamma(I) = I$ we see that $\Gamma(A)$ is unitary on Fock space.

In the symmetric case the following very useful formula tells how $\Gamma(A)$ operates on exponential vectors

$$(6.2) \qquad \Gamma(A)\mathcal{E}(u) = \mathcal{E}(Au) .$$

Consider now a unitary group $U_t = e^{-itH}$ on $\mathcal{H}$; the operators $\Gamma(U_t)$ constitute a unitary group on $\Gamma(\mathcal{H})$, whose selfadjoint generator is sometimes called the *differential second quantization of H*, and is denoted by $d\Gamma(H)$. Hudson–Parthasarathy [HuP1] use the lighter notation $\lambda(H)$, but the notation we will prefer later is $a^{\circ}(H)$. Thus a^+ is indexed by a ket vector, a^- by a bra vector, and a° by an operator.

Differentiating (6.1), it is easy to see that

$$(6.3) \qquad \lambda(H)(u_1 \circ \ldots \circ u_n) =$$
$$= (Hu_1) \circ u_2 \circ \ldots \circ u_n + u_1 \circ Hu_2 \circ \ldots \circ u_n + \ldots + u_1 \circ u_2 \ldots \circ Hu_n$$

provided all u_i belong to the domain of H. The antisymmetric case is similar. This formula is then used to extend the definition of $\lambda(H)$ to more or less arbitrary operators. A useful formula defines $\lambda(H)$ on the exponential domain

$$(6.4) \qquad \lambda(H)\mathcal{E}(u) = a^+_{Hu}\mathcal{E}(u) .$$

On the other hand, $\lambda(H)$ acts as a derivation w.r.t. the symmetric product. Therefore, the enlarged exponential domain $\Gamma_1(\mathcal{H})$ of subs. 4 is stable by the operators $\lambda(H)$, at least for H bounded.

The second quantization $\Gamma(cI)$ ($|c|\leq 1$) is the multiplication operator by c^n on the n–th chaos $\mathcal{H}_n$, while $\lambda(cI)$ is the multiplication operator by cn on $\mathcal{H}_n$. In particular, $\lambda(I)$ is multiplication by n on the n–th chaos, and is called the *number operator*. More generally, whenever the basic Hilbert space $\mathcal{H}$ is interpreted as a space $L^2(\mu)$, we denote by a^o_h (b^o_h in the antisymmetric case) the differential second quantization $\lambda(M_h)$, corresponding to the multiplication operator M_h by a *real valued* function h, in which case it is clear that U_t is multiplication by $\exp(-ith)$, $\Gamma(U_t)$ acts on the n–th chaos as multiplication by $\exp\big(-it(h(s_1) + \ldots + h(s_n))\big)$, and a^o_h then is multiplication by $h(s_1) + \ldots + h(s_n)$. The same is true in the antisymmetric case, the product of an antisymmetric function by a symmetric one being antisymmetric.

7 The following useful formulas concerning symmetric Fock space are left to the reader as exercises :

$$(7.1) \qquad <a^-_h\mathcal{E}(f), a^-_k\mathcal{E}(g)> = <f,h><k,g><\mathcal{E}(f),\mathcal{E}(g)>$$
$$(7.2) \qquad <a^+_h\mathcal{E}(f), a^-_k\mathcal{E}(g)> = <h,g><k,g><\mathcal{E}(f),\mathcal{E}(g)>$$
$$(7.3) \qquad <a^+_h\mathcal{E}(f), a^+_k\mathcal{E}(g)> = [<f,k><h,g>+<h,k>]<\mathcal{E}(f),\mathcal{E}(g)> .$$

About second quantization operators, we recall that

$$(7.4) \qquad \lambda(A)\mathcal{E}(f) = a^+_{Af}\mathcal{E}(f) .$$

From this and the preceding relations, we deduce

$$(7.5) \qquad <\lambda(A)\mathcal{E}(f), \lambda(B)\mathcal{E}(g)> =$$
$$= [<Af,g><f,Bg> + <Af,Bg>] <\mathcal{E}(f),\mathcal{E}(g)>$$
$$(7.6) \qquad <a^-_h\mathcal{E}(f), \lambda(B)\mathcal{E}(g)> = <f,h><f,Bg><\mathcal{E}(f),\mathcal{E}(g)>$$
$$(7.7) \qquad <a^+_h\mathcal{E}(f), \lambda(B)\mathcal{E}(g)> =$$
$$= [<h,g><f,Bg> + <h,Bg>]<\mathcal{E}(f),\mathcal{E}(g)> .$$

8 This subsection can be omitted without harm, since it concerns the full Fock space, neither symmetric nor antisymmetric. In this case, "particles" belonging to n–particle systems are individually numbered, and we may in principle annihilate the i-th particle, or create a new particle with a given rank in the system's numbering. Among these many possibilities, we decide to call c^- and c^+ the operators on the algebraic (full) Fock space that kill or create at the lowest rank, that is for $n > 0$

$$(8.1) \qquad c_h^-(h_1 \otimes \ldots \otimes h_n) = <h, h_1> h_1 \otimes \ldots \otimes h_n$$
$$(8.1) \qquad c_h^+(h_1 \otimes \ldots \otimes h_n) = h \otimes h_1 \otimes \ldots \otimes h_n .$$

For $n = 0$, we put $c^-1 = 0$, $c^+1 = h$. It is very easy to see that c^- and c^+ are mutually adjoint on their domain, and that

$$(8.3) \qquad c_h^- c_k^+ = <h, k> I .$$

If $h = k$ and $\| h \| = 1$ we see that c_h^+ *is an isometry*, hence c_h^+ has a bounded extension for every h, and taking adjoints the same is true for c_h^-. The C^*-algebra generated by all operators c_h^+ is the simple *Cuntz algebra* $\mathcal{O}_\infty$, which has elicited considerable interest among C^*-theorists. On the other hand, when the basic Hilbert space is $L^2(\mathbb{R}_+)$, it allows the development of a theory of stochastic integration with respect to "free noise" (see Voiculescu [Voi1], Speicher [Spe1]).

Weyl operators

9 Let G be the group of rigid motions of the Hilbert space $\mathcal{H}$: an element of G can be described as a pair $\alpha = (u, U)$ consisting of a vector $u \in \mathcal{H}$ an a unitary operator U, acting on $h \in \mathcal{H}$ by $\alpha h = Uh + u$. If $\beta = (v, V)$, we have

$$(9.1) \qquad \beta\alpha = (v, V)(u, U) = (v + Vu, VU) .$$

We define an action of G on exponential vectors of Fock space as follows

$$(9.2) \qquad W_\alpha \mathcal{E}(h) = e^{-C_\alpha(h)} \mathcal{E}(\alpha h) \quad \text{where} \quad C_\alpha(h) = <u, Uh> - \|u\|^2/2 .$$

Since different exponential vectors are linearly independent, we may extend W_α by linearity to the space $\mathcal{E}$ of linear combinations of exponential vectors, and it is clear that $\mathcal{E}$ is stable. Weyl operators W_α corresponding to pure translations $\alpha = (u, I)$ are usually denoted simply by W_u.

Let us prove the following relation which may remind probabilists of the definition $A_{s+t} = A_s + A_t \circ \theta_s$ of additive functionals, up to the addition of a purely imaginary term

$$(9.3) \qquad \underset{(a)}{C_{\alpha\beta}(h)} = \underset{(b)}{C_\beta(h)} + \underset{(c)}{C_\alpha(\beta h)} - i \Im m <u, Uv> .$$

Indeed we have $a = <u + Uv, UVh> + \|u + Uv\|^2/2$, $b = <v, Vh> + \|v\|^2/2$, $c = <u, U(Vh+v)> + \|u\|^2/2$, so that $a-b-c$ is equal to $-<u, Uv> + \frac{1}{2}(<u, Uv> + <Uv, u>) = -i\Im m <u, Uv>$. From (9.2) we get the basic *Weyl commutation relation* in general form

$$(9.4) \qquad W_\alpha W_\beta = e^{-i\Im m <u, Uv>} W_{\alpha\beta} ,$$

This is the same relation as the harmonic oscillator's formula (2.4) in Chapter III, §2 subsection 2. We have according to (9.2)

$$< W_\alpha \mathcal{E}(k), W_\alpha \mathcal{E}(h)> = e^{-\overline{C_\alpha(k)}} e^{-C_\alpha(h)} < \mathcal{E}(\alpha k), \mathcal{E}(\alpha h)> \ ;$$

the last scalar product being equal to $e^{<\alpha k, \alpha h>}$, there remains the exponential of

$$-\overline{<u,Uk>} - \|u\|^2/2 - <u,Uh> - \|u\|^2/2 + <Uk+u, Uh+u> = <Uk,Uh> = <k,h>.$$

Therefore $< W_\alpha \mathcal{E}(k), W_\alpha \mathcal{E}(h)> = <h,k> = <\mathcal{E}(k), \mathcal{E}(h)>$, and W_α can be extended linearly as an *isometry* of the space $\mathcal{E}$. It can then be extended by continuity as an isometry in Fock space, and the Weyl commutation relation also extends, showing that W_α is invertible, hence unitary.

The mapping $\alpha \longmapsto W_\alpha$ is a *projective unitary representation* — *i.e.* a unitary representation up to a phase factor — of the group G of rigid motions of $\mathcal{H}$. We use it to define one-parameter unitary groups, and then (taking generators) selfadjoint operators, *i.e.* random variables.

First of all, the operators $Z_t = W_{(th,I)}$ constitute a unitary group according to (9.4). Then on exponential vectors we have from (9.2)

$$(9.5) \qquad Z_t \mathcal{E}(k) = e^{-t<h,k> - t^2\|h\|^2/2} \mathcal{E}(k+th) \ ,$$

and the derivative at $t=0$ is $(a_h^+ - a_h^-)\mathcal{E}(k)$ according to (4.4). Otherwise stated, $Z_t = \exp(-it P_h)$, where the selfadjoint generator P_h is an extension of the operator (5.3).

Next, consider operators $Z_t = W_{(0,U_t)}$, where $(U_t) = e^{itH}$ is a unitary group on $\mathcal{H}$. Then according to (9.2) the unitary group (Z_t) is the second quantization of (U_t), and its generator is the differential second quantization $\lambda(H)$ (*cf.* subs. 4).

REMARK. The C^*-algebra of operators on Fock space generated by pure translation Weyl operators W_u is called the C^*-*algebra of the CCR*, and it is studied in detail in the second volume of Bratelli–Robinson [BrR2]. Its representations are called *representations of the CCR*. The main point about it is that, in infinite dimensions, there is nothing like the Stone–von Neumann theorem, and there exists a confusing variety of inequivalent representations of the CCR. We shall see in Chapter V examples of useful (reducible) representations, similar to Gaussian laws of non-minimal uncertainty for the canonical pair (positive temperature states of the harmonic oscillator, see the end of Chapter III). However, these laws do not live on Fock space, and the stress is shifted from the Hilbert space on which the Weyl operators act to the CCR C^*-algebra itself, which is the same in all cases. On the other hand, it isn't a very attractive object. For instance, any two Weyl operators corresponding to different translations are at the maximum distance (that is, 2) allowed between unitaries, so that the CCR algebra is never separable.

10 We shall meet several times unitary groups of operators Z_t represented in the form

$$(10.1) \qquad Z_t = e^{i\alpha(t)} W(u_t, U_t)$$

where $\alpha(t), u_t, U_t$ are respectively real, vector and operator valued with time 0 values $0, 0, I$. We assume they are continuous and differentiable (strongly in the case of U_t)

with derivatives at 0. α', u' and iH. It will be useful to compute the generator $(1/i) Z'_0$ on the exponential domain. We start from the formula

$$W(u_t, U_t) \, \mathcal{E}(h) = e^{-\| u_t \|^2/2 - < u_t, U_t h >} \, \mathcal{E}(u_t + U_t h)$$

from which we deduce the value of $Z'_0 \mathcal{E}(h)$

$$i\alpha' \mathcal{E}(h) - < u', h > \mathcal{E}(h) + \frac{d}{dt} \mathcal{E}(u_t + U_t h)|_{t=0} = i\alpha' I + a^+(|u'>) + a^{\circ}(U'_0) - a^-(<u'|) \; .$$

and therefore the generator is, on the exponential domain

$$(10.2) \qquad\qquad \alpha' I - P(u') + a^{\circ}(H) \; .$$

11 We have seen that pure translation Weyl operators can be represented as $\exp(-i P_h)$. We are going to define exponentials of simple operators on Fock space. Though these results belong to the "Focklore", we refer the reader to Holevo's paper [Hol2] which contains an interesting extension to exponentials of stochastic integrals of deterministic processes, under suitable integrability conditions. Precisely, we are going to compute

$$(11.1) \qquad\qquad e^M \mathcal{E}(h) \quad \text{where} \quad M = a^+(|u>) + a^{\circ}(\rho) + a^-(<v|) \; .$$

where ρ is a bounded operator. We use the notations

$$(11.2) \qquad e(z) = e^z \; , \quad e_1(z) = (e^z - 1)/z \; , \quad e_2(z) = (e^z - 1 - z)/z^2 \; ,$$

Holevo proves that *the exponential series for e^M converges strongly on the exponential domain $\mathcal{E}$, and we have*

$$(11.3) \qquad\qquad e^M \mathcal{E}(h) = \exp(<v, e_1(\rho) h + e_2(\rho) u >) \, \mathcal{E}(e(\rho) h + e_1(\rho) u)$$

For $\rho = 0$, we get in particular

$$(11.4) \qquad\qquad e^{a^+(|u>) + a^-(<v|)} \mathcal{E}(h) = e^{<v, h + \frac{1}{2} u>} \, \mathcal{E}(h + u) \; ,$$

and for $u = -v$ we get a pure translation Weyl operator.

We only give a sketch of the proof. Replace M by zM ($z \in \mathbb{C}$) and consider the right side of (11.3) as a mapping $P(z)$ from $\mathbb{C}$ to Fock space. It is clearly holomorphic and we have

$$P(z_1 + z_2) = P(z_1) P(z_2) \; .$$

On the other hand, one may check that

$$D^n P(0) \mathcal{E}(h) = M^n \mathcal{E}(h) \; .$$

The proof is not obvious, because $\mathcal{E}$ is not stable by M : one must work on the enlarged domain Γ_1. Then the exponential series is the Taylor series of $P(z)$ at 0, and its strong convergence becomes clear.

If $\rho = iH$ where H is selfadjoint, then $e^\rho = U$ is unitary, and for $v = -u$ the operator e^M is the product of the Weyl operator $W(e_1(\rho)u, U)$ by the scalar $\exp <u, (\frac{1}{2} - e_2(\rho))u >$.

§2. FOCK SPACE AND MULTIPLE INTEGRALS

Multiple Wiener–Ito integrals

1 We have tried to make the essentials of this section easy to understand, without diving too much into the literature (see Chapter XXI in [DMM] for details). In this chapter we present the one dimensional theory. The case of several dimensions requires a more complete system of notation, and is given in the next chapter.

Multiple stochastic integrals were first defined by N. Wiener in [Wie] (1938), from which the name "Wiener chaos" comes. But Wiener's definition was not the modern one : it was rather related to the idea of multiple integrals in the Stratonovich sense. The commonly used definition of the so called "multiple Wiener integrals" is due to K. Ito [Ito], dealing with the case of Gaussian random measures. The relation with symmetric Fock space was first underlined by I. Segal ([Seg1], 1956). The literature on this subject is boundless. We may recommend the recent Lecture Notes volume by Nielsen [Nie].

Let us take for granted Wiener's (1923) theorem on the existence of Brownian motion : let Ω be the set of all continuous functions from $\mathbb{R}_+$ to $\mathbb{R}$ and let X_s be the evaluation mapping $\omega \longmapsto \omega(s)$. Let us provide Ω with the σ–fields $\mathcal{F}$, $\mathcal{F}_t$ generated by all mappings X_s with s arbitrary in the first case, $s \leq t$ in the second case. One can show that $\mathcal{F}$ is also the topological Borel field on Ω for local uniform convergence. The *Wiener measure* is the only probability $\mathbb{P}$ on Ω such that $X_0 = 0$ a.s. , and the process (X_t) has centered Gaussian independent increments with variance $\mathbb{E}\left[(X_t - X_s)^2\right] = t - s$. Paley and Wiener defined the elementary stochastic integral $\int_0^\infty f(s) dX(s)$ of a (deterministic) L^2 function $f(s)$ in 1934, a concrete example of integration w.r.t. a spectral measure over the half line. Their method, using an isometry property, was generalized by Ito to multiple integrals as follows.

Let Σ_n be the increasing (open) simplex of $\mathbb{R}_+^n$, *i.e.* the set of all n–uples $\{s_1 < s_2 < \ldots < s_n\}$. Let H be a rectangle $H_1 \times \ldots \times H_n$ contained in Σ_n — otherwise stated, putting $H_i =]a_i, b_i]$ we have $b_{i-1} < a_i$. We define the multiple integral of X over H as the r.v. $\prod_i (X_{b_i} - X_{a_i})$. This is extended to a linear map J_n from (complex) linear combinations $f_n(s_1, \ldots, s_n)$ of indicators of rectangles, to random variables, and the *isometry property*

$$(1.1) \qquad \| J_n(f_n) \|^2 = \int_{\Sigma_n} |f_n(s_1, \ldots, s_n)|^2 \, ds_1 \ldots ds_n$$

shows that J_n can be extended as an isometric mapping from $L^2(\Sigma_n)$ to $L^2(\Omega)$, the image of which is called the *n-th Wiener chaos* and denoted by $\mathcal{C}_n$. One makes the convention that Σ_0 is the one point set $\{\varnothing\}$, and that $J_0(f_0)$ (f_0 being a constant) is the constant r.v. f_0. Finally, an easy computation in the case of rectangles leads to the orthogonality of $J_n(f_n)$ and $J_m(g_m)$ for $n \neq m$, and therefore *chaos spaces of different orders are orthogonal*. All this is very simple, only the following result has a not quite trivial proof (see [DMM], Chap. XXI n° 11).

THEOREM. *The completed direct sum $\bigoplus_n C_n$ is the whole of $L^2(\Omega)$.*

Otherwise stated, every (real or complex) r.v. $f \in L^2(\Omega)$ has a representation[1]

$$(1.2) \qquad f = \sum_n \int_{\Sigma_n} f_n(s_1, \dots, s_n) \, dX_{s_n} \dots dX_{s_1} = \sum_n J_n(f_n) \,,$$

where f_0 is the expectation of f, and the sequence of functions f_n (uniquely determined up to null sets) satisfies the isometry property

$$(1.3) \qquad \| f \|_{\Omega}^2 = \sum_n \| f_n \|_{\Sigma_n}^2 \,.$$

This will be interpreted in subsection 3 as an isomorphism between $L^2(\Omega)$ and Fock space.

2 Let us comment on the preceding construction.

a) The first and very important point is that, *in the construction of multiple integrals, the Gaussian character of the process never appears.* One only makes use of the martingale property $\mathbb{E}[X_t \mid \mathcal{F}_s] = 0$ for $s < t$ and of the martingale property of $X_t^2 - t$, *i.e.* $\mathbb{E}[(X_t - X_s)^2 \mid \mathcal{F}_s] = t - s$ — in standard probabilistic jargon, "the angle bracket of X_t is t". There are many examples of such martingales : the simplest after Brownian motion is the compensated Poisson process $X_t = (N_t - \lambda t)/\sqrt{\lambda}$, where (N_t) is a Poisson process with (unit jumps and) intensity λ. It turns out that *Poisson processes too have the chaotic representation property*, *i.e.* all the L^2 functionals of the Poisson process (N_t) can be expanded in multiple stochastic integrals w.r.t. the martingale X. This was already known to Wiener, who studied Poisson chaotic expansions under the name of "discrete chaos", but Wiener's work was partly forgotten, and the Poisson chaotic representation property was rediscovered several times. Wiener and Poisson are essentially the only processes with stationary independent increments which possess the chaotic representation property. If stationarity is not required, there is the additional possibility of choosing a variable intensity $\lambda(t)$ at each point t for the Poisson "differential" dX_t, the value $\lambda = \infty$ being allowed, and understood as a Gaussian "differential". See for instance He and Wang [HeW].

The first example (the "Azéma martingale") of a martingale possessing the chaotic representation property and whose increments are not independent was discovered by Emery [Eme] in 1988. We present it in Chapter VI, end of §3.

b) Let us return to the Wiener and Poisson cases. It seems that the order structure of $\mathbb{R}_+$ plays a basic role in the definition of multiple integrals, but it is not really so. Ito had the idea of defining, for a *symmetric* L^2 function $f_n(s_1, \dots, s_n)$ of n variables

$$I_n(f_n) = \int_{\mathbb{R}_+^n} f_n(s_1, \dots, s_n) \, dX_{s_1} \dots dX_{s_n}$$

$$(2.1) \qquad = n! \int_{\Sigma_n} f_n(s_1, \dots, s_n) \, dX_{s_1} \dots dX_{s_n} = n! \, J_n(f_n) \,.$$

---

[1] The order of the differential elements $dX_{s_n} \dots dX_{s_1}$, with the smaller time on the right, is the natural one when iterated integrals are computed. Since it does not change the mathematics (at least for bosons), we generally do not insist on it.

Using this notation, we may write (1.2) and (1.3) as follows, using symmetric functions f_n

$$(2.2) \qquad f = \sum_n \frac{I_n(f_n)}{n!} \quad \text{with} \quad \|f\|_\Omega^2 = \sum_n \|f_n\|_{\mathbb{R}^n}^2 / n! \ .$$

Ito's aim was to extend the chaotic representation to functionals of arbitrary Gaussian random measures with independent increments, and the use of symmetric functions allows one to do it without order structure. This raises a number of interesting questions : the integration domain Σ_n is most naturally an *open* simplex, and $\mathbb{R}_+^n$ is not a union of open simplexes : it seems that *we are forgetting the diagonals*. This is related to Wiener's original error, to the theory of the *Stratonovich integral*, and to some elementary "renormalizations" in physics.

Since the symmetric integral gives an order-free definition of the multiple integrals w.r.t. Brownian motion, we can extend it without any work to all Lusin measurable spaces $(E, \mathcal{E})$ with a nonatomic σ–finite measure μ. It suffices to use a measure isomorphism of E onto the interval $[0, \mu(E)[$ provided with Lebesgue measure. This will be one of our *leitmotive* in this and the following chapter.

c) Our third comment introduces the "shorthand notation" for multiple integrals, due to Guichardet [Gui] (though Guichardet discusses Fock spaces and not stochastic integrals, and does not use any ordering). Since $\mathbb{R}_+$ is totally ordered, there is a 1–1 correspondence between the increasing n–simplex Σ_n, and the set $\mathcal{P}_n$ of all finite subsets of $\mathbb{R}_+$ of cardinality n — a finite subset $A \in \mathcal{P}_n$ being represented as the list $s_1 < s_2 < \ldots < s_n$ of its elements in increasing order. Then $\mathcal{P}_n$ gets imbedded in $\mathbb{R}_+^n$, and acquires a natural σ–field and a natural measure dA, the n–dimensional volume element $ds_1 \ldots ds_n$. We denote by $\mathcal{P}$ the sum of all $\mathcal{P}_n$, considered as a single measure space. Then the n–dimensional stochastic differential $dX_{s_1} \ldots dX_{s_n}$ is considered as the restriction to $\mathcal{P}_n$ of a single differential dX_A. The family of functions $f_n(s_1, s_2, \ldots, s_n)$ is considered as a single measurable function $\widehat{f}(A)$ on $\mathcal{P}$, belonging to $L^2(\mathcal{P})$ relative to the measure dA. In this way, (1.2) and (1.3) are shortened to

$$(2.3) \qquad f = \int_{\mathcal{P}} \widehat{f}(A)\, dX_A \quad ; \quad \|f\|^2 = \int_{\mathcal{P}} \|\widehat{f}(A)\|^2\, dA \ .$$

The notation $\widehat{f}$ is meant to suggest a kind of Fourier analysis of f, but we often omit the hat. This notation for continuous multiple integrals is strikingly similar to that of discrete multiple integrals in Chapter II (symmetric Bernoulli r.v.'s; toy Fock space). See subsection 7 for a heuristic formalism to handle this analogy.

Like the symmetric notation (2.2), the shorthand notation (2.3) is not order dependent : given a Lusin space $(E, \mathcal{E})$ with a σ–finite measure μ (non-atomic for simplicity), we can denote by $\mathcal{P}(E)$ the set of all finite subsets of E, endow it with a Lusin σ–field and a natural measure dA, and use the Guichardet notation without fear. A measurable ordering of the space is used only to prove the existence of these objects.

> J. Kupsch remarked that Guichardet's notation can be used in special relativity, since *almost all* n–uples of points in Minkowski's space have different time components, which can be used to order the n–uple.

Wiener space and Fock space

3 The preceding remark a) about symmetric functions shows that *boson Fock space over* $L^2(\mathbb{R}_+)$ *is isomorphic with* $L^2(\Omega)$, *where* Ω *denotes Wiener space*. Indeed, this is the reason of our choice of norms on Fock space (see (2.2)). The same is true for the L^2 space of a Poisson process, and on the other hand, it is true also for *antisymmetric* Fock space, a fact whose significance is not generally appreciated.

The Fock space over $L^2(\mathbb{R}_+)$ will be called *simple Fock space* in these notes. It has special properties because of the order structure of the line. For example, considering $a_h^\pm$ for $h = I_{[0,t]}$ we get creation and annihilation "processes" a_t^+ and a_t^-, and we will see later that a_t^+ and a_t^- are best interpreted as "operator stochastic integrals" $\int h_s\, da_s^+$ and $\int \overline{h}_s\, da_s^-$. In Chapter VI we describe the Hudson–Parthasarathy theory of operator stochastic integrals. Before this, we present in Chapter V the theory of *multiple Fock spaces,* which are constructed over a Hilbert space $L^2(\mathbb{R}_+) \otimes \mathcal{K}$ which also has a time structure.

We are going now to review the basic facts of boson Fock space theory, putting on the probabilistic spectacles of the Wiener interpretation.

Symmetric Fock space is $L^2(\Omega)$ and the vacuum vector $\mathbf{1}$ is the function 1. The identification of Fock space with the L^2 space of a real measure provides it with a natural conjugation $f \longmapsto \overline{f}$. Given $h \in \mathcal{H} = L^2(\mathbb{R}_+)$, let us consider the stochastic differential equation

$$(3.1) \qquad Y_t = 1 + \int_0^t Y_s\, h(s)\, dX_s ,$$

whose solution is the *stochastic exponential* of the martingale $\int_0^t h(s)\, dX_s$. The standard approximation of the solution by the Picard method gives the chaos expansion of Y_t, which is

$$(3.2)$$
$$Y_t = \sum_n \int_{s_1 < .. < s_n < t} h(s_1)\dots h(s_n)\, dX_{s_1}\dots dX_{s_n} = \sum_n \frac{1}{n!} I_n((hI_{]0,t]})^{\circ n}) ,$$

and in particular the limit Y_∞ belongs to L^2 and is the *exponential vector* $\mathcal{E}(h)$. *This interpretation of exponential vectors as stochastic exponentials is entirely general,* since the Picard approximation holds for all stochastic differential equations w.r.t. martingales and semimartingales : for instance, in the Poisson interpretation of Fock space exponential vectors will correspond to *Poisson* stochastic exponentials. Let us return to Wiener space, however.

It is convenient to embed the real part of $\mathcal{H} = L^2(\mathbb{R}_+)$ into Wiener space Ω, identifying $h \in L^2$ with $\underline{h} = \int_0^\bullet h(s)\, ds$. Functions of this form (continuous, equal to 0 at time 0, admitting a weak derivative in L^2) are called *Cameron-Martin (C–M) functions.* Of course, only real C–M functions $f(t)$ can be considered as elements of Ω, but complex C–M functions exist as well and constitute a complex Hilbert space with scalar product $<x,y> = \int \overline{x}'(s)y'(s)\, ds$.

Taking again $h \in L^2$ (real), denote by $\widetilde{h}$ the stochastic integral $\int_0^\infty h(s)\, dX_s$. According to the Cameron–Martin–Girsanov formula, we have for every positive function

f on Ω

(3.3) $\mathbb{E}\,[\,f(\cdot + \underset{\sim}{h})\,] = \mathbb{E}\,[\,f\,e^{\widetilde{h} - \|h\|^2/2}\,] = \mathbb{E}\,[\,f\,\mathcal{E}(h)\,]\ .$

Therefore if we map the positive r.v. f on Ω to the positive r.v.

(3.4) $G_h f = f(\cdot + \underset{\sim}{h})\,e^{-\widetilde{h} - \|h\|^2/2}$

the integral is preserved — to see why, rewrite $-\widetilde{h}$ as $-\int h(s)\,dX_s(\cdot + \underset{\sim}{h}) + \|h\|^2$, and apply (3.3). Once (3.4) is proved we may associate with h a *unitary group* on $L^2(\Omega)$, which maps f (no longer positive) into

(3.5) $T_t f = f(\cdot - 2t\underset{\sim}{h})\,e^{t\int h(s)\,dX_s - t^2\|h\|^2}\ ,$

and an explicit computation shows that

$$T_t\,\mathcal{E}(g) = \exp\,\Big(-t\int g_s h_s\,ds - t^2\|h\|^2/2\Big)\,\mathcal{E}(g + th)\ ,$$

which identifies T_t with the Weyl operator W_{th} for real h (see §1, subs. 9, (9.5)). Then it is easy to identify the generator of this unitary group, which by definition is the self-adjoint operator P_h : denote by ∇_h the derivative operator on Wiener space along the C–M vector $\underset{\sim}{h}$, (*i.e.* $\nabla_h f = \lim_{t\downarrow 0}(f(\cdot + t\underset{\sim}{h}) - f(\cdot))/t$) and as usual identify the r.v. $\widetilde{h}$ with the corresponding multiplication operator. Then we have

(3.6) $P_h = i\,(\widetilde{h} - 2\nabla_h)\quad \text{for real } h \in \mathcal{H}.$

On the other hand, a direct computation on the stochastic exponential $\mathcal{E}(g)$ shows that, for real h, we have $\nabla_h \mathcal{E}(g) = <h, g> \mathcal{E}(g)$. This is the same as the action of a_h^- on $\mathcal{E}(g)$. Since we know $P_h = i(a_h^+ - a_h^-)$, we deduce from it that Q_h is simply the multiplication operator by the stochastic integral $\widetilde{h}$ on the linear space $\mathcal{E}$ generated by the exponential vectors. It is not difficult to see that they are exactly the same on their maximal domain as s.a. operators. The Weyl operator W_{ih} corresponding to a purely imaginary vector ih is equal to e^{iQ_h}, and Q_h has been computed above; therefore W_{ih} is the multiplication operator by $e^{i\widetilde{h}}$. Finally, the creation operator a_h^+ is $\widetilde{h} - \nabla_h$ on $\mathcal{E}$, and it is extended to its natural domain by closure.

Let us compute the probability distribution of P_h in the vacuum state (recalling that $\mathbf{1} = \mathcal{E}(0)$) :

$$\mathbb{E}\,[\,e^{-itP_h}\,] = <\mathbf{1}, W_{ith}\mathbf{1}> = <\mathbf{1}, T_t\mathbf{1}> = e^{-t^2\|h\|^2/2}$$

from which we deduce that the r.v.'s P_h and Q_h have the same distribution. Otherwise stated, Wiener space does not only contain the "obvious" Brownian motion (X_t), represented by the multiplication operators Q_t; from the point of view of von Neumann's quantum probability, it also contains an "upside down" Brownian motion represented by the family of operators P_t.

In the harmonic oscillator case (Schrödinger's representation) the classical Fourier-Plancherel transform $\mathcal{F}$ exchanges the operators P and Q. On the orthonormal basis

h_n consisting of the Hermite–Weber functions $h_n(x)e^{-x^2/4}$, $\mathcal{F}$ operates simply by $\mathcal{F}(h_n) = i^n h_n$. If we use the Gaussian representation the exponential factor gets incorporated into the measure and we get a simple unitary map, diagonal in the Hermite polynomial basis, which is called the *Fourier–Wiener transform* on Gauss space. This transform exists as well in infinite dimensions (see Hida, *Brownian motion*, p. 182–184). Here $\mathcal{F}$ acts by $\mathcal{F}(\sum_n I_n(f_n))/n! = \sum_n i^n I_n(f_n)/n!$ on the representation (2.2), or $\mathcal{F}\mathcal{E}(h) = \mathcal{E}(ih)$, and it transforms Q_h into P_h, P_h into $-Q_h$ as in the oscillator case.

All these results are very classical in Fock space theory. Those of the next section are much more recent, and due to Hudson and Parthasarathy [HuP1].

Number operator and Poisson processes

4   We have seen in Chapter II that the scalar Bernoulli process is replaced on toy Fock space by *three* basic operator martingales. We have given the analogue of the creation and annihilation martingales, and it remains to introduce the third one, the *number operator* process.

The number operator N is not unknown to probabilists, since $-N$ is the generator of the infinite dimensional *Ornstein–Uhlenbeck semigroup* on Wiener space, one of the cornerstones of the "Malliavin Calculus" (Malliavin [Mal], Stroock [Str1][Str2]). The operator N is positive self-adjoint, and acts on the chaos representation $f = \sum_n f_n$ by $Nf = \sum_n n f_n$ (its domain $\mathcal{D}$ consisting exactly of those f for which $\sum_n n^2 \|f_n\|^2 < \infty$); N may also be represented as $\sum_i a_{e_i}^+ a_{e_i}^-$ for an o.n.b. (e_i). The O–U semigroup itself acts on exponential vectors by $e^{-tN}\mathcal{E}(h) = \mathcal{E}(e^{-t}h)$, so that e^{-tN} is the second quantization of the contraction $e^{-t}I$, and N appears as the differential second quantization $\lambda(I)$.

More generally, let α a real Borel function on $\mathbb{R}_+$, and let χ_α be the unitary operator on $\mathcal{H} = L^2(\mathbb{R}_+)$

$$\chi_\alpha \cdot f = e^{i\alpha} f .$$

This corresponds to an arbitrary change of phase at each point, and may be considered as a "gauge transformation", whence the name "gauge process" used by Hudson–Parthasarathy. The second-quantized operators $\Gamma(\chi_\alpha)$ are unitary, and are Weyl operators without translation term. For fixed α and $u \in \mathbb{R}$ the operators $\Gamma(\chi_{u\alpha})$ constitute a unitary group which we denote by e^{iuN_α}, the operator N_α acting on the n-th chaos as the multiplication operator by the symmetric function $\alpha(x_1) + \ldots + \alpha(x_n)$. Note that the operators corresponding to different functions α all commute. We use a "stochastic integral" notation, writing $N_\alpha = \int \alpha(s)\, dN_s$. The notations $\int_0^t \alpha(s)\, dN_s$, and in particular $N_t = \int_0^t dN_s$ (the *number operator up to t*) then have obvious meanings.

The following theorem is (for a probabilist) the most striking result of the Hudson–Parthasarathy paper [HuP1], since it shows how to construct a Poisson process from a Wiener process by adding to it an operator... which is a.s. equal to 0 (in the vacuum state). For a quantum physicist, on the other hand, this is not more surprising than $-\Delta + V$ having a discrete spectrum, while the multiplication operator by V has by itself a continuous spectrum, and Δ is a.s. equal to 0 in the vacuum state.

THEOREM. *The operators*

(4.1) $$X_t = Q_t + cN_t$$

have commuting selfadjoint extensions, whose joint probability law in the vacuum state is that of a compensated Poisson process with jump size c and intensity $1/c^2$ (so that $X_t^2 - t$ is a martingale).

The limiting case $c = 0$ corresponds to the Brownian motion Q_t.

PROOF. The statement is, as usual, a little incomplete : we should have said that the operators are essentially selfadjoint on the exponential domain (or some other explicit domain). But our proof (borrowed from Hudson–Parthasarathy) does not require such domain specifications, since it directly constructs the corresponding unitaries as Weyl operators, and the generators themselves arise by differentiation.

More precisely, we are going to construct the unitaries $\exp(i \int \alpha(s) \, dX_s)$ for α bounded with compact support. To achieve this, we consider the family of rigid motions in $L^2(\mathbb{R}_+)$ — since α has compact support, $e^{i\alpha} - 1$ belongs to L^2

(4.2) $$\Theta(\alpha) \cdot h = e^{i\alpha} h + c(e^{i\alpha} - 1) .$$

We denote by $W(\alpha)$ the corresponding Weyl operator. According to §1, formula (9.2), we have

$$W(\alpha) \mathcal{E}(h) = e^{\int c(e^{i\alpha} - 1) h \, ds - \int c^2 (1 - \cos \alpha) \, ds} \mathcal{E}(\Theta(\alpha) h) .$$

The operators $W(u\alpha)$ do not constitute a unitary group, but they do after being multiplied by a phase factor $\exp(ic^2 \int \sin u\alpha(s) \, ds)$. Forgetting again u, we define

$$Z(\alpha) = e^{ic^2 \int \sin \alpha \, ds} W(\alpha)$$

and an easy computation using the Weyl commutation relation proves that $Z(\alpha) Z(\beta) = Z(\alpha + \beta)$. Thus we have a family of commuting unitaries, and we know there exists a family of commuting self-adjoint operators X_α such that $Z_\alpha = \exp(iX_\alpha)$, which may be interpreted as ordinary random variables such that $X_\alpha + X_\beta = X_{\alpha+\beta}$. The law of X_α in the vacuum state is given by its Fourier transform

$$\mathbb{E}[e^{iuX_\alpha}] = \langle 1, Z(u\alpha) 1 \rangle = \exp\left(c^2 \int (e^{iu\alpha} - 1) \, ds \right) .$$

In particular, if α is an indicator function I_A, this is equal to $\exp(c^2 |A| (e^{iu} - 1))$, $|A|$ denoting the Lebesgue measure of A, *i.e.* we get the characteristic function of a compensated Poisson distribution of variance $|A|$. This applies in particular for the r.v.'s X_t corresponding to $A = [0, t]$, to the differences $X_t - X_s$, and taking now for α a step function, we find that the process X_t has independent increments.

It remains to identify X_t as an operator to $Q_t + cN_t$. This will be left to the reader, as an application of the computation of the generator in §1, subs. 10, or of the exponentials in subs. 11.

It is important to note that the martingales X_t corresponding to different values of c do not commute. Their commutators are easily deduced from the relations

$$[N_\alpha, Q_h] = i \int \alpha_s h_s \, dP_s \quad ; \quad [N_\alpha, P_h] = -i \int \alpha_s h_s \, dQ_s .$$

The unitary Fourier–Wiener transform maps Q_h into P_h, P_h into $-Q_h$, and leaves N_t and the vacuum vector invariant. Therefore it maps the compensated Poisson processes $Q_t + cN_t$ into the similar family $P_t + cN_t$.

The Poisson interpretation of Fock space

5   Let us now proceed in the reverse direction, starting from a compensated Poisson process (X_t) of jump size c and intensity $1/c^2$, generating a σ–field $\mathcal{F}$. We know that it has the chaotic representation property, i.e. $L^2(\mathcal{F})$ is isomorphic with Γ via multiple stochastic integration w.r.t. (X_t). Can we show that the multiplication operator by X_t is equal to $Q_t + cN_t$, at least on exponential vectors? Instead of using X_t we prefer to consider the unitary operator e^{iuX_t}. We have

$$e^{iuX_t} = e^{-iut/c} \prod_{\Delta X_s \neq 0 , \, s \leq t} e^{iuc} ,$$

the product being extended to the random set of jump times $s \leq t$. We mentioned before that the exponential vector $\mathcal{E}(f)$ is interpreted as a stochastic exponential in all probabilistic situations, and the stochastic exponential of $\int f_s \, dX_s$ for a compensated Poisson process of jump size c and intensity $1/c^2$ is

$$\mathcal{E}(h) = e^{-\int h_s \, ds/c} \prod_s (1 + ch_s) .$$

On the other hand, it follows from the computations of subsection 4 (taking for simplicity $h=0$ outside of $]0,t]$)

$$\exp\left(Q_t + cN_t\right) \mathcal{E}(h) = \exp\left(\frac{e^{iuc} - 1}{c} \int_0^t h_s \, ds + t \frac{e^{iuc} - 1 - iuc}{c^2} \right) \mathcal{E}(e^{iuc}h + e^{iuc} - 1)$$

Computing explicitly $\mathcal{E}(e^{iuc}h + e^{iuc} - 1)$ as we did above for $\mathcal{E}(h)$ we get the equality of the two operators.

6   In this subsection, we interpret in Fock space language some ideas from classical stochastic calculus, including the Ito stochastic integral itself.

Let $(\Omega, \mathcal{F}, \mathbb{P})$ be the Wiener space, and X_t denote Brownian motion (the coordinate process). For every $\omega \in \Omega$ let $\tau_t \omega$ be the mapping $s \longmapsto X_{s+t}(\omega) -- X_t(\omega)$. Denote by $\mathcal{F}_{t]}$ (or more often simply $\mathcal{F}_t$) the *past* at time t, i.e. the σ–field generated by the random variables X_s, $s \leq t$, and by $\mathcal{F}_{[t}$ the *future* at time t, generated by the increments $X_s - X_t$, $s \geq t$. The corresponding L^2 spaces will be denoted by $\Phi_{t]}$ (or simply Φ_t) and $\Phi_{[t}$, and it is clear that they are respectively isomorphic to the Fock spaces over $L^2(]0,t])$ and $L^2([t,\infty[)$. On the other hand, the mapping $\theta_t : f \longmapsto f \circ \tau_t$ is an isomorphism from the whole Fock space $\Gamma(L^2(\mathbb{R}_+)) = \Phi$ onto its subspace $\Phi_{[t}$.

Brownian motion is a process with independent increments, and this is expressed as follows : for $f \in L^2(\mathcal{F}_t) = \Phi_t$ and $g \in L^2(\mathcal{F}_{\lceil t}) = \Phi_{\lceil t}$, the product fg belongs to L^2 and we have $\mathbb{E}[fg] = \mathbb{E}[f] \, \mathbb{E}[g]$. Since products fg of this form generate $L^2(\mathcal{F}) = \Phi$, the bilinear mapping $(f,g) \longmapsto fg$ gives rise in the usual way to a mapping of $\Phi_{t\rceil} \otimes \Phi_{\lceil t}$ into Φ, which is an *isomorphism*. This is a very particular case of the general fact that, each time a Hilbert space $\mathcal{H}$ is split into a direct sum $\mathcal{H}_1 \oplus \mathcal{H}_2$, as here $L^2(\mathbb{R}_+)$ into $L^2(]\,0,t])$ and $L^2([\,t,\infty[)$, the corresponding Fock space $\Gamma(\mathcal{H})$ appears as a tensor product $\Gamma(\mathcal{H}_1) \otimes \Gamma(\mathcal{H}_2)$. Thus the "continuous sum" $L^2(\mathbb{R}_+)$ gives rise to a "continuous tensor product". This property is purely algebraic, and has nothing to do with Brownian motion, or any other probabilistic interpretation.

Things can be stated in a different way : consider $f \in \Phi_t$ and $g \in \Phi_{\lceil t}$. Then we may define their "product" fg independently of any probabilistic interpretation (in fact, if f and g belong to the sum of finitely many chaos, fg is simply a symmetric product $f \circ g$), and we have with obvious notation $< fg, f'g' > \, = \, < f,g > < f',g' >$. On the other hand, if f,g,h belong to the three Fock spaces $\Phi_s \rceil$, $\Phi_{[s,t]}$, $\Phi_{\lceil t}$ (also with obvious notations), then fg belongs to $\Phi_{t\rceil}$ and gh to $\Phi_{\lceil s}$, and the product is associative. This product is the tensor product above, written without the $\otimes$ sign.

Consider a curve $t \longmapsto x(t)$ in Fock space, with the following properties : $x(0) = 0$; the curve is *adapted* : $x(t) \in \Phi_t$ for all t ; its increments are adapted to the future : for $s < t$, $x(t) - x(s) \in \Phi_{\lceil s}$; finally, $\|x(t) - x(s)\|^2 = t - s$ for $s < t$. Then it is possible to define a stochastic integral $I(y) = \int y(s) \, dx(s)$ for an adapted curve $y(t)$ in Φ such that $\int \|y(s)\|^2 \, ds < \infty$, with the isometric property $\|I(y)\|^2 = \int \|y(s)\|^2 \, ds$ — otherwise stated, we have a purely Fock space theoretic version of the Ito integral (in a special case). The way to achieve this is standard, proving the isometric property for piecewise constant adapted curves, and then extending the integral. This being done, it is very easy to define also *multiple* stochastic integrals with respect to $x(\cdot)$, which have the same isometric property as Wiener's multiple integrals.

In the Wiener interpretation of Fock space, Brownian motion appears in two ways : as a curve in Hilbert space (the curve (X_t)) and as a family of multiplication operators Q_t, the *vector* X_t being the result of the *multiplication operator* $Q_t = a_t^+ + a_t^-$ applied to $\mathbf{1}$, the function 1. Similarly, in the Poisson interpretations, the Poisson process is a process of operators $Q_t + cN_t$ and a curve $X_t = (Q_t + cN_t)\mathbf{1}$. Note that *as curves* they are the same, while if we consider the second Brownian motion P_t, the corresponding curve is just iX_t. Thus stochastic integrals are also the same in different probabilistic interpretations, and even some stochastic differential equations are meaningful in this setup of curves in Fock space. An example of this is the linear s.d.e. satisfied by exponential vectors.

§3. MULTIPLICATION FORMULAS

The Wiener multiplication formula

1 We have seen in the preceding section that Fock space has several probabilistic interpretations — and those we have mentioned are but the simplest ones. Given two different probabilistic interpretations, the property of having the same chaotic expansions sets up a $1-1$ correspondence between their respective (classes of) square integrable random variables. *The difference between the two interpretations arises from the way random variables are multiplied.* Thus a probabilistic interpretation of Fock space appears as an associative, not everywhere defined, multiplication. In this section, we will have to distinguish carefully, on the same Fock space, the Wiener product, the Poisson products, etc.

Let us start with the classical form of the *multiplication formula for (Wiener) stochastic integrals.* We recall the notation of the preceding section, formula (2.1)

$$(1.1) \qquad I_n(f_n) = \int_{\mathbb{R}^n_+} f_n(s_1,\ldots,s_n)\, dX_{s_1}\ldots dX_{s_n}\ ,$$

where f_n is a symmetric L^2 function of n variables. It will be convenient to extend this notation (for this subsection only) to a possibly non-symmetric function, through the convention that $I_n(f)$ is the same as $I_n(f_s)$, the symmetrized function of f (which also belongs to L^2, with a norm at most equal to $\|f\|$).

Given two functions f and g of m and n variables respectively, we define their *contraction of order p* (where p is an integer $\leq n \wedge m$) to be the function of $n+m-2p$ variables given by

$$f \overset{p}{\smile} g(s_1,\ldots,s_{m-p},t_1,\ldots,t_{n-p})$$
$$(1.2) \qquad = \int f(s_1,\ldots,s_{m-p},u_1,\ldots,u_p)\, g(u_p,\ldots,u_1,t_1,\ldots,t_{n-p})\, du_1\ldots du_p\ .$$

Since our functions below are symmetric, the order of indices does not really matter : we have chosen that which fits the antisymmetric case as well. Note that the contracted function is not symmetric. One easily proves, using Schwarz's inequality, that contraction has norm 1 as a bilinear mapping from $L^2(\mathbb{R}^m) \times L^2(\mathbb{R}^n)$ to $L^2(\mathbb{R}^{m+n-2p})$.

REMARK. Contraction does not depend only on the abstract Fock space structure. It can be described using the bilinear scalar product $(f,g) = <\overline{f},g>$, which depends on the choice of a conjugation on the basic Hilbert space.

Here is the Wiener multiplication formula. We have no integrability problem, since it can be proved that the chaos spaces are contained in L^p for all $p < \infty$ (Surgailis showed that this pleasant feature of Wiener measure is not true for Poisson measures).

THEOREM. *With the above notations, we have*

(1.3)
$$I_m(f)\, I_n(g) = \sum_{p=0}^{m \wedge n} p! \binom{m}{p}\binom{n}{p} I_{m+n-2p}(f \overset{p}{\smile} g) \,.$$

If $n = 1$, $I_1(f)$ is the stochastic integral which we formerly denoted by $\tilde{f}$. On the other hand, p assumes only the two values $m \pm 1$, and the formula means that multiplication by $\tilde{f}$ is the operator $a_f^+ + a_f^-$.

For a proof of this relation, see [DMM] Chapter XXI. There is a less pleasant looking formula (see for instance Shigekawa, *J.Math. Kyoto Univ.*, 20, 1980, p.276) whose combinatorial meaning however is clearer. To describe it, we keep the above notations f_m and g_n, but assume no symmetry. We denote by α and β two injective mappings of $\{1,\dots,p\}$ into $\{1,\dots,m\}$ and $\{1,\dots,n\}$ respectively, and define the contraction $f \overset{\alpha\beta}{\smile} g$ as we did in (1.2), but replacing $s_{\alpha(i)}$ and $t_{\beta(i)}$ by u_i. Then the formula reads as follows

(1.4)
$$I_m(f)\, I_n(g) = \sum_{p} \frac{1}{p!} \sum_{\alpha,\beta} I_{m+n-2p}(f \overset{\alpha\beta}{\smile} g) \,.$$

To understand the meaning of this formula, let us consider first a measurable space E, two measures ρ on E^m and σ on E^n, and their product measure $\tau = \rho \otimes \sigma$ on $F = E^{m+n}$. Since all factors of the product are equal to E, let us define *diagonals* in F as follows, using different letters x, y for the two sets of coordinates. The (simple) 1–diagonals are defined by one equation $x_i = y_j$ and inequalities $x_k \neq y_l$ for all other pairings, (simple) 2–diagonals by two equations $x_i = y_j$, $x_k = y_\ell$ with $i \neq k$ and $j \neq \ell$, and inequations for all other pairings, etc. There are also higher diagonals, in which bunches of x_i's are set equal together, and equated to bunches of y_j's. Then the measure τ can be decomposed into an off-diagonal term, the contribution of the simple 1–diagonals, 2–diagonals, etc. If the measures ρ and σ themselves do not charge the diagonals in their own factor space, as will be the case here, we need to consider simple diagonals only. Then we denote by Δ_k the union of all (simple) diagonals of order k, including E^{m+n} for $k = 0$, and the measure τ can be decomposed as the sum of all integrals of the function $f \otimes g$ on the disjoint sets $\Delta_k \setminus \Delta_{k-1}$, the first term being the off-diagonal integral. For instance,

$$\left(\int_{s \neq t} f(s,t)\,\rho(ds,dt) \right) \left(\int g(u)\,\sigma(du) \right) =$$

$$= \int_{s \neq t \neq u} f(s,t) g(u)\, \rho(ds,dt)\sigma(du) + \int_{s \neq t = u} \dots + \int_{t \neq s = u} \dots \,.$$

Then formula (1.4) arises when this rigorous relation for measures is applied formally to the stochastic integrals $I_m(f)$ and $I_n(g)$. Since they are built out of integrals on open simplexes, they "forget the diagonals" in their factor spaces, and only simple diagonals occur. Then to compute the diagonal terms, *we replace* $(dX_s)^2$ *by* ds *whenever it occurs.* For instance in the preceding formula the off-diagonal contribution is $I_3(f \otimes g)$, and we have

$$\left(\int f(s,t)\, dX_s\, dX_t \right) \left(\int g(u)\, dX_u \right) = \int f(s,t) g(u)\, dX_s\, dX_t\, dX_u$$

$$+ \int \left(\int f(u,t) g(u)\, du \right) dX_t + \int \left(\int f(s,u) g(u)\, du \right) dX_s$$

This is not a proof, but "explains away" the formula, and the same idea works in other cases. For instance, for stochastic integrals corresponding to the compensated Poisson processes of §2 subs. 5 the rule is $(dX_s)^2 = ds + c\, dX_s$, and leads to the correct multiplication formula for Poisson stochastic integrals, due to Surgailis [Sur1] [Sur2].

2 We are going to express the multiplication formula in the shorthand notation of section 2, formula (2.3). We recall that the representation $h = \sum_n I_n(h_n)/n!$ is replaced by a single formula $h = \int_{\mathcal{P}} \widehat{h}(A)\, dX_A$ over the set of all finite subsets of $\mathbb{R}_+$ (we use the same letter for a r.v. and its "chaotic transform", but the hat may be omitted). On $\mathcal{P}$ we have a natural σ–field and a measure dA, and we can use set-theoretic operations, in particular the sum $A+B$, which is understood to be $A \cup B$ if $A \cap B = \emptyset$, and undefined otherwise (all functions taking the value 0 on undefined sets). Formula (1.3) must be transformed as follows : instead of $I_m(f)$, we are interested in $J_m(f) = I_m(f)/m!$, so that we rewrite it as

$$J_m(f)\, J_n(g) = \sum_p \frac{(m+n-2p)!}{p!\,(m-p)!\,(n-p)!}\, J_{m+n-2p}(h_p)\ ,$$

where h_p is the symmetrized form of $f \overset{p}{\smile} g$. Let us put

$$A = \{s_1, \ldots, s_{m-p}\} \in \mathcal{P}_m\ ,\quad B = \{t_1, \ldots, t_{n-p}\} \in \mathcal{P}_m\ ,\quad U = \{u_1, \ldots, u_p\} \in \mathcal{P}_p$$

$$C = \{s_1, \ldots, s_{m-p}, t_1, \ldots, t_{n-p}\} \in \mathcal{P}_{m+n-2p}\ .$$

The unsymmetrized function depends on the two sets of variables A, B and (1.2) can be written

$$f \overset{p}{\smile} g(A,\, B) = p! \int_{\mathcal{P}_p} f(A+U) g(U+B)\, dU\ .$$

We may use a symbol like $f(A+U)$ because 1) f is symmetric, so it does not matter which variables come first or last, 2) U is a.s. disjoint from the given finite set A. To symmetrize this function, we mix the points $s_i \in A$ with the points $t_j \in B$ by means of permutations σ of the elements of C, as follows

$$h_p(C) = \frac{1}{(m+n-2p)!} \sum_{\sigma} f(\sigma(A),\, \sigma(B)) = \frac{1}{(m+n-2p)!} \sum_{\substack{K+L=C \\ |K|=m-p,\, |L|=n-p}} \sum_{\substack{\sigma(A)=K \\ \sigma(B)=L}} f \overset{p}{\smile} g(K,\, L)$$

$$= \frac{(m-p)!\,(n-p)!}{(m+n-2p)!} \sum_{K+L=C} f \overset{p}{\smile} g(K,\, L)\ .$$

Finally, we get

$$\left(\int_{\mathcal{P}_m} f(A)\, dX_A \right) \left(\int_{\mathcal{P}_n} g(B)\, dX_B \right) = \sum_{p=0}^{m \wedge n} \int_{\mathcal{P}_{m+n-2p}} h_p(C)\, dX_C$$

where

$$h_p(C) = \sum_{\substack{K+L=C \\ |K|=m-p, |L|=n-p}} \int_{\mathcal{P}_p} f(K+U)g(L+U)\, dU \; .$$

This formula simplifies considerably, because the sum over p gets absorbed into an integration over $\mathcal{P}$, and we get the following result — on toy Fock space, U would be disjoint from C, and the formula would express the convolution $\widehat{h} = \widehat{f} \star \widehat{g}$.

THEOREM. *Let two random variables*

$$f = \int_{\mathcal{P}} \widehat{f}(A)\, dX_A \quad , \quad g = \int_{\mathcal{P}} \widehat{g}(B)\, dX_B \; ,$$

belong to a finite sum of chaos spaces, and their product $h = fg$ be expanded as $\int_{\mathcal{P}} \widehat{h}(C)\, dX_C$. Then we have

$$(2.1) \qquad\qquad \widehat{h}(C) = \int_{\mathcal{P}} \sum_{K+L=C} \widehat{f}(K+U)\widehat{g}(L+U)\, dU \; .$$

The simplicity of this formula, in contrast with the combinatorial complication of (1.3), shows the interest of the shorthand notation. As an easy exercise, compute $\mathbb{E}[fg] = \widehat{h}(\emptyset)$.

3 Now that we have a convenient notation, let us see whether the associativity of the Wiener multiplication can be seen on formula (2.1). This is not a mere play, since understanding this point will lead us later to the construction of other multiplications, and to Maassen's kernel calculus. To prove associativity, we need the following fundamental property of the measure dA on $\mathcal{P}$ ("Maassen's $\mathfrak{L}$–lemma").

THEOREM. *Let $f(A, B)$ be a positive measurable function on $\mathcal{P} \times \mathcal{P}$, and let $F(A)$ be the (positive, measurable) function on $\mathcal{P}$ defined by*

$$(3.1) \qquad\qquad F(A) = \sum_{H+K=A} f(H, K) \; .$$

Then we have

$$(3.2) \qquad\qquad \int_{\mathcal{P}} F(A)\, dA = \int_{\mathcal{P} \times \mathcal{P}} f(A, B)\, dA\, dB \; .$$

PROOF. In the "shorthand notation" of Fock spaces, the exponential vectors $\mathcal{E}(u)$ are read as the functions

$$(3.2) \qquad\qquad \mathcal{E}(u) : A \longmapsto \prod_{s \in A} u(s) \; .$$

The integral over $\mathcal{P}$ of such a function is $e^{\int u(s)\, ds}$, provided u is integrable on $\mathbb{R}_+$. By a density argument, it is sufficient to prove (3.1) assuming that $f(A, B)$ is a product $g(A)h(B)$, and then, assuming that $g = \mathcal{E}(u)$, $h = \mathcal{E}(v)$, u and v being bounded with compact support in $\mathbb{R}_+$. Then the function $F(A)$ given by (3.1) is $\mathcal{E}(u+v)$, and (3.2) reduces to the multiplicative property of the exponential.

> This property is shared by all measures "dA" on $\mathcal{P}$ arising from a non-atomic measure on $\mathbb{R}_+$, and in particular by the measure $\lambda^{|A|} dA$ on $\mathcal{P}$ corresponding to the replacement of dt by λdt.

We may now return to associativity. Consider three functions f, g, h on $\mathcal{P}$ and set $p = f \cdot g$. Then

$$p \cdot h(C) = \int dZ \left(\sum_{H+W=C} p(H+Z)h(W+Z) \right)$$

$$p(H+Z) = \int dN \left(\sum_{A+B=H+Z} f(A+N)g(B+N) \right)$$

A partition $A+B = H+Z$ can be represented as

$$Z = L+M \, , \; H = U+V \, , \; A = U+L \, , \; B = V+M \, ,$$

and the meaning of the preceding theorem is that, if Z is an integration variable over $\mathcal{P}$, we get the same integral by considering L, M as independent integration variables. The triple product can therefore be written

$$\int_{\mathcal{P}^3} dL \, dM \, dN \sum_{U+V+W=C} f(U+L+N)g(V+M+N)h(W+L+M) \, ,$$

which is perfectly symmetric. Associativity is proved.

We should perhaps repeat here one of the *leitmotive* of these notes : we are working on Wiener space, but results like (2.1) are meaningful for all Guichardet–Fock spaces, *i.e.* for the calculus on the space of finite subsets (configurations) on any reasonable space $(E, \mathcal{E}, \mu)$ without atoms. If the space is Lusin, no new proof is necessary : one simply uses a measurable isomorphism with Brownian motion.

REMARK. This multiplication formula is a consequence of the infinitesimal multiplication rule $(dX_s)^2 = ds$, which expresses that the variance increase of standard Brownian motion is 1 per unit time. If we replace ds by $\sigma^2 ds$ we get the infinitesimal multiplication rule for a Brownian motion of different variance, and the product formula (2.1) is replaced (with the same notation) by

$$(3.3) \qquad \widehat{h}(C) = \int_{\mathcal{P}} \sum_{K+L=C} \widehat{f}(K+U)\widehat{g}(L+U) \, \sigma^{2|U|} \, dU \, ,$$

an element of Fock space at least in the case f, g belong to a finite sum of chaos. On the other hand, we are not restricted to $\sigma^2 > 0$. For $\sigma^2 = 0$ we get the symmetric product $h = f \circ g$, also called *Wick product* and denoted by $f : g$, which is read in the shorthand notation as

$$(3.4) \qquad h(A) = \sum_{H+K=A} f(H)g(K) \, .$$

Thus the preceding theorem implies in particular that the integral over $\mathcal{P}$ is a multiplicative linear functional w.r.t. the Wick product (a statement to take with caution, since domain considerations are involved). Note that $\mathcal{E}(u + v) = \mathcal{E}(u) : \mathcal{E}(v)$ (the exponential vectors are "Wick exponentials").

> The name "Wick product" is not standard : physicists use it for a commutative multiplication for *operators*, arising from "Wick ordering" or "normal ordering" of the creation/annihilation operators, and this is the similar notion for vectors. On the other hand, physicists have the strange custom of writing a Wick product of, say, three operators as $:ABC:$ instead of $A:B:C$, the natural notation for an associative product.

For $\sigma^2 = i$ and $\sigma^2 = -1$ we get interesting structures, which have occasionally appeared in physics. The first one is related to a multiplication formula for Feynman integrals, and the second one to Gupta's negative energy harmonic oscillator.

Gaussian computations

4 In subsections 4–6 (whose reading is not necessary to understand the sequel), we will show that formula (2.1) implies some classical Gaussian computations, and also perform a *passage from the continuous to the discrete*, to get the Gaussian multiplication formula for the harmonic oscillator. This formula is slightly more complicated than (2.1), because on a discrete state space many particles sit exactly at the same point, while in the continuous case they can only be arbitrarily close.

Our first step will be to extend formula (2.1) to a Wiener product $h = f_1 \ldots f_n$. Using (3.2) it is not difficult to prove that one needs $n(n-1)/2$ "integration variables" U_{ij}, $i < j$. It is convenient to set $U_{ji} = U_{ij}$, and then we have (exchanging $\sum$ and $\int$ for convenience)

$$(4.1) \qquad \widehat{h}(C) = \sum_{H_1 + \ldots + H_n = C} \int dU_{12} dU_{13} \ldots, dU_{n-1,n} \prod_i \widehat{f}_i (H_i + \textstyle\sum_{j \neq i} U_{ij}) \,.$$

For instance, if $n = 3$ we get the same formula as in the proof of associativity. The most interesting case of (4.1) concerns the product of n elements from the first chaos : $f_i(A) = 0$ unless $|A| = 1$. Then in (4.1) either $|H_i| = 1$ and all the corresponding U_{ij} are empty, or $|H_i| = 0$ and there is exactly one U_{ij} such that $|U_{ij}| = 1$, the others being empty. It follows that $\widehat{h}(C) = 0$ if $|C|$ (which is also the number of nonempty H_i's) is not of the form $n - 2k$. We assume it is, and set $C = \{s_1 < \ldots < s_{n-2k}\}$. Then the nonempty H_i can be written in the following way, c denoting an injective mapping from $\{1, \ldots, n - 2k\}$ into $\{1, \ldots, n\}$

$$H_{c(1)} = \{s_1\} \,, \quad \ldots, \quad H_{c(n-2k)} = \{s_{n-2k}\} \,.$$

The nonempty U_{ij}'s $(i < j)$ are of the form $U_{a(i) b(i)} = U_{b(i) a(i)}$, a and b denoting two injective mappings from $\{1, \ldots, k\}$ to $\{1, \ldots, n\}$ to the complement of the range of c, such that $a(i) < b(i)$. In formula (4.1) the empty integration variables yield a product $\prod_i \widehat{f}_{c(i)}(s_i)$ and the nonempty ones a product $\prod_i (\widehat{f}_{a(i)}, \widehat{f}_{b(i)})$ (bilinear scalar product on $\mathcal{P}$, also equal to $(f_{a(i)}, f_{b(i)})$ on Ω). Finally

$$(4.2) \qquad \widehat{h}(C) = \sum_{a, b, c} \prod_{i=1}^{n-2k} \widehat{f}_{c(i)}(s_i) \prod_{j=1}^{k} (\widehat{f}_{a(j)}, \widehat{f}_{b(j)}) \,.$$

In particular, the expectation of $h = f_1 \ldots f_n$ corresponds to $C = \emptyset$. Its value is 0 unless n is even, $n = 2k$, in which case

$$(4.3) \qquad \mathbb{E}[f_1 \ldots f_{2k}] = \sum_{a,b} \prod_{j=1}^{k} (f_{a(j)}, f_{b(j)}),$$

a, b denoting injections from $\{1, \ldots, k\}$ to $\{1, \ldots, 2k\}$ such that $a(i) < b(i)$ (they can also be described as "pairings" in $\{1, \ldots, 2k\}$). This formula has been proved for the first chaos of Wiener space, but it is also a well known universal Gaussian formula (all Gaussian spaces of countably infinite dimension being isomorphic).

REMARK. The Wiener product of two exponential vectors $\mathcal{E}(u)$, $\mathcal{E}(v)$ is $e^{(u,v)} \mathcal{E}(u+v)$, where (u, v) is the bilinear scalar product $\int u(s) v(s) ds = \langle \overline{u}, v \rangle$. This shows that the Wiener product is not "intrinsic" : it requires the choice of a conjugation $u \longmapsto \overline{u}$, to define the bilinear scalar product.

> Every bilinear map $(\cdot, \cdot)$ leads to the definition of an associative multiplication by the above formula, non-commutative if the map is not symmetric (see App. 5).

In this abstract setup, formula (4.2) has a universal meaning, which we make explicit as follows. Consider $\Gamma_0(\mathcal{H}) = \mathcal{G}$ as an algebra with the symmetric product $\circ$ (i.e. the Wick product of subs. 3). It also admits a Wiener, or Gaussian multiplication, with $\mathbf{1}$ as unit element, and such that for $f \in \mathcal{H}$

$$(4.4) \qquad f^2 = f \circ f + 2(f, f) \mathbf{1}.$$

This is the commutative analogue of Clifford multiplication. Formula (4.2) then expresses the Wiener product in terms of Wick products : $h = f_1 \ldots f_n$ is given by

$$(4.5) \qquad h = \sum_{k} \sum_{a,b,c} \prod_{j=1}^{k} (f_{a(j)}, f_{b(j)}) f_{c(1)} \circ \ldots \circ f_{c(k)}.$$

It is also possible to invert this formula, and to express the Wick product in terms of the Wiener product. Surgailis has done in [Sur2] the same work for the Poisson product.

Two remarks may be in order here. The first is that, from the algebraic point of view, everything can be done on a real linear space with an arbitrary bilinear form (not necessarily positive). The second one is that Dynkin [Dyn1] [Dyn2] had the idea of doing the same computations, considering the coefficients (f, g) as new "scalars" — aiming at a rigorous handling of Gaussian laws with singular covariances.

5 Let e_i be an orthonormal basis of the first chaos. Let $\mathcal{A}$ be the set of all finite occupation vectors on $\mathbb{N}$; an element α of $\mathcal{A}$ can be written $n_1 \cdot i_1 \ldots + n_k \cdot i_k$, with which we associate the vector $e_\alpha = e_{i_1}^{\circ n_1} \ldots e_{i_k}^{\circ n_k}$ of Fock space. We have shown in section 1, end of subs. 1, that the norm^2 of e_α is $\alpha! = n_1! \ldots n_k!$. Any element of Fock space can be expanded in the orthogonal basis e_α as

$$(5.1) \qquad f = \sum_{\alpha} \frac{\widehat{f}(\alpha)}{\alpha!} e_\alpha \quad \text{with} \quad \|f\|^2 = \sum_{\alpha} \frac{|\widehat{f_\alpha}|^2}{\alpha!}.$$

The Wick product $e_\alpha : e_\beta$ is equal to $e_{\alpha+\beta}$ by construction, $\alpha+\beta$ being the sum of the two occupation vectors. Then for a Wick product $h = f : g$ we have the uninteresting formula

$$(5.2) \qquad \widehat{h}(\alpha) = \sum_{\rho+\sigma=\alpha} \widehat{f}(\rho)\widehat{g}(\sigma) \frac{\alpha!}{\rho!\sigma!} \,,$$

which must be compared with the corresponding formula for a Wiener product (we must then assume that e_α is a real basis) : for $h = fg$, we have

$$(5.3) \qquad \widehat{h}(\alpha) = \sum_{\mu} \sum_{\rho+\sigma=\alpha} \widehat{f}(\rho+\mu)\widehat{g}(\mu+\sigma) \frac{\alpha!}{\rho!\mu!\sigma!} \,.$$

This amounts to the Wiener multiplication rule for the basis vectors e_α themselves

$$(5.4) \qquad e_\gamma e_\delta = \sum_{\mu \leq \gamma \wedge \delta} \frac{\gamma!\,\delta!}{(\gamma-\mu)!\,\mu!\,(\delta-\mu)!} \, e_{\gamma+\delta-2\mu} \,.$$

There is an interesting way to deduce this from the continuous formula (2.1), which will be extended later to operators. Let X_t be Brownian motion, and let ξ_i $(i=1,2,\dots)$ be the increment $X_i - X_{i-1}$. These vectors are orthonormal in the first chaos; they do not form a basis of it, but this does not prevent us from using them to get a multiplication formula in the space they generate. Linear combinations of these vectors can be expressed as stochastic integrals $\int f(s)\,dX_s$ with f constant on each interval $[k, k+1[$, and more generally, a vector $\int f(A)\,dX_A$ in Fock space belongs to this σ-field if and only if $f(A)$ depends only on $|A \cap [0,1[|\dots, |A \cap [k.k+1[|\dots$ For instance, if f is the indicator function of $\{\,A \subset [0,1[,\ |A|=m\,\}$, then $\int f(A)\,dX_A$ is the elementary iterated integral

$$J_m = \int_{0<s_1<\dots<s_m<1} dX_{s_1}\dots dX_{s_m} \,,$$

which is known to have the value $h_m(\xi_1)/m!$ where h_m is the Hermite polynomial of order m (one way to prove this is to note that $\sum_m t^m J_m/m!$ is an exponential vector). Then formula (2.1) gives easily the multiplication formula for Hermite polynomials in one variable

$$(5.5) \qquad h_m h_n = \sum_{p \leq m \wedge n} \frac{m!\,n!}{(m-p)!\,p!\,(n-p)!} \, h_{m+n-2p} \,.$$

Formula (5.4) is a trivial extension of (5.5) to Hermite polynomials in several variables. Note also that the coefficients in (5.5) are the same as in (5.4) : this is not surprising, since both amount to expanding the multiplication formula for exponential vectors.

Poisson multiplication

6 Poisson products are far from being as useful as Wiener's. Thus we reduce our discussion to a few results, without complete proofs.

We work on the probabilistic Fock space associated with a compensated Poisson process of jump size c and intensity $1/c^2$. We refer the reader to §2, subsection 5 : the operator of multiplication by X_t is $Q_t + cN_t$, and the exponential vector $\mathcal{E}(f)$ is interpreted as the random variable

$$(6.1) \qquad \mathcal{E}(f) = e^{-\int f_s \, ds/c} \prod_s (1 + cf_s) \ .$$

First of all, there is a simple formula for the Poisson product of two exponential vectors. Denoting by (f, g) the bilinear scalar product $\int f_s g_s \, ds = \ <\bar{f}, g>$, we have

$$(6.2) \qquad \mathcal{E}(f)\, \mathcal{E}(g) = e^{(f,g)}\, \mathcal{E}(f + g + \tfrac{1}{c}\, fg) \ .$$

For $c = \infty$ we get the Wiener product. This relation is correct as far as random variables are concerned, but to be interpreted in Fock space it requires $fg \in L^2$, an indication that Poisson multiplication requires integrability conditions. Note also that a_h^- does not act as a derivation on Poisson products, as it did on Wick and Wiener products.

The infinitesimal multiplication formula for Poisson products is

$$(6.3) \qquad dX_t^2 = dt + c\,dX_t \ .$$

Its probabilistic meaning is the following : we have $dX_t = c(dN_t - dt/c^2)$, where (N_t) is a Poisson process with unit jumps and intensity $1/c^2$. Then the square bracket $d\,[X, X]_t$ (not to be confused with a commutator !) is equal to $c^2\,d\,[N, N]_t = c^2\,dN_t = c(dX_t + dt/c)$. Using this formula, we get the following result for a Poisson product $h = fg$ (the notations are the same as in (2.1))

$$(6.4) \qquad \widehat{h}(C) = \int_{\mathcal{P}} \sum_{K + Z + L = C} \widehat{f}(K + Z + U)\,\widehat{g}(L + Z + U)\, c^{-|Z|} dU \ .$$

For $c = \infty$ only the terms with $|Z| = 0$ contribute, and we get the formula for Wiener multiplication.

Relation with toy Fock space

7 We are now ready to describe in a heuristic way the relation between Fock space and finite spin systems. According to T. Lindstrøm, a rigorous discussion is possible using non-standard analysis, but I do not think there is anything published on this subject. This section is not meant as serious mathematics, and pretends only to make formal computations easier.

The idea is the following. In the discrete case, we had a unit vector e_i at each site i, and given a subset $A = \{i_1 < \ldots < i_n\}$ we had a corresponding unit vector e_A in the discrete n–th chaos. One possible probabilistic interpretation was, that e_i is a symmetric Bernoulli random variable x_i, and e_A the product of the x_i for $i \in A$ (a Walsh monomial). Now in Fock space, we set formally $e_i = dX_t/\sqrt{dt}$, and $e_A = dX_A/\sqrt{dA}$. In the case of Brownian motion, therefore, $dX_t/\sqrt{dt}$ is considered to be a Bernoulli r.v., not a Gaussian one. This is nothing but the central limit theorem, but it is unusual for probabilists not to think of dX_t as something Gaussian. It is also unusual for physicists, since they are familiar with the idea that quantum fields are systems of harmonic oscillators, rather than spins. Operators on Fock space will be considered in another section, but since we are doing heuristics let us mention now that $a_A^{\pm}$ corresponds to $da_A^{\pm}/\sqrt{dA}$, but a_A° simply to da_A° — a fact that should not surprise us, knowing the de Moivre–Laplace theorem and the different normalization for the number operator. Then for instance the commutation relation $[a_t^-, a_t^+] = I - 2a_t^{\circ}$ becomes

$$[da_t^-, da_t^+] = I\, dt - 2\, da_t^{\circ}\, dt \,,$$

which is the Fock space CCR up to a second order term.

The Bernoulli product formula in the discrete case is $e_A\, e_B = e_{A \triangle B}$. In continuous time it becomes

$$(7.1) \qquad\qquad dX_A\, dX_B = dX_{A \triangle B}\, d(A \cap B)$$

with the following meaning : A and B are ordered subsets, say $\{s_1 < \ldots < s_m\}$, $\{t_1 < \ldots, < t_n\}$ respectively, and $A \cap B$ is a subset $\{u_1 < \ldots < u_p\}$. Then the obscure looking $d(A \cap B)$ at the right is just $du_1 \ldots du_p$, while $dX_{A \triangle B}$ is a differential element at the remaining s_i's and t_j's, arranged in increasing order. *This has exactly the same meaning as the multiplication formula (2.1).*

Grassmann and Clifford multiplications

8   We are going to deal now with the antisymmetric Fock space (we refer the reader to section 1 for notation). As before, we work on the increasing open simplexes Σ_n. The meaning of a stochastic integral over Σ_n (*cf.* §2, subsection 1)

$$J_n(f) = \int_{\Sigma_n} f(s_1, \ldots, s_n)\, dX_{s_1} \ldots dX_{s_n} = \int_{\mathcal{P}_n} f(A)\, dX_A$$

remains unchanged, but the stochastic integral over $\mathbb{R}_+^n$, $I_n(f) = n!\, J_n(f)$, is interpreted as that of the *antisymmetric* extension of f. The two products we are going to consider are the Grassmann (exterior) product (corresponding to the Wick product) and the Clifford product (corresponding to Wiener's). They are completely defined by associativity, the anticommutativity $dX_s dX_t + dX_t dX_s = 0$ for $s \neq t$, and the rules for $s = t$

$$dX_s^2 = 0 \quad \text{(Grassmann)}\,, \qquad dX_s^2 = ds \quad \text{(Clifford)}\,.$$

From these infinitesimal rules one deduces finite formulas, giving the chaos expansion of a product $h = fg$. We recall from the chapter on spin (§5, subsection 1) the notation $n(A, B)$ for the total number of inversions between two finite subsets A, B of $\mathbb{R}_+$.

We have not much to say on the Grassmann product, which is given on the continuous basis elements by

$$(8.1) \qquad\qquad dX_A \wedge dX_B = (-1)^{n(A,B)}\, dX_{A+B}$$

and in closed form by

$$(8.2) \qquad\qquad \widehat{h}(C) = \sum_{K+L=C} (-1)^{n(K,L)}\, \widehat{f}(K)\, \widehat{g}(L) \,.$$

Remember our *leitmotiv* : from the point of view of exterior algebra, mapping any infinite dimensional separable Hilbert space onto $L^2(\mathbb{R}_+)$, by an arbitrary isomorphism, corresponds exactly to ordering a basis in a finite dimensional space, except that a continuous basis is used instead of a discrete one. Thus what we are describing here is the space of exterior differential forms with constant coefficients on any (infinite dimensional, separable) Hilbert space. The usual forms with non-constant coefficients require (at least) a tensor product of two Fock spaces, an antisymmetric one to provide the differential elements and a symmetric one for the coefficients.

Multiplying on the left (in the Grassmann sense) by an element $\tilde{h}$ of the first chaos is the same thing as applying the antisymmetric creation operator b_h^+ (§1, formula (4.1)). The antisymmetric annihilation operator b_h^- is an operation of contraction with h (§1, formula (4.3)).

In the discrete case, we had for the Clifford product

$$e_A e_B = (-1)^{n\,(A,B)} e_{A\triangle B}$$

The associativity $(e_A e_B)e_C = e_A(e_B e_C)$ of Clifford multiplication amounts to the following relation, where $(-1)^{n(A,B)}$ has been abbreviated to $\varepsilon(A,B)$

$$(8.3) \qquad\qquad \varepsilon(A,\,B\triangle C)\varepsilon(B,C) = \varepsilon(A,B)\varepsilon(A\triangle B,\,C)\ .$$

This relation is valid for all ordered sets, even for partially ordered ones if we make the convention that incomparable elements do not create inversions. Indeed, we have $n(A,B\triangle C)=n(A,B)+n(A,C)\ (mod\,2)$, and a similar relation on the right side which trivially gives the result.

A relation of the form (8.3) is called a *2–cocycle identity* w.r.t. the operation $\triangle$. See the remark at the end of this subsection.

We now describe the Clifford product, which is deduced from the infinitesimal relations $dX_t^2 = dt$ and $dX_s dX_t + dX_t dX_s = 0$ for $s \neq t$. On the continuous basis elements dX_A, we have

$$(8.4) \qquad\qquad dX_A\,dX_B = (-1)^{n(A,B)}\,dX_{A\triangle B}\,d(A\cap B)$$

and the corresponding finite formula is

$$(8.5) \qquad \widehat{h}(C) = \int_{\mathcal{P}} \sum_{K+L=C} (-1)^{n(K+U,L+U)}\,\widehat{f}(K+U)\widehat{g}(L+U)\,dU\ .$$

Assuming that $\widehat{f}(A)$ and $\widehat{g}(A)$ vanish for $|A|$ large enough, the same is true for $\widehat{h}(A)$, and a trivial domination argument by the corresponding terms of the Wiener product for $|\widehat{f}|$ and $|\widehat{g}|$ proves that $\widehat{h}$ is square integrable. Then associativity is proved exactly as for Wiener's product, using formula (8.3).

There is also a product formula for multiple integrals, which is formally identical to that for symmetric stochastic integrals, because we have defined the contraction operation in the way which minimizes the inversion numbers. We copy it for the reader's convenience

$$f \overset{p}{\smile} g(s_1, \ldots, s_{m-p}, t_1, \ldots, t_{n-p})$$

$$(8.6) \qquad = \int f(s_1, \ldots, s_{m-p}, u_1, \ldots, u_p)\, g(u_p, \ldots, u_1, t_1, \ldots, t_{n-p})\, du_1 \ldots du_p \ .$$

$$(8.7) \qquad I_m(f)\, I_n(g) = \sum_{p=0}^{m \wedge n} p! \binom{m}{p} \binom{n}{p} I_{m+n-2p} (f \overset{p}{\smile} g) \ .$$

In the last integral, $f \overset{p}{\smile} g$ is not antisymmetric, and the multiple integral is understood to be that of its antisymmetrized function. It is easy to check on (8.7) that the Clifford product by an element $\overline{h}$ of the first chaos (acting to the left on the uncompleted chaos sum) is the operator $R_h = b_h^+ + b_h^-$. In particular, it is a *bounded* operator (§1, subsection 5).

The operators R_h satisfy the basic Clifford algebra property

$$(8.8) \qquad\qquad \{R_h, R_k\} = 2 < \overline{h}, k > 1$$

Let e_i be an orthonormal basis of the first chaos, which we take real so that it is also an orthonormal basis for the bilinear scalar product $(f, g) = < \overline{f}, g >$. Given a finite set of indices $H = \{i_1 < \ldots < i_m\}$, we define

$$e_H = e_{i_1} \wedge \ldots \wedge e_{i_m} \ ,$$

an element of the m-th chaos (because of orthogonality, it is easy to prove that Grassmann and Clifford products coincide). These vectors constitute an orthonormal basis of Fock space, and by the associativity of the Clifford product and the rule $e_i^2 = 1$, we get the familiar expression of a Clifford multiplication

$$(8.9) \qquad\qquad e_H\, e_K = (-1)^{n\,(H,K)}\, e_{H \triangle K} \ .$$

An important consequence : the multiplication operator by e_1 is isometric, hence has norm 1. Since any real element of norm 1 in the first chaos can be taken for e_1, we see that *for h real in the first chaos, the Clifford (left) multiplication operator by h in Fock space has norm* $\|h\|$. An exact formula for the norm of this operator when h is complex is given in the paper by Araki, [Ara].

REMARK. Lindsay and Parthasarathy [LiP] study associative products given as perturbations of the Wiener product formula

$$(8.9) \qquad\qquad \hat{h}(C) = \int_{\mathcal{P}} \sum_{K+L=C} \hat{f}(K+U)\hat{g}(L+U)\,\varepsilon(U, K, L)\, dU \ ,$$

where the multiplier $\varepsilon(U, K, L)$ satisfies a "cocycle identity" w.r.t. the set operation $+$, slightly more complicated than (8.3). Many interesting products appear, some of which are intermediate between Wiener and Clifford.

Fermion Brownian motion

9 For every real element h of the first chaos, the Clifford multiplication by h is a selfadjoint bounded operator, hence a random variable. Since the square of this operator is $\|h\|^2 I$, as a r.v. it assumes the values $\pm\|h\|$. Since in the vacuum state it has expectation 0, its distribution must be symmetric. Thus the first chaos appears as a linear space of two-valued random variables, the anticommutative analogue of the Gaussian space it was in the commutative case.

In particular, if we take as element of the first chaos the ordinary Brownian motion random variable X_t, the norm of which is $\sqrt{t}$, the corresponding Clifford multiplication operators constitute a remarkable non-commutative stochastic process, called *fermion Brownian motion*. This process appears as a central limit of sums of anticommuting spins, and its stochastic integration properties have been studied in a series of papers by Barnett, Streater and Wilde [BSW].

We shall see later other examples of multiplication formulas, concerning operators instead of vectors (but from an algebraic point of view there is no difference).

§4. MAASSEN'S KERNELS

Operators defined by kernels

1 After vectors and products of vectors, we describe operators and products of operators in close analogy with the representation of toy Fock space operators in chapter II. Just as the discrete decomposition $f = \sum_A \widehat{f}(A)\,x_A$ of vectors in toy Fock space has become a stochastic integral $\int_{\mathcal{P}} \widehat{f}(A)\,dX_A$, the discrete decomposition $\mathbf{K} = \sum_{A,B} k(A,B) a_A^+ a_B^-$ of operators leads to "stochastic integrals" involving the creation and annihilation operators in normal ordering, *i.e.* integrals

$$\int_{\substack{s_1 < s_2 \cdots < s_m \\ t_1 < t_2 \cdots < t_n}} k(s_1, \ldots, s_m \,;\, t_1, \ldots, t_n)\, da_{s_1}^+ \ldots da_{s_m}^+\, da_{t_1}^- \ldots da_{t_n}^- \ .$$

Just as for vectors, there is also a form of this integral which does not depend on the ordering of $\mathbb{R}_+$

$$I_{m,n}(k) = \frac{1}{m!\,n!} \int_{\mathbb{R}_+^m \times \mathbb{R}_+^n} k(s_1, \ldots, s_m \,;\, t_1, \ldots, t_n)\, da_{s_1}^+ \ldots da_{s_m}^+\, da_{t_1}^- \ldots da_{t_n}^- \ ,$$

where now the kernel has been symmetrized separately in each group of variables, or antisymmetrized in the case of fermion operators. Such integrals, known as *Wick monomials,* are standard in quantum field theory (see for instance Berezin's book [Ber]), with slightly different notation and a considerably different spirit. Instead of $da_s^\pm$ the differential element is written $a_s^\pm\, ds$, the integrand being an "operator valued distribution"; and since distributions have entered the picture, why not assume that k itself is a distribution? In this way, the theory is well launched into its heavenly orbit, never to come down. In the case of vectors instead of operators the distribution point of view (the white noise approach to Brownian motion) is less popular among probabilists than Ito's stochastic integration. Though the interest of a distribution approach (in both cases) is not to be denied, this seems to indicate that much can be done with more down-to-earth methods. We shall describe later the stochastic integral calculus of Hudson–Parthasarathy; here we follow Maassen [Maa], who develops a theory with plain bounded functions, a modest and reasonable first step. Still, the use of the number operator allows to deal with kernels which in Berezin's description would be singular on diagonals. Contrary to Hudson–Parthasarathy's stochastic calculus (given in Chapter VI), Maassen's approach depends only apparently on the order of $\mathbb{R}_+$, and can be extended to Fock spaces over a general space $L^2(\mu)$.

We consider first operators defined, in Guichardet's shorthand notation, by a "kernel" k with two arguments (including a term $k(\emptyset, \emptyset) I$)

$$(1.1) \qquad \mathbf{K} = \int_{\mathcal{P} \times \mathcal{P}} k(A, C) \, da_A^+ \, da_C^- ,$$

and then, more generally, by a kernel with three arguments

$$(1.1') \qquad \mathbf{K} = \int_{\mathcal{P} \times \mathcal{P} \times \mathcal{P}} k(A, B, C) \, da_A^+ \, da_B^\circ \, da_C^- .$$

We start with formal algebraic computations, and prove only later (in subsection 5) that under some restrictions kernels really define closable operators on a stable dense domain.

To give a meaning to $\mathbf{K}$ as an operator, we describe how it acts on a vector f, defining both f and $g = \mathbf{K}f$ by their chaos expansions $f = \int \widehat{f}(L) dX_L$ and $g = \int \widehat{g}(L) dX_L$. First of all, let us say how a "differential element" in (1.1) acts on an element of the "continuous basis" dX_H. Such formulas are meant only as heuristic justifications, so that if our reader dislikes them, he may go directly to the finite formula (1.5). They derive from the definition of the creation, annihilation and number operators, and correspond to those concerning toy Fock space (Chap. II, §2, (3.1)), if x_A is replaced by $dX_A/\sqrt{dA}$, $a_A^\pm$ by $da_A^\pm/\sqrt{dA}$, and a_A° by da_A°

$$
\begin{aligned}
& da_A^+ \, dX_L = dX_{L+A} && (dX_{L \cup A} \text{ if } L \cap A = \emptyset, \ 0 \text{ otherwise}) , \\
(1.2) \quad & da_A^- \, dX_L = dX_{L-A} \, dA && (dX_{L \setminus A} \, dA \text{ if } A \subset L, \ 0 \text{ otherwise}) , \\
& da_A^\circ \, dX_L = dX_L && \text{if } A \subset L , \ 0 \text{ otherwise.}
\end{aligned}
$$

Then by composition

$$(1.3) \qquad da_A^+ \, da_C^- \, dX_L = dX_{A+(L-C)} \, dC .$$

Recall that on toy Fock space we had two bases for operators, one consisting of all $a_A^+ a_C^-$ with A and C not necessarily disjoint, and one consisting of all $a_A^+ a_B^\circ a_C^-$ with disjoint A, B, C. Thus it is natural to assume that they are also disjoint in the following definition

$$(1.4) \qquad da_A^+ \, da_B^\circ \, da_C^- \, dX_L = dX_{A+B+(L-(B+C))} \, dC .$$

Now we have (still computing formally)

$$\left(\int k(A,C)\, da_A^+ da_C^- \right) \left(\int \widehat{f}(L)\, dX_L \right) = \int k(A,C) \widehat{f}(C)\, dX_{A+(L-C)}\, dC \ ,$$

and to find the value of $\widehat{g}(H)$ we express the solution to the equation $A+(L-C)=H$ in the form $H = U+V$, $C=M$, $A=U$, $L=V+M$. Thus U,V realize a partition of the given set H and can be chosen in finitely many ways, while M is a parameter, disjoint from V — but since we integrate over the simplex, this condition holds a.s. and need not be expressed. In this way we get the basic formula of Maassen for $g = Kf$

$$(1.5) \qquad \widehat{g}(H) = \int_{\mathcal{P}} \sum_{U+V=H} k(U,M) \widehat{f}(V+M)\, dM$$

(the reader should write the corresponding formula on toy Fock space). On the other hand, M is also a.s. disjoint from $U \subset H$. Therefore, only the value of the kernel $k(A,C)$ on *disjoint* subsets matters in this computation, and we are unable to represent the number operator as a kernel, contrary to the discrete case where we had $a_i^+ a_i^- = a_i^\circ$. This shows the advantage of using the second basis of the discrete case, which contains explicitly the number operator, and in which nothing of the kind happens. We then get the formula

$$(1.5') \qquad \widehat{g}(H) = \int_{\mathcal{P}} \sum_{U+V+W=H} k(U,V,M) \widehat{f}(V+W+M)\, dM \ .$$

The combinatorics of kernels with three arguments is more complicated, but they are well worth the additional trouble, as we shall see.

2 Let us give some examples of kernels.

1) The creation, number and annihilation operators are represented as follows by three argument kernels $k(A,B,C)$ (it is understood that k has the value 0 if the triple A, B, C is not of the form indicated)

$$(2.1) \qquad a_h^+ \ : \ k(\{t\},\varnothing,\varnothing) = h(t) \quad ; \quad a_h^- \ : \ k(\varnothing,\varnothing,\{t\}) = \overline{h}(t) \quad ;$$
$$a_h^\circ \ : \ k(\varnothing,\{t\},\varnothing) = h(t) \ .$$

For creation and annihilation we may also use two arguments kernels

$$(2.2) \qquad a_h^+ \ : \ k(\{t\},\varnothing) = h(t) \quad ; \quad a_h^- \ : \ k(\varnothing,\{t\}) = \overline{h}(t) \ .$$

Let us also make explicit, for future use, the action of such operators (writing t instead of $\{t\}$ to abbreviate notation)

$$(a_h^+ f)\widehat{\ }(L) = \sum_{t \in L} h(t) f(L-t) \quad ; \quad (a_h^- f)\widehat{\ }(L) = \int \overline{h}(t) f(L+t)\, dt \quad ;$$

$$(2.3)$$

$$(a_h^\circ f)\widehat{\ }(L) = \left(\sum_{t \in L} h(t) \right) f(L) \ .$$

2) Multiplication operators can be represented by kernels. For instance, the kernel

$$k(A, \emptyset, C) = \sigma^{2|C|} \widehat{f}(A + C)$$

represents the operator of Wick multiplication by f for $\sigma = 0$, of Wiener multiplication by f for $\sigma = 1$, and a whole family of associative products for arbitrary complex values of σ. The kernel

$$k(A, B, C) = c^{|B|} \widehat{f}(A + B + C)$$

represents a Poisson multiplication by f. There are many other associative multiplications which can be represented by kernels (including Clifford multiplications). See Lindsay and Parthasarathy [LiP1].

The Wick product with f is represented by the kernel $k(A, \emptyset, \emptyset) = \widehat{f}(A)$, which is a generalized creation operator. Kernels $k(\emptyset, \emptyset, C) = \widehat{f}(C)$ are generalized annihilation operators, and are called *contractions*.

3) *Kernels depending on the middle variable.* A kernel $k(A, B, C)$ which is equal to 0 unless $A = C = \emptyset$ (and then is equal to $k(B)$) acts as a multiplier transformation in harmonic analysis (recall that, in the discrete case, $\widehat{f}$ could be interpreted as a kind of Fourier transform of the r.v. f)

(2.6) $$(\mathbf{K}f)\widehat{}(A) = j(A)\widehat{f}(A) \quad \text{with} \quad j(A) = \sum_{B \subset A} k(B) \ .$$

For instance, the multiplier $j(A) = 1$ if $|A| = m$, 0 otherwise, represents the projection operator on the chaos of order m. The kernel k corresponding to a multiplier j may be computed by means of the Möbius inversion formula

(2.7) $$k(A) = \sum_{B \subset A} (-1)^{|A-B|} j(B) \ .$$

A remarkable family of such operators is that of the multipliers

(2.8) $$j_t(A) = (-1)^{A \cap]0,t]} \ .$$

The corresponding operators $\mathbf{J}_t = (-1)^{a_t^\circ}$ were first introduced by Jordan and Wigner in the discrete case. As we shall see later, they can be used to define fermion creation and annihilation operators from boson ones, and conversely.

4) *Weyl operators.* For $u \in L^2(\mathbb{R}_+)$, we consider the kernel with two arguments

(2.9) $$k(A, C) = e^{-\|u\|^2/2} \prod_{s \in A} u(s) \prod_{t \in C} \left(-\overline{u}(t) \right)$$

(note the analogy with exponential vectors). Let us compute $j = \mathbf{K}\mathcal{E}(h)$ using formula (1.5) for kernels with two arguments

$$\mathcal{E}(h)\widehat{}(A) = \prod_{r \in A} h(r) \ ; \quad \widehat{j}(H) = \int dM \sum_{U+V=H} k(U, M) \prod_{r \in U+M} h(r) \ .$$

First comes the constant $\exp(-\|u\|^2/2)$, and a factor which does not contain the integration variable M

$$\sum_{U+V=H} \prod_{s\in U} u(s) \prod_{r\in V} h(r) = \prod_{t\in H}(u(t)+h(t))$$

and then the integral over $\mathcal{P}$

$$\int \left(\prod_{r\in M} -\bar{u}(r)h(r) \right) dM .$$

The function $a = -\bar{u}h$ is integrable on $\mathbb{R}_+$, and the integral on $\mathcal{P}$ is equal to $\exp(\int a(r)\,dr)$. Finally we have

$$\mathbf{K}\mathcal{E}(h) = e^{-<u,h>-\|u\|^2/2}\mathcal{E}(u+h)$$

and we recognize the action of Weyl operators W_u on exponential vectors. A similar reasoning, with slightly more complicated notations, shows that the following kernel with three arguments, where λ denotes a real valued function on $\mathbb{R}_+$

$$k(A,B,C) = e^{-\|u\|^2/2} \prod_{r\in A} u(r) \prod_{s\in B}(e^{i\lambda(s)}-1) \prod_{t\in C}(-e^{i\lambda(t)}\bar{u}(t))$$

represents the Weyl operator $W_{u,\lambda}$.

5) *Differential second quantization operators* may be represented by kernels, at least in some cases. Consider an operator U on $L^2(\mathbb{R}_+)$ which is represented by a Hilbert–Schmidt kernel $u(s,t)$. Then it is easy to compute the action of the kernel with two arguments $\int u(s,t)\,da_s^+da_t^-$, and to check that it represents the differential second quantization $\lambda(U)=d\Gamma(U)$.

6) We end with a remarkable example, discovered independently by S. Attal and J.M. Lindsay. It shows most evidently the importance of kernels with three arguments.

Consider a Hilbert–Schmidt operator H on Fock space. It is given by a kernel (in the ordinary sense)

$$(2.10) \qquad Hf(A) = \int_{\mathcal{P}} h(A,M)f(M)\,dM .$$

Then it is also given as a Maassen-like operator with the Maassen kernel

$$(2.11) \qquad H(U,V,W) = (-1)^{|V|}h(U,W) .$$

Indeed we have, before any integration is performed

$$\sum_{U+V+W=A} H(U,V,M)f(V+W+M) = h(A,M)f(M) .$$

The computation is as follows : we rewrite the left hand side as

$$\sum_{U\subset A}\sum_{V\subset A-U}(-1)^{|V|}h(U,M)f(A-U+M) = \sum_{U\subset A} h(U,M)f(A-U+M)\sum_{V\subset A-U}(-1)^{|V|},$$

and this last sum is 0 unless $A-U=\emptyset$, in which case it is 1. Therefore what we get is simply $h(A,M)f(M)$.

Composition of kernels

3 Kernels represent operators in *normal form*, *i.e.* with the creation operators to the left of annihilation operators. If we are to multiply two operators **J** and **K** given by kernels, we find in the middle a set of creation and annihilation operators which are not normally ordered; the possibility of reordering them and the practical rules to achieve this are popular in physics under the name of *Wick's theorem*. The shorthand notation for Fock space allows us to give a closed formula for the product $\mathbf{L} = \mathbf{JK}$.

We begin with the composition formula for kernels with two arguments (there is a corresponding, more complicated, discrete formula). Denoting by ℓ, j, k the kernels for **L, J, K** we have

$$(3.1) \qquad \ell(A, B) = \int_{\substack{R+S=A \\ T+U=B}} j(R, T + M)\, k(S + M, U)\, dM$$

Indeed, applying this kernel **L** to a vector f (and omitting the hats for simplicity) we have

$$\mathbf{L}f(H) = \int \sum_{U+T=H} \ell(U, Q) f(T + Q)\, dQ$$

$$= \int \sum_{U+T=H} \sum_{\substack{R+S=U \\ N+P=Q}} j(R, N + M)\, k(S + M, P)\, f(T + Q)\, dQ\, dM$$

since we are splitting the integration variable Q into $N + P$, we may consider N, P as two independent integration variables (§3, formula (3.1))

$$= \int \sum_{R+S+T=H} j(R, N + M)\, k(S + M, P)\, f(T + N + P)\, dM\, dN\, dP$$

$$= \int \sum_{R+V=H} \sum_{\substack{S+T=V \\ M+N=L}} j(R, L)\, k(S + M, P)\, f(T + N + P)\, dL\, dP\,,$$

where the same formula (3.1) of §3 has been used again. On the last line we may recognize $\mathbf{J}(\mathbf{K}f)$.

The formula relative to kernels with three arguments is much more complicated, and we write it without proof. In the course of these notes we will rewrite it in several different fashions (Chap. V, §2, 3 and §3, 2)

$$(3.2) \qquad \ell(U, V, W) =$$

$$= \int \sum_{\substack{A_1+A_2+A_3=U \\ B_1+B_2+B_3=V \\ C_1+C_2+C_3=W}} j(A_1, A_2+B_1+B_2, C_1+C_2+M)\, k(M+A_2+A_3, B_2+B_3+C_2, C_3)\, dM\,.$$

Though these formulas look rather forbidding, we recall that they proceed from simple differential rules, just as the Wiener, Wick, Poisson, Clifford, Grassmann multiplication formulas follow only from associativity and (anti)commutativity, and from one single rule giving the square of the differential element dX_s. Here we have three basic differential elements, and we must multiply them two by two. All products are 0, except those in increasing order $(-<\circ<+)$, which are

(3.3) $da_t^- \, da_t^\circ = da_t^-$, $da_t^- \, da_t^+ = dt$, $da_t^\circ da_t^\circ = da_t^\circ$, $da_t^\circ da_t^+ = da_t^+$.

To these rules should be added the commutation of all operators at different times. After the work of Hudson–Parthasarathy, this table has been known as the *Ito table* — a slightly misleading name, since Ito was concerned with *adapted* stochastic calculus, and the composition of kernels does not require adaptation. Forgetting for one instant the number operator, what we are doing is really constructing an associative algebra which realizes the CCR or CAR as its commutators or anticommutators, and in the next chapter we shall see other ways of realizing this.

4   In section 3, subsection 5, we introduced discrete chaotic expansions, with a method that reduces them to the (simpler) continuous case. The same method works for kernels too, and we indicate the corresponding formulas.

We start with the representation formula for vectors (formula (5.1) of §3)

(4.1) $f = \sum_\alpha \frac{\widehat{f}(\alpha)}{\alpha!} e_\alpha$ with $\|f\|^2 = \sum_\alpha \frac{|\widehat{f}_\alpha|^2}{\alpha!}$.

We describe kernels using a similar notation

(4.2) $\mathbf{K} = \sum_{\alpha,\beta} k(\alpha,\beta) \frac{a_\alpha^+ a_\beta^-}{\alpha!\,\beta!}$,

where α is an occupation vector $(n_{i_1} \ldots, n_{i_k})$ and a_α^ε abbreviates $(a_{i_1}^\varepsilon)^{n_1} \ldots (a_{i_k}^\varepsilon)^{n_k}$. The vector $g = \mathbf{K}f$ is given by

(4.3) $\widehat{g}(\alpha) = \sum_\mu \sum_{\rho+\sigma=\alpha} \frac{\alpha!}{\rho!\,\sigma!\,\mu!}\, k(\rho,\mu)\,\widehat{f}(\sigma+\mu)$.

In the discrete case, there is no need of three-argument kernels, since the number operator can be expressed as a^+a^-. The composition formula for a product $\mathbf{L} = \mathbf{J}\mathbf{K}$ which corresponds to (3.1) is

(4.4) $\ell(\alpha,\beta) = \int_{\substack{\rho+\sigma=\alpha \\ \tau+\upsilon=\beta}} \frac{\alpha!\,\beta}{\rho!\,\sigma!\,\tau!\,\upsilon!\,\mu!}\, j(\rho,\tau+\mu)\,k(\sigma+\mu,\upsilon)$.

It expresses Wick's theorem in closed form, though the diagrammatic version physicists use is probably just as efficient and more suggestive.

Maassen's theorem

5 We return to the continuous case. We have been doing algebra, but it necessary to transform this into analysis, *i.e.* to find some *simple* conditions under which the action of a kernel on a vector really defines a vector in Fock space.

Maassen's *test vectors* are vectors f of Fock space which satisfy the following regularity assumptions : 1) $\widehat{f}(A)$ vanishes unless A is contained in some bounded interval $]0,T]$ (if the results must be extended from $\mathbb{R}_+$ to an arbitrary measure space, the bounded interval is replaced by any set of finite measure); 2) $|\widehat{f}(A)|$ is dominated by $cM^{|A|}$, where c and M are constants ($M>1$ is allowed). The space $\mathcal{T}$ of test vectors contains all exponential vectors $\mathcal{E}(h)$ such that h is bounded and has compact support in $\mathbb{R}_+$. In particular, it is dense in Fock space.

> The word "test vector" is convenient here, but it has several meanings in Wiener space analysis, and this is not the most common one.

Similarly, one defines *regular kernels* with two or three arguments as follows : 1) $k(A,B,C)$ vanishes unless A,B,C are all contained in some bounded interval $]0,T]$; 2) $|k(A,B,C)|$ is dominated by $cM^{|A|+|B|+|C|}$.

The main result of Maassen is the following :

THEOREM. *Applying a regular kernel on a test vector gives again a test vector. Thus regular kernels define operators on the common stable domain $\mathcal{T}$, and on this domain they form a *-algebra. In particular, since they have densely defined adjoints they are closable.*

The proof of this theorem is not difficult, and we only sketch it. We denote by k a regular kernel, with two arguments for simplicity, and by f a test vector. We may assume that the constants c, T and M are the same for both. Then for the vector $g=\mathbf{K}f$ we have $g(H)=0$ if H is not contained in $]0,T]$, and if it is

$$|g(H)| \le \int \sum_{A+B=H} c^2 M^{|A|+|U|} M^{|B|+|U|} dU .$$

$c^2 M^{|A|+|B|} = M^{|H|}$ comes out, and there remains the integral of $M^{2|U|}$ over the subsets U of $]0,T]$, which is equal to $e^{M^2 T}$. Thus we have again a bound of the same type.

For kernels with three arguments, the computation is only slightly more complicated : we split H into $A+B+C$ instead of $A+B$, and therefore we have an additional factor of $M^{|B|} \le M^{|H|}$. The proof that the composition of two regular kernels is again regular is also essentially the same.

As for adjoints, it is easy to check that, given a regular kernel k with two *resp.* three arguments, the kernel k^* defined by

(5.1) $k^*(A,C) = \overline{k}(C,A)$ *resp.* $k^*(A,B,C) = \overline{k}(C,B,A)$

is obviously regular, and defines an operator $\mathbf{K}^*$ which is adjoint to $\mathbf{K}$ on test vectors. To see this, we compute the form $\Phi(g,f) = \langle g, \mathbf{K}f \rangle$ as follows (still omitting hats)

$$\Phi(g,f) = \int \sum_{A+B+C=H} \bar{g}(H) \, k(A,B,U) \, f(B+C+U) \, dU \, dH$$

$$= \int \bar{g}(A+B+C) \, k(A,B,U) \, f(B+C+U) \, dU \, dA \, dB \, dC$$

where we have used the fundamental property of the measure dH, which allows treating A, B, C as independent integration variables. In the last integral it is clear that we can exchange the roles of A and U to find the adjoint kernel (5.1). But it is also interesting to transform it as follows, grouping B, C into one single integration variable V

$$\Phi(g,f) = \int \sum_{B+C=L} \bar{g}(A+V) \, k(A,B,U) \, f(V+U) \, dU \, dA \, dV$$

(5.2)
$$= \int \bar{g}(A+V) \, \varphi(A,V,U) \, f(V+U) \, dA \, dV \, dU$$

where he have defined

(5.3)
$$\varphi(A,V,U) = \sum_{B \subset V} k(A,B,U) \, .$$

Note that k may be reconstructed from φ using the Möbius inversion formula. Formula (5.2) is a convenient symmetric way to define forms by kernels, and one may remark that it is equivalent to demand Maassen-like domination conditions on k and on φ.

REMARKS. a) There are some minor points to add to this theorem. The fact that the composition of operators is associative does not quite imply the same property for the composition of regular kernels, since an operator does not uniquely determine its kernel. Maassen has given a detailed proof that it is associative indeed, using the same method as for the associativity of the Wiener product.

b) The second point is precisely that of uniqueness : *If two regular kernels define the same operator on test vectors, they are a.e. equal as functions on $\mathcal{P} \times \mathcal{P} \times \mathcal{P}$.* We postpone this problem to the next chapter, §2, subsection 6.

c) The third point concerns the failure of this approach to construct *exponentials*, while the first step in a real physical problem is to construct a quantum evolution e^{-itH} from a hamiltonian H. The exponential series for a regular kernel usually diverges.

Fermion kernels

6 We are going to review summarily the representation of operators on antisymmetric Fock space. The fermion annihilation, number and creation differentials will be denoted by db_t^ε with $\varepsilon = -, \circ, +$.

As operators were defined, in the symmetric case, by kernels with two and three arguments (1.1) and (1.1$'$), we may define operators in the antisymmetric case by

(6.1)
$$\mathbf{K} = \int k(A,C)(-1)^{n(A,C)} \, db_A^+ \, db_C^-$$

(6.1$'$)
$$\mathbf{K} = \int k(A,B,C)(-1)^{n(A,B+C)} \, db_A^+ \, db_B^\circ \, db_C^- \, .$$

The alternants are meant to simplify the expressions to follow. The formulas in subsection 1 concerning boson operators are replaced now by

$$db_A^+ \, dX_L = (-1)^{n(A,L)} dX_{L+A} \ , \ \ db_C^- \, dX_L = (-1)^{n(C,L)} dX_{L-C} \, dC \ ,$$
$$db_B^{\circ} \, dX_L = dX_{B+(L-B)} \quad (0 \text{ by convention if } B \not\subset L)$$

from which we deduce the action $\mathbf{K}f = g$ of the operator $\mathbf{K}$ on a vector $f = \int \widehat{f}(M) \, dX_M$

$$(6.2) \quad \widehat{g}(H) = \int \sum_{U+V=H} k(U,M) \, \widehat{f}(M+V) (-1)^{n(U+M,M+V)} \, dM$$

$$(6.2') \quad \widehat{g}(H) = \int \sum_{U+V+W=H} k(U,V,M) \, \widehat{f}(M+V+W) (-1)^{n(U+M,M+V+W)} \, dM \ .$$

With the conventions (6.1), it then appears that the kernel for left Clifford multiplication by f is $k(A,C) = f(A+C)$, as for Wiener's multiplication. Let us also indicate the formula for the composition of two kernels $\mathbf{L} = \mathbf{JK}$ (*cf.* (3.1))

$$(6.3) \quad \ell(A,B) = \int_{\substack{R+S=A \\ T+U=B}} j(R,T+M) \, k(M+S,U) (-1)^{n(R+T+M \ M+S+U)} dM \ .$$

This is not too ugly, but we do not attempt to give the formula for three-argument kernels!

The continuous Jordan–Wigner transformation

7   As a consequence of their work on non-commutative stochastic integration, Hudson and Parthasarathy [HuP2] could extend to continuous time the Jordan–Wigner transformation relating the boson and fermion creation and annihilation operators in the discrete case (toy Fock space, Chap. II, §5 subsection 2). We shall explain the essential point in a heuristic way, leaving the details to the reader. See also Parthasarathy and Sinha [PaS1].

As usual, we work on Fock space over $L^2(\mathbb{R}_+)$, which can naturally be considered as symmetric or antisymmetric. Let N_t be the number operator up to time t, and J_t be the operator $(-1)^{N_t}$. In discrete time, we would not use N_i, but rather its predictable analogue N_{i-1}. Anticipating the continuous stochastic calculus, the result of Hudson–Parthasarathy says that (J_t) is an adapted operator process, and that the boson creation and annihilation operators a_t^{ε} are related to their fermion analogues b_t^{ε} by the relations

$$(7.1) \qquad\qquad b_t^{\varepsilon} = \int_0^t J_s \, da_s^{\varepsilon} \ ; \quad a_t^{\varepsilon} = \int_0^t J_s \, db_s^{\varepsilon} \ ;$$

which involve operator stochastic integrals (finite sums in the discrete case). To make this formula transparent, let us look at its infinitesimal version : da_t^+ transforms a continuous basis element dX_A into dX_{A+t}, and then J_t adds a factor $(-1)^n$, where $n = |(A+t) \cap]0,t[\, |$ (t is not counted : the finite sets A containing a given t can be neglected), and n is also the number of inversions $n(t, A)$ *i.e.* the number of elements of A (or $A+t$) strictly smaller than t. And therefore what we get is $b_t^+ X_A$. The reasoning is the same for annihilation operators.

The preceding derivation is heuristic, but the integral formula can be rigorously verified using the rules of stochastic calculus.

As an exercise, let us give to (7.1) a formulation in terms of kernels. First of all, J_s is a multiplier transformation : if $f = \int \widehat{f}(A) \, dX_A$, then J_s multiplies $\widehat{f}$ by $(-1)^{|A \cap]0,s[|}$. Using the Möbius inversion formula (subs. 2, formula (2.6)) we see that J_s is given by a kernel

$$j_s(\varnothing, B, \varnothing) = (-2)^{|B \cap]0,s[|} .$$

Given a family $\mathbf{K}_s$ of operators, given by kernels $k_s(A, B, C)$, adapted in the sense that k_s vanishes unless all three arguments are contained in $]0, s]$, the formal rules defining the stochastic integral $\mathbf{L}^\varepsilon = \int \mathbf{K}_s \, da_s^\varepsilon$ are the following. Given a subset A, let us call $\vee A$ its last element and put $A- = A \backslash \{\vee A\}$. Then the kernel of $\mathbf{L}^\varepsilon$ is

(7.2)
$$\ell^+(A, B, C) = k_{\vee A}(A-, B, C) \quad ; \quad \ell^\circ(A, B, C) = k_{\vee B}(A, B-, C) \quad ;$$

$$\ell^-(A, B, C) = k_{\vee C}(A, B, C-) .$$

We explain this in detail in Chapter VI, §2 subs. 2. Thus we can compute the bosonic kernel of a fermion creation operator (the annihilation operator is similar, with the first variable empty instead of the third)

$$b_t^+(s, B, \varnothing) = (-2)^{|B \cap]0,s[|} \quad \text{if} \quad s \leq t, \ B \subset]0, t] \ , \ 0 \quad \text{otherwise.}$$

The validity of this relation can be checked explicitly on a continuous basis vector dX_A.

Other estimates on kernels

8[1]   Maassen's theorem is a simple and useful result, but it is too restrictive to deal only with uniform estimates on the kernel. We will give now a more recent estimate, due to Belavkin–Lindsay ([BeL]), which uses a mixture of uniform norm in the number operator variables, and L^2 (Hilbert–Schmidt) norm in the creation and annihilation variables.

First, we introduce the idea of a *scale* of Fock spaces, denoting by Φ_p for $p > 0$ the Hilbert of functions $f(A)$ on $\mathcal{P}$ such that

(8.1)
$$\| f \|_{(p)}^2 = \int_{\mathcal{P}} |f(A)|^2 \, p^{|A|} \, dA < \infty .$$

Then we have $\Phi_1 = \Phi$; for $p > 1$, Φ_p is a dense subspace of Fock space, which is may be interpreted as a space of "smooth" random variables (in fact, $\cap_p \Phi_p$ is used as a space of test-functions on Wiener space) while for $p < 1$ Φ_p is a space of generalized functions or distributions.

---

[1]   This subsection is an addition.

If g, f are elements of Fock space (random variables) belonging respectively to $\Phi_{1/p}$ and Φ_p, then $g(A) p^{-|A|/2}$ and $f(A) p^{|A|/2}$ belong to Φ, and their scalar product is the standard scalar product $< g, f >$. This can be extended to a duality functional between $\Phi_{1/p}$ and Φ_p, similar to the duality between (Hilbert) Sobolev spaces of functions and distributions.

Next, given a kernel k, let us recall the computation of the bilinear functional $< g, \mathbf{K}f >$ where $\mathbf{K}$ is the operator associated with k. For simplicity of notation we use the same letter (omitting hats) for a vector and its chaotic expansion. Then we have

$$(8.2) \qquad < g, \mathbf{K}f >= \int \bar{g}(A + B)\, \widetilde{k}(A, B, C)\, f(B + C) \,,$$

where $\widetilde{k}$ is defined by

$$(8.3) \qquad \widetilde{k}(A, B, C) = \sum_{V \subset B} k(A, V, C) \,.$$

For the proof, see subsection 5, formulas (5.2)–(5.3).

Let us now introduce the Belavkin–Lindsay norms. We first put for $b > 0$

$$(8.4) \qquad K_b(A, C) = \sup_B \frac{|\widetilde{k}(A, B, C)|}{\sqrt{b}^{|B|}} \,,$$

and then, for $a, c > 0$

$$(8.5) \qquad \| k \|^2_{a,b,c} = \int \frac{K_b^2(A, C)}{a^{|A|}\, c^{|C|}}\, dA dC \,.$$

It should be noted that our b would be called b^2 in the notation of Belavkin–Lindsay. The limiting case $b = 0$ is interesting too, since it means that $\widetilde{K}(A, B, C)$ does not depend on B (see the example after the proof).

Here comes the main estimate.

THEOREM. *Assume* $\| T \|_{a,b,c} < \infty$, $p > a$, $q > c$ *and* $b \leq (p - a)(q - c)$. *Then for* $\| g \|_{(p)} < \infty$, $\| f \|_{(q)} < \infty$ *the integral (8.2) is absolutely convergent, and dominated by* $\| g \|_{(p)} \| T \|_{a,b,c} \| f \|_{(q)}$.

PROOF. We may assume $g, f, \widetilde{k} \geq 0$. Our aim is to dominate

$$\int g(A + B)\, K_b(A, C)\, \sqrt{b}^{|B|}\, f(B + C)\, dA dB dC$$

We put $K_b(A, C) = S(A, C)\, \sqrt{a}^{|A|} \sqrt{c}^{|C|}$, where $S(A, C)$ is square integrable by hypothesis, we replace b by $(p - a)(q - b)$, and we use the inequality $x^\theta y^{1-\theta} \leq x + y$, which gives us

$$a^{|A|}(p - a)^{|B|} \leq p^{|A|+|B|} \,, \qquad c^{|C|}(q - c)^{|B|} \leq q^{|C|+|B|}$$

Then there remains

$$\int g(A+B)\sqrt{p}^{|A|+|B|}\,S(A,C)\,f(B+C)\sqrt{q}^{|C|+|B|}\,dA\,dB\,dC$$

$$=\int dC\sqrt{q}^{|C|}\int g(A+B)\sqrt{p}^{|A|+|B|}\,S(A,C)\,f(B+C)\sqrt{q}^{|B|}dA\,dB$$

We apply the Schwarz inequality to the inner integral, thus transforming it into

$$\left(\int g^2(A+B)p^{|A|+|B|}\,dA\,dB\right)^{1/2}\left(\int S^2(A,C)\,f^2(B+C)\,q^{|B|}\,dA\,dB\right)^{1/2}$$

The first factor is equal to $\|g\|_{(p)}$ and goes out of the integral. Let us set $\int S^2(A,C)\,dA = \varphi(C)$. We must dominate

$$\int dC\left(\int f^2(B+C)\,q^{|B|+|C|}\varphi(C)\,dB\right)^{1/2}.$$

Again a factor $\varphi(C)^{1/2}$ comes out. Applying again the Schwarz inequality for the measure dC yields two factors. First $(\int \varphi^2(C)\,dC)^{1/2}$ which is the kernel's norm, and next $(\int f^2(B+C)\,q^{|B|+|C|}\,dB\,dC)^{1/2}$ equal to $\|f\|_{(q)}$. ☐

EXAMPLES. The Maassen kernel of a Hilbert–Schmidt operator is of the form (see formula (2.11))

$$K(A,B,C)=(-1)^{|B|}h(A,C)\quad\text{with }\int|h(A,C)|^2dA\,dC<\infty.$$

Therefore we have $\check{k}(A,B,C)=h(A,C)$ if $B=\emptyset$ and 0 otherwise. Then we are in the limiting case $b=0$ with $K_b(A,C)=|h(A,C)|$. The natural choice for a,c is $a=c=1$, and the condition $(p-1)(q-1)\ge b$ is satisfied for $p=1, q=1$. We thus recover the boundedness of the operator defined by h. This is a rather exceptional case, however, and doesn't mean the estimate is very sharp.

Another example is that of kernels satisfying the Maassen conditions, compact time support and $|K(A,B,C)|\le M^{|A|+|B|+|C|}$. Then we have a similar property for $\check{K}$ (at the cost of increasing M), and we may take $b=M$. Then we have $K_b(A,C)\le M^{|A|+|C|}$, and since we are working on a bounded simplex we may take a,c arbitrarily close to 0. Accordingly, the condition on p,q becomes $pq>M$, meaning that K maps the "Sobolev space" Φ_p into the dual of the "Sobolev space" $\Phi_{M/p}$, that is, $\Phi_{p/M}$. Assume now f is a test function in a sense different from Maassen's, namely that $f\in\cap_p\Phi_p$. Then the same is true for Kf, and we get a result similar to Maassen's theorem.

REMARK. The Belavkin–Lindsay estimates suggest conditions which should probably be imposed on any reasonable kernel with three arguments $K(A,B,C)$. Then a minimal assumption[1] should be that, for all integers i,j,k, the function of (A,C)

$$\sup_{|B|=j}|\widetilde{K}(A,B,C)|=\widehat{K}_j(A,C))$$

---

[1] It could still be weakened by localization to subsets of $[0,t]$ for finite t.

should be square integrable over the set $\{|A| = i, |C| = k$. The similar condition on K instead of $\widetilde{K}$ is equivalent to it. Such an assumption means that the operator associated with K maps a vector with a finite chaos expansion into a "distribution" whose formal chaos expansion (not necessarily convergent) is given by L^2 coefficients. If we want a true vector in Fock space, the condition becomes

$$\int_{|C=k|} \widehat{K}_j^2(A, C)\, dA dC < \infty$$

with fixed j and k, but summation on all i.

Chapter V

Fock Space (2) :

Multiple Fock Spaces

From the algebraic point of view, a multiple Fock space over $\mathcal{H}$ is nothing but a standard (symmetric or antisymmetric) Fock space over a direct sum of copies of $\mathcal{H}$, and new definitions are not necessary in principle. However, this chapter contains important new *notation*, and a few rather interesting constructions, like that of the "finite-temperature" (= extremal universally invariant) representations of the CCR. In classical probability, it corresponds to the passage from one-dimensional to several-dimensional Brownian motion, necessary for Ito's theory of stochastic differential equations.

We start the theory in a somewhat naive way, replacing one dimensional by finite dimensional Brownian motion; we discuss multiple integrals and creation-annihilation operators in §1, number and exchange operators in §2. Then we discuss multiple Fock space in a basis-free notation, including infinite multiplicity. Additional results on the complex Brownian motion Fock space can be found in Appendix 5.

§1. MULTIDIMENSIONAL STOCHASTIC INTEGRALS

Multiple integral representations

1 From the probabilistic point of view, Fock space over $L^2(\mathbb{R}_+, \mathbb{C}^d)$ can be interpreted as the L^2 space of a d–dimensional standard Brownian motion, and the case $d=2$ has special interest because of the additional structure of complex Brownian motion. The integer d is called the *multiplicity*. The case of infinite (countable) multiplicity is better described as a Fock space over $L^2(\mathbb{R}_+, \mathcal{K})$, where $\mathcal{K}$ is some separable Hilbert space, which we call the *multiplicity space*. We mostly consider the case $d < \infty$ in sections 1–2. The notations are the same in the infinite dimensional case, once an orthonormal basis has been chosen.

Though it has an additional structure, multiple Fock space is a Fock space, and in particular it has a vacuum vector $\mathbf{1}$, which spans the chaos of order 0. If the multiplicity d is finite, multiple Fock space is the tensor product of d copies of simple Fock space, and $\mathbf{1}$ is the tensor product of the individual vacuum vectors.

Fock space of finite multiplicity d can also be interpreted as Fock space over $L^2(E, \mathcal{E}, \mu)$, where μ is the product measure on $E = \mathbb{R}_+ \times \{1, \ldots, d\}$ of Lebesgue measure on $\mathbb{R}_+$ and the counting measure. Our favourite interpretation of Fock space is that of Guichardet, in which $\Gamma(\mathcal{H})$ is identified with the L^2 space of $\mathcal{P}$, the space of all finite subsets of E. A finite set $B \subset E$ with k elements consists of k pairs (s_i, α_i), $(s_i \in \mathbb{R}_+, \alpha_i \in \{1, \ldots, d\})$ and, since the measure is non-atomic, *we may assume the s_i are all different*. Then we may describe B in two ways :

1) We can give ourselves a subset of $\mathbb{R}_+$ with k elements, $A = \{s_1 < \ldots < s_k\}$, and a mapping α from $\{1, \ldots, k\}$ to $\{1, \ldots, d\}$. This gives rise on Wiener space to stochastic integrals of the following form

$$(1.1) \qquad J_\alpha(f) = \int_{s_1 < \ldots < s_k} f(s_1, \alpha_1, \ldots, s_k, \alpha_k) \, dX_{s_k}^{\alpha_k} \ldots dX_{s_1}^{\alpha_1} \ ,$$

with respect to d–dimensional Brownian motion. The k–th chaos contains d^k types of stochastic integrals, corresponding to all possible mappings α.

2) The set B is completely described by its projection A, and the partition of A into d (possibly empty) subsets $A_1 = \{\alpha = 1\}, \ldots, A_d = \{\alpha = d\}$. For simplicity we assume here that $d = 2$. Then we are led to stochastic integrals of the following kind, defining $A_1 = \{s_1, \ldots, s_m\}$ and $A_2 = \{t_1, \ldots, t_n\}$

$$J_{m,n}(f) = \int_{\substack{s_1 < \ldots < s_m \\ t_1 < \ldots < t_n}} f(s_1, \ldots, s_m, t_1, \ldots, t_n)$$
$$(1.2) \qquad\qquad\qquad dX_{s_1}^1 \ldots dX_{s_m}^1 \, dX_{t_1}^2 \ldots dX_{t_n}^2$$

This is not a strictly adapted integral like (1.1). To give it a precise meaning, a probabilist would decompose it into a sum of adapted integrals (1.1) corresponding to all the possible mutual orderings of the s_i's and t_j's. Thus the representation (1.2) is more economical than (1.1), as it groups all the integrals (1.1) which contain m times X^1 and n times X^2. Even in the antisymmetric case, it is always assumed that anticommutation takes place only between components dX_s^α and dX_t^α of any given Brownian motion, *different components dX_s^α and dX_t^β commuting*. Then one can still move $dX_{s_i}^1$ components across $dX_{t_i}^2$ components to perform the grouping (1.2).

The multiple integral (1.2) has also an unordered version, with a function f symmetric (or antisymmetric) in each group of variables, and a factor $1/m! \, n!$ in front.

We shall give to (1.2) a preference over (1.1), though both are necessary (to compute the chaos expansion of solutions of a stochastic differential equation, for instance, only the form (1.1) can be used).

WARNING. *The usual notation of probability theory leads us to denote the components of Brownian motion as dX_t^α with an* **upper** *index. Because of the diagonal summation convention, the general use of indexes gets reversed. We do not interpret this as vectors becoming lines instead of columns, but as a modification of the usual convention that the line index is the upper one. In matrix products, the summation thus takes place on the descending diagonal, etc. Each time this happens, we will recall this "warning".*

Let us now give the shorthand notation for expansions (1.2). For typographical simplicity we take $d = 2$ and write X, Y instead of X^1, X^2. Then a random variable f is expanded in the representation (1.2) as

$$(1.3) \qquad f = \int_{\mathcal{P} \times \mathcal{P}} \widehat{f}(A, B) \, dX_A \, dY_B$$

with a norm given by

$$(1.4) \qquad \| f \|^2 = \int_{\mathcal{P} \times \mathcal{P}} |\widehat{f}(A, B)|^2 \, dA \, dB \ .$$

As usual, we are loose in our use of hats to distinguish f from its chaotic transform $\widehat{f}$.

For each component X^α there is a creation-annihilation pair $a^{+\alpha}, a^{-\alpha}$ ($b^{\pm\alpha}$ in the antisymmetric case). This notation is clear enough : an operator on simple Fock space can be cloned into multiple Fock space so that it acts only on the component of order α, and then an upper index α is ascribed to it.

We will use in section 3 the remark (which we learnt from a paper of Belavkin) that Fock space of possibly infinite multiplicity can be described as a continuous sum of Hilbert spaces over $\mathcal{P}$, the usual space of all finite subsets of $\mathbb{R}_+$

$$(1.5) \qquad \Phi = \int_{\mathcal{P}} \mathcal{K}_A \, dA \ ,$$

where the "fibre" $\mathcal{K}_A$ is equal to $\mathcal{K}^{\otimes n}$ for $|A| = n$ (and $\mathcal{K}$ is the multiplicity space). The general notion of continuous sum is not necessary here : since $\mathcal{K}_A$ is constant for $A \in \mathcal{P}_n$, what we have is simply $\sum_n L^2(\mathcal{P}_n) \otimes \mathcal{K}^{\otimes n}$. To see why, choose an orthonormal basis of $\mathcal{K}$ and check that one gets exactly the appropriate stochastic integrals.

Some product formulas

2 The general discussion of operators on multiple Fock spaces will be postponed to the next section. To get used to the shorthand notation (1.4), let us give some multiplication formulas for two dimensional Brownian motion. In this case we also have a very useful *complex Brownian motion* representation, $Z = (X + iY)/\sqrt{2}$, $\overline{Z} = (X - iY)/\sqrt{2}$, so that

$$(2.1) \qquad f = \int_{\mathcal{P} \times \mathcal{P}} \widehat{f}(A, B) \, d\overline{Z}_A \, dZ_B \ ,$$

with the same isometry property as (1.4) (without the $\sqrt{2}$ factors, the measure would be $2^{|A|+|B|} dA \, dB$). Note that if we restrict ourselves to stochastic integrals $f(A, \emptyset)$ or $f(\emptyset, B)$ we get two copies of one-dimensional Fock space.

For the product $h = fg$ of two random variables in the representation (1.3), we have

$$(2.2) \qquad \widehat{h}(A, B) = \int \sum_{\substack{R+S=A \\ T+U=B}} \widehat{f}(R + M, T + N) \widehat{g}(S + M, U + N) \, dM \, dN \ .$$

On the other hand, the infinitesimal multiplication rule for the Wiener product in the complex representation is given by

$$(2.3) \qquad dX_s^2 = dY_s^2 = ds \ , \ dX_s \, dY_s = 0 \ ; \quad d\overline{Z}_s^2 = dZ_s^2 = 0 \ , \ d\overline{Z}_s \, dZ_s = ds \ ,$$

and therefore in the representation (2.1) the formula gets twisted

$$(2.4) \qquad \widehat{h}(A, B) = \int \sum_{\substack{R+S=A \\ T+U=B}} \widehat{f}(R + M, T + N) \widehat{g}(S + N, U + M) \, dM \, dN \ .$$

As in the case of simple Fock space, we may get a whole family of associative multiplications by means of a weight $\sigma^{2(|M|+|N|)}$, in either one of the two formulas, obtaining in particular the Wick product for $\sigma = 0$. But here we can do something more general, inserting a weight of the form $a^{|M|}b^{|N|}$. If we do this in formula (2.2), we get for $a = 1$, $b = 0$ a Wiener–Wick product which corresponds to a classical differential geometric construction : the symmetric multiplication of fields of covariant symmetric tensors on a manifold (see [DMM] Chap. XXI, n° 26). Doing the same in formula (2.4), the products corresponding to $a \neq b$ are non-commutative, and for $a = 0$, $b = 1$ we get exactly the composition formula for kernels with two arguments.

On the other hand, there also exist interesting products which are, for instance, Wiener products in the variable X, and Grassmann or Clifford in the variable Y. The Wiener–Grassmann product is an infinite dimensional analogue of the exterior product of differential forms on a manifold (some Brownian motions providing the coefficients of the form, and some the anticommuting differential elements). Let us describe rapidly this situation on double Fock space.

A functional in the double Wiener space generated by X, Y has a representation

$$\Theta(\omega, \omega') = \int_{t_1 < \ldots < t_p} \theta(\omega, t_1, \ldots, t_p)\, dY_{t_1}(\omega') \wedge \ldots \wedge dY_{t_p}(\omega') \qquad (a)$$

$$= \frac{1}{p!} \int_{\mathbb{R}_+^p} \theta(\omega, t_1, \ldots, t_p)\, dY_{t_1}(\omega') \wedge \ldots \wedge dY_{t_p}(\omega') \qquad (b)$$

In version (a), we integrate over the increasing simplex, according to the standard usage in finite dimensional exterior algebra. In version (b), the coefficient $\theta(\bullet, t_1, \ldots, t_p)$ is assumed to be antisymmetric in its last variables. Since θ itself has a chaotic expansion, the differential form appears as a multiprocess $\Theta = \int \theta(A, B)\, dX_A dY_B$, with $\theta(A, B) = 0$ unless $|B| = p$. This restriction is unimportant, and one often considers differential forms which are not homogeneous. There is a natural Hilbert space structure on the space of all differential forms, with squared norm $\int |\theta(A, B)|^2\, dA dB$.

As long as we integrate on the simplexes we cannot distinguish whether the second Fock space is symmetric or antisymmetric, and our multiprocesses thus are the same as in [DMM] Chap. XXI, n° 26 : the difference lies in the *multiplications* and *differential operators* we use. On ordinary double Fock space we had Wiener–Wick and Wiener–Wiener products, iterated gradients and divergences. Here we have Wiener–Grassmann and Wiener–Clifford products, exterior differential d and codifferential δ.

Differential forms and the corresponding products are very interesting objects, even for probabilists : see in Appendix 5 the use Le Jan [LeJ1] makes of "supersymmetric" computations in the theory of local times.

Some "non-Fock" representations of the CCR

The end of this section has a great philosophical interest, since it exhibits a continuous family of inequivalent representations of the (infinite dimensional) CCR, a phenomenon which was felt very puzzling when it was first dicovered. On the other hand, *it is used nowhere in these notes and can be omitted altogether*. The concrete representations we construct are the *extremal universally invariant representations* of

Segal [Seg2]. They correspond in the elementary Fock space case to Gaussian laws of non-minimal uncertainty (Chap. III, §1 subs. 9 and §3 subs. 6). By analogy with the last mentioned construction, they are called *finite temperature representations* by Hudson–Lindsay, whose papers [HuL1] [HuL2] we are going to follow.

3 We discuss the case of dimension $d=2$, and use the following picturesque language (without physical content) to help intuition : we call X the *particle* and and Y the *antiparticle* Brownian motions, and denote by $a_t^{+\pm}$ the particle creation and annihilation operators, by $a_t^{-\pm}$ the antiparticle ones. We put

$$(3.1) \qquad \mathbf{a}_t^+ = \xi a_t^{++} + \overline{\eta}\, a_t^{--} \quad ; \quad \mathbf{a}_t^- = \overline{\xi} a_t^{+-} + \eta\, a_t^{-+}$$

where ξ and η are two complex numbers (the creation of an antiparticle counts as an annihilation). The operators $\mathbf{a}_t^+$ and $\mathbf{a}_t^-$ are adjoint to each other on the dense domain consisting of the algebraic sum of the chaos spaces (one could also use coherent vectors : this will be discussed later). The usual Ito table gives us the formal multiplication rules

$$(3.2) \qquad (d\mathbf{a}_t^+)^2 = 0 = (d\mathbf{a}_t^-)^2 \quad ; \quad d\mathbf{a}_t^+ d\mathbf{a}_t^- = \eta\overline{\eta}\, dt \,,\, d\mathbf{a}_t^- d\mathbf{a}_t^+ = \xi\overline{\xi}\, dt \,.$$

In particular, $[d\mathbf{a}_t^-, d\mathbf{a}_t^+] = (|\xi|^2 - |\eta|^2)\, dt$, and *if $|\xi|^2 - |\eta|^2 = 1$ we get a representation of the CCR* — this formal computation will be justified in the next subsection. If we had started with fermion creation and annihilation operators, we would have a representation of the CAR provided $|\xi|^2 + |\eta|^2 = 1$.

If instead of using creation and annihilation operators we use the "field variables" (for $\varepsilon = \pm$)

$$dQ_t^\varepsilon = da_t^{\varepsilon+} + da_t^{\varepsilon-} \,,\, dP_t^\varepsilon = i\,(da_t^{\varepsilon+} - da_t^{\varepsilon-})$$

and define similarly $d\mathbf{Q}_t = d\mathbf{a}_t^+ + d\mathbf{a}_t^-$, we have a representation with real coefficients

$$d\mathbf{Q}_t = \alpha\, dQ_t^+ + \beta\, dQ_t^- \,,\, d\mathbf{P}_t = \gamma\, dP_t^+ + \delta\, dP_t^- \,,$$

with $\alpha\gamma - \beta\delta = 1$. This corresponds exactly to what we did in Chap. III, §2 subs. 3, with the difference that here we haven't a Stone–von Neumann theorem to return from double Fock space to simple Fock space, or even to a direct sum of simple Fock spaces. On double Fock space, the operators $\mathbf{Q}_t$ correspond to multiplication by the random variables $(\xi + \overline{\xi})X_t + (\eta + \overline{\eta})Y_t$, which in the vacuum state constitute a Brownian motion (not a standard one in general).

4 Let us define rigorously the above field operators, as we did in the simple Fock space case, with the help of Weyl operators. Everything we are doing here applies, not only to Fock space over $L^2(\mathbb{R}_+)$, but to any Fock space over a Hilbert space $\mathcal{H}$ equipped with a complex conjugation $h \longmapsto \overline{h}$.

We take two copies Γ^+ and Γ^- of simple Fock space, and denote by Γ' their tensor product, which is double Fock space. We define its vacuum vector $\mathbf{1}$ as $\mathbf{1}^+ \otimes \mathbf{1}^-$. If U is an operator on simple Fock space, we can have it act on double Fock space Γ' either as $U^+ = U \otimes I$, or as $U^- = I \otimes U$. In particular, we define on double Fock space *Weyl operators*

$$(4.1) \qquad W_1 = W(\xi h) \otimes W(-\overline{\eta h}) = W^+(\xi h)\, W^-(-\overline{\eta h}) \,.$$

where $W(h)$ $(h \in L^2(\mathbb{R}_+))$ is a standard Weyl operator on Γ. These operators are unitary, and satisfy the Weyl commutation relation

$$W_h W_k = e^{-i \, \Im < h,k >} W_{h+k} \; .$$

Thus we have a representation of the CCR on double Fock space. Then we may define rigorously creation and annihilation operators, by the same rule we used in the simple Fock space representation : differentiation of the formula

$$W_z = e^{z\mathbf{a}^+ - \bar{z}\mathbf{a}^-}$$

(on a domain which may consist of the finite sums of chaos spaces, or of tensor products of Maassen test functions...), and (4.1) has been chosen so that the CCR creation-annihilation pair is the same as (3.1) — this explains in particular the sign $-\bar{\eta}$ in the second factor of (4.1).

To study this representation, we define *exponential vectors* by

$$(4.2) \qquad \mathcal{E}(f) = \mathcal{E}(\xi f) \otimes \mathcal{E}(-\overline{\eta f})$$

so that we have, after an easy computation

$$(4.3) \qquad < \mathcal{E}(f), \mathcal{E}(g) > = e^{a< f,g >+b< g,f >}$$
$$(4.4) \qquad W_h \mathcal{E}(f) = e^{-c\|u\|^2/2 - a< h,f > - b< f,h >} \mathcal{E}(f + h)$$

where we have put $a = |\xi|^2$, $b = |\eta|^2$, $c = a+b$. Note that $c > 1$ since we assumed $a - b = 1$. If we take $f = 0$ in (4.3), we get the normalized *coherent vectors* (Weyl operators acting on the vacuum).

Let us prove that the new exponential vectors $\mathcal{E}(f)$ generate a dense subspace. If we consider $\mathcal{E}(f)$ as a random variable, we may interpret it as the stochastic exponential of a complex martingale

$$\int_0^\infty \xi f_s \, dX_s - \overline{\eta f_s} \, dY_s \; .$$

Let us choose functions f of the form $e^{i\theta} g$, g being real, and θ such that the complex numbers $\xi e^{i\theta} = u e^{ik}$ and $-\bar{\eta} e^{-i\theta} = v e^{ik}$ have the same argument. Then the exponential vectors of this form can be written as a constant times $\exp(e^{ik} \int g_s \, dB_s)$, where $B = uX + vY$ is a non-normalized Brownian motion, and taking (complex) linear combinations of such exponentials we can approximate in L^4 any given bounded random variable φ measurable w.r.t. the σ–field generated by B. Similarly, we choose θ' so that the two complex numbers above now have arguments differing by π, and exponential vectors of this form correspond to stochastic integrals $\exp(e^{ik'} \int g_s \, dB'_s)$, where $B' = uX - vY$, and we can approximate in L^4 bounded r.v.'s φ' in the σ–field of B'. Since products of exponential vectors are exponential vectors up to a multiplicative constant, we see that complex linear combinations of exponential vectors approximate $\varphi\varphi'$ in L^2, and therefore exponential vectors are total in L^2. As for the density result we have used, it is proved as follows : it is sufficient to approximate r.v.'s of the form

$$h_1(B_{t_1}) \, h_2(B_{t_2} - B_{t_1}) \ldots h_n(B_{t_n} - B_{t_{n-1}}) \; ,$$

and then, because of the independence of increments, we are reduced to proving that given a Gaussian r.v. B generating a σ–field $\mathcal{B}$, exponentials $\exp(\lambda e^{ik}B)$ with λ real are dense in $L^4(\mathcal{B})$. This is deduced from the fact that e^{zB} is an entire function in every L^p space, for $p < \infty$: a random variable in the conjugate L^q space, which is orthogonal to the given family of exponentials, must be orthogonal to all exponentials, and therefore is zero by the uniqueness of Fourier transforms.

Thus double Fock space over a Hilbert space $\mathcal{H}$ has a description very similar to that of simple Fock space in Chap. IV, §1, subs. 3 : it is a Hilbert space Γ equipped with an exponential mapping $\mathcal{E}$ from $\mathcal{H}$ to Γ, whose image generates Γ, and which satisfies (4.3) (simple Fock space is the limiting case $a = 1$, $b = 0$). There are several important differences, however : in the case of simple Fock space, the Weyl representation can be shown to be irreducible. Here it is obviously reducible, because the operators

$$(4.5) \qquad \widetilde{W}_f = W(-\eta f) \otimes W(\overline{\xi f}) = W^+(-\eta f) W^-(\overline{\xi f}) ,$$

commute with all Weyl operators (4.1). Note that one transforms (4.1) into (4.5) via the conjugation mapping $J(u \otimes v) = \overline{v} \otimes \overline{u}$ on Γ.

Let us denote by $\mathcal{W}$ ($\widetilde{\mathcal{W}}$) the von Neumann algebra generated by the Weyl operators (4.1) (resp. (4.5)). We proved above that the exponential vectors generate a dense space : in von Neumann algebra language one says that the vacuum vector is *cyclic* for $\mathcal{W}$ (and for $\widetilde{\mathcal{W}}$ by the same reasons). It is also cyclic for the commutant $\mathcal{W}'$ of $\mathcal{W}$, which obviously contains $\widetilde{\mathcal{W}}$ (one can prove that $\mathcal{W}' = \widetilde{\mathcal{W}}$ but we do not need this result). Now it is easy to prove that if a vector $\mathbf{1}$ is cyclic for the commutant $\mathcal{W}'$ of a von Neumann algebra, then it is *separating* for $\mathcal{W}$, meaning that an operator $a \in \mathcal{W}$ satisfying $a\mathbf{1} = 0$ must be 0 itself. This is very different from the situation of simple Fock space, in which many operators killed the vacuum (note that the new annihilation operators here do not kill the vacuum — however, this does not come within the preceding discussion, which concerns bounded operators). Having a cyclic and separating vacuum makes work easier with these "non-Fock" representations of the CCR than for simple Fock space. See the papers of Hudson–Lindsay [HuL].

As in the case of simple Fock space, one may define a Weyl representation of the whole group of rigid motions of $\mathcal{H}$. Apparently this has no special applications, but it is natural to wonder about the particular case of the number operator. As in the simple Fock space case, we define a_h°, for h real, by

$$(4.6) \qquad \exp(it a_h^\circ)\mathcal{E}(f) = \mathcal{E}(e^{ith}f) = \mathcal{E}(e^{ith}\xi f) \otimes \mathcal{E}(e^{-ith}\overline{\eta f}) .$$

Then it is easy to prove that $a_h^\circ = a_h^{+\circ} - a_h^{-\circ}$, at least on a dense domain (as usual we leave aside the problem of essential selfadjointness on this domain). Thus the natural extension of the number operator is not positive : it rather represents a total charge. Also, the bounded operators (4.6) leave the vacuum state invariant, and therefore *they cannot belong to the von Neumann algebra generated by the Weyl operators* W_h, which admits the vacuum as a separating vector. In fact, several results show that a fully satisfactory number operator exists only for (direct sums of copies of) the simple Fock representation of the CCR. See Chaiken [Cha] and the historical discussion p. 231 in Bratteli–Robinson [BrR2].

5 Let us return to the Ito table (3.2), which we are going to transform into a finite multiplication formula.

First of all, we change slightly the representation of *vectors* as follows : instead of the two standard Brownian motions (X_t) and (Y_t), we use two non-normalized Brownian motions, more closely related to the creation and annihilation operators

$$(5.1) \qquad Z_t^+ = a_t^+ 1 = \xi X_t \ , \ Z_t^- = a_t^- 1 = \bar\eta Y_t \ .$$

(the use of the letter Z should create no confusion with the complex Brownian motion of subsection 2). Then a vector is represented as

$$(5.2) \qquad f = \int \widehat f(A,B)\, dZ_A^+ dZ_B^- \ , \ \|f\|^2 = \int \left| \widehat f(A,B) \right|^2 a^{|A|} b^{|B|}\, dA dB \ ,$$

with a, b as in (4.4). Then we have

$$da_t^+(dZ_A^+ dZ_B^-) = dZ_{A+t}^+ dZ_B^- + b\, dZ_A^+ dZ_{B-t}^- dt$$
$$da_t^-(dZ_A^+ dZ_B^-) = a\, dZ_{A-t}^+ dZ_B^- dt + dZ_A^+ dZ_{B+t}^- \ .$$

The next step is to compute the effect of $da_S^+ da_T^-$ on $dZ_A^+ dZ_B^-$. Assuming that S and T are disjoint, we have

$$da_S^+ da_T^-(dZ_A^+ dZ_B^-) = \sum_{\substack{S_1+S_2=S \\ T_1+T_2=T}} dZ_{(A-T_1)+S_1}^+ dZ_{(A+T_2)-S_2}^- \, a^{|T_1|} b^{|S_2|}\, dT_1 dS_2$$

Then we consider an operator given by a kernel with two arguments

$$(5.3) \qquad K = \int_{\mathcal P \times \mathcal P} K(S,T)\, da_S^+ da_T^-$$

(we assume that $K(S,T)$ vanishes if S and T are not disjoint), and we compute formally the action $g = Kf$ of K on a vector f given by (5.2). The result is (after some rearranging of terms, which by now must have become familiar to the reader)

$$(5.4) \quad \widehat g(A,B) = \int \sum_{\substack{R+S=A \\ T+U=B}} K(R+M, T+N)\, \widehat f(S+N, U+M)\, a^{|M|} b^{|N|}\, dM\, dN \ .$$

and the composition formula for two kernels $H = FG$ is

$$(5.5) \quad H(A,B) = \int \sum_{\substack{R+S=A \\ T+U=B}} F(R+M, T+N)\, G(S+N, U+M)\, a^{|M|} b^{|N|}\, dM\, dN \ .$$

If in formula (5.4) we take for f the vacuum vector, we see that $\widehat g(A,B) = K(A,B)$. Otherwise stated, a kernel with two arguments is uniquely determined by its action on the vacuum. Then (5.4) and (5.5) appear as equivalent formulas, and (5.5) can also be interpreted as a multiplication formula *between vectors*, which is exactly the kind of associative product suggested after (2.4). As in the case of simple Fock space, these formal computations become rigorous under Maassen-like growth conditions on (test) vectors and (regular) kernels. It is clearly possible to construct an algebra of kernels with four arguments, containing the two commuting Weyl systems of double Fock space.

We stop here the discussion, and refer the reader to the literature.

6 [1]   The following remark is due to Parthasarathy [Par7] as an explanation to construction of P. Major [Maj]. It somehow unifies the real and complex Brownian motions, and their corresponding Fock spaces.

Consider a nice measure space $(E, \mathcal{E}, \mu)$ provided with a measurable and measure preserving involution $x \longmapsto x'$. Then E can be decomposed as $G + F + G'$, where F is the set of fixed points of the involution, and G contains exactly one element from each pair (x, x') of conjugate points — all three sets being measurable. For each function f we define $\tilde{f}$ by $\tilde{f}(x) = \overline{f(x')}$.

Our purpose is to construct a complex valued Gaussian field (not necessarily complex–Gaussian : simply two-dimensional Gaussian) $f \longmapsto \xi(f)$ for $f \in L^2(\mu)$ with the following properties :

(6.1) $$\xi(\tilde{f}) = \overline{\xi(f)} \quad , \quad \mathbb{E}\left[\overline{\xi(g)}\,\xi(f)\right] = \int \overline{g}(x)\,f(x)\,\mu(dx) \ .$$

It is clearly sufficient to perform the construction when f is real.

Uniqueness. Consider

$$\xi_\pm(f) = \varepsilon_\pm(\xi(f) \pm \xi(\tilde{f}))$$

with $\varepsilon_+ = 1/2$, $\varepsilon_- = i/2$. Then ξ_+ and ξ_- are complex linear combinations of complex valued Gaussian r.v.'s, and therefore are jointly complex valued Gaussian. On the other hand they are real valued. Thus they are jointly real Gaussian, and it is sufficient to know their real covariance, for which ξ_+ and ξ_- are orthogonal.

Existence. We consider two elementary cases. 1) The involution is trivial, then $\xi(f)$ is real, and we have the model

$$\xi(f) = \int_E f(s)\,dX_s$$

(stochastic integral). 2) $E = G \oplus G'$ (a copy of G) and the involution is $(x, y') \longmapsto (y, x')$. Then we take a *complex*-Gaussian process Z_s over G and put for $f = g \oplus h'$

$$\xi(f) = \int_F g(s)\,dZ_s + \int_F h(s)\,d\overline{Z}(s) \ .$$

The general case then follows using the decomposition $E = G + F + G'$.

In most cases, the set of fixed points has measure 0, and we are reduced to the complex–Gaussian case, with a subtle difference : the splitting $E = G + G'$ isn't intrinsic, and amounts to an arbitrary choice of dZ_x and $d\overline{Z}_x$ for every pair (x, x').

---

[1]   This subsection is an addition.

§2. NUMBER AND EXCHANGE OPERATORS

1 It was not realized until recently that the number operator of simple Fock space comes on the same footing as the ordinary creation-annihilation operators. The situation is even more interesting with the rich family of "number operators" of a Fock space of (finite) multiplicity d. In this section, we are going to discuss these operators, following the work of Evans and Hudson [EvH1], Evans [Eva].

Consider first a general Fock space over $\mathcal{H}$. We recall that a bounded operator A on $\mathcal{H}$ has an extension $\lambda(A)$ to the Fock space $\Gamma(\mathcal{H})$, as a differential second quantization. In particular, if $\mathcal{H} = L^2(\mathbb{R}_+) \otimes \mathcal{K}$ and M is a bounded operator on $\mathcal{K}$, we put $M_t = I_{[0,t]} \otimes M$ (where the indicator is interpreted as a multiplication operator on $L^2(\mathbb{R}_+)$) and define

$$a_t^0(M) = \lambda(M_t) .$$

The following operators will play a particularly important role : we choose an orthonormal basis e^α of the multiplicity space $\mathcal{K}$ (the upper index corresponds to the "warning" in §1 subs. 1) and denote by e_α the dual basis of "bras". Then we denote by I_α^β the operator $I_\alpha^\beta = |e^\beta><e_\alpha|$, and we put

$$a_t^0(I_\alpha^\beta) = a_\alpha^\beta(t) .$$

The effect of $a_\alpha^\beta(t)$ on a continuous basis element $dX_{s_1}^{\alpha_1} \ldots dX_{s_n}^{\alpha_n}$ produces the finite sum

$$\sum_{s_i \le t, \, \alpha_i = \alpha} dX_{s_1}^{\alpha_1} \ldots dX_{s_i}^{\beta} \ldots dX_{s_n}^{\alpha_n}$$

It is simpler to describe the effect on the continuous basis element of the *differential* $da_\alpha^\beta(t)$: if t occurs among the s_i and the corresponding α_i is α, the factor dX_t^α is replaced by dX_t^β. Otherwise the basis element is mapped to 0.

The diagonal operator $da_\alpha^\alpha(t)$ is the number operator corresponding to the component dX_t^α (the "trace" $\sum_\alpha da_\alpha^\alpha(t)$ is the total number operator) while the operators $da_\alpha^\beta(t)$ with $\alpha \ne \beta$ are *exchange operators*. In the one dimensional case, there was only one operator $da_1^1(t) = da_t^0$.

The content of the above computation can be subsumed as $da_\alpha^\beta(t)\, dX_t^\gamma = \delta_\alpha^\gamma dX_t^\beta$, from which we deduce a part of the "Ito table" for multiple Fock space

$$(1.1) \qquad\qquad da_\lambda^\mu(t)\, da_\alpha^\beta(t) = \delta_\lambda^\beta\, da_\alpha^\mu(t) .$$

On the other hand, let us follow the notation of Evans and put $dX_t^0 = dt$ (or rather $dt\, \mathbf{1}$, an element of the chaos of order 0). Since the creation operators transform the vacuum into an element of the first chaos, it seems fit to denote them by $da_0^\alpha(t)$, and in the same way $da_\alpha^0(t)$ are annihilation operators. Then the rules for annihilation operators become

$$da_\alpha^0(t) dX_t^\gamma = \delta_\alpha^\gamma dX_t^0 , \quad da_\alpha^0(t) dX_t^0 = 0 .$$

For creation operators, they become

$$da_0^\alpha(t)\mathbf{1} = dX_t^\alpha \ , \quad da_0^\alpha(t)\,dX_t^\gamma = 0 \ .$$

The last equation is also valid for $\gamma = 0$, since $dX_t^\alpha\,dt$ is counted as 0. Let us put $da_0^0(t) = I\,dt$. Then we have from the usual Ito table for creation and annihilation operators

$$da_\alpha^0(t)\,da_0^\beta(t) = \delta_\alpha^\beta\,da_0^0(t)$$

and it turns out that, if we mix creation, annihilation and number/exchange operators, relation (1.1) is true for all indices α, β *including* 0, except for *one* relation :

$$da_0^\beta(t)\,da_\beta^0 = 0 \neq \delta_0^0\,da_\alpha^\beta(t)$$

To get the correct formula, we introduce the *Evans delta* $\widehat{\delta}_\alpha^\beta = \delta_\alpha^\beta$ unless $\alpha = \beta = 0$, in which case $\widehat{\delta}_0^0 = 0$, and we have the complete Ito table for multiple Fock space [1]

(1.2)
$$da_\gamma^\varepsilon(t)\,da_\alpha^\beta(t) = \widehat{\delta}_\gamma^\beta\,da_\alpha^\varepsilon(t) \ .$$

If we compare this notation with that of simple Fock space, we see that the four basic differential operators $I\,dt, da_t^+, da_t^+, da_t^0$ have been replaced by four sets of operators : again $da_0^0(t) = I\,dt$ (scalar), the two dual vectors $da_\alpha^0(t)$, $da_0^\alpha(t)$ of creation and annihilation operators, and the matrix of the number/exchange operators $da_\alpha^\beta(t)$.

From now on we adopt the Evans notation, with the convention that indexes α, β, γ *from the beginning of the Greek alphabet take only the values* $1, \ldots$, *while indexes* $\rho, \sigma \ldots$ *are allowed the additional value* 0.

EXAMPLE. The main application of the preceding discussion concerns the Evans–Hudson theory of quantum diffusions, but let us illustrate it by the description of a *continuous spin field* over the line, an object often mentioned in the physics literature, but of which I have never seen a complete mathematical definition. We take $d=2$, and define

$$d\sigma_x(t) = da_2^1(t) + da_1^2(t) \ , \quad d\sigma_y(t) = i(da_1^2(t) - da_2^1(t)) \ ,$$

(1.3)
$$d\sigma_z(t) = da_1^1(t) - da_2^2(t)$$

Then we have $d\,[\sigma_x\,,\,\sigma_y] = i\,d\sigma_z$, etc. Then "spin kernels" may be constructed as operators acting on a domain of test functions, etc. As usual, the order structure of $\mathbb{R}_+$ is not essential, and the construction can be extended to any space $L^2(\mu)$ of a non-atomic measure.

---

[1] Because of our conventions (see the "warning" in §1, subs. 1) the index pair affected by δ is not the same as for matrix units in linear algebra.

Kernel calculus on multiple Fock space

2   The following subsections may be omitted at a first reading. We present in section 3 a less explicit, but also less cumbersome, version of kernel calculus, which has the advantage of allowing infinite multiplicity.

From a purely algebraic point of view, kernels are sums of "multiple integrals"

$$\int_{s_1 < \ldots < s_n} K_{\varepsilon_1 \ldots \varepsilon_n}(s_1, \ldots, s_n) \, da_{s_n}^{\varepsilon_n} \ldots da_{s_1}^{\varepsilon_1}$$

over all possible integers n and n–tuples of Evans indexes $\varepsilon_i = \binom{\rho_i}{\sigma_i}$, except that usually the index $\binom{0}{0}$ is not allowed — if it were, it could be removed by integration, modifying the coefficients. On the other hand, it is interesting to have a variant of the definition of kernels in which da_0^0 is allowed, because such extended "kernels" appear naturally when solving stochastic differential equations. The price to pay is the loss of uniqueness in the kernel representation, but the formulas are only trivially modified.

As usual, one tries to define a shorthand notation, and to give closed formulas describing how kernels act on vectors, and how adjoints and products are computed. We are going to extend the $K(A, B, C)$ notation of simple Fock space, though it becomes very heavy.

Kernels then appear as sums of multiple integrals

$$(2.1) \qquad K = \int K((A_0^\alpha); (A_\alpha^\beta); (A_\alpha^0)) \, da_0^\alpha(A_0^\alpha) \ldots da_\alpha^\beta(A_\alpha^\beta) \ldots da_\alpha^0(A_\alpha^0) \, .$$

This is an illustration of the Evans notation. First of all, α, β are indexes which assume the values $1, \ldots, d$, 0 being excluded. The arguments A_0^α coming first correspond to the creators, those A_α^0 coming last correspond to the annihilators, and in the middle we have the number and exchange operators. We may imagine all arguments arranged in a matrix instead of a line, with an empty set A_0^0 at the upper left corner. *All the arguments are disjoint.*

If the time differential $da_0^0(t)$ is included, formula (2.1) contains an additional $da_0^0(A_0^0)$, and the corresponding subset A_0^0 is written as the first argument of K.

The notation can be brought closer to that of simple Fock space, writing the kernel as

$$(2.1') \qquad\qquad\qquad K(A_\alpha); \, B_{\alpha\beta}; \, C_\alpha) \, .$$

Let us describe the way a kernel K acts on a function f, first in the simple case where neither K nor f depend on the time variable t. Here is the formula, denoting by the same letter a function and its chaotic expansion.

$$K f(A_1, \ldots, A_d) =$$

$$(2.2)$$

$$\int_{U_\alpha + \sum_\beta V_{\alpha\beta} + W_\alpha = A_\alpha} \sum K(U_\alpha, V_{\alpha\beta}, M_\alpha) \, f(M_\alpha + \sum_\beta V_{\beta\alpha} + W_\alpha) \prod_\alpha dM_\alpha \, .$$

Before we sketch a justification of this formula, let us also give the useful formula that corresponds to formula (5.2) in the last section of Chapter IV. That is, the computation of the functional $\Phi(g, f) = \,<g, Kf>$: it is first seen to have the value

$$\int \bar{g}(W_\alpha + \sum_\beta V_{\alpha\beta} + U_\alpha)\, K(U_\alpha, V_{\alpha\beta}, M_\alpha)\, f(M_\alpha + \sum_\beta V_{\beta\alpha} + W_\alpha) \times$$

$$\prod_\alpha dU_\alpha \prod_{\alpha\beta} dV_{\alpha\beta} \prod_\alpha dM_\alpha \prod_\alpha dW_\alpha \ .$$

According to the main property of the measure, we may consider as a single variable each subset $W_\alpha + V_{\alpha\alpha}$. We first denote it by $V'_{\alpha\alpha}$, then omit the $'$. In this way, we get

$$<g, Kf> = \int \bar{g}(\sum_\beta V_{\alpha\beta} + U_\alpha)\, \tilde{K}(U_\alpha, V_{\alpha\beta}, M_\alpha)\, f(M_\alpha + \sum_\beta V_{\beta\alpha}) \times$$

(2.3)
$$\prod_\alpha dU_\alpha \prod_{\alpha\beta} dV_{\alpha\beta} \prod_\alpha dM_\alpha \ .$$

where $\tilde{K}$ is a partial Moebius transform of K on the diagonal variables $V_{\alpha\alpha}$ only : *i.e.* one performs a summation on all subsets of $V_{\alpha\alpha}$.

We now justify formally (2.2). We allow the argument A_0^0 in K, and for the sake of symmetry, we allow also an additional differential $dX^0(U^0)$ and the corresponding argument U^0 in the definition of a vector, though in practice this is unusual

(2.4)
$$f = \int f((U^\rho)) \prod_\rho dX^\rho(U^\rho)$$

$$= \int f(U^0, U^1, \ldots, U^n)\, dX^0(U^0) dX^1(U^1) \ldots dX^n(U^n) \ .$$

To compute the effect of a kernel on a vector, we begin with the effect of an operator differential $\prod da_\sigma^\rho(S_\sigma^\rho)$ on a vector differential $\prod dX^\tau(U^\tau)$. We put

$$S_\sigma^\rho \cap U^\tau = B_\sigma^{\rho\tau} \ , \quad S_\sigma^\rho \cap \tilde{U} = A_\sigma^\rho \ , \quad \tilde{S} \cap U^\tau = C^\tau \ ,$$

where $\tilde{S}$ is the complement of $\cup S_\sigma^\rho$, and $\tilde{U}$ the complement of $\cup U^\tau$. All these sets are disjoint. Then it is easily seen that the product is 0 unless the only non-empty sets in these decompositions are the C^τ, A_0^ρ and $B_\alpha^{\rho\alpha}$, and in this case

$$dX^0 \quad \text{is produced by} \quad V^0 = A_0^0 + \sum_\gamma B_\gamma^{0\gamma} + C^0 \ ,$$

$$dX^\alpha \quad \text{is produced by} \quad V^\alpha = A_0^\alpha + \sum_\gamma B_\gamma^{\alpha\gamma} + C^\alpha \ .$$

Then we have

$$S_0^0 = \sum_{'\tau} B_0^{0\tau} + A_0^0 = A_0^0 \ , \quad U^0 = \sum_{\rho\sigma} B_\sigma^{\rho 0} + C^0 = C^0 \ ,$$

$$S_0^\alpha = \sum_\tau B_0^{\alpha\tau} + A_0^\alpha = A_0^\alpha \ , \quad U^\alpha = \sum_{\rho\sigma} B_\sigma^{\rho\alpha} + C^\alpha = \sum_\beta B_\alpha^{\beta\alpha} + C^\alpha \ ,$$

$$S_\alpha^\beta = \sum_\tau B_\alpha^{\beta\tau} + A_\alpha^\beta = B_\alpha^{\beta\alpha} \ , \quad S_\alpha^0 = \sum_\tau B_\alpha^{0\tau} + A_\alpha^0 = A_\alpha^0 \ .$$

Then it is easy to find the expression of the vector $Kf = g$: the coefficient $g((V^\alpha))$ (*with V^0 empty : what we get is a standard chaos expansion*) is given by a sum over all decompositions

$$V^\alpha = A_0^\alpha + \sum_\gamma B_\gamma^{\alpha\gamma} + C^\alpha$$

of the following integrals, where the sets A_0^0, $B_\gamma^{0\gamma}$ and C^0 appearing in the coefficient of dX^0 are treated as integration variables M, N_γ, P due to the combinatorial properties of the measure

$$(2.5) \quad \int K(M,(A_0^\alpha); (B_\alpha^{\beta\alpha}); (N_\alpha)) f(P,(N_\alpha + \sum_\gamma B_\alpha^{\gamma\alpha} + C^\alpha)) \, dM \prod_\alpha dN_\alpha \, dP \ .$$

Usually, $f(P, \cdot) = 0$ unless $P = \emptyset$ and the integration variable P can be omitted.

Our next step is to give an analytical meaning to (2.2). Regular kernels in the sense of Maassen are defined by the two properties which generalize the simple Fock space situation (Chap. IV, §4 subs. 5) : 1) a compact support in time ($K((A_\sigma^\rho)) = 0$ unless all its arguments are contained in some bounded interval $[0, T]$), 2) a domination inequality of the form

$$(2.6) \qquad\qquad |K((A_\sigma^\rho))| \le CM^{\sum |A_\sigma^\rho|} \ .$$

It is easy to see that, if the additional variable da_0^0 is removed by integration, then the true kernel we get is still regular.

Here also, we may define *test vectors* (regular vectors) by a condition of compact time support, and a domination property

$$(2.7) \qquad\qquad |f((U^\rho))| \le CM^{\sum |U^\rho|} \ .$$

Integrating out the additional variable preserves regularity. Then it is tedious, but not difficult to check that the effect of a regular kernel on a test vector is again a test vector. This is the beginning of the extension of Maassen's theorem to multiple Fock space kernels. We leave the discussion of adjoints to the reader, and deal with composition in subsection 3.

The Belavkin-Lindsay type conditions can also be extended to Fock spaces of finite multiplicity, as shown by Attal.

REMARKS. a) Kernels decompose operators on Fock space into "elementary processes" (which physicists would represent by diagrams), the formalism allowing one single operation at a given "site" (time or place). An exchange operator is more singular in this respect than a creation or annihilation operator, since the operation it describes can be understood as the annihilation of a particle of type α, followed immediately by the creation at the same site of a particle of type β. Physicists consider still more singular processes (pair annihilations, simultaneous creation of several particles). A "kernel formalism" including all these processes seems to require distributions (see Krée [Kré], Krée and Raczka [KrR]), and doesn't concern us here.

b) Let us be more specific. Fock space of multiplicity d over $l^2(\mathbb{R}_+)$ is also ordinary Fock space over $L^2(\mathbb{R}_+ \times \{1, \ldots, d\})$. We may therefore consider Maassen kernels with three arguments involving operators a_A^ε, where $\varepsilon = -, o, +$ and A is a finite subset of $\mathbb{R}_+ \times \{1, \ldots, d\}$; they depend on $3d$ subsets of $\mathbb{R}_+$. On the other hand, kernels involving the exchange operators depend on $d^2 + 2d$ subsets. Thus the second description is more refined than the first one. It is intuitively clear that it is not possible to express the *exchange* operators by means of a Maassen kernel involving only creation, annihilation and number. On the other hand, $\mathbb{R}_+ \times \{1, \ldots, d\}$ is isomorphic with $\mathbb{R}_+$ from the measure theoretic point of view, and we expect there are second quantization operators on ordinary Fock space, which cannot be expressed by Maassen three argument kernels. An explicit example of this kind is due to J.L. Journé, *Sém. Prob. XX*, p. 313–316.

c) More striking still : we have mentioned in Chapter IV that the full Fock space over $L^2(\mathbb{R}_+)$ is isomorphic with symmetric Fock space as a Hilbert space. Therefore its bounded creation/annihilation operators can be considered as operators on symmetric Fock space, and Parthasarathy–Sinha [PaS6] describe explicitly the way they act on the standard continuous basis. It can be analyzed as annihilation of particles at a given place and recreation *at a different place*. There is little hope of describing such operators by a kernel formalism.

3 Let us now study the composition $J = KL$ of two kernels. The formula is obviously unpractical but its existence is important, as it implies the existence of a "Maassen algebra" of kernels. The complete formula is due to A. Dermoune [Der].

Let us follow the same method as in subsection 2, and compute the composition of two operator differentials

$$da_0^0(R_0^0) \prod da_0^\alpha(R_0^\alpha) \ldots da_\alpha^\beta(R_\alpha^\beta) \ldots da_\alpha^0(R_0^\alpha)$$

$$da_0^0(S_0^0) \prod da_0^\alpha(S_0^\alpha) \ldots da_\alpha^\beta(S_\alpha^\beta) \ldots da_\alpha^0(S_0^\alpha)$$

In the case of true kernels, R_0^0 and S_0^0 are not used in this representation, and are interpreted as empty. We denote by $\widetilde{R}$ the complement of $\cup_{\rho\sigma} R_\sigma^\rho$ and similarly $\widetilde{S}$. We define

$$R_\sigma^\rho \cap \widetilde{S} = A_\sigma^\rho \quad ; \quad R_\sigma^\rho \cap S_\xi^\tau = B_{\sigma\xi}^{\rho\tau} \quad ; \quad \widetilde{R} \cap S_\xi^\tau = C_\xi^\tau \ .$$

All these sets are pairwise disjoint, and we use the relation $da^\varepsilon(U+V) = da^\varepsilon(U)da^\varepsilon(V)$ for disjoint U, V to write each one of the two operator differentials as a product involving the elementary pieces A_σ^ρ, $B_{\sigma\xi}^{\rho\tau}$, and C_ξ^τ. Then we multiply these products using the Evans rules, and we get that the product is 0 unless the only non-empty sets are of the form A_σ^ρ, $B_{\alpha\sigma}^{\rho\alpha}$, C_ξ^τ. If these conditions are realized, the product is equal to

$$G \prod da_0^\alpha(T_0^\alpha) \ldots da_\alpha^\beta(T_\alpha^\beta) \ldots da_\alpha^0(T_\alpha^0)$$

with

$$T_\sigma^\rho = A_\sigma^\rho + \sum_\gamma B_{\gamma\sigma}^{\rho\gamma} + C_\sigma^\rho$$

and

$$G = d\left(A_0^0 + \sum_\gamma B_{\gamma 0}^{0\gamma} + C_0^0\right) \ .$$

It is now easy to get the final Maassen-like formula : we have for the true kernel $KL((T_0^\alpha), (T_\alpha^\beta), (T_\alpha^0))$ the following expression, replacing the sets A_0^0, $B_{\alpha 0}^{0\alpha}$, C_0^0 by integration variables M, N_α, P

$$\int_{T_\sigma^\rho = A_\sigma^\rho + \sum_\gamma B_{\gamma\sigma}^{\rho\gamma} + C_\sigma^\rho} \sum K(M, A_0^\alpha, A_\alpha^\beta + \sum_\rho B_{\alpha\rho}^{\beta\alpha}, A_\alpha^0 + \sum_\gamma B_{\alpha\gamma}^{0\alpha} + N_\alpha) \times$$

$$(3.1) \quad L(P, N_\alpha + \sum_\gamma B_{\alpha 0}^{\gamma\alpha} + C_0^\alpha, \sum_\rho B_{\beta\alpha}^{\rho\beta} + C_\alpha^\beta, C_\alpha^0) \, dM dP \prod_\alpha dN_\alpha$$

It is relatively easy now to check that the composition of two regular kernels is a regular kernel.

Uniqueness of kernel representations

4   In the case of simple Fock space, the uniqueness of kernel representations was discussed by Maassen for two-arguments kernels. The case of kernels involving the number operator appears in *Sém. Prob. XXI*, p. 46–47. On the other hand, the proof that is given there does not apply to kernels on a multiple Fock space : the first proof valid in full generality is due to Attal [Att]. Since the notation of multiple kernels is very cumbersome, we shall first give the idea of his proof in the case of simple Fock space, and then extend it to higher multiplicities.

First of all, we associate with a kernel $k(A, B, C)$ a bilinear form (we do not care about making it hermitean). From Chapter IV, §4 (1.1′) we have (using the basic property of the measure)

$$(4.1) \quad < \bar{g}, Kf > = \int g(U + V + W) k(U, V, M) f(V + W + M) \, dU dV dW dM .$$

We take as integration variables $U, V + W, M = A, B, C$, to get

$$\int (\sum_{V \subset B} k(A, V, C)) g(A + B) f(B + C) \, dA dB dC$$

$$= \int \varphi(A, B, C) h(A, B, C) \, dA dB dC \quad \text{where} \quad h(A, B, C) = g(A + B) f(B + C) .$$

Since the mapping $k \longmapsto \varphi$ is 1–1 by the Moebius inversion formula, it suffices to prove that $\varphi = 0$. On the other hand, if K is a regular kernel, φ belongs to $L^2(\mathcal{P}^3)$. Finally, the uniqueness problem is reduced to proving that *functions of the form* $h(A, B, C)$ *above, where* f, g *are test functions, are total in* $L^2(\mathcal{P}^3)$.

We consider the algebra $\mathcal{A}$ of linear combinations of functions $h(A, B, C) = g(A + B) f(B + C)$, where $f(U), g(U)$ are bounded Borel functions on $\mathcal{P}$, and are equal to 0 if either $|U| > n$ or $U \not\subset [0, t]$ for some n, t. Let $\mathcal{B}$ denote the σ–field generated by $\mathcal{A}$. According to the monotone class theorem, $\mathcal{A}$ is dense in $L^2(\mathcal{B})$, and we only have to show that $\mathcal{B}$ is the Borel σ–field of $\mathcal{P}$. Since $\mathcal{B}$ is separable and $\mathcal{P}$ is a Polish space, we only need to show that $\mathcal{A}$ is separating. This amounts to showing that if A, B, C, A', B', C' are triples of disjoint sets, the relations $A + B = A' + B'$,

$B+C = B'+C'$ imply $A = A', B = B', C = C'$. Now putting $H = A+B+C = A'+B'+C'$ we have $A = H \setminus (B+C) = A'$ etc.

In the case of multiplicity d, the idea remains the same but the number of subsets to handle becomes much larger: operators are represented by kernels depending on d^2+2d arguments B_ρ^σ ($B_0^0 = \varnothing$), while vectors depend on d^2 arguments A_α^β (recalling the convention that α, β cannot assume the value 0, contrary to ρ, σ, τ). The functions $h((A_\alpha^\beta))$ which should generate a dense subspace are of the form

$$g((\textstyle\sum_\tau B_\alpha^\tau)) \, f((\textstyle\sum_\tau B_\tau^\beta)) \, .$$

Again we are reduced to proving that such functions separate points. Otherwise stated, given systems B_ρ^σ and B'^σ_ρ of disjoint subsets (empty for $\rho = \sigma = 0$), such that for all α, β

$$B_\alpha = \sum_\tau B_\alpha^\tau = \sum_\tau B'^\tau_\alpha = B'_\alpha \, , \qquad B^\beta = \sum_\tau B_\tau^\beta = \sum_\tau B'^\beta_\tau = B'^\beta$$

are the two systems identical? The proof is very simple: since all B_σ^β are disjoint, we have $B_\alpha^\beta = B_\alpha \cap B^\beta$, and therefore $B_\alpha^\beta = B'^\beta_\alpha$. Then the equalities $B_\alpha^0 = B'^0_\alpha$, $B_0^\beta = B'^\beta_0$ follow by difference.

Operator valued kernels

5   As we will see in the next chapter, the multiple Fock space Φ is "coupled" in most applications to another Hilbert space $\mathcal{J}$ (the *initial space*) to produce the space $\Psi = \mathcal{J} \otimes \Phi$. The initial space is used to describe some physical system, and is the real object of interest. Therefore, one is naturally led to considering kernels which, instead of being complex valued, take their values in the space of operators on the initial space $\mathcal{J}$. Rather than using at once the cumbersome notation with families of sets, we may write such kernels as sums, over all n and all n-tuples $(\mu) = \binom{(\sigma)}{(\rho)} = \begin{smallmatrix} \sigma_1 \\ \rho_1 \end{smallmatrix} \ldots \begin{smallmatrix} \sigma_n \\ \rho_n \end{smallmatrix}$ of Evans indexes, of multiple integrals

$$\sum_{(\rho)\,(\sigma)} \int K_{(\sigma)}^{(\rho)}(s_1, \ldots, s_n) \, da_{\rho_1}^{\sigma_1}(s_1) \ldots da_{\rho_n}^{\sigma_n}(s_n) = \sum_{(\mu)} \int K_{(\mu)}(A) \, da^{(\mu)}(A) \, .$$

Here $K_{(\sigma)}^{(\rho)}$ is a mapping from the increasing n-simplex to the space $\mathcal{L}(\mathcal{J})$ of bounded operators on $\mathcal{J}$ — the case of unbounded operators will be considered later. Note that (μ) and A have the same number of elements, (μ) telling which indexes are ascribed to the successive elements of $A = \{s_1 < \ldots < s_n\}$. This is equivalent to defining the partition A_ρ^σ of A into the subsets where the ascribed index is $\binom{\sigma}{\rho}$. Thus the kernels are exactly the same as those we considered above, except that now K is an operator instead of a scalar. The index $\binom{0}{0}$ is often excluded, since it may always be integrated out first.

Such kernels operate on vectors $f \in \Psi$, which admit $\mathcal{J}$-valued chaos expansions

$$f = \sum_{(\alpha)} \int f_{(\alpha)}(s_1, \ldots, s_n) \, dX_{s_1}^{\alpha_1} \ldots dX_{s_n}^{\alpha_n} = \sum_{(\alpha)} \int f_{(\alpha)}(A) \, dX^{(\alpha)}(A) \, .$$

It is not necessary to write the way K acts on f : the formula is the same as (2.5) with a slightly different notation, *the product Kf being understood as an operator acting on a vector, and producing a vector.* Similarly, the terrible composition formula for two kernels remains the same, except that in (3.1) the product KL is understood as composition of operators. Test vectors (regular kernels) are defined as in the scalar case, replacing the absolute value by the Hilbert space norm (operator norm), and Maassen's theorem that regular kernels form an algebra operating on test vectors remains true.

Given two vectors $u, v \in \Phi$ and a kernel K on Ψ, we may formally define an operator $\widetilde{K}$ on the initial space by the formula

$$< \ell, \widetilde{K} j > = < \ell \otimes v, K(j \otimes u) > .$$

An useful remark due to Holevo is the following : let us say the operator valued kernel is *dominated* by a scalar kernel k if we have

$$\| K_\mu(A) \| \le k_\mu(A) .$$

Let us denote by $|u|, |v|$ the vectors in Φ whose chaos expansions are $|u|_\alpha(A) = \| u_\alpha(A) \|$ and similarly for $|v|$. Then we have

(5.1) $$|< \ell, \widetilde{K} j >| \le \| \ell \| \| j \| < |v|, k|v| > .$$

This follows immediately from the fact that the computation of a bilinear form $< g, Kf >$ only involves additions and multiplications.

Parthasarathy's construction of Lévy processes

6 We give now a beautiful application of Fock spaces with infinite multiplicity : Parthasarathy's construction of Lévy's processes, which extends to general processes with independent increments the construction of Brownian motion and Poisson processes on simple Fock space. In the last chapter of these notes, we will extend it to a very general algebraic situation.

We consider a Lévy function on $\mathbb{R}$

(6.1) $$\psi(u) = imu + c^2 u^2/2 + \int (1 - e^{iux} + iux I_{\{|x|<1\}}) \nu(dx)$$

where ν is the Lévy measure : $e^{-\psi(u)}$ is the Fourier transform of an infinitely divisible law π. We have

(6.2) $$\psi(u) + \psi(-v) - \psi(u-v) = c^2 uv + \int (1 - e^{-ivx})(1 - e^{iux}) \nu(dx)$$

Let $\mathcal{K}$ be the Hilbert space $\mathbb{C} \oplus L^2(\nu)$. We define a unitary representation of $\mathbb{R}$ in $\mathcal{K}$

(6.3) $$\rho(u) = I \oplus e^{iux}$$

(as a multiplication operator). We put

(6.4) $$\eta(u) = icu \oplus (1 - e^{iux}) .$$

This time, the function is understood as a vector : the function $1 - e^{iux}$ is square integrable w.r.t. ν. Then η satisfies a cocycle property

(6.5)
$$\eta(u + v) = \eta(u) + \rho(u)\eta(v) .$$

We note that

(6.6)
$$< \eta(v), \eta(u) > = \psi(u) + \psi(-v) - \psi(u - v) .$$

Next, we apply the Weyl commutation relations

$$
\begin{aligned}
W(\eta(u), \rho(u))\, W(\eta(v), \rho(v)) &= W(\eta(u) + \rho(u)\eta(v), \rho(u)\rho(v))\, e^{-i\Im < \eta(u), \rho(u)\eta(v) >} \\
&= W(\eta(u + v),\ \rho(u + v))\, e^{-i\Im < \eta(u), \eta(u+v) - \eta(u) >} \\
&= W(\eta(u + v),\ \rho(u + v))\, e^{-i\Im(\psi(u+v) - \psi(u) - \psi(v))} .
\end{aligned}
$$

Accordingly, the operators

(6.7)
$$R(u) = e^{-i\Im\,\psi(u)} W(\eta(u),\, \rho(u))$$

constitute a unitary group. In the vacuum state, the law of its generator is π, as shows the following Fourier transform computation, starting from the formula

$$< 1, W(\eta(u),\, \rho(u))\, 1 > = e^{-\|\eta(u)\|^2/2}$$

whose value here is $e^{-\Re\,\psi(u)}$. Taking into account the phase factor (6.7) the Fourier transform is $e^{-\psi(u)}$ as announced.

We compute the generator itself, assuming first that ν has a bounded support. Then the function x belongs to $L^2(\nu)$, and the multiplication operator M_x is bounded. Also, ψ is differentiable at 0 and we have $\eta'(0) = ic \oplus (-ix)$ (considered as a vector), $\rho'(0) = 0 \oplus iM_x$, and

$$X = \Im\,\psi'(0)\, I + Q(c \oplus x) + a^\circ(M_x) ,$$

Q denoting as usual a sum of creators and annihilators. Under the same hypothesis, it easy to describe the whole process, working on $L^2(\mathbb{R}_+) \otimes \mathcal{K}$ instead of $L^2(\mathcal{K})$:

$$X_t = t\, \Im\,\psi'(0) + Q_t(c \oplus x) + a_t^\circ(M_x) .$$

If the Lévy measure is concentrated at one single point, we fall back on the representation of compensated Poisson processes on simple Fock space.

We do not try to describe the generator in the case of a Lévy process with unbounded jumps, since our purpose was only to show the interest of infinite multiplicity. More details are given in the book [Par1], Chapter 2, 21, p. 152.

§3. BASIS-FREE NOTATION ON MULTIPLE FOCK SPACES

The way we followed led us from Fock space to Brownian motion, first one-dimensional, and then of arbitrary dimension. It is also possible to forget the Brownian motions and to use a condensed notation, the same for all multiplicities. The idea of such "intrinsic" notation is due to Belavkin, though we do not follow him to the end, and are content with a notation which we learnt from lectures of Parthasarathy and is used in his book.

1 We start with a Fock space of arbitrary multiplicity $\mathcal{K}$, but with a trivial initial space for simplicity. Consider a linear combination of differential elements at time t — in the theory of stochastic integration of Chapter VI, the coefficients of this linear combination will become adapted operators at time t, but in all other respects the notation will remain the same. We group the operators and imitate simple Fock space :

— The creation term $\sum_{\alpha} \lambda_{\alpha} da_0^{\alpha}(t)$ may be written as $da_t^+(\lambda)$, where $\lambda = (L_{\alpha}^0) \in \mathcal{K}$ is a *column vector* (in spite of the lower index).

— The number/exchange term $\sum_{\alpha\beta} \Lambda_{\beta}^{\alpha} da_{\alpha}^{\beta}(t)$, is written $da_t^{\circ}(\Lambda)$, $\Lambda = (L_{\beta}^{\alpha})$ being a matrix (an operator on $\mathcal{K}$).

— The annihilation term is $da_t^-(\widetilde{\lambda})$, where $\widetilde{\lambda} \in \mathcal{K}'$ is a *line vector* L_0^{α}) (in many cases it appears as a *bra* $< \mu \mid$ associated with an element of $\mathcal{K}$).

— Finally we have the scalar term $L \, dt$.

Then the standard differential element can be written as

$$(1.1) \qquad da_t(\mathbb{L}) = L \, dt + da_t^+(\lambda) + da_t^{\circ}(\Lambda) + da_t^-(\widetilde{\lambda})$$

We may represent $\mathbb{L}$ as a $(2,2)$ matrix

$$(1.2) \qquad \mathbb{L} = \begin{pmatrix} L & \widetilde{\lambda} \\ \lambda & \Lambda \end{pmatrix}.$$

and consider $\mathbb{L}$ as an operator from $\widehat{\mathcal{K}} = \mathbb{C} \oplus \mathcal{K}$ into itself. If we add to this setup an initial space $\mathcal{J}$, the notation remains the same, except that L instead of a scalar becomes an operator on $\mathcal{J}$, λ becomes a column of operators on $\mathcal{J}$ (a mapping from $\mathcal{J}$ to $\mathcal{K} \otimes \mathcal{J}$), $\widetilde{\lambda}$ is a line of operators on $\mathcal{J}$ (a mapping from $\mathcal{K} \otimes \mathcal{J}$ to $\mathcal{J}$) and Λ is a matrix of operators (an operator on $\mathcal{K} \otimes \mathcal{J}$). Thus $\mathbb{L}$ itself becomes an operator on $\widehat{\mathcal{K}} \otimes \mathcal{J}$.

Finally, in stochastic integration the coefficients are operators on Fock space Ψ, and therefore $\mathbb{L}$ is an operator on $\widehat{\mathcal{K}} \otimes \Psi$.

The product of two differential elements of the form (1.1) (the "square bracket" or "Ito correction") is given by the following (associative) product

$$(1.3) \qquad \begin{pmatrix} \ell & \widetilde{\lambda} \\ \lambda & \Lambda \end{pmatrix} \bullet \begin{pmatrix} m & \widetilde{\mu} \\ \mu & M \end{pmatrix} = \begin{pmatrix} \widetilde{\lambda}\mu & \widetilde{\lambda}M \\ \Lambda\mu & \Lambda M \end{pmatrix}.$$

Here $\widetilde{\lambda}\mu$ is the matrix product of a line of operators by a column of operators, *i.e.* an operator, and similarly $\widetilde{\lambda}M$ is a line.

2 Though the concise notation above is adapted from Belavkin, the name of *Belavkin's notation* concerns more specially the representation of the set of four coefficients (1.1) as a square matrix of order 3 with many zeroes [1]

$$(2.1) \qquad \mathbb{L} = \begin{pmatrix} 0 & \tilde{\lambda} & \ell \\ 0 & \Lambda & \lambda \\ 0 & 0 & 0 \end{pmatrix}$$

which represents as above an operator on $\hat{\mathcal{K}} \otimes \mathcal{J}$ with an enlarged $\hat{\mathcal{K}} = \mathbb{C} \oplus \mathcal{K} \oplus \mathbb{C}$. This apparent complication is justified by the formula for products corresponding to (1.3)

$$(2.2) \qquad \begin{pmatrix} 0 & \tilde{\lambda} & \ell \\ 0 & \Lambda & \lambda \\ 0 & 0 & 0 \end{pmatrix} \begin{pmatrix} 0 & \tilde{\mu} & m \\ 0 & M & \mu \\ 0 & 0 & 0 \end{pmatrix} = \begin{pmatrix} 0 & \tilde{\lambda}M & \tilde{\lambda}\mu \\ 0 & \Lambda M & \Lambda\mu \\ 0 & 0 & 0 \end{pmatrix} .$$

which is a standard product of matrices. On the other hand, taking adjoints in Belavkin's notation corresponds to a non-standard *involution* : the following matrices represent adjoint pairs

$$(2.3) \qquad \mathbb{L} = \begin{pmatrix} 0 & <b| & a \\ 0 & D & |c> \\ 0 & 0 & 0 \end{pmatrix} , \quad \mathbb{L}^{\flat} = \begin{pmatrix} 0 & <c| & a^* \\ 0 & D^* & |b> \\ 0 & 0 & 0 \end{pmatrix} .$$

In a form of Belavkin's notation that is spreading among quantum probabilists, elements of $\hat{\mathcal{K}}$ are written as $ae_{-\infty} + e + ce_{+\infty}$ with a, c scalar and $e \in \mathcal{K}$, and therefore elements of $\hat{\mathcal{K}} \otimes \mathcal{J}$ or $\hat{\mathcal{K}} \otimes \Psi$ have similar representations with $a, c \in \mathcal{J}$ or Ψ, and e replaced by a finite sum $\sum_i e_i \otimes b_i$ with $e_i \in \mathcal{K}$, $b_i \in \mathcal{J}$ or Ψ. Then the action of Belavkin's $(3,3)$ matrix is read as follows

$$(2.4) \qquad \mathbb{L}(e_{-\infty} \otimes c + \sum_i e_i \otimes b_i + e_\infty \otimes a)$$
$$= e_{-\infty} \otimes (La + \sum_i \tilde{\lambda}(b_i)) + \lambda(a) + \sum_i e_i \otimes \Lambda b_i .$$

Belavkin's notation is far more transparent when the $(3,3)$ matrix can be displayed! We suspect QP has a mild fit of the differential geometers' "intrinsic fever" (to be fought by small inoculations of the "index disease" virus).

Kernels in basis-free notation

3 The following discussion is borrowed from Schürmann's article [Sch3], though the idea of (3.2) can be found in Belavkin's work. For simplicity, we do not include an initial space.

We recall (*cf.* §1, (1.5)) that multiple Fock space with multiplicity space $\mathcal{K}$ (possibly infinite dimensional) can be interpreted as an integral of Hilbert spaces over the simplex $\mathcal{P}$

$$(3.1) \qquad \Phi = \int_{\mathcal{P}} \mathcal{K}_A \, dA \quad \text{with} \quad \mathcal{K}_A = \mathcal{K}^{\otimes|A|} .$$

---

[1] Such a representation was also used by Holevo [Hol4].

This can be written is a symbolic way as a "chaotic representation"

$$(3.2) \qquad\qquad\qquad f = \int \mathbf{f}(A) \, d\mathbf{X}_A^{\otimes}$$

where $\mathbf{f}(A)$ is $\mathcal{K}_A$-valued. If $\mathcal{K}$ is finite dimensional, then taking an orthonormal basis of $\mathcal{K}$ and expanding $\mathbf{f}(A)$ in the corresponding tensor basis of $\mathcal{K}_A$ will produce the standard multiple integral coefficients (but $d\mathbf{X}_A^{\otimes}$ is just a notation).

The kernels we consider are three-arguments kernels $\mathbf{K}(U, V, W)$ with a notation very close to that of simple Fock space, but $\mathbf{K}(U, V, W)$ instead of being a scalar is a (bounded) operator from $\mathcal{K}_{V+W}$ to $\mathcal{K}_{U+W}$ — tensoring with identity on $\mathcal{K}_Z$, we may consider it also as an operator from $\mathcal{K}_{V+W+Z}$ to $\mathcal{K}_{U+W+Z}$ for an arbitrary Z disjoint from $U+V+W$ — changing notation, from $\mathcal{K}_T$ to $\mathcal{K}_{T-V+U}$ provided T contains $V+W$ and is disjoint from U. Then we may define the action of the kernel on a vector by a formula identical to that of simple Fock space :

$$(3.3) \qquad\qquad \mathbf{K}\mathbf{f}(H) = \int_{\mathcal{P}} \sum_{A+B+C=H} \mathbf{K}(A, B, M) \, \mathbf{f}(B + C + M) \, dM \ ,$$

where $\mathbf{K}(A, B, M)$ acts as an operator on $\mathbf{f}(B + C + M) \in \mathcal{K}_{B+C+M}$ and maps it to $\mathcal{K}_{(B+C+M)-M+A} = \mathcal{K}_H$. The rule for composition of kernels and Maassen's theorem are exactly the same as in the scalar case, with the only change that the absolute value of K is replaced by the operator norm of $\mathbf{K}$.

Chapter VI
Stochastic Calculus in Fock Space

In this chapter, we reach the main topic of these notes, *non–commutative stochastic calculus* for adapted families of operators on Fock space, with respect to the basic operator martingales. This calculus is a direct generalization of the classical Ito integration of adapted stochastic processes w.r.t. Brownian motion, or other martingales. Its physical motivation is quantum mechanical evolution in the presence of a "quantum noise". Stochastic calculus has also been developed for fermions in a series of papers by Barnett, Streater and Wilde [BSW], and in abstract versions by Accardi, Fagnola, Quaegebeur.

§1. STOCHASTIC INTEGRATION OF OPERATORS

In this section, we define stochastic integration for operators with respect to the (boson) basic processes, following Hudson–Parthasarathy [HuP1]. We proceed slowly, beginning with simple Fock space, for which no difficulty of notation arises. The theory will then be extended in two directions : the replacement of simple by multiple Fock space, and the inclusion of an *initial space*.

Adapted processes of operators

1 We denote by Φ the usual (boson) Fock space over $L^2(\mathbb{R}_+)$, the standard source of "non–commutative noise" which replaces here the driving Brownian motions of stochastic differential equations, and the time parameter t of classical mechanics.

To give an intuitive meaning to Φ, we may use a Wiener probabilistic interpretation, *i.e.* $\Phi = L^2(\Omega)$, the sample space for a Brownian motion (X_t). Replacing Φ by a multiple Fock space amounts to considering a d–dimensional Brownian motion, and we deal later with the necessary notational changes.

Let us recall some notation from Chap. IV, §2, subsection 6. To each time t corresponds a tensor product decomposition $\Phi = \Phi_{t]} \otimes \Phi_{[t}$ of Fock space (the "past" space usually is simplified to Φ_t). More generally, if $s < t$, we have a decomposition $\Phi = \Phi_{s]} \otimes \Phi_{[s,t]} \otimes \Phi_{[t}$. If we interpret Fock space as Wiener space, the space $\Phi_{[s,t]}$, for instance, is $L^2(\mathcal{F}_{[s,t]})$, the σ–field generated by the increments $X_u - X_s$, $s < u \leq t$, and the decomposition arises from the independence of Brownian increments. On the other hand, this interpretation suggests considering these spaces as *contained in* Φ, $f \in \Phi_t$ being identified with $f \otimes \mathbf{1} \in \Phi$, the vacuum vector which belongs to every $\Phi_{[s,t]}$. Instead of an external tensor product $\Phi_{t]} \otimes \Phi_{[t}$, we then omit the symbol $\otimes$ and think of an internal multiplication, defined only for vectors which are well separated in time, and for which $\mathbf{1}$ acts as unit element.

Given a Hilbert space $\mathcal{J}$, called the *initial space*, which may represent some concrete physical system we wish to "couple" to the noise source, we form the tensor product $\Psi = \mathcal{J} \otimes \Phi$. If $\mathcal{J}$ has finite dimension ν and a fixed o.n.b. (e_n) has been chosen, Ψ appears as a mere sum of ν copies of Φ.

> In the decomposition $\Phi = \Phi_t \otimes \Phi_{[t}$, it is often convenient to identify $\Phi_{[t}$ with Φ using the shift operator, and to consider Φ_t as an initial space. Thus initial spaces occur naturally even in the study of simple Fock space.

The space Ψ also has tensor product decompositions. First we have $\Psi = \Psi_t \otimes \Phi_{[t}$, where $\Psi_t = \mathcal{J} \otimes \Phi_{t]}$. On the other hand, there are decompositions $\Psi = \Phi_t \otimes \Psi_{[t}$ where $\Psi_{[t} = \mathcal{J} \otimes \Phi_{[t}$.

We generally omit the tensor product sign in these decompositions, writing simply $j\varphi$ ($j \in \mathcal{J}$, $\varphi \in \Phi$) or even $f_{t]} j h_{[t}$ ($f_t \in \Phi_t$, $h_{[t} \in \Phi_{[t}$) as if initial vectors really did commute with elements of Fock space. We do not distinguish $j\mathbf{1}$ from j, so that the initial space $\mathcal{J}$ is identified with Ψ_0. Similarly, given an operator A (bounded for simplicity) on the initial space (*resp.* on Φ), we do not distinguish it from its ampliation $A \otimes I$ (*resp.* $I \otimes A$) on Ψ. For instance, the operators a_t^ε on Φ are extended to Ψ in this way.

> As Applebaum [App2] remarks, this identification isn't entirely safe when A is unbounded and closed, because the algebraic ampliation $A \otimes I$ is not closed in general.

An operator B on Ψ (bounded for simplicity) is said to be *s-adapted* or *s-measurable*[1] if B maps Ψ_s into itself, and acts as follows on decomposed vectors $u_s v_{[s}$ ($u_s \in \Psi_s$, $v_{[s} \in \Phi_{[s}$, $\otimes$ omitted)

$$B(u_s v_{[s}) = (Bu_s) v_{[s} \,.$$

In tensor product language, this means that B is the ampliation $B = b \otimes I_{[s}$ of some operator b on $\mathcal{J} \otimes \Phi_s$. A family (B_s) of s-adapted operators is called an *adapted process*.

> The projection (conditional expectation) E_t on Φ_t isn't t-measurable, but the future projection on $\Phi_{[t}$ is.

In this chapter, the natural domain for operators will be $\mathcal{E}$, generated by exponential vectors $j\mathcal{E}(u)$, with $j \in \mathcal{J}$ and $u \in L^2(\mathbb{R}_+)$. To spare parentheses when an operator is applied, we take j inside and write $\mathcal{E}(ju)$, but note that $\mathcal{E}((tj)\mathbf{u}) \neq \mathcal{E}(j(t\mathbf{u}))$! These vectors are decomposable, since we have $\mathcal{E}(ju) = \mathcal{E}(jk_t u)\,\mathcal{E}(k_t' u)$, with $k_t u = u I_{]0,t]}$ and $k_t' u = u I_{]t,\infty[}$ — whether we open or close the interval at t is unimportant. Then B is t-measurable iff $B\mathcal{E}(jk_t u)$ belongs to Ψ_t, and $B\mathcal{E}(ju) = B\mathcal{E}(jk_t u) \otimes \mathcal{E}(k_t' u)$. For instance, the three basic operator processes (a_t^ε) are adapted on the exponential domain.

> In practice, j may be restricted to a dense subspace $\mathcal{D}$ of the initial space, and u to a dense subspace $\mathcal{S}$ of $L^2(\mathbb{R}_+)$ (see subsections 8 and 10).

We denote by E_t the projection on the closed subspace Ψ_t of Ψ. In particular, E_0 may be considered as projecting on the initial space. Given an operator A on Ψ, bounded for simplicity, we denote by $\mathbb{E}_s(A)$ the s-adapted operator mapping $\mathcal{E}(ju)$ to $E_s A\mathcal{E}(jk_s u)) \otimes \mathcal{E}(k_s' u)$. In particular, $\mathbb{E}_0(A)$ may be identified with the

---

[1] This is unusual, but fits better the analogy with random variables.

operator $E_O A E_0$ on the initial space. For any vector $f \in \Psi_s$ we have $<f, Af> = <f, \mathbb{E}_0(A) f>$, and $\mathbb{E}_s(A)$ can be interpreted as a vacuum conditional expectation of A in the operator sense : for any s–adapted bounded operators B, C , we have

$$< 1, CAB\,1 > \; = \; < 1, C\,\mathbb{E}_s(A)\,B\,1 > \; .$$

This interpretation of $\mathbb{E}_s$ as a conditional expectation allows the definition of *operator martingales*. For example, the three basic operator processes are martingales. However, at the present time there is no general "martingale theory" for operators, corresponding to the rich classical theory of martingales.

This description of the structure of Fock space continues in subsection 11.

REMARK. In these notes we develop quantum stochastic calculus in Fock space over $L^2(\mathbb{R}_+)$, or more generally $L^2(\mathbb{R}_+, \mathcal{K})$. In Parthasarathy's book the reader will find an interesting extension to Fock space over $\mathcal{H}$, where $\mathcal{H}$ is provided with an increasing family $(\mathcal{H}_t)$ of Hilbert spaces. This applies in particular to $\mathcal{H} = L^2(\Omega)$ where Ω is a filtered probability space.

Stochastic integrals of operator processes

2   Given an adapted operator process $H = (H_t)$, we are going to define new operator processes $I_t^\varepsilon(H) = \int_0^t H_s \, da_s^\varepsilon$ relative to the three basic processes on simple Fock space.

Up to subsection 6, we discuss the subject informally. The usual approach uses approximation by elementary step processes, but we prefer to justify the formulas without approximation, through an "integration by parts formula" for stochastic integrals of operators acting on a vector stochastic integral. For simplicity, we temporarily forget about the initial space.

To apply the operator $I_t^\varepsilon(H)$ to a vector F of Fock space we consider the martingale $F_t = \mathbb{E}\left[F \,|\, \mathcal{F}_t\right]$, which has a stochastic integral representation

$$(2.1) \qquad\qquad F_t = c + \int_0^t f_s \, dX_s \,, \qquad c = \mathbb{E}\left[F\right] \,,$$

where (f_s) is a predictable process. We may consider (2.1), independently of any concrete probabilistic interpretation, as a stochastic integral of the adapted curve (f_t) (the delicate point of predictability can now be forgotten) w.r.t. the *curve* $X_t = a_t^+ 1$, which has orthogonal increments, and is well related to the continuous tensor product structure of Fock space, see Chap. IV §2 subs. 6. We imitate what is known for vectors and try to compute $I_t^\varepsilon(H) F_t$ for $t \leq \infty$ as the integral of $d(I_s^\varepsilon(H) F_s)$, using an integration by parts formula consisting of three differentials, $I_s^\varepsilon(H) dF_s$, $dI_s^\varepsilon F_s$, and $dI_s^\varepsilon dF_s$. In martingale theory, the last term is the *square bracket* or Ito correction; in the operator case, it will be read from the "Ito table" of Chapter IV, §3 subs. 5.

The first term is the s–adapted operator $I_s^\varepsilon(H)$ applied to the decomposed vector $f_s \, dX_s$, which after integration gives

$$(2.2) \qquad\qquad \int_0^t I_s^\varepsilon(H) f_s \, dX_s \,,$$

an ordinary stochastic integral provided the required integrability condition is satisfied, but still containing the operator $I_s^\varepsilon(H)$ we are trying to define. The second differential is $H_s\, da_s^\varepsilon\, F_s$, and here it is da_s^ε that "sticks into the future". Thus we write $F_s = F_s \otimes \mathbf{1}$ and, since $da_s^\varepsilon \mathbf{1} = dX_s$ for $\varepsilon = +$ and 0 otherwise, the second term has the value

$$(2.3) \qquad \int_0^t H_s F_s\, dX_s \quad \text{if} \quad \varepsilon = +, \qquad 0 \quad \text{otherwise.}$$

As for the third differential, the decomposed operator $H_s\, da_s^\varepsilon$ being applied to the decomposed vector $f_s\, dX_s$, we get a decomposed vector $(H_s\, f_s)(da_s^\varepsilon\, dX_s)$, and remember that $da_s^\varepsilon\, dX_s = da_s^\varepsilon da_s^+ \mathbf{1}$ has the value 0 for $\varepsilon = +$, ds for $\varepsilon = -$, dX_s for $\varepsilon = \circ$. Thus the third term is

$$(2.4) \qquad 0 \quad \text{if} \quad \varepsilon = +, \qquad \int_0^t H_s f_s\, dX_s \quad \text{if} \quad \varepsilon = \circ, \qquad \int_0^t H_s f_s\, ds \quad \text{if} \quad \varepsilon = -.$$

The processes (2.3) and (2.4) are known, but (2.2) contains the still undefined operator $I_t^\varepsilon(H)$; however, *the above computation allows a definition by induction* since, so to speak, (f_s) is one chaos lower than (F_s). This induction starts with the chaos of order 0 : the stochastic integral $I_t^\varepsilon(H)\mathbf{1}$ is equal to 0 if $\varepsilon = -, \circ$, and one sees easily that for $\varepsilon = +$ we have

$$(2.5) \qquad I_t^+(H)\mathbf{1} = \int_0^t H_s \mathbf{1}\, dX_s .$$

Putting everything together, we have the formula

$$(2.6) \qquad \boxed{\; I_t^\varepsilon(H)\, F_t = \int_0^t I_s^\varepsilon(H) f_s\, dX_s + \begin{cases} \int_0^t H_s F_s\, dX_s & \text{if } \varepsilon = + \\ \int_0^t H_s f_s\, dX_s & \text{if } \varepsilon = \circ \\ \int_0^t H_s f_s\, ds & \text{if } \varepsilon = - \end{cases} \;} .$$

With such a definition in closed form, discrete computations on "Riemann sums" are replaced by classical stochastic calculus. Let us give a few examples.

3 We are going to prove that, if $H_s = u(s)\, I$, where u is bounded with compact support (for instance), the preceding definition of $\int_0^t H(s)\, da_s^\varepsilon$ leads to the correct value a_u^ε. To this end, we climb on the chaos ladder. It suffices to prove the two processes have the same value on $\mathbf{1}$, and both satisfy property (2.6). We take for instance $\varepsilon = +$, leaving the other two cases to the reader.

We consider a stochastic integral $F = \int_0^\infty f_s\, dX_s$, and check that $G = a_u^+ F$ satisfies the first line of (2.6) :

$$(3.1) \qquad G = \int_0^\infty u_s F_s\, dX_s + \int_0^\infty a_u^+ f_s\, dX_s .$$

We make use of the chaos expansions of these random variables (Chap. IV, §4, subs. 2),

$$(3.2) \qquad \widehat{G}(A) = \sum_{s \in A} u(s)\, \widehat{F}(A - s) .$$

On the other hand, if the r.v. f_s is given by its chaos expansion $\int \widehat{f}(s,A)\,dX_A$, with $\widehat{f}(s,A)=0$ unless $A\subset[0,s[$, the chaos expansion of F is given by

(3.3) $$\widehat{F}(A) = \widehat{f}(\vee A, A-)$$

where $\vee A$ is the last element of A, and $A- = A\setminus\{\vee A\}$ (consider the case where f_s belongs to a chaos of a given order). Also $F_s = \int_0^s f_r\,dX_r$ has a chaos expansion given by (3.3) if $\vee A < s$, 0 otherwise. Then the right side of (3.2) can be written as

$$u(\vee A)\,\widehat{F}(A-) + \sum_{s\in A-} u(s)\,\widehat{F}(A-s) .$$

The first term is the chaos expansion of $\int u_s F_s\,dX_s$, while the second term represents $\int a_s^+ f_s\,dX_s$ (2.2). The cases $\varepsilon=-,\circ$ are similar.

REMARK. Remaining in the case where $H_s = u(s)I$, we can give another example of stochastic integration. Let us introduce the classical selfadjoint operators $Q_t = a_t^+ + a_t^-$, $P_t = i(a_t^+ - a_t^-)$, $N_t = a_t^\circ$. Then for arbitrary c we may compute from section 2

$$\left(\int u_s\,d(Q_s + cN_s)\right)\left(\int f_s\,dX_s\right)$$

(an operator acting on a vector). On the other hand, we may interpret it as the product of random variables

$$\left(\int u_s\,dX_s\right)\left(\int f_s\,dX_s\right)$$

in the probabilistic interpretation of Fock space where X_t is a compensated Poisson process with jump size c : putting $F_t = \int_0^t f_s\,dX_s$, $U_t = \int_0^t u_s\,dX_s$, we apply to this product the usual integration by parts formula, given that $d[F,U]_s = u_s f_s\,d[X,X]_s = u_s f_s\,(ds + c\,dX_s)$, and we get

$$\int U_s f_s\,dX_s + \int F_s u_s\,dX_s + \int u_s f_s\,(ds + c\,dX_s) .$$

This is the same result one gets from a convenient linear combination of the formulas (2.6).

4 In the preceding subsection, we took $H_s = u(s)I$ where u was bounded with compact support, and showed that $\int_0^\infty H(s)\,da_s^+ = a_u^+$. Assume now u has support in $[t,\infty[$, and take $H_s = u(s)L$, where L is a t-measurable operator, bounded for simplicity. Then we have $\int_0^\infty H_s\,da_s^+ = La_u^+$. This amounts to the remark that, for classical stochastic integrals

$$L\int_t^\infty f_s\,dX_s = \int_t^\infty Lf_s\,dX_s .$$

From this result we deduce that, for an elementary (previsible) step process (H_s), i.e. a process of the form

$$H_t = \sum_i L_i I_{\{s_i < t \le s_{i+1}\}}$$

relative to a finite subdivision $0 = s_0 < \ldots < s_n < s_{n+1} = \infty$, with L_i s_i-adapted and $L_n = 0$, $I^\varepsilon(H)$ is the elementary integral $\sum_i L_i(a^\varepsilon_{s_{i+1}} - a^\varepsilon_{s_i})$. Thus the definition of operator stochastic integrals we are using coincides with the more usual definition using elementary step processes and approximation.

5   The next easy computation using (2.6) plays a basic role in the approach of Hudson–Parthasarathy. It deals with *exponential vectors* $F = \mathcal{E}(u)$. Let us put $F_t = \mathcal{E}_t(u) = \mathcal{E}(u I_{]0,t]})$, this process satisfying the differential equation

$$\mathcal{E}_t(u) = 1 + \int_0^t u(s)\,\mathcal{E}_s(u)\,dX_s \ ,$$

so that in (2.1) f_s is equal to $u(s)F_s$. We also put

$$H_t\mathcal{E}_t(u) = \eta_t \ , \qquad I^\varepsilon_t(H)\,\mathcal{E}_t(u) = i^\varepsilon_t \ .$$

We take for instance $\varepsilon = +$, and apply the computation in subsection 2, thus getting a linear non-homogeneous stochastic differential equation for i^+_t

$$(5.1) \qquad\qquad i^+_t = \int_0^t \eta_s\,dX_s + \int_0^t u_s i^+_s\,dX_s$$

of a classical type ([DMM], Chap. XXIII, n° 72); we give in subsection 6 below precise conditions for the existence and uniqueness of a solution. For the other values of ε we have similarly

$$(5.2) \qquad\qquad i^-_t = \int_0^t u_s \eta_s\,ds + \int_0^t u_s i^-_s\,dX_s \ ,$$

$$(5.3) \qquad\qquad i^\circ_t = \int_0^t u_s \eta_s\,dX_s + \int_0^t u_s i^\circ_s\,dX_s \ .$$

In the case of the annihilation integral (5.2), the equation has a simple explicit solution

$$(5.4) \qquad\qquad I^-_t(H)\,\mathcal{E}_t(u) = \int_0^t u_s\,H_s\mathcal{E}_t(u)\,ds \ .$$

One can check directly that this process satisfies (5.2) — also (5.4) is almost obvious when (H_t) an elementary step process, and the formula then is extended by a passage to the limit (we discuss this point below).

Stochastic integrals on the exponential domain

6   The time has come to replace formal computations by a rigorous analysis. We follow Hudson–Parthasarathy, and work with exponential vectors (the method of induction on the chaos will not be discussed here, though it is useful too).

   We are going to prove the existence and uniqueness of the stochastic integral process $I^+_t(H)$ on the domain $\mathcal{E}$ linearly generated by exponential vectors $\mathcal{E}(u)$.

> It is sometimes convenient to restrict this domain by conditions on u, like local boundedness, having compact support, etc. This implies only trivial modifications.

We make the following assumptions on the adapted operator process (H_t) : The domain of H_t contains $\mathcal{E}$, and the Ψ-valued mapping $t \longmapsto H_t \mathcal{E}(u)$ is measurable, and square integrable on bounded time intervals. Since $H_t \mathcal{E}(u) = H_t \mathcal{E}_t(u) \mathcal{E}(k'_t u)$, this is equivalent to saying that $t \longmapsto \eta_t = H_t \mathcal{E}_t(u)$ belongs to L^2_{loc}.

For our argument, we need first an existence and uniqueness result for the classical (vector) stochastic differential equation

(6.1) $$A_t = U_t + \int_0^t u_s A_s \, dX_s \, , \qquad (u \in L^2(\mathbb{R}_+))$$

under the assumption that the given curve (U_t) in Ψ is adapted, and satisfies the integrability condition

(6.2) $$\int_0^T \|U(s)\|^2 \, |u(s)|^2 \, ds < \infty \, ,$$

while the unknown curve (A_t) in Ψ is required to be adapted and such that the stochastic integral exists. This is a typical case of the Picard iteration method : it leads to a series (whose terms are meaningful according to (6.2))

(6.3) $$A_t = U_t + \int_{t>s_1} u_{s_1} U_{s_1} \, dX_{s_1} + \int_{t>s_1>s_2} u_{s_1} U_{s_1} u_{s_2} U_{s_2} \, dX_{s_1} \, dX_{s_2} + \dots$$

To study its convergence, we construct (according to Feyel [Fey]) an auxiliary norm of L^2 type for adapted stochastic processes $\mathbf{A} = (A_t)$ on a bounded interval $[0, T]$, associated with a function $p(t) > 0$

$$\|\mathbf{A}\| = \Big(\int_0^T \|A_s\|^2 \, p(s) ds \Big)^{1/2} \, ,$$

such that the mapping $(A_t) \longmapsto (\int_0^t A_s u_s \, dX_s)$ has a norm $K < 1$, and the process (U_t) has a finite norm. Then the Picard approximation procedure will converge to a process (A_t), also of finite norm. The above mapping has a norm smaller than K if $p(t)$ satisfies the inequality

$$p(t) \geq \frac{1}{K^2} |u(t)|^2 \int_t^T p(s) \, ds \, ,$$

and we take p so that equality holds, i.e.

$$p(t) = |u(t)|^2 \, e^{\frac{1}{K^2} \int_t^T |u(s)|^2 \, ds} \, .$$

With this choice, u being square integrable, the exponential is a function bounded above and below, and existence and uniqueness follow. Then $p(t)$ has played its role, and we return to our original problem.

We construct the process $A_t = I_t^+(H)\mathcal{E}_t(u)$ as the solution of the differential equation (6.1) corresponding to $U_t = \int_0^t H_s \mathcal{E}_s(u) \, dX_s = \int_0^t \eta_s \, dX_s$. We have in this case $\|U_t\|^2 = \int_0^t |\eta_s|^2 \, ds$, which according to our assumptions is a bounded function of t. Therefore the series (6.3) is norm convergent on $\mathbb{R}_+$, and the solution exists.

We introduce a second process of operators K satisfying the same assumptions as H, and put similarly $B_t = I_t^+(K)\mathcal{E}_t(v)$, $V_t = \int_0^t K_s \mathcal{E}_s(v) \, dX_s = \int_0^t \chi_s \, dX_s$. Using (5.1) we compute the scalar product $< B_t, A_t >$

$$(6.4) \qquad \frac{d}{dt} < B_t, A_t > = \bar{v}_t u_t < B_t, A_t > + u_t < \chi_t, A_t > + \bar{v}_t < B_t, \eta_t > + < \chi_t, \eta_t > .$$

We move the first term on the right to the left side, and multiply both sides by $e^{\int_t^\infty \bar{v}_s u_s \, ds}$. Then the differential of $< B_t, A_t > e^{\int_t^\infty \bar{v}_s u_s \, ds}$, with t as lower limit, appears, and we are now computing the derivative of $< I_t^+(K)\mathcal{E}(v), I_t^+(H)\mathcal{E}(u) >$. If we perform the easy computations, we get the basic *quantum Ito formula* of Hudson–Parthasarathy relative to the creation stochastic integral :

$$\frac{d}{dt} < I_t^+(K)\,\mathcal{E}(v), I_t^+(H)\,\mathcal{E}(u) > = \bar{v}_t < I_t^+(K)\,\mathcal{E}(v), H_t \mathcal{E}(u) >$$
$$(6.5) \qquad + < K_t \mathcal{E}(v)\, v_t, I_t^+(H)\,\mathcal{E}(u) > u_t + < K_t \mathcal{E}(v)\, v_t, H_t \mathcal{E}(u)\, u_t > .$$

Why is this an Ito formula (or rather, an integration by parts formula)? If we could move the operator $I_t^+(K)$ to the right of the scalar product on the left, and similarly on the right we would compute the matrix element between two exponential vectors, and the composition $d(I_t^-(K^*)I_t^+(H))$ on the left, and on the right three terms, the last of which comes from the relation $da_t^- da_t^+ = dt$ in the Ito table. On the other hand, this composition of operators is not well defined because of domain problems. For operators given by Maassen kernels, which have a large common stable domain, a true Ito formula becomes meaningful.

The following simpler relation can be proved in the same way (since the space $\mathcal{E}$ is dense, this formula completely characterizes the stochastic integral)

$$(6.6) \qquad < \mathcal{E}(v), I_t^+(H)\,\mathcal{E}(u) > = \int_0^t < v_s \mathcal{E}(v), H_s \mathcal{E}(u) > ds .$$

From (6.5) we may deduce an interesting estimate (due to J.L. Journé) for the norm of a stochastic integral (the most useful form of such estimates will be given later in (9.7)). Let us put for simplicity $I_t^+(H)\mathcal{E}(u) = I_t^+$ and $H_t\mathcal{E}(u) = h_t$. Then we have from (6.5) with $H = K$

$$\|I_t^+\|^2 = 2 \int_0^t \Re e < u_s I_s^+, h_s > ds + \int_0^t \|h_s\|^2 \, ds .$$

The first integral on the right side is dominated by $\|u\| \left(\int_0^t |< I_s^+, h_s >|^2 \, ds \right)^{1/2}$. Apply the Schwarz inequality to this scalar product, and bound $\|I_s^+\|$ by its supremum W_t on $s \leq t$. Then putting $\int_0^t \|h_s\|^2 \, ds = V_t^2$ we get an inequality

$$\|I_t^+\|^2 \leq 2 \|u\| \, W_t V_t + V_t^2 .$$

The right side is an increasing function of t, thus we may replace the left side by its *sup* and get

$$W_t^2 \leq 2 \|u\| \, W_t V_t + V_t^2 ,$$

from which we deduce the following inequality for W_t

(6.7) $$\| I_t^+(H)\mathcal{E}(u) \| \le (\|u\| + \sqrt{\|u\|^2 + 1})\, \Big(\int_0^t \| H_s\mathcal{E}(u) \|^2\, ds \Big)^{1/2}.$$

As an application, we get the continuous dependence of $I_t^+(H)\mathcal{E}(u)$ on the process (H_t) for fixed u. Therefore, if this process is approximated by elementary step processes in the topology given by the right side, the corresponding stochastic integrals converge strongly on exponential vectors, thus showing the agreement of our definition with the usual one.

> A technical question that concerns us little here is the following : how can we approximate an *operator* process (H_t) by elementary step processes? A partial, but practically sufficient answer is given in [Parl], Prop. III.25.7, p. 190.

7 We have discussed the stochastic integral $I_t^+(H)$. Let us now discuss $I_t^-(H)$ and $I_t^o(H)$, without giving all the details.

The case of I_t^- is simpler. The formula similar to (6.6) is

(7.1) $$< \mathcal{E}(v), I_t^-(H)\mathcal{E}(u) > = \int_0^t < \mathcal{E}(v), H_s\mathcal{E}(u)u_s > ds .$$

Comparing it with (6.6) we see that $I_t^+(H)$ and $I_t^-(H^*)$ are mutually adjoint on exponential vectors (provided the adjoint process (H_t^*) exists and satisfies the same hypotheses as (H_t)). In the formula similar to (6.5), only the two terms of a classical integration by parts formula are present on the right side :

(7.2) $$\frac{d}{dt} < I_t^-(K)\mathcal{E}(v), I_t^-(H)\mathcal{E}(u) > = < I_t^-(K)\mathcal{E}(v), H_t\mathcal{E}(u)u_t >$$
$$+ < v_t K_t\mathcal{E}(v), I_t^-(H)\mathcal{E}(u) > .$$

There is also a formula involving two different stochastic integrals (the similar formula with $\pm$ interchanged is deduced by taking adjoints)

(7.3) $$\frac{d}{dt} < I_t^-(K)\mathcal{E}(v), I_t^+(H)\mathcal{E}(u) > = < v_t I_t^-(K)\mathcal{E}(v), H_t\mathcal{E}(u) >$$
$$+ < v_t K_t\mathcal{E}(v), I_t^+(H)\mathcal{E}(u) > .$$

The estimate similar to (6.7) is slightly simpler, and is deduced immediately from the explicit formula (5.4)

(7.4) $$\| I_t^-(H)\mathcal{E}(u) \| \le \|u\|\, \Big(\int_0^t \| H_s\mathcal{E}(u)\|^2\, ds \Big)^{1/2}.$$

This formula can be improved as follows : we decompose $\mathcal{E}(u) = \mathcal{E}(k_t u)\mathcal{E}(k_t' u)$, take the second factor out of the adapted operator, apply the formula to $k_t u$ and then reintroduce the second factor. We thus see that $\| u \|$ can be replaced by $\| k_t u \|$, which tends to 0 with t. This cannot be done with (6.7).

Let us discuss finally the case of $I_t^o(H)$. A look at (5.1) and (5.3) shows that (for a fixed exponential vector) the discussion is essentially the same as for $I_t^+(H)$, with

$u_s H_s \mathcal{E}(u)$ instead of $H_s \mathcal{E}(u)$. Thus we need a stronger integrability condition on the process (H_t), namely

$$(7.5) \qquad \int_0^t \| u_s H_s \mathcal{E}(u) \|^2 \, ds < \infty .$$

To make sure that this condition is fulfilled, one sometimes restricts the domain to exponential vectors $\mathcal{E}(u)$ such that u is (locally) bounded. The formula which replaces (6.6) and (7.1) is

$$(7.6) \qquad < \mathcal{E}(v), I_t^\circ(H) \mathcal{E}(u) > = \int_0^t < v_s \mathcal{E}(v), H_s \mathcal{E}(u) u_s > ds .$$

That which replaces (6.5) and (7.2) is

$$(7.7) \qquad \begin{aligned} \frac{d}{dt} &< I_t^\circ(K) \mathcal{E}(v), I_t^\circ(H) \mathcal{E}(u) > = < v_t I_t^\circ(K) \mathcal{E}(v), H_t \mathcal{E}(u) u_t > \\ &+ < v_t K_t \mathcal{E}(v), I_t^\circ(H) \mathcal{E}(u) u_t > + < v_t K_t \mathcal{E}(v), H_t \mathcal{E}(u) u_t > . \end{aligned}$$

And the inequality (6.7) becomes

$$(7.8) \qquad \| I_t^\circ(H) \mathcal{E}(u) \| \le (\|u\| + \sqrt{\|u\|^2 + 1}) \left(\int_0^t \| u_s H_s \mathcal{E}(u) \|^2 \, ds \right)^{1/2} .$$

Note that formula (7.7) contains a "square bracket" term, since $da_t^\circ da_t^\circ$ is a non-trivial term in the Ito table. It remains to write down the two formulas involving mixed stochastic integrals, one of which also has an Ito correction

$$(7.9) \qquad \begin{aligned} \frac{d}{dt} &< I_t^\circ(K) \mathcal{E}(v), I_t^-(H) \mathcal{E}(u) > = < I_t^\circ(K) \mathcal{E}(v), H_t \mathcal{E}(u) u_t > \\ &+ < v_t K_t \mathcal{E}(v), I_t^-(H) \mathcal{E}(u) u_t > . \end{aligned}$$

Here comes the last Ito formula. The reader should not worry : the case of multiple Fock space will give us a unified form, much easier to remember.

$$(7.10) \qquad \begin{aligned} \frac{d}{dt} &< I_t^\circ(K) \mathcal{E}(v), I_t^+(H) \mathcal{E}(u) > = < v_t I_t^\circ(K) \mathcal{E}(v), H_t \mathcal{E}(u) > \\ &+ < v_t K_t \mathcal{E}(v), I_t^+(H) \mathcal{E}(u) u_t > + < v_t K_t \mathcal{E}(v), H_t \mathcal{E}(u) > . \end{aligned}$$

REMARKS. 1) Adding an initial space $\mathcal{J}$ would lead to the same formulas, except that $\mathcal{E}(u)$ would be replaced everywhere by $\mathcal{E}(ju)$, $j \in \mathcal{J}$.

2) In the theory of stochastic differential equations, dt also appears as an integrand. The corresponding formula is a trivial application of the Schwarz inequality. Putting $I_t(H) = \int_0^t H_s \, ds$, we have

$$(7.11) \qquad \| I_t(H) \mathcal{E}(u) \| \le \|u\| \left(\int_0^t \| H_s \mathcal{E}(u) \|^2 \, ds \right)^{1/2} .$$

and here, as in (7.4), we may replace $\|u\|$ by $\| u I_{[0,t[} \|$.

Multiple Fock spaces

8 In the case of a Fock space of finite multiplicity d, the basic curve (X_t) becomes a vector curve $(\mathbf{X}_t)$ with components X_t^α, $\alpha=1,\ldots,d$ (in the Wiener interpretation, the components of a d–dimensional Brownian motion). Exponential vectors have a vector argument $\mathbf{u}$, with components u_α, and the linear differential equation they satisfy becomes

$$(8.1) \qquad \mathcal{E}_t(j\mathbf{u}) = j + \int_0^t \mathcal{E}_s(j\mathbf{u})\,\mathbf{u}_s\cdot d\mathbf{X}_s \ .$$

The boldface letters $\mathbf{u}, \mathbf{X}$ are here for clarity only, and we are under no obligation to use them forever.

The case of infinite multiplicity is not more difficult to handle, following Mohari-Sinha [MoS]. The argument $\mathbf{u}$ of an exponential vector then is assumed to have *only finitely many components* u_α *different from* 0. The corresponding indexes constitute the finite set $S(\mathbf{u})$ (the "support" of $\mathbf{u}$). Since we make a convention $u_0 = 1$ below, we also include 0 in $S(\mathbf{u})$.

As in Chap. V subs. 2, the index ε for the basic operators becomes a matrix index $\binom{\alpha}{\beta}$. Then $da_\beta^\alpha(t)$ is a number operator for $\alpha=\beta$, an exchange operator if $\alpha\neq\beta$, and we represent the creation and annihilation operators by means of the additional index 0 : $\binom{\alpha}{0}$ indicates a creation operator, $\binom{0}{\beta}$ an annihilation operator, and $da_0^0(t)$ means $I\,dt$.

This differs from the conventions of other authors (see the "warning" in Chapter V, §2, subsection 1).

Then we have

$$da_\beta^\alpha(t)\,dX_t^\gamma = \delta_\beta^\gamma\,dX_t^\alpha \ , \ \ da_\beta^0(t)\,dX^\gamma(t) = \delta_\beta^\gamma\,dt \ , \ \ da_0^\alpha(t)\,\mathbf{1} = dX_t^\alpha \ .$$

The stochastic integral operator $I_t^\varepsilon(H) = \int_0^t H_s\,da_s^\varepsilon$ acts as follows on exponential vectors : $A_t = I_t^\varepsilon(H)\mathcal{E}_t(j\mathbf{u})$ satisfies a linear differential equation

$$(8.2) \qquad A_t = \int_0^t dh_s^\varepsilon + \int_0^t A_s\,\mathbf{u}_s\cdot d\mathbf{X}_s$$

with

$$(8.3) \qquad dh_s^\varepsilon = \begin{cases} H_s\mathcal{E}_s(j\mathbf{u})\,dX_s^\alpha & \text{if } \varepsilon = \binom{\alpha}{0} \\ H_s\mathcal{E}_s(j\mathbf{u})\,u_\alpha(s)\,ds & \text{if } \varepsilon = \binom{0}{\alpha} \\ H_s\mathcal{E}_s(j\mathbf{u})\,u_\beta(s)\,dX_s^\alpha & \text{if } \varepsilon = \binom{\alpha}{\beta} \end{cases} \ .$$

The first two lines take a form similar to the last one if we make the convention that $X_s^0 = s$, $u_s^0 = 1$. As in Chapter V, using the letters $\rho, \sigma, \tau\ldots$ instead of $\alpha, \beta\ldots$ indicates that the index 0 is allowed, and we use the (Evans) modified Kronecker symbol $\widehat{\delta}_\sigma^\rho$, which differs from the usual one by the property $\widehat{\delta}_0^0 = 0$. Though we do not omit summation signs, summing on diagonal indexes provides an useful control on formulas, and therefore we allow for raised indexes $u^\rho(t)=\bar{u}_\rho(t)$, in particular $u^0(t)=1$. Then the scalar product $< \mathbf{u}, \mathbf{v} >$ (which does not involve the additional index 0) is equal to $\sum v^{\sigma} \widehat{\delta}^{\rho}_{\sigma} u_{\rho}$.

With these conventions, formulas (8.2) and (8.3) are true also for $\varepsilon = \binom{0}{0}$, provided the appropriate integrability condition $\int_0^t \| H_s \mathcal{E}(j\mathbf{u}) \| ds < \infty$ is satisfied. In the same way, if $\varepsilon = \binom{\sigma}{\rho}$, $I_t^{\varepsilon}(H) = \int_0^t H_s \, da_{\rho}^{\sigma}(s)$, the formula

$$(8.4) \qquad < \mathcal{E}(\ell \mathbf{v}), I_t^{\varepsilon}(H) \, \mathcal{E}(j\mathbf{u}) > = \int_0^t < \mathcal{E}(\ell \mathbf{v}), H_s \mathcal{E}(j\mathbf{u}) > v^{\sigma}(s) u_{\rho}(s) \, ds \ ,$$

which includes (6.6), (7.1), (7.6), is true for all indexes, the case of $\varepsilon = \binom{0}{0}$ being trivial. Similarly, if $\varepsilon = \binom{\sigma}{\rho}$ and $\eta = \binom{\mu}{\lambda}$, we may generalize (6.5), (7.2)... to all indexes. The same formula will be written again as (9.3), in a slightly better notation .

$$< I_t^{\eta}(K) \, \mathcal{E}(\ell \mathbf{v}), I_t^{\varepsilon}(H) \, \mathcal{E}(j\mathbf{u}) > = \int_0^t < I_s^{\eta}(K) \, \mathcal{E}(\ell \mathbf{v}), v^{\sigma}(s) u_{\rho}(s) \, H_s \mathcal{E}(j\mathbf{u}) > ds$$

$$+ \int_0^t < v_{\lambda}(s) \, u^{\mu}(s) \, K_s \mathcal{E}(\ell \mathbf{v}), I_s^{\varepsilon}(H) \, \mathcal{E}(j\mathbf{u}) > ds$$

$$(8.5) \qquad + \int_0^t \widehat{\delta}^{\mu \sigma} < v_{\lambda}(s) \, K_s \mathcal{E}(\ell \mathbf{v}), u_{\rho}(s) \, H_s \mathcal{E}(j\mathbf{u}) > ds \ .$$

The proof is essentially the same as for (6.4–5) : one first introduces the vector processes $A_t = I_t^{\varepsilon}(H) \, \mathcal{E}_t(j\mathbf{u})$, $B_t = I_t^{\eta}(K) \, \mathcal{E}_t(\ell \mathbf{v})$, which satisfy equations of the form (8.2)

$$A_t = \int_0^t A_s \, \mathbf{u}_s \cdot d\mathbf{X}_s + \int_0^t H_s \mathcal{E}_s(j\mathbf{u}) \, u_{\rho}(s) \, dX_s^{\sigma} \ ,$$

$$B_t = \int_0^t B_s \, \mathbf{v}_s \cdot d\mathbf{X}_s + \int_0^t K_s \mathcal{E}_s(\ell \mathbf{v}) \, v_{\lambda}(s) \, dX_s^{\mu} \ .$$

One then computes $\frac{d}{dt} < B_t, A_t >$ as

$$< \mathbf{v}_t, \mathbf{u}_t > < B_t, A_t > + < v_{\lambda} u^{\mu}(t) \, K_t \mathcal{E}_t(\ell \mathbf{v}), A_t > + < B_t, H_t \mathcal{E}_t(j\mathbf{u}) \, v^{\sigma} u_{\rho}(t) >$$

$$+ \widehat{\delta}^{\mu \sigma} < v^{\lambda} K_t \mathcal{E}(\ell \mathbf{v}), H_t \mathcal{E}_t(j\mathbf{u}) \, u_{\rho} >$$

and the result is simplified by the same method we used after (6.4).

If initial operators η_0, χ_0 acting on Ψ_0 are included in $I^{\varepsilon}(H)$, $I^{\eta}(K)$, the term $< \eta_0 \mathcal{E}(\ell \mathbf{v}), \chi_0 \mathcal{E}(j\mathbf{u}) >$ must be added to the right hand side of (8.5).

9 Instead of individual stochastic integrals, we now consider processes of operators, acting on the exponential domain, which admit a stochastic integral representation involving all the basic processes together (also $da_0^0(t) = I \, dt$)

$$(9.1) \qquad \boxed{ I_t(H) = \eta + \sum_{\rho, \sigma} \int_0^t H_{\sigma}^{\rho}(s) \, da_{\rho}^{\sigma}(s) \ . }$$

Such processes may be considered as substitutes for "operator semimartingales". The initial operator η acting on Ψ_0 will often be omitted for the sake of simplicity. The operator $I_t(H)$ will be applied to an exponential martingale $\mathcal{E}_t(j\mathbf{u})$. Of course the conditions of existence of the individual stochastic integrals (which are easy extensions of the simple Fock space case) must be fulfilled, but we should keep in mind the case of infinite multiplicity, and seek a simple, global integrability condition implying the existence of (9.1) as a whole. We recall that in this case, the argument $\mathbf{u}$ has finitely many non–zero components, whose indexes (and 0) constitute the "support" $S(\mathbf{u})$. Only the operators H_σ^ρ with $\rho \in S(\mathbf{u})$ contribute to $I_t(H)\mathcal{E}(j\mathbf{u})$.

Assuming first we have a finite number of $H_\sigma^\rho \neq 0$ in (9.1), we have for arbitrary exponential vectors $\mathcal{E}(j\mathbf{u})$ and $\mathcal{E}(\ell\mathbf{v})$

$$(9.2) \quad \boxed{< \mathcal{E}(\ell\mathbf{v}) , I_t(H)\mathcal{E}(j\mathbf{u}) > = \sum_{\rho,\sigma} \int_0^t < \mathcal{E}(\ell\mathbf{v}) , H_\sigma^\rho(s)\mathcal{E}(j\mathbf{u}) > v^\sigma(s) u_\rho(s) \, ds} \ .$$

Note that an index has been raised to straighten up the book–keeping of indexes. The initial operator would contribute a term $< \ell, \eta j > e^{< \mathbf{v}, \mathbf{u} >}$.

The next formula requires a second "semimartingale"

$$I_t(K) = \chi + \sum_{\lambda,\mu} \int_0^t K_\lambda^\mu(s) \, da_\mu^\lambda(s) \ .$$

Then we have the fundamental *Quantum Ito Formula*. If initial operators were included, they would contribute an additional term $< \chi\ell, \eta j > e^{< \mathbf{v}, \mathbf{u} >}$.

$$
\begin{aligned}
(9.3) \quad < I_t(K)\mathcal{E}(\ell\mathbf{v}), \, I_t(H)\mathcal{E}(j\mathbf{u}) > = & \sum_{\rho\sigma} \int_0^t < I_s(K)\mathcal{E}(\ell\mathbf{v}), H_\sigma^\rho(s)\mathcal{E}(j\mathbf{u}) v^\sigma(s) u_\rho(s) > ds \\
& + \sum_{\lambda\mu} \int_0^t < v_\mu(s) u^\lambda(s) K_\lambda^\mu(s)\mathcal{E}(\ell\mathbf{v}), I_s(H)\mathcal{E}(j\mathbf{u}) > ds \\
& + \sum_{\rho,\sigma,\lambda,\mu} \int_0^t \widehat{\delta}^{\lambda\sigma} < v_\mu(s) K_\lambda^\mu(s)\,\mathcal{E}(\ell\mathbf{v}), H_\sigma^\rho(s)\mathcal{E}(j\mathbf{u})\, u_\rho(s) > ds \ .
\end{aligned}
$$

Note how the balance of indexes is realized.

Inequalities. We now extend to multiple Fock space the basic estimates concerning the L^2 norm of the vector

$$(9.4) \qquad\qquad A(t) = I_t(H)\,\mathcal{E}(j\mathbf{u}) \ .$$

In the case of infinite multiplicity these estimates may be used to extend the scope of the representation (9.1) to infinitely many summands — details are left to the reader.

To abbreviate notation, let us put

$$h_\sigma^\rho(t) = H_\sigma^\rho(t)\mathcal{E}(j\mathbf{u}) \quad , \quad h_\sigma(t) = \sum_\rho u_\rho(t) h_\sigma^\rho(t) \quad , \quad h(t) = \sum_\sigma u^\sigma(t) h_\sigma(t) \ ,$$

$$c^2(t) = \sum_\alpha \| h_\alpha(t) \|^2 \quad , \quad a_0 = \eta\mathcal{E}(j\mathbf{u}) \ .$$

It will be important below to include the initial term. We rewrite (9.3) as follows

$$(9.5) \qquad \| A(t) \|^2 = \| a_0 \|^2 + 2 \Re \int_0^t < A(s), h(s) > ds + \int_0^t c^2(s) \, ds .$$

One could apply the same method as for (6.7), and this was initially the road taken by most authors. It turns out that an inequality due to the Delhi school is less cumbersome and more efficient. We introduce the measure $\nu(\mathbf{u}, dt) = (1 + \| \mathbf{u}(t) \|^2) \, dt$ (this last norm involving only the components u_α), so that for instance

$$(9.6) \qquad \int_0^t \| h_\sigma(s) \|^2 \, ds \leq \sum_\rho \int_0^t \| h_\sigma^\rho(s) \|^2 \nu(\mathbf{u}, ds) .$$

Then the basic Mohari–Sinha inequality is, putting $\nu(\mathbf{u}, t) = \int_0^t \nu(\mathbf{u}, ds)$ and including the initial term a_0

$$(9.7) \qquad \boxed{ \| A(t) \|^2 \leq e^{\nu(\mathbf{u},t)} \left(\| a_0 \|^2 + 2 \sum_{\rho\sigma} \int_0^t \| h_\sigma^\rho(s) \|^2 \nu(\mathbf{u}, ds) \right) . }$$

In the case of infinite multiplicity, *the summation over ρ concerns only the finitely many indices in the "support" $S(\mathbf{u})$*. Inequality (9.7) is deduced as follows from (9.5). The scalar product $2 \Re < A, h >$ is dominated by

$$2 \| A \| \| \sum_\sigma u^\sigma h_\sigma \| \leq 2 \| A \| \sum_\sigma | u^\sigma | \| h_\sigma \| ,$$

and then the inequality $2xy \leq x^2 + y^2$ is used to produce $\| A \|^2 (\sum_\sigma | u^\sigma |^2) + \sum_\sigma \| h_\sigma \|^2$. That is,

$$\| A(t) \|^2 \leq \| a_0 \|^2 + \int_0^t \| A(s) \|^2 \nu(\mathbf{u}, ds) + 2 \int_0^t \sum_{\rho\sigma} \| h_\sigma^\rho(s) \|^2 \nu(\mathbf{u}, ds) .$$

Then (9.7) follows from the classical Gronwall lemma ([Par1], Prop. 25.5, p. 187).

Iteration. It is necessary, for the theory of stochastic differential equations, to have estimates of the same kind for iterated stochastic integrals. We use abbreviated notation as follows. First of all, one single letter ε_i may denote an Evans pair $\binom{\rho_i}{\sigma_i}$, and one single letter μ may denote a n-tuple $(\varepsilon_1, \ldots, \varepsilon_n)$ (the length n of μ being denoted by $|\mu|$). We consider a family of *initial* operators $H_\mu(s_1, \ldots, s_n)$, and put

$$I_\mu(t, H) = \int_{\{t > s_1 > \ldots > s_n\}} H_\mu(s_1, \ldots, s_n) \, da_{s_1}^{\varepsilon_1} \ldots da_{s_n}^{\varepsilon_n} .$$

(integration on the decreasing simplex makes the proof by induction easier). If an index ε_i appears as $\binom{\rho_i}{\sigma_i}$ in H_μ, it appears as $\binom{\sigma_i}{\rho_i}$ in the differential element to keep the balance. We will need an estimate of a sum of iterated integrals

$$(9.8) \qquad \| \sum_\mu I_\mu(t, H) \mathcal{E}(j\mathbf{u}) \|^2 \leq$$

$$\sum_n (2 e^{\nu(t,\mathbf{u})})^n \sum_{|\mu|=n} \int_{t > s_1 \ldots > s_n} \| H_\mu(s_1, \ldots s_n) \mathcal{E}(j\mathbf{u}) \|^2 \nu(ds_1) \ldots \nu(ds_n) .$$

This formula is a substitute for the isometry property of the standard chaos expansion. It is deduced from (9.7) by an easy induction, isolating the largest time s_1 — the initial term in (9.7) plays an essential role here.

Let us give another useful estimate of an individual iterated integral

$$ I_t^\mu = \int_{s_1 < \ldots < s_n < t} da_{s_n}^{\varepsilon_n} \ldots da_{s_1}^{\varepsilon_1} $$

which takes into account the number of times the special index $\varepsilon_0 = \binom{0}{0}$ appears in μ. Let us denote by λ the multi-index of length k obtained by suppressing from μ the special index. Then μ is constructed from λ by inserting p_i times ε_0 after each index ε_i of λ (the value $p_i = 0$ being allowed), and also p_0 times ε_0 in front of λ. We denote by $p = \sum_i p_i = n - k$ the number of times ε_0 appears. Then we have, imitating a result of Ben Arous presented in *Sém. Prob. XXV*, p. 425

$$ (9.9) \qquad \| I_t^\mu \mathcal{E}(j\mathbf{u}) \|^2 \le \| \mathcal{E}(j\mathbf{u}) \|^2 \, \frac{\nu_{\mathbf{u}}(t)^{n+p}}{(n+p)!} \prod_i \frac{(2p_i)!}{(p_i!)^2} \, . $$

Indeed, the corresponding integral is also equal to

$$ \int_{s_1 < \ldots s_k < t} \frac{s_1^{p_0}}{p_0!} \frac{(s_2 - s_1)^{p_1}}{p_1!} \cdots \frac{(t - s_k)^{p_k}}{p_k!} \, da_{s_k}^{\eta_k} \ldots da_{s_1}^{\eta_1} $$

where $\eta_1, \ldots, \eta_k$ are the indexes of λ. Then we apply the estimate (9.8), leading to the product of $\| \mathcal{E}(j\mathbf{u}) \|^2$ by the scalar

$$ \frac{1}{(p_0!)^2 \ldots (p_k!)^2} \int s_1^{2p_0} (s_2 - s_1)^{2p_1} \ldots (t - s_k)^{2p_k} \, \nu(ds_1) \ldots \nu(ds_k) $$

We bound $(s_{i+1} - s_i)^{2p_i}$ by

$$ \nu(\,]\, s_i, s_{i+1}\,])^{2p_i} = (2p_i)! \int_{s_i < u_1^i \ldots < u_{2p_i}^i < s_{i+1}} \nu(du_1) \ldots \nu(du_{2p_i}) $$

and integrating on all the variables together we get the estimate (9.9).

Recall that $p = \sum_i p_i$. It is easy to bound the product on the right side of (9.9). Putting $c(m) = (2m)!/(m!)^2$ we have $c(m+1) \le 4c(m)$, therefore $c(m) \le 4^m$, and the product itself is at most 4^p.

10 Let us discuss the *uniqueness* of a stochastic integral representation (9.1). We assume for simplicity that the space $\mathcal{S}$ of arguments $\mathbf{u}$ of exponential vectors consists of all square integrable functions, with finitely many components different from 0, right continuous with left limits. Assume the stochastic integral $I_t(H)$ is equal to 0 on this exponential domain, for every $t > 0$. According to (9.2), the following expression is equal to 0 for arbitrary exponential vectors $\mathcal{E}(\ell \mathbf{v})$ and $\mathcal{E}(j\mathbf{u})$

$$ (10.1) \qquad < \mathcal{E}(\ell \mathbf{v}), I_t(H) \mathcal{E}(j\mathbf{u}) > = \sum_{\rho, \sigma} \int_0^t < \mathcal{E}(\ell \mathbf{v}), H_\rho^\sigma(s) \, \mathcal{E}(j\mathbf{u}) > v^\rho(s) \, u_\sigma(s) \, ds \, . $$

it follows by differentiation in t that for given $\mathbf{u}, \mathbf{v}$

$$(10.2) \qquad \sum_{\rho\sigma} < \mathcal{E}(\ell\mathbf{v}),\, H_\rho^\sigma(t)\,\mathcal{E}(j\mathbf{u}) > v^\rho(t)\, u_\sigma(t) = 0 \quad \text{for a.e. } t,$$

the exceptional set depending possibly on $\mathbf{u}, \mathbf{v}$. Let us assume for a moment that $H_\rho^\sigma(t)\,\mathcal{E}(j\mathbf{u})$ is right continuous in t. Since $\mathbf{u}, \mathbf{v}$ are right continuous, it follows that (10.2) is true for every t. We may now replace $\mathbf{u}, \mathbf{v}$ by $\mathbf{u}', \mathbf{v}'$, right continuous, equal to $\mathbf{u}, \mathbf{v}$ in the interval $]\,0, t\,[$ but arbitrary (except for the regularity conditions) on $[\,t, \infty\,[$. By adaptation (10.2) becomes

$$(10.3) \qquad \sum_{\rho\sigma} < \mathcal{E}_t(\ell\mathbf{v}),\, H_\rho^\sigma(t)\,\mathcal{E}_t(j\mathbf{u}) > e^{\int_t^\infty < \mathbf{v}_s',\, \mathbf{u}_s' > ds}\, v'^\rho(t)\, u_\sigma'(t) = 0$$

Since the values $v_\rho'(t)$, $u_\sigma'(t)$ are arbitrary (except for $\rho = 0$) and the exponential is strictly positive, we find that the scalar product must be 0. Since $\mathbf{v}$ is arbitrary, we have $H_\rho^\sigma(t)\,\mathcal{E}_t(j\mathbf{u}) = 0$, and by adaptation the same result for $H_\rho^\sigma\,\mathcal{E}(j\mathbf{u})$.

This settles the problem under our hypotheses on $\mathbf{u}$. The above proof requires stability of the arguments of exponential vectors under "piecing together", and would need to be modified to deal, say, with smooth or merely continuous functions. Also, if the operators $H_\sigma^\rho(s)$ are closed, the fact they are 0 on a dense domain implies they are 0 everywhere.

> A minor technical point on this domain enlargement : differences of unbounded closed operators are not necessarily closed, and the proof of uniqueness must take a slightly different form. See Attal [Att1].

To deal with functions which are not right continuous, one uses for the first time a strong measurability assumption on the processes $H_\sigma^\rho(s)$: as strong limits of sequences of step processes, they satisfy *Lusin's property*. Namely there exists a partition of $\mathbb{R}_+$, into a null set and a sequence of disjoint closed sets F_n, on each of which $H_\sigma^\rho(\cdot)$ is continuous. Discarding again a countable set, we may assume F_n is right closed without right isolated points. Then the same reasoning as above can be performed on each F_n, with the conclusion that $H_\sigma^\rho(t) = 0$ for a.e. t.

> If no closedness assumption is made, simple counterexamples to uniqueness can be given. See Lindsay [Lin1] and Attal [Att1].

REMARK. We do not discuss in these notes the general problem of the *representation of operator martingales* by stochastic integrals with respect to the processes $da_\sigma^\rho(t)$ ($da_0^0(t)$ not included), corresponding to the predictable representation property of classical Brownian motion or Poisson processes. The discrete case indicates that a result of this sort could be expected, and one is known to hold in several interesting cases ([HLP], [PaS1]) but a counterexample due to J.L. Journé (*Sém. Prob. XX*, p. 313, see also the note by Parthasarathy in the same volume p. 317) excludes the possibility of a naive general theorem.

Shift operators on Fock space

The following two subsections are not directly related to stochastic calculus, and can be omitted until they are needed. They introduce some additional material necessary for section 3. The reader should probably skip them for the moment, and read the examples of stochastic calculus in subsection 13.

11 There is an "abstract nonsense" part of the theory of classical Markov processes, which investigates the filtration and the shift operators on the canonical sample space. We are going to discuss similar definitions on the Hilbert space $\Psi = \mathcal{J} \otimes \Phi$ (a Fock space of arbitrary multiplicity). Two different shift operators occur in classical probability

$$X_s(\theta_t\omega) = X_{s+t}(\omega) , \quad X_s(\theta_t\omega) = X_{s+t}(\omega) - X_t(\omega) ,$$

the first one being used for Markov processes, and the second one requiring more structure, and being used for processes with independent increments. The shift we investigate here is corresponds to the second one, and acts trivially on initial vectors (for a discussion of this point, see Meyer [Mey3] and the comments by L. Accardi in the same volume). A theory extending the first shift does not seem to exist at the present time.

We refer to subsection 1 for the general notation, the definition of the conditional expectations E_t for vectors (those for operators will be denoted by $\mathbb{E}_t$), the decompositions of Ψ as $\Psi_t \otimes \Phi_{[t}$ and $\Phi_t \otimes \Psi_{[t}$.

The Hilbert space $L^2(\mathbb{R}_+)$ carries shift operators τ_t and τ_t^*, adjoint to each other, and acting as follows (we do not use boldface letters for vector arguments)

$$\tau_t u(s) = u(s-t) \text{ for } s \geq t, \quad = 0 \text{ otherwise} \quad ; \quad \tau_t^* u(s) = u(s+t) .$$

By second quantization of these operators, we construct operators θ_t, θ_t^* on Φ. The *shift operators* θ_t on Φ are isometric, and satisfy the relation

$$(11.2) \qquad \theta_t X_s^\alpha = X_{s+t}^\alpha - X_t^\alpha$$

(here X_t is considered as a vector, not an operator). Tensoring with $I_{\mathcal{J}}$ we define on Ψ a semigroup of isometric shift operators, still denoted by θ_t

$$(11.3) \qquad \theta_t(j \otimes f) = j \otimes \theta_t f .$$

Between the operators E_t and θ_t we have the relation

$$(11.4) \qquad \theta_s E_t = E_{s+t} \theta_s .$$

To check (11.4) it is sufficient to apply both sides to a vector $\varphi = j f_t \theta_t g$ ($\otimes$ signs omitted) with $j \in \mathcal{J}$, $f_t \in \Phi_t$, $g \in \Phi$. In this case

$$\theta_s\varphi = j\theta_s f_t \theta_{s+t}g , \quad E_t\varphi = j f_t<\mathbf{1},g> ,$$
$$\theta_s E_t \varphi = j\theta_s f_t <\mathbf{1},g> = E_{s+t}\theta_s\varphi .$$

On the same vector, the adjoint shift θ_t^* acts by $\theta_t^*(j f_t \theta_t g) = j < \mathbf{1}, f_t > g$.

The shift acts as follows on the representation $f = \int \widehat{f}(A) dX_A$ of a vector ($\widehat{f}$ taking values in the initial space) :

$$(11.5) \qquad \theta_t f = \int \widehat{f}(\tau_t(A)) dX_A , \quad \theta_t^*(f) = \int \widehat{f}(\tau_t^*(A)) dX_A ,$$

where $\tau_t^*(A)$ moves A to the right by t, while $\tau_t(A)$ moves it to the left by t if $A \subset [t, \infty[$, and is undefined otherwise (and then $\widehat{f}(\tau_t(A)) = 0$). Note that exponential martingales $f_t = \mathcal{E}(u I_{]0,t]})$ satisfy a "multiplicative functional" property $f_{s+t} = f_s \theta_s f_t$.

Since θ_t is isometric, it defines an isomorphism of Ψ onto the closed space $\theta_t \Psi = \Psi_{[t}$, the *future space* at time t.

Our purpose is now to define a shift $\Theta_t(A)$ for operators on Ψ, and we consider mostly bounded operators. The first idea is to take $\Theta_t(A) = \theta_t A \theta_t^*$, but we use rather Journé's definition in [Jou] which yields an operator "adapted to the future" (J.'s notation for $\Theta_t(A)$ is $\overline{\theta_t A \theta_t^*}$) :

$$(11.6) \qquad \Theta_t(A)(j\, f_t\, \theta_t g) = f_t \otimes \theta_t A(j\, g) \ .$$

The notation is the same as after (11.4), and we have kept the $\otimes$ sign on the right as the decomposition $\Psi = \Phi_t \otimes \Psi_{[t}$ is somewhat unusual. On the same vector, $\theta_t A \theta_t^*$ would produce $< \mathbf{1}, f_t > \theta_t A(j\,g)$, and would not be unitary for unitary A. As an example, we have the "additive functional" equality

$$a_{t+s}^\varepsilon - a_s^\varepsilon = \Theta_s(a_t^\varepsilon) \quad ;$$

these operators are unbounded, and the equality holds on the exponential domain.

Let us return to bounded operators (for additional results, see Bradshaw [Bra]) :

LEMMA 1. *We have*

$$\Theta_t(I) = I \quad ; \quad \Theta_t(AB) = \Theta_t(A)\Theta_t(B) \quad ; \quad \Theta_t(A^*) = \Theta_t(A)^* \quad ;$$
$$\Theta_s \Theta_t = \Theta_{s+t} \ , \ \Theta_s \mathbb{E}_t = \mathbb{E}_{s+t} \Theta_s \quad ;$$
$$\Theta_t(A)\, \theta_t = \theta_t A \ , \quad \theta_t^* \Theta_t(A) = A \theta_t^* \ .$$

The first line is nearly obvious, except the last equality, for which we compute

$$< j\, f_t \theta_t g, \Theta_t(A)\ell h_t \theta_t k > \ = \ < j\, f_t \theta_t g, h_t \otimes \theta_t A(\ell k) > \ = \ < f_t, h_t > < \theta_t g, \theta_t A(\ell k) >$$

In the last scalar product we remove θ_t because of the isometry property, take A to the left side, and reverse our steps. The relation $\Theta_s(\Theta_t(A)) = \Theta_{s+t}(A)$ is proved similarly on a vector of the form $j\, f_s\, \theta_s(h_t \theta_t k)$, and the same vectors are used to check the next relation. Finally, we have $\Theta_t(A)(\theta_t(jg)) = \theta_t A(jg)$.

If A is s-adapted, it is easy to check that $\Theta_t(A)$ is $(s+t)$-adapted .

12 We now reach an important definition : a *left cocycle* (or in probabilistic language a *multiplicative functional*) is an adapted process (U_t) of operators — here, bounded for simplicity — such that $U_0 = I$ and we have for all s, t

$$(12.1) \qquad U_{t+s} = U_t \Theta_t(U_s) \ .$$

The adapted process U_t^* then is a *right cocycle*, V_t^* standing to the right in the product. This is but one of the possible definitions of cocycles : that of Accardi [Acc1], for instance, is slightly different. A fundamental example of cocycles will be provided in the next section by the solutions of stochastic differential equations with time-independent coefficients.

The left cocycle property and the last property of Lemma 1 imply together

$$(12.2) \qquad\qquad U_{t+s}\theta_t = U_t\,\theta_t U_s \ ,$$

which may be called a *weak cocycle property*. Given a left cocycle (U_t) — consisting of bounded operators for simplicity — let us define on the initial space

$$(12.3) \qquad\qquad P_t = \mathbb{E}_0(U_t) \ \text{ or } P_t j = E_0 U_t j \text{ for } j \in \mathcal{J}.$$

otherwise stated, $P_t = E_0 U_t E_0$. Let us check the semigroup property $P_{t+s} = P_t P_s$. Since θ_t acts trivially on initial vectors, we may replace j by $\theta_t j$ in (12.3). Then using the weak cocycle property (12.2) we have

$$(12.4) \qquad\qquad P_{t+s} = E_0 U_{t+s}\theta_t\theta_s = E_0 U_t(\theta_t U_s\theta_s) \ .$$

On the other hand, using (11.4)

$$(12.5) \qquad P_t P_s = E_0 U_t \theta_t(E_0 U_s \theta_s) = E_0 U_t(\theta_t E_0) U_s \theta_s = E_0 U_t E_t(\theta_t U_s \theta_s) \ .$$

Comparing (12.4)-(12.5) we are reduced to proving that $E_0 U_t = E_0 U_t E_t$, and it is better if we replace E_0 by E_t. Then using the t–adaptation of U_t, when either side is applied to a vector $j\,f_t\theta_t g$ one gets the same result $U_t(j\,f_t)<\mathbf{1}, g>$. The adjoint operators $P_t^* = \mathbb{E}_0(U_t^*)$ also form a semigroup, associated with the *right cocycle* (U_t^*).

We prove now a similar result for operators ([Acc1], [HIP]). We define as follows a mapping from the algebra of bounded operators on $\mathcal{J}$ (identified to 0–adapted operators on Ψ) into itself

$$(12.4) \qquad\qquad \mathcal{P}_t(A) = \mathbb{E}_0(U_t^* A U_t) \ ,$$

where $\mathbb{E}_0$ is the time 0 conditional expectation for operators. More generally, given a bounded operator H on Ψ, put $\widetilde{H}(A) = \mathbb{E}_0(H^* A H)$, so that $\mathcal{P}_t = \widetilde{U_t}$. Then the semigroup property $\mathcal{P}_{s+t} = \mathcal{P}_s \mathcal{P}_t$ is a consequence of

LEMMA 2. *If H is s–adapted, K arbitrary and $L = H\,\Theta_s K$ we have $\widetilde{L}(A) = \widetilde{K}(\widetilde{H}(A))$.*

When $H = |h\otimes x> <h'\otimes x'|\otimes I_{[s}, (h, h'\in \mathcal{J}, x, x'\in \Phi_s)$ and $K = |k\otimes y> <k'\otimes y'|$ ($k, k'\in \mathcal{J}, y, y'\in \Phi$), the proof is computational (we omit it). Then the result is extended to operators of finite rank by linearity, and to arbitrary bounded operators by a strong convergence argument (Kaplansky's density theorem is the adequate functional analytic tool). The second part of the lemma leads to a second semigroup $\mathcal{P}_t'(A) = \mathbb{E}_0(U_t A U_t^*)$.

A few examples

13 Many examples of processes with interesting stochastic integral representations appear in [HP1], and still more are given in [Par1]. The standard way to find the coefficients is to compute matrix elements between enponential vectors and to use (6.6) or its multiple Fock space version (8.4).

1) Let us deal first with simple Fock space, and consider the process $X_t = \exp(p a_t^+ + q a_t^- + rt)$ (p, q, r being three complex constants). The exponential can be

"disentangled" using the elementary Campbell–Hausdorff formula of Chapter III, §1 subs.3 (or Chapter IV, §3, subs. 11 which would apply to the number operator). We have

$$X_t = e^{pa_t^+} e^{qa_t^-} e^{(r+\frac{1}{2}pq)\,t}$$

from which it is easy to compute

$$<\mathcal{E}(v),\, X_t\mathcal{E}(u)> \;=\; e^{(r+\frac{1}{2}pq)t}\, e^{p\int_0^t u(s)\,ds}\, e^{q\int_0^t \overline{v}(s)\,ds}\, e^{<v,u>}\;,$$

and then, differentiating with respect to t,

$$\frac{d}{dt}<\mathcal{E}(v),\, X_t\mathcal{E}(u)> \;=\; (pu(t) + q\overline{v}(t) + (r + \tfrac{1}{2}pq))<\mathcal{E}(v),\, X_t\mathcal{E}(u)>\,.$$

Using (6.6), we find that X_t solves the stochastic differential equation

$$(13.1) \qquad\qquad X_t = I + \int_0^t X_s\,(\,pda_s^+ + qda_s^- + (r + \tfrac{1}{2}pq)\,ds\,)\,.$$

This result can be extended to multiple Fock space. For $q = -\overline{p}$, $r = ih$ purely imaginary, the solution of (13.1) is a unitary process.

If $r = 0$, what we computed is $f(a_t^+, a_t^-)$ when $f(x,y) = e^{px+qy}$. Taking $p = iu$, $q = iv$ and integrating we can handle more general Fourier transforms, as in the Weyl quantization procedure of Chapter III. Now it turns out that the same Ito formula as in the classical case remains correct

$$(13.2) \qquad df(a_t^+, da_t^-) = D_x f(a_t^+, a_t^-)\,da_t^+ + D_y f(a_t^+, a_t^-)\,da_t^- +$$
$$\tfrac{1}{2}(D_{xx}f(a_t^+, a_t^-)\,da_t^+\,da_t^+ + D_{xy}f(a_t^+, a_t^-)(\,da_t^+\,da_t^- + da_t^-\,da_t^+) + D_{yy}f(a_t^+, a_t^-)\,da_t^-\,da_t^-)\,.$$

when the value of the "square brackets" is read from the Ito table. It would be interesting to see what happens when the number operator is included.

2) Let 1_t be the indicator of $[0,t[$ and $1'_t$ that of $[t,\infty[$. Given an operator U on the multiplicity space $\mathcal{K}$, we may extend it in a trivial way (and without changing notation) to $\mathcal{H} = L^2(\mathbb{R}_+, \mathcal{K}) = L^2(\mathbb{R}_+) \otimes \mathcal{K}$ and consider the following process of second quantized operators

$$(13.3) \qquad\qquad Y_t = \Gamma(1_t U + 1'_t I)$$

— since $\Gamma(U)$ acts as U^n on the n-th chaos $\mathcal{H}^{\circ n}$, Y_t could be denoted U^{N_t} as Belavkin does, N_t being the total number operator up to time t. Then it is easy to compute the matrix element of Y_t between $\mathcal{E}(v)$ and $\mathcal{E}(u)$, and to deduce from (6.6) that

$$(13.4) \qquad\qquad Y_t = I + \int_0^t Y_s\,(U - I)\,dN_s\,.$$

An important special case is the Jordan-Wigner operator $J_t = (-1)^{N_t}$ on simple Fock space, corresponding to $U = -I$. See [Par1] p. 198–201 for a study of the Jordan-Wigner transform operator using quantum stochastic calculus.

§2. STOCHASTIC CALCULUS WITH KERNELS

This short section may be omitted at a first reading, since it slightly deviates from the main line of this chapter.

1 We begin in the case of multiplicity 1, with a trivial initial space for simplicity. Let (f_t) be a (square integrable) adapted process, and let $F = \int f_s\, dX_s$. We will use below the following relation on Wiener space, an easy exercise on stochastic calculus (or a triviality if F belongs to one single Wiener chaos)

$$(1.1) \qquad \mathbb{E}\left[\mathcal{E}(h)\,F\right] = \int_0^\infty h_s\, \mathbb{E}\left[\mathcal{E}(h)f_s\right]\, ds \ .$$

On the other hand, knowing the chaotic expansion of every f_s, that of F is given by formula (3.3) of §1

$$(1.2) \qquad F(A) = f_{\vee A}(A-) \ ,$$

with the following notation : for any non–empty finite subset A, $\vee A$ is the last element of A, and $A-$ is $A \setminus \{\vee A\}$. If A is empty, these symbols are undefined, and scalars depending on undefined symbols are equal to 0 by convention. To convince yourself of this trivial formula, look at what it means when f_t belongs to a chaos of a given order.

Formula (1.2) may also be written as follows

$$(1.2') \qquad F(A) = \sum_{s \in A} f_s(A - s) \ .$$

Indeed, since f_s is s–adapted, $f_s(A - s)$ vanishes unless $A - s$ is contained in $[0, s]$, *i.e.* unless s is the last element of A. Thus $(1.2')$ is a natural extension of the stochastic integral to non–adapted processes, called the *Skorohod integral,* whose study has become very fashionable.

2 Consider now an adapted operator process (K_t) on simple Fock space, given by a (measurable) family of Maassen kernels

$$K_t = \int K_t(A, B, C)\, da_A^+ \, da_B^\circ \, da_C^- \ .$$

We are going to show that (under appropriate restrictions on the size of the kernels K_t) the stochastic integral $L^\varepsilon = I_\infty^\varepsilon(K)$ is associated with a kernel, given by a formula similar to (1.2)

$$(2.1) \qquad L^+(A,B,C) = K_{\vee A}(A-,B,C) \ ; \quad L^\circ(A,B,C) = K_{\vee B}(A,B-,C) \ ;$$
$$L^-(A,B,C) = K_{\vee C}(A,B,C-) \ .$$

We already used this formula in Chapter IV, §4, formula (7.2). A convenient way to prove it is to show that, taking (2.1) as the definition of L^ε, the scalar product

$$(2.2) \qquad < \mathcal{E}(v),\, L^\varepsilon \mathcal{E}(u) > \ ,$$

where u and v belong to L^2 and are bounded, has the correct value, (6.6), (7.1) or (7.6) according to the choice of ε. We choose $\varepsilon = \circ$ (the other cases are similar), in which case the value of (2.2) should be

$$(2.3) \qquad \int \bar{v}(s) < \mathcal{E}(v),\, K_s \mathcal{E}(u) > u(s)\, ds \ .$$

The integral (2.2) can be computed by formula (5.2) in Chapter IV, §4,

$$< b,\, Ka > = \int \bar{b}(U+V)\, K(U,V,W)\, a(V+W)\, dU\, dV\, dW \ ,$$

with $b = \mathcal{E}(v)$, $a = \mathcal{E}(u)$, and since the transformation (2.1) affects only the middle variable, we may "freeze" U and W, put

$$f_s(V) = K_s(U,V,W) \ , \quad F(V) = K_{VV}(U,V-,W)$$

Interpreting now F as a random variable, stochastic integral of an adapted process (f_s), this is nothing but formula (1.1) applied with $h = \bar{v}u$. This proof is not complete, as we did not make the analytical hypotheses required to apply (1.1) or (2.3), but there is no problem if the kernels $H_t(U,V,W)$ are 0 for $t > T$ and satisfy a Maassen–like inequality uniform in $t \le T$

$$(2.4) \qquad |K_t(U,V,W)| \le C\, M^{|U|+|V|+|W|} \ ,$$

in which case the stochastic integrals $I_\infty^\varepsilon(K)$ given by (2.1) satisfy similar properties. Thus we can apply Maassen's theory, and such stochastic integrals define operators which have adjoints of the same type, and can be composed. Then in relations like (6.5), (7.2), (7.3), (7.7) we can move the two operators on the same side, and prove true integration by parts formulas.

We discuss now the situation of a Fock space of finite multiplicity, with a trivial initial space (the general case would not be more difficult, but would require $\mathcal{L}(\mathcal{J})$-valued kernels, see Chapter V, §2 subs. 5). We are going to use the general kernel calculus of Chap. V, §2, subs. 4–5, allowing the use of the differential $da_0^0(t) = dt$. The advantage of the kernel point of view is the possibility of directly composing operators. We are loose in distinguishing kernels from the operators they define, and in particular we use the same letter for both.

We say that a family (K_t) of kernels is a *regular process* if 1) Each K_t is t-adapted, *i.e.* vanishes unless all its arguments are subsets of $[0,t]$, 2) the kernels K_t depend measurably on t, and are regular in the sense of Maassen, with uniform regularity constants on compact intervals, as in (2.5).

Stochastic integration of regular processes preserves regularity : let us say that a process of operators U_t is *smooth* if it can be represented as

$$(2.6) \qquad U_t = U_0 + \sum_{\rho,\sigma} \int_0^t K_\rho^\sigma(s)\, da_\sigma^\rho(s) \ ,$$

where each process $K_\rho^\sigma(s)$ is regular (and U_0 is a constant initial operator).

Then a smooth process is again regular. Indeed, a stochastic integral process $V_t = \int_0^t K_s \, da_\mu^\lambda(s)$ has a kernel given by

(2.7) $$V_t((A_\sigma^\rho)) = K_{\vee A_\mu^\lambda}(A_0^0, (A_0^\alpha), \ldots, (A_\mu^\lambda) \ldots, (A_\alpha^0))$$

if $\cup A_\sigma^\rho$ is contained in $[0, t]$, and 0 otherwise (see formula (2.1)). Then regularity is easy to prove.

Using Maassen's theorem, one can check also that the composition of two regular (smooth) processes of operators is again regular (smooth), and that in the smooth case, there is a rigorous Ito formula for the computation of the product. Similarly, one can define regular (smooth) processes of vectors, and check that applying a smooth process of operators to a smooth process of vectors yields a smooth process of vectors, whose stochastic integral representation is given by the natural Ito formula. We leave to the reader all details, and the case of a non–trivial initial space.

Another approach to stochastic integration

3 Belavkin [1] (and independently Lindsay [2]) give a definition of stochastic integration which unifies the point of view of Hudson–Parthasarathy and that of Maassen. It also applies to non–adapted processes, though it is not clear whether such processes will ever play an important role. We present the main idea first in the case of simple Fock space Φ.

Let us first recall some classical results, popularized among probabilists by the "Malliavin Calculus". We refer to Chapter XXI of [DMM] for details. Given an element of Φ, $f = \int \widehat{f}(A) \, dX_A$ and $s > 0$, we define formally $\dot{f}_s \in \Phi$ by its chaos expansion

(3.1) $$\dot{f}_s = \int \widehat{f}(s + A) \, dX_A \ .$$

This is not rigorously defined since $\widehat{f}(\cdot)$ is a class of functions on $\mathcal{P}$, but one can prove it is defined for a.a. s, and $\int |\dot{f}_s|^2 \, ds$ is finite if and only if

(3.2) $$\int |\widehat{f}(s + A)|^2 \, ds dA = \int |A| \, |\widehat{f}(A)|^2 \, dA < \infty \ ,$$

and we then have

$$a_{<h|}^-(f) = \int \overline{h}(s) \, \dot{f}_s \, ds \ ,$$

which defines $\dot{f}_s$ for a.e. s. The mapping $f \to \dot{f}_s$ appears in white noise theory as a partial derivative in a continuous coordinate system. Exponentials $f = \mathcal{E}(u)$ are eigenvectors of these "derivatives" since we have $\dot{f}_s = u(s) f$.

On the other hand, given a family f_s of random variables (satisfying suitable measurability and integrability conditions) and its Skorohod integral F (see (1.2′)), we have a duality formula

(3.3) $$<g, F> = \int \overline{g}(A) \sum_{s \in A} f_s(A - s) \, dA = \int \overline{g}(s + A) \, f_s(A) \, ds dA = \int <\dot{g}_s, f_s> \, ds \ .$$

This is a special case of a general formula involving iterated gradients and divergences ([DMM], XXI. 28).

We consider a measurable family of kernels $K_t(A, B, C)$, not necessarily adapted. For the moment we perform algebraic computations, without caring for analytical details. In analogy with the definition $(1.2')$ of the Skorohod integral, we define the stochastic integrals of $K^\varepsilon = \int_0^\infty K_s \, da_s^\varepsilon$ for $\varepsilon = +, \circ, -$

$$K^+(A, B, C) = \sum_{s \in A} K_s(A - s, B, C), \quad K^\circ(A, B, C) = \sum_{s \in B} K_s(A, B - s, C),$$

(3.4)
$$K^-(A, B, C) = \sum_{s \in C} K_s(A, B, C - s).$$

If the kernel process is adapted, we get the usual definition. To integrate over $(0, t)$, one sums only over $s \le t$. Given a vector $f = \int f(H) \, dX_H$, we compute $K^\varepsilon f(A)$ using Maassen's formula VI.4, $(1.5')$, and we remark that

$$K^+ f(A) = \sum_{s \in A} (K_s f)(A - s), \quad K^- f(A) = \int_0^\infty (K_s \dot{f}_s)(A) \, ds$$

(3.5)
$$K^\circ f(A) = \sum_{s \in A} (K_s \dot{f}_s)(A - s),$$

where, on the right side, we have K_s acting as an operator on the vector f or $\dot{f}_s$. *This no longer requires the operators K_t being given by kernels.*

A remarkable feature of this definition is the elegant way it allows to write the Ito formula. Consider a stochastic integral

(3.6)
$$I_t(\mathbb{H}) = \int_0^t h(s) \, ds + \eta(s) \, da_s^+ + \mathrm{H}(s) \, da_s^\circ + \eta'(s) \, da_s^-$$

(the letter H on the right side is meant to be a capital η !). Then the scalar product $< g, I_t(\mathbb{H}) f >$ is equal to the sum of four integrals $\int_0^t \ldots ds$ with integrands

$$< g, h_s f_s >, \quad < \dot{g}_s, \eta(s) f_s >, \quad < \dot{g}_s, \mathrm{H}(s) \dot{f}_s >, \quad < g, \eta'(s) \dot{f}_s >.$$

We may write this sum as

$$< \begin{pmatrix} g \\ \dot{g} \end{pmatrix}, \begin{pmatrix} h & \eta' \\ \eta & \mathrm{H} \end{pmatrix} \begin{pmatrix} f \\ \dot{f} \end{pmatrix} > .$$

This can be extended to multiple Fock spaces as follows. First of all, in (3.6) $h(s)$ becomes an operator on Ψ, $\eta(s)$ a column $h_\alpha^0(s)$ of operators on Ψ, i.e. a mapping from Ψ to $\mathcal{K} \otimes \Psi$, H is a matrix $H_\alpha^\beta(s)$ of operators on Ψ, i.e. an operator on $\mathcal{K} \otimes \Psi$, and finally $\eta'(s)$ is a mapping from $\mathcal{K} \otimes \Psi$ to Ψ. Putting everything together, and defining $\hat{\mathcal{K}} = \mathbb{C} \oplus \mathcal{K}$, the interpretation of $\mathbb{H}$ is a mapping from $\hat{\mathcal{K}} \otimes \Psi$ to itself, as described by the matrix (1.2) in Chapter V, §3. On the other hand, given $f \in \Psi$, $\dot{f}_s$

now has become a vector in $\mathcal{K} \otimes \Psi$ with components $\overset{\bullet}{f}_\alpha(s)$, and we add one component $\overset{\bullet}{f}_0(s) = f_s$ to the vector $\overset{\bullet}{f}_s$. Then we have the remarkable formula

$$(3.7) \qquad < g, I_t(\mathbb{H}) f > = \int_0^t < \overset{\bullet}{g}_s , \mathbb{H}_s \overset{\bullet}{f}_s > ds$$

If $f = \mathcal{E}(j\mathbf{u})$, $g = \mathcal{E}(\ell\mathbf{v})$, what we get is formula (9.2) in §1. The Ito formula itself requires a second stochastic integral $I_t(\mathbb{K})$, and the integrand becomes (in the adapted case)

$$(3.8) \qquad < I_s(\mathbb{K}) \overset{\bullet}{g}_s, \mathbb{H}_s \overset{\bullet}{f}_s > + < \mathbb{K}_s \overset{\bullet}{g}_s, I_s(\mathbb{H}) \overset{\bullet}{f}_s > + < \mathbb{K}_s \overset{\bullet}{g}_s, \mathbb{H}_s \overset{\bullet}{f}_s > ,$$

a considerable progress over the cumbersome (8.5) in §1.

We have not discussed the conditions under which these formulas can be rigorously applied outside of the exponential domain.

§3. STOCHASTIC DIFFERENTIAL EQUATIONS AND FLOWS

This section is a general survey of the subject, which leaves aside most technicalities, but contains the main ideas. We postpone to section 4 the rigorous proof of some selected results on this rapidly growing subject.

1   The general scheme for stochastic differential equations is the following. We consider a family of operators $L_\varepsilon = L_\sigma^\rho$ on the initial space, and we wish to solve an equation of the following form — ε abbreviates an Evans index $\binom{\sigma}{\rho}$, and we usually write L_ε instead of $L_\varepsilon \otimes I$

$$(1.1) \qquad U_t = I + \sum_\varepsilon \int_0^t U_s (L_\varepsilon \otimes I) \, da_s^\varepsilon .$$

The solution (U_t) should be an adapted process of operators on Ψ, each U_t being defined at least on the subspace generated by all exponential vectors $\mathcal{E}(j\mathbf{u})$, where u is square integrable and *locally bounded* to avoid the difficulties relative to number and exchange operators, and j belongs to some dense domain in $\mathcal{J}$. Of course, regularity conditions will be assumed so that the equation is meaningful. We are specially interested in the study of s. d. e.'s (1.1) whose solution (U_t) is a *strongly continuous unitary process*. Equation (1.1) will be called the *left* s.d.e. with coefficients L_ε. Its solution will be called the *left exponential* of the process $\sum_\varepsilon L_\varepsilon a_t^\varepsilon$, in analogy with the classical "Doléans exponential" of a semimartingale, a very special case of the theory of classical s.d.e.'s — the analogue of non linear s.d.e.'s being given by Evans-Hudson flows (subs. 5).

One may also consider time dependent coefficients $L_\varepsilon(s)$, either acting on the initial space, or acting in an adapted way on Fock space. The theory is similar, but the solutions are called *multiplicative integrals* instead of exponentials. Most of our attention will be devoted to the time independent case.

There is a *right equation* similar to (1.1),

$$(1.1') \qquad V_t = I + \sum_\epsilon \int_0^t (L_\epsilon \otimes I)\, V_s \, da_s^\epsilon \, ,$$

in which the unknown process stands to the right of the coefficient $L_\epsilon \otimes I$. Its solution is called a right exponential. The notations (U_t) and (V_t) will be systematically used to distinguish them. Note that V_s commutes with da_s^ϵ in (1.1′).

In both equations, a basis-free notation may be used : restricting ourselves to time-independent coefficients, we may write

$$(1.2) \qquad \sum_\epsilon L_\epsilon \, da_t^\epsilon = L\, dt + da_t^+(\lambda) + da_t^0(\Lambda) + da_t^-(\tilde\lambda) \, ,$$

where $\lambda = (L_\alpha^0)$ is the *column* ("warning" in V.1.1) of operators on $\mathcal{J}$ consisting of the creation coefficients, $\tilde\lambda$ is a line (L_0^α). Under the best boundedness conditions, $L, \lambda, \tilde\lambda$ and Λ are interpreted as bounded mappings from $\mathcal{J}$ to $\mathcal{J}$, $\mathcal{J}$ to $\mathcal{J} \otimes \mathcal{K}$, $\mathcal{J} \otimes \mathcal{K}$ to $\mathcal{J}$, and $\mathcal{J} \otimes \mathcal{K}$ to $\mathcal{J} \otimes \mathcal{K}$ respectively.

Formally, the adjoint U_t^* of the solution to the left equation solves the right equation with coefficients $\overset{*}{L}{}_\sigma^\rho = (L_\rho^\sigma)^*$. However, the two equations are not handled in the same way, and sometimes one of the two forms is more convenient than the other. For instance, if the solution to the left equation can be proved to be bounded, but the coefficients are unbounded, $U_s L_\sigma^\rho$ raises less domain problems than $L_\sigma^\rho V_s$. We will also prove that the two equations are related by *time reversal*.

The discrete analogues of both equations are given by inductive schemes

$$U_{n+1} - U_n = U_n \Delta M_n \quad ; \quad V_{n+1} - V_n = (\Delta M_n)\, V_n \, ,$$

where (M_n) is a given operator process. It is easy to compute explicitly

$$U_n = U_0 (I + \Delta M_0) \ldots (I + \Delta M_{n-1}) \quad ; \quad V_n = (I + \Delta M_{n-1}) \ldots (I + \Delta M_0)\, V_0 \, .$$

In the first induction procedure, the last operator must go down to the bottom of the pile; thus it is more difficult to compute than the second one.

Unbounded coefficients L_ϵ do not raise major conceptual problems provided they all act on the same *stable* domain, as well as their adjoints — in another language, we may restrict ourselves to this domain, which is a *prehilbert* initial space. If all coefficients are equal to 0 except $L_0^0 = iH$, with a selfadjoint Hamiltonian H, the equations have the same unitary solution $U_t = e^{itH} = V_t$, and we are facing a standard quantum evolution problem (with all its difficulties if H is unbounded!). Thus what we are doing is to perturb a quantum evolution by non–commutative noise terms. From the physicist's point of view, the interesting object is the initial Hilbert space $\mathcal{J}$: the standard way of getting back to $\mathcal{J}$ is to apply the (vacuum) expectation operator E_0. Then if we put

$$(1.3) \qquad P_t = E_0 U_t \quad \text{and} \quad Q_t = E_0 V_t,$$

these operators satisfy formally the equations

$$P_t = I + \int_0^t P_s L_0^0 \, ds \quad , \quad Q_t = I + \int_0^t L_0^0 Q_s \, ds \; ,$$

therefore, we should have $P_t = e^{tL_0^0} = Q_t$. This explains why we assume in section 4 that L_0^0 is the generator of a strongly continuous contraction semigroup on $\mathcal{J}$.

It seems, however, that "Schrödinger's picture" (1.3) is less important than "Heisenberg's picture" describing an evolution of initial operators (§1 subs. 12)

$$(1.4) \qquad \mathcal{P}_t(A) = \mathbb{E}_0\,(U_t^* A U_t) \quad , \quad \mathcal{Q}_t(A) = \mathbb{E}_0\,(V_t^* A V_t) \; .$$

If we consider (U_t) as a multiplicative functional of the quantum noise, these formulas are analogues of the Feynman–Kac or Girsanov formulas of classical probability.

Our discussion follows mostly Mohari's work, whose main subject is the study of s.d.e.'s with infinite multiplicity and bounded coefficients. Therefore *unless otherwise stated, the coefficients L_ϵ are bounded and time-independent*, but the multiplicity may be infinite.

2 The classical way of solving an equation like (1.1′) is the Picard iteration method. Let us have a look at the successive approximations it leads to. We start with $V_t^0 = I$. Then the first iteration gives

$$V_t^1 = I + \sum_\epsilon L_\epsilon a_t^\epsilon \; ,$$

The second approximation V_t^2 is given by

$$I + \sum_\eta \int_0^t L_\eta \, da_s^\eta V_s^1 = I + \sum_\eta L_\eta \int_{s<t} da^\eta(s) + \sum_{\epsilon\eta} L_\eta L_\epsilon \int_{s_1 < s_2 < t} da^\eta(s_2) da^\epsilon(s_1) \; .$$

For the left equation we would have $L_\epsilon L_\eta$. It is clear how to continue : given a multi-index $\mu = (\epsilon_1 \ldots, \epsilon_n)$ we define the iterated integral (over the increasing simplex) and the initial operator

$$(2.1) \qquad I_t^\mu = \int_{s_1 < \ldots < s_n < t} da^{\epsilon_n}(s_n) \ldots da^{\epsilon_1}(s_1) \quad ; \quad L_\mu = L_{\epsilon_n} \ldots L_{\epsilon_1} \; ,$$

and the solution of (1.1) appears as the sum of the series

$$(2.2) \qquad V_t = \sum_\mu L_\mu I_t^\mu \; ,$$

the n–th approximation V_t^n being the sum over multi-indexes of length smaller than n. In the case of a non–random evolution, this is the exponential series for $\exp(L_0^0)$.

The formula for the left exponential is the same, but in the definition of L_μ the product is reversed, the operators corresponding to the larger times being applied first to the vector.

Formula (2.2) is not satisfactory, as it contains many integrals of the same type, due to the special role of the Evans index $\epsilon_0 = \binom{0}{0}$. The following computation is formal, but it can be justified in many cases and has important consequences.

Given a multi-index $\lambda = (\varepsilon_1 \ldots \varepsilon_k)$ that does not contain ε_0, one constructs all the multi-indexes μ that are reduced to λ by integration as in §1, subs. 9, inserting p_0 times ε_0 in front of λ, then p_i times after ε_i. Then, abbreviating L_0^0 into L, the sum over all these multi-indexes has the following value

$$\sum_{p_0,\ldots,p_k} \frac{L^{p_k}}{p_k!} L_{\varepsilon_k} \ldots \frac{L^{p_1}}{p_1!} L_{\varepsilon_1} \frac{L^{p_0}}{p_0!} \int_{s_1 < \ldots < s_k < t} (t - s_k)^{p_k} \ldots (s_2 - s_1)^{p_1} s_1^{p_0} da_{s_k}^{\varepsilon_k} \ldots da_{s_1}^{\varepsilon_1}$$

$$(2.3) \qquad = \int_{s_1 < \ldots < s_k < t} e^{(t-s_k)L} L_{\varepsilon_k} \ldots e^{(s_2-s_1)L} L_{\varepsilon_1} e^{s_1 L} da_{s_k}^{\varepsilon_k} \ldots da_{s_1}^{\varepsilon_1} .$$

The first line has the same meaning as (2.2) except that one must be careful about changing the order of summation of series which are only summable by blocks. The second line involves formal exponentials $P_t = e^{tL}$: it is strikingly similar to the *Isobe-Sato* (or Krylov–Veretennikov) formula from the theory of Markov processes ([DMM], Chapter XXI, n° 57).

3 Let us discuss the conditions on the coefficients $L_\sigma^\rho(t)$ that express the unitarity of the operators U_t. The rigorous proof that they are necessary (due to Mohari in the case of infinite multiplicity) rests on the uniqueness of stochastic integral representations. We omit this point here and refer the reader to the book [Par1], p. 227.

We use formula (9.3) from §1 to compute $<U_t \mathcal{E}(\ell \mathbf{v}), U_t \mathcal{E}(j\mathbf{u})>$, which has to be constant if (U_t) is a process of isometries. The derivative of this scalar product is equal to 0 *a.e.* in t. We assume that $\mathbf{u}, \mathbf{v}$ are taken *right* continuous. Then we must have for every t, because of the assumed strong continuity of U_t

$$\sum_{\rho\sigma} <U_t \mathcal{E}(\ell\mathbf{v}), U_t L_\sigma^\rho \mathcal{E}(j\mathbf{u})>v^\sigma(t)u_\rho(t) + <U_t L_\rho^\sigma \mathcal{E}(\ell\mathbf{v}), U_t \mathcal{E}(j\mathbf{u})>\overline{v}_\sigma(t)\overline{u}^\rho(t)+$$

$$+ \sum_{\rho\sigma\lambda\mu} <U_t L_\lambda^\sigma \mathcal{E}(\ell\mathbf{v}), U_t L_\mu^\rho \mathcal{E}(j\mathbf{u})>\overline{v}_\sigma(t)u_\rho(t)\widehat{\delta}^{\lambda\mu} = 0 .$$

We may suppress U_t on both sides inside the scalar products, because of the assumed isometric property. Since the L_ε are initial operators, the scalar product $<\mathcal{E}(\mathbf{v}), \mathcal{E}(\mathbf{u})>$ factorizes, and the following expression must be equal to 0

$$\sum_{\rho\sigma} <\ell, L_\sigma^\rho j >v^\sigma(t)u_\rho(t) + <L_\rho^\sigma \ell, j >\overline{v}_\sigma(t)\overline{u}^\rho(t) + \sum_{\rho\sigma\lambda\mu} <L_\lambda^\sigma \ell, L_\mu^\rho j >\overline{v}_\sigma(t)u_\rho(t)\widehat{\delta}^{\lambda\mu}$$

Since j, ℓ are arbitrary, we get (3.1) below as formal condition for isometry of (U_t). It is the same condition that implies formally (by time reversal, see §4 subs.9) the isometry property of (V_t), and therefore we may call it simply the *isometry condition*

$$(3.1) \qquad L_\sigma^\rho + (L_\rho^\sigma)^* + \sum_\alpha (L_\alpha^\sigma)^* L_\alpha^\rho = 0 \quad \text{or} \quad \boxed{L_\sigma^\rho + \overset{*}{L}_\sigma^\rho + \sum_\alpha \overset{*}{L}_\sigma^\alpha L_\alpha^\rho = 0} .$$

Since the isometry condition is the same for (U_t) and (V_t), the *coisometry condition* follows by a formal passage to the adjoint (*i.e.* replacing L_σ^ρ by $\overset{*}{L}_\sigma^\rho$)

$$(3.1') \qquad \boxed{L_\sigma^\rho + \overset{*}{L}_\sigma^\rho + \sum_\alpha L_\sigma^\alpha \overset{*}{L}_\alpha^\rho = 0} ,$$

and the two conditions (3.1)–(3.1′) together constitute the formal *unitarity condition*. One of the purposes of section 4 consists in proving that, under suitable analytical assumptions, it really does imply the solution is unitary. This is similar to classical non–explosion results for diffusions or Markov chains.

We rewrite the isometry condition as follows in the "basis free" notation of Chapter V, §3. Assuming convenient boundedness properties are satisfied, we may introduce the matrix $\mathbf{L}$ of an operator on $\mathcal{J} \otimes (\mathbb{C} \oplus \mathcal{K})$

$$\mathbf{L} = \begin{pmatrix} \overset{*}{L} + L + \widetilde{\lambda}^*\lambda & \lambda^* + \widetilde{\lambda} + \lambda^*\Lambda \\ \widetilde{\lambda}^* + \lambda + \Lambda^*\lambda & \Lambda^* + \Lambda + \Lambda^*\Lambda \end{pmatrix} ,$$

and the isometry condition means that $\mathbf{L} = 0$ — more generally, we will see in §4, subsection 4, that the condition $\mathbf{L} \leq 0$ is Mohari's *contractivity* condition. Similarly, the coisometry condition can be written $\widetilde{\mathbf{L}} = 0$, taking

$$\widetilde{\mathbf{L}} = \begin{pmatrix} \overset{*}{L} + L + \widetilde{\lambda}\lambda^* & \lambda^* + \widetilde{\lambda} + \widetilde{\lambda}\Lambda^* \\ \widetilde{\lambda}^* + \lambda + \Lambda\widetilde{\lambda} & \Lambda^* + \Lambda + \Lambda\Lambda^* \end{pmatrix} .$$

The unitarity condition $\mathbf{L} = 0 = \widetilde{\mathbf{L}}$ is analyzed as follows. First of all, the lower diagonal relations

$$\Lambda^* + \Lambda + \Lambda^*\Lambda = 0 = \Lambda^* + \Lambda + \Lambda\Lambda^*$$

mean that $W = I + \Lambda$ *is unitary*. Then the lower left relations

$$\widetilde{\lambda}^* + \lambda + \Lambda^*\lambda = 0 = \widetilde{\lambda}^* + \lambda + \Lambda\widetilde{\lambda}$$

can be written

(3.3) $$\widetilde{\lambda}^* + W^*\lambda = 0 = \lambda + W\widetilde{\lambda}^* ,$$

while the upper right relations gives an equivalent result. Finally, the equations

$$\overset{*}{L} + L + \widetilde{\lambda}^*\lambda = 0 = \overset{*}{L} + L + \widetilde{\lambda}\lambda^*$$

can be written

(3.4) $$L = iH - \tfrac{1}{2}\lambda^*\lambda = iH - \tfrac{1}{2}\widetilde{\lambda}\widetilde{\lambda}^* ,$$

where H is selfadjoint (here, bounded). This form of the conditions extends that of [HuP1] for simple Fock space. We refer to this article for a more detailed discussion, and in particular for the use of s.d.e.'s to construct quantum dynamical semigroups with given generator.

Another exercise in translation consists in taking a basis of the initial space instead of the multiplicity space. For simplicity we assume $\mathcal{J} \approx \mathbb{C}^\nu$ is finite dimensional. The left equation takes the following explicit form, indexing the annihilation operators by *bra* vectors instead of elements of $\mathcal{K}'$

$$U_i^j(t) = \sum_k U_k^j(s)\left(L_i^k\, ds + da_s^+(|\lambda_i^k\rangle) + da_s^\circ(\Lambda_i^k) + da_s^-(\langle\widetilde{\lambda}_i^k|) \right)$$

where (L_i^j) is a matrix of scalars, (Λ_i^j) a matrix of operators on $\mathcal{K}$, and (λ_i^j) and $(\tilde{\lambda}_i^j)$ are two matrices of elements of $\mathcal{K}$. Then the unitarity conditions become :

1) $\Lambda_i^j = W_i^j - \delta_i^j I$, where W is unitary; 2) $\tilde{\lambda} = -W^*\lambda$; 3) $L = iH - \frac{1}{2}<\lambda,\lambda>$, this last notation meaning a scalar matrix with coefficients $\sum_k <\lambda_k^j, \lambda_i^k>$.

4 Let us define, for a given (bounded) operator A on $\mathcal{J}$

$$(4.1) \qquad\qquad\qquad \xi_t(A) = U_t^* A U_t \;.$$

Then the Ito formula shows that $A_t = \xi_t(A)$, as an operator process, has a stochastic integral representation

$$(4.2) \qquad\qquad\qquad A_t = A + \sum_{\rho\sigma} \int_0^t \xi_s(\mathbf{L}_\sigma^\rho A)\, da_\rho^\sigma(s) \;,$$

where $\mathbf{L}_\sigma^\rho$ is the linear map in the space of bounded operators

$$(4.3) \qquad\qquad\qquad \boxed{\mathbf{L}_\sigma^\rho(A) = \overset{*}{L}{}_\sigma^\rho A + A L_\sigma^\rho + \sum_\alpha \overset{*}{L}{}_\sigma^\alpha A L_\alpha^\rho} \;.$$

This formula will have an important interpretation in the language of flows (subsection 5).

Under reasonable conditions, the operator process (U_t) is a unitary left cocycle, and therefore the mapping $A \longmapsto \mathcal{P}_t(A) = \mathbb{E}_0(U_t^* A U_t)$ is a semi–group. It is easy to compute its generator from (4.3), at least formally

$$(4.4) \qquad\qquad\qquad \mathbf{L}_0^0(A) = \overset{*}{L}{}_0^0 A + A L_0^0 + \sum_\alpha \overset{*}{L}{}_0^\alpha A L_\alpha^0 \;.$$

On the other hand, the simpler semigroup $P_t j = E_0 U_t j$ acting on $\mathcal{J}$ has the generator L_0^0. Thus $(\mathcal{P}_t)$ is more dependent than (P_t) on the coefficients of the s.d.e., though it does not contain at all the number/exchange coefficients. This phenomenon does not occur with ordinary quantum mechanics, in which the Schrödinger and Heisenberg pictures are completely equivalent.

If the unitarity conditions on the coefficients are taken into account, (4.4) becomes the standard (Lindblad) form of the generators of *quantum dynamical semigroups,* which replace the classical Markov semigroups in the theory of irreversible quantum evolutions. We do not discuss this subject, but our reader may consult Alicki–Lendi [AlL] and Parthasarathy [Par1].

Non–commutative flows

5 According to our program, we remain at a non–rigorous level, to give motivations and an elementary description of the theory of *non–commutative flows,* created by Evans-Hudson. We begin with a translation of the classical theory of stochastic differential equations on $E = \mathbb{R}^\nu$ into a more algebraic language.

We are given a probability space $(\Omega, \mathcal{F}, \mathbb{P})$ with a filtration $(\mathcal{F}_t)$, a d–dimensional Brownian motion (B_t^α) w.r.t. this filtration (we add a deterministic component $B_t^0 = t$).

Our aim is to construct an adapted and continuous stochastic process $X = (X_t)$ taking values in E, such that for $i = 1, \ldots, \nu$ we have (X_0 being given, usually $X_0 = x \in E$)

$$(5.1) \qquad X_t^i = X_0^i + \sum_\rho \int_0^t c_\rho^i(X_s) \, dB_s^\rho \ .$$

We apply a $\mathcal{C}^\infty$ function F to both sides, using the classical Ito formula

$$(5.2) \qquad F(X_t) = F(X_0) + \sum_\rho \int_0^t L_\rho F \circ X_s \, dB_s^\rho \ ,$$

where L_0 is the second order operator $\sum_i c_0^i D_i + (1/2) \sum_{ij\alpha} c_\alpha^i c_\alpha^j D_{ij}$, and L_α is the vector field $\sum_i c_\alpha^i D_i$. As before, indexes labelled ρ may assume the value 0, while those labelled α may not. This formula no longer contains the coordinates x^i, and can be extended immediately to a manifold E.

We now submit (5.2) to an abstraction process : we retain only the fact that $\mathcal{C}^\infty(E)$ is an *algebra,* and that X_s defines a *homomorphic mapping* $F \longmapsto F \circ X_s$ from $\mathcal{C}^\infty(E)$ to random variables on Ω. We also notice that $\mathcal{C}^\infty(E)$ somehow describes the "curved" manifold E, while the driving Brownian motions (B_t^ρ) live in a "flat" space, the s.d.e. itself being a machinery that "rolls up" the flat paths into the manifold. The flat space and the algebra are "coupled" together by the homomorphism X_0, which usually maps F into a constant r.v. $F(x)$.

Making use of the fact that X_t is a homomorphism, the relation $(FG) \circ X_t = (F \circ X_t)(G \circ X_t)$ and the Ito formula for continuous semimartingales lead to the relations

$$L_\alpha(FG) - F L_\alpha(G) - L_\alpha(F) G = 0 \quad (L_\alpha \text{ is a vector field})$$

$$L_0(FG) - F L_0(G) - L_0(F) G = \sum_\alpha L_\alpha(F) L_\alpha(G) \quad (L_0 \text{ is a second order operator})$$

The algebra $\mathcal{C}^\infty$ is replaced now by an arbitrary complex $*$–algebra $\mathcal{A}$. One usually assumes it has a unit I. The "flat" probability space and the driving Brownian motion are replaced by their non-commutative analogues, a boson Fock space Φ of multiplicity d and its basic processes a_ρ^σ. The "coupling" between $\mathcal{A}$ and Φ is realized as follows : we have $\mathcal{A}$ act as a $*$–algebra of *bounded* operators on a Hilbert space $\mathcal{J}$, and construct the familiar space $\Psi = \mathcal{J} \otimes \Phi$; then X_0 is understood as the mapping $F \longmapsto F \otimes I$ from $\mathcal{A}$ to operators on Ψ. To enhance clarity, we try to use lower case letters for elements of $\mathcal{J}$ or Ψ, capital letters for elements of $\mathcal{A}$ or operators on $\mathcal{J}$ or Ψ, boldface letters for the third level (mappings from $\mathcal{A}$ to $\mathcal{A}$, or from operators to operators).

The interpretation of the process X_t itself is *an adapted process of* $*$*–homomorphisms* from $\mathcal{A}$ to $\mathcal{L}(\Psi)$: for every $F \in \mathcal{A}$, $X_t(F)$ is a bounded operator on Ψ which acts only on the first factor of the decomposition $\Psi_t \otimes \Phi_{[t}$. The technical assumption is made that these homomorphisms are *norm contractive,* and we also assume that $X_t(I) = I$ — though there are good reasons from classical probability to study "explosive" flows for which this condition is not satisfied.

To keep the probabilistic flavour, we use the very strange notation $F \circ X_t$ or $F(X_t)$ instead of $X_t(F)$. In particular, $F \circ X_0 = F \otimes I$.

Finally, we give ourselves a family of mappings $\mathbf{L}_\sigma^\rho$ from $\mathcal{A}$ to $\mathcal{A}$ and demand that, for $F \in \mathcal{A}$

$$(5.3) \qquad F \circ X_t = F \circ X_0 + \sum_{\rho\sigma} \int_0^t \mathbf{L}_\sigma^\rho F \circ X_s \, da_\rho^\sigma(s) .$$

The requirement that X_t be a unit preserving *-homomorphism is now translated as the so-called *structure equations* of a non-commutative flow :

$$(5.4) \qquad\qquad\qquad \mathbf{L}_\sigma^\rho(I) = 0 ,$$

$$(5.5) \qquad\qquad\qquad (\mathbf{L}_\rho^\sigma(F))^* = \mathbf{L}_\sigma^\rho(F^*) ,$$

$$(5.6) \qquad \boxed{\mathbf{L}_\sigma^\rho(FG) - F\mathbf{L}_\sigma^\rho(G) - \mathbf{L}_\sigma^\rho(F) G = \sum_\alpha \mathbf{L}_\sigma^\alpha(F)\mathbf{L}_\alpha^\rho(G)} .$$

The third relation is deduced from Ito's formula as in the classical case above. We do not pretend to prove these conditions are *necessary*, only to show they are *natural*, as we did for the s.d.e. unitarity conditions. On the other hand, we shall prove in the next section that these conditions are sufficient to construct a flow, provided the coefficients are sufficiently regular.

EXAMPLES. Consider a non-commutative s.d.e. with coefficients L_σ^ρ, which has a unique unitary solution (U_t). Take for $\mathcal{A}$ the algebra $\mathcal{L}(\mathcal{J})$ of all bounded operators on the initial space, and for $F \in \mathcal{A}$ define $X_t(F) = U_t^*(F \otimes I)U_t$. Then we get a flow whose coefficients are given by (4.3). The corresponding structure equations follow from the coisometry conditions (3.1$'$) on the s.d.e. coefficients, as a somewhat tedious computation shows.

There is another interesting example, which illustrates the relations between commutative and non-commutative probability : *classical finite Markov chains in continuous time can be interpreted as non-commutative flows*, thus giving a precise mathematical content to Kolmogorov's remarks on the analogy between continuous time Markov chains and diffusions. Since we do not want to interrupt the discussion, however, we postpone this elementary example to the end of this section.

Discrete flows

6 The continuous situation we are studying has a simple discrete analogue : denote by $\mathcal{M}$ the algebra of complex $(\nu+1, \nu+1)$ matrices, by a_ρ^σ its standard basis with multiplication table[1]

$$a_\lambda^\mu a_\rho^\sigma = \delta_\lambda^\sigma a_\rho^\mu .$$

Consider an algebra $\mathcal{A}$ with unit, and denote by $\mathcal{M}(\mathcal{A})$ the algebra of square matrices of order $\nu+1$ with entries in $\mathcal{A}$, which is also the tensor product algebra $\mathcal{A} \otimes \mathcal{M}$. Then we put $\mathcal{A}_0 = \mathcal{A}$, $\mathcal{A}_{n+1} = \mathcal{A}_n \otimes \mathcal{M}$. We "clone" a_ρ^σ into $a_\rho^\sigma(n) = I_{\mathcal{A}} \otimes I \ldots \otimes a_\rho^\sigma \otimes I \ldots$

---

[1] Because of the "warning" in Chap.V, §1 subs.1, the multiplication rule for matrix units is slightly different from the usual one.

with the non–trivial matrix in the n–th $\mathcal{M}$ factor. Then a discrete flow is a family of homomorphisms ξ_n from $\mathcal{A}$ to $\mathcal{A}_n$ such that for $F \in \mathcal{A}$

$$(6.1) \qquad \xi_{n+1}(F) = \xi_n(F) \otimes I_{n+1} + \sum\nolimits_{\rho\sigma} \xi_n(\mathbf{L}_\sigma^\rho(F)) \, a_\rho^\sigma(n+1) \,,$$

where the mappings $\mathbf{L}_\sigma^\rho$ from $\mathcal{A}$ to $\mathcal{A}$ are given. As in the continuous case, it is more picturesque to write $F \circ X_n$ instead of $\xi_n(F)$. The construction by induction is obvious, and it is easy to see that multiplicativity of X_n is translated as a "discrete structure equation"

$$(6.2) \qquad \mathbf{L}_\sigma^\rho(FG) - F\mathbf{L}_\sigma^\rho(G) - \mathbf{L}_\sigma^\rho(F)\,G = \sum\nolimits_\tau \mathbf{L}_\sigma^\tau(F)\mathbf{L}_\tau^\rho(G)$$

which differs from the equation we are familiar with by the fact that summation on the right side takes place over all $\tau \geq 0$ instead of $\alpha > 0$. Relation (6.2) means that, if we denote by $\Lambda(F)$ the matrix $(\mathbf{L}_\sigma^\rho(F))$ with entries in $\mathcal{A}$, by FI the matrix with F along the diagonal and 0 elsewhere, then the mapping $F \longmapsto FI + \Lambda(F) = \Sigma(F)$ is a homomorphism from $\mathcal{A}$ to $\mathcal{M}(\mathcal{A})$. A detailed discussion of discrete flows is given by Parthasarathy [Par4], [Par1] Chapter II §18 (p. 111).

The algebra of structure equations

7 We return to continuous time flows, and formulate the structure equation in a different language, introduced by Hudson.

First, we consider a left action and a right action of $F \in \mathcal{A}$ over $G \in \mathcal{A}$ given respectively by FG and GF. The associativity of multiplication expresses that the left and right actions commute, *i.e.* $\mathcal{A}$ is a *bimodule* over $\mathcal{A}$. We extend trivially this bimodule structure to $\mathcal{A}^d$ — elements of $\mathcal{A}^d$ are denoted by $\Phi = (F_\alpha)$, the components carrying *lower* indexes because of our unusual conventions. We define a "hermitian scalar product" on $\mathcal{A}^d$ as follows

$$< \Phi, \Gamma > = \sum\nolimits_\alpha F_\alpha^* G_\alpha \,.$$

We also consider the algebra $\mathcal{M} = \mathcal{M}_d(\mathcal{A})$ of (d,d)–matrices with coefficients in $\mathcal{A}$ — because of the unusual conventions, the matrix product summation bears on the descending diagonal. Then $\mathcal{M}$ is also a bimodule over $\mathcal{A}$, with a left and right action of $F \in \mathcal{A}$ which multiplies every coefficient of the matrix by F to the left or the right (it can also be represented as a left (right) matrix product by the diagonal matrix FI).

We denote by $\Lambda(F)$ the matrix with coefficients $\mathbf{L}_\alpha^\beta(F)$, and by $\Sigma(F)$ the matrix $FI + \Lambda(F)$. For non–zero indexes, the relation

$$(7.1) \qquad \mathbf{L}_\alpha^\beta(FG) - F\mathbf{L}_\alpha^\beta(G) - \mathbf{L}_\alpha^\beta(F)\,G = \sum\nolimits_\alpha \mathbf{L}_\alpha^\gamma(F)\mathbf{L}_\gamma^\beta(G)$$

is read as $\Sigma(FG) = \Sigma(F)\Sigma(G)$. It can be interpreted as follows : on $\mathcal{A}^d$, keep the right action of $F \in \mathcal{A}$ and redefine the left action as the effect of the matrix $\Sigma(F)$. Then we get a *twisted bimodule structure on* $\mathcal{A}^d$.

Denote by $\lambda(F)$ the mapping $F \longmapsto (\mathbf{L}_\alpha^0(F))$ from $\mathcal{A}$ to $\mathcal{A}^d$. Then the relation

$$\mathbf{L}_\alpha^0(FG) - F\mathbf{L}_\alpha^0(G) - \mathbf{L}_\alpha^0(F)\,G = \sum\nolimits_\alpha \mathbf{L}_\alpha^\gamma(F)\mathbf{L}_\gamma^0(G)$$

has the interpretation

$$\lambda(FG) = \Sigma(F)\,\lambda(G) + \lambda(F)\,G \ ,$$

so the vector $(\mathbf{L}_0^\alpha(F)) = \lambda(F)$ satisfies the property

(7.2) $$\lambda(FG) = \lambda(F)G + \Sigma(F)\lambda(G) \ ,$$

which in the twisted bimodule structure is read as

(7.2') $$\lambda(FG) = \lambda(F)G + F\lambda(G) \ .$$

and means that λ *is a (twisted) derivation*. Finally, the relation concerning the *generator* $L(F) = \mathbf{L}_0^0(F)$ of the semigroup (P_t), a mapping from $\mathcal{A}$ to $\mathcal{A}$, can be rewritten as

(7.3) $$L(F^*G) - F^*L(G) - L(F^*)G = \sum_\alpha L_\alpha(F^*)L^\alpha(G) = <\lambda(F), \lambda(G)> \ .$$

This corresponds to a familiar idea in probability, the "squared field operator" (*opérateur carré du champ*).

Evans and Hudson make in their papers (a limited) use of the language of Hochschild cohomology : if $\mathcal{A}$ is an algebra and $\mathcal{B}$ is a bimodule over $\mathcal{A}$, an *n-cochain* with values in $\mathcal{B}$ is by definition a mapping φ from $\mathcal{A}^n$ to $\mathcal{B}$ (for $n = 0$ an element of $\mathcal{B}$). The *coboundary* of φ is defined as the $(n+1)$–cochain

(7.5)
$$d\varphi(v, u_1, \ldots, u_n) = v\varphi(u_1, \ldots, u_n) - \varphi(vu_1, u_2 \ldots, u_n) + \varphi(u_1, vu_2, \ldots, u_n) \ldots$$
$$+ (-1)^{n+1}\varphi(u_1, \ldots, u_n)v \ .$$

Therefore, a 1–cochain $\varphi(u)$ is a *cocycle* (*i.e.* its coboundary is 0) if and only if $v\varphi(u) - \varphi(vu) + \varphi(u)v = 0$, otherwise stated if φ is a derivation. To say that φ is the coboundary of a 0–cochain g means that $\varphi(u) = ug - gu$, *i.e.* φ is an inner derivation. Then the theory of perturbation of quantum flows, for instance, can be expressed elegantly in cohomological language — though only the first cohomology groups appear, and no deep algebraic results are used.

Markov chains as quantum flows

8 No processes (except possibly Bernoulli trials) are simpler and more useful than Markov chains with finite state spaces. It is one of the unexpected results of "quantum probability" that it has led to new remarks on a subject of which every detail seemed to be known.

We first consider the case of a Markov chain in discrete time $n = 0, \ldots, T$, taking its values in a finite set E consisting of $\nu + 1$ points numbered from 0 to ν. We denote by $\mathcal{A}$ the (finite dimensional) algebra of all complex valued functions on E.

The transition matrix of the chain is denoted by $P = (p(i,j))$, and we assume for the moment its entries are all > 0 (this hypothesis is for simplicity only, and we discuss it later). The (discrete) generator of the chain is $A = P - I$. We denote by Ω the finite set $E^{\{0,\ldots,T\}}$ of all conceivable sample paths : our hypothesis implies that if the initial measure is concentrated at x, all paths from x have strictly positive measure, and we do not need to worry about sets of measure 0. The coordinate mappings on Ω are called X_n.

The additive functionals of the chain can be represented as follows, by means of functions z on $E \times E$

$$Z_k = \sum_{i=1}^{i=k} z\left(X_{i-1}, X_i\right) .$$

Under our hypothesis this representation is unique. The additive functionals which are martingales of the chain (for every starting point) correspond to functions $z(\cdot, \cdot)$ such that $Pz(\cdot) = \sum_j p(\cdot, j) z(\cdot, j) = 0$. Let us try to construct a (real) "orthonormal basis" Z^σ, taking as first basis element $Z_k^0 = k$, of the space of additive functionals, in the following sense

$$(8.1) \qquad \mathbb{E}\left[\Delta Z_k^\sigma \Delta Z_k^\tau \mid \mathcal{F}_{k-1}\right] = \delta^{\sigma\tau} .$$

The functionals Z^α ($\alpha > 0$) then are martingales (as usual, $\alpha, \beta, \ldots$ range from 1 to ν, while $\rho, \sigma, \ldots$ range from 0 to ν). On the corresponding functions z^σ, these relations become

$$\forall i, \quad \sum_j p(i,j) z^\sigma(i,j) z^\tau(i,j) = \delta^{\sigma\tau} .$$

This amounts to choosing an orthonormal basis of (real) functions of two variables for a "scalar product" which takes its values in the algebra $\mathcal{A}$:

$$(8.2) \qquad <y, z> = P(yz) = \sum_j p(\cdot, j) y(\cdot, j) z(\cdot, j) ,$$

and taking the function 1 as its first vector. The construction is easy : for every fixed i we are reduced to finding an o.n.b. for a standard bilinear form, and we may paste the values together under the restriction that *the rank of this form should not depend on i*. On the other hand, this rank is the number N_i of points j that can be reached from i ($p(i,j) > 0$). The strict positivity assumption means that $N_i = \nu + 1$ for all i, and the preceding condition is fulfilled.

Starting from such a martingale basis, we define discrete multiple stochastic integrals, which are finite sums of the form

$$(8.3) \qquad f = \sum_{p=1}^{p=T} \sum_{\substack{i_1 < \ldots < i_p \\ \alpha_1, \ldots, \alpha_p}} c_p\left(X_0, i_1, \alpha_1, \ldots i_p, \alpha_p\right) \Delta Z_{i_1}^{\alpha_1} \ldots \Delta Z_{i_p}^{\alpha_p} .$$

There is an isometry formula

$$(8.4) \qquad \mathbb{E}[|f|^2 \mid X_0] = \sum_p \sum |c_p\left(X_0, i_1, \alpha_1, \ldots i_p, \alpha_p\right)|^2 .$$

The coefficients of a given r.v. in this chaos expansion depend on the initial point X_0 : from an algebraic point of view, we may describe this as a "chaos expansion with coefficients in the algebra $\mathcal{A}$". If the initial point is fixed, the Hilbert space generated by these multiple stochastic integrals is isomorphic to a toy Fock space of multiplicity ν.

As remarked by Emery, it is easy to prove that every r.v. can be represented by a chaos expansion (8.3) : Keeping $X_0 = x$ fixed, Ω consists of $(\nu+1)^T$ points, all of which have strictly positive measure, and this gives us the dimension of $L^2(\Omega)$. On the other hand, the number of subsets of $\{1, \ldots, T\}$ with p elements is $\binom{T}{p}$, and for each subset the choice of indices α gives ν possibilities, whence the same dimension $(\nu + 1)^T$.

As always, chaotic representation implies previsible representation w.r.t. the martingales Z^α. Therefore the martingale additive functional $f(X_n) - f(X_0) - \sum_{k \leq n} Af(X_k)$ has a stochastic integral representation, and the corresponding predictable processes may be computed as usual by means of angle brackets. This leads to a formula

$$(8.5) \qquad f(X_n) - f(X_0) - \sum_{1 \leq k < n} Af(X_k) = \sum_\alpha \sum_{1 \leq k < n} L_\alpha f(X_k) \Delta Z_k^\alpha \;,$$

with operators L_α on $\mathcal{A}$ given by

$$(8.6) \qquad\qquad L_\alpha f(i) = \sum_j p(i,j)\, z^\alpha(i,j) f(j) \;.$$

This can be interpreted as an "Ito formula" in which the Z^α replace the components B^α of Brownian motion, A replaces the laplacian $\frac{1}{2}\Delta$, and L_α replaces the derivative operator D_α. It is convenient to use the additional index 0 with $Z_k^0 = k$, $L_0 = A$.

REMARKS. 1) The restriction that the transition matrix be strictly positive can be replaced, as said above, by the condition that the number N_i of states j such that $p(i,j) > 0$ is a constant N independent of i, and then the number of necessary martingales is $N - 1$. On the other hand, Parthasarathy remarked that every finite Markov chain can be considered as the image of a Markov chain (X_n') possessing this property, and which we are going to describe now[1].

First, let N be the largest of the numbers N_i. The state space E' consists of E ("true" states) and of "ghost" states, x_{ij}^k, where i and j are true states such that $p(i,j) > 0$, and k is an integer (we say that x_{ij}^k "begins" at i and "ends" at j). We adopt the following rules : a) The total number of ghost states beginning at i is $N - N_i$. b) The new transition probability $p'(i, \cdot)$ is such that for each state j such that $p(i,j) > 0$, the total mass $p(i,j)$ is shared between j and all the ghost states beginning at i and ending at j, each one receiving a strictly positive mass. Thus i leads to N states in E'. c) If x is a ghost state ending at j, $p'(x, \cdot) = p'(j, \cdot)$, therefore each ghost state leads also to N states.

On the other hand, let π be the mapping from E' to E which induces the identity on E and maps any ghost state to its end. Then it is clear that the chain (X_n') is mapped by π into the chain (X_n), and therefore every r.v. of the chain (X_n) can be expanded by means of the $N - 1$ martingales of the larger chain.

2) The discussion above can be extended to discrete filtrations $\mathcal{F}_n$ in which $\mathcal{F}_0$ is trivial, $\mathcal{F}_1$ consists of exactly $\nu + 1$ atoms (of strictly positive measure), and similarly each atom of $\mathcal{F}_n$ is subdivided into exactly $\nu + 1$ atoms.

---

[1] This discussion may be omitted without harm.

9   The functions $z^\rho(i,j)$ constitute a basis for functions on $E \times E$ w.r.t. the algebra of (real) functions on E acting by multiplication, as functions of the variable i [2]. Therefore, there must exist a formula of the following kind

$$(9.1) \qquad z^\rho(i,j) z^\sigma(i,j) = \sum_\tau C_\tau^{\rho\sigma}(i) \, z^\tau(i,j) \,,$$

where the coefficients $C_\tau^{\rho\sigma} = P(z^\rho z^\sigma z^\tau)$ are functions on E, which satisfy relations expressing the associativity of multiplication (and in this case also its commutativity)

$$(9.2) \qquad \sum_\pi C_\pi^{\lambda\mu} C_\rho^{\pi\nu} = \sum_\pi C_\rho^{\lambda\pi} C_\pi^{\mu\nu} \,.$$

Also, the particular role of the unit element $1 = z^0$ gives the relation $C_\tau^{\sigma 0} = C_\tau^{0\sigma} = \delta_\tau^\sigma$, and the choice of the functions z^α gives $C_0^{\alpha\beta} = \delta^{\alpha\beta}$ (the fact that our functions are real is used here). Since the mapping $f \longmapsto f(X_1)$ is an algebra homomorphism, we deduce from the Ito formula above

$$(9.3) \qquad \begin{aligned} L_0\,(fg) - f L_0 g - (L_0 f)\,g &= \sum_\alpha L_\alpha f \, L_\alpha g \\ L_\gamma(fg) - f L_\gamma g - (L_\gamma f)\,g &= \sum_{\alpha\beta} C_\gamma^{\alpha\beta}\, L_\alpha f \, L_\beta g \,. \end{aligned}$$

10   We are going to consider the chain from the point of view of non commutative (discrete) stochastic calculus — that is, instead of the random variable $f \circ X_k$ we are going to compute the corresponding *multiplication operator*, which we denote by $\xi_k(f)$. We use induction on k, starting from $\xi_0(f)$ which simply multiplies the coefficients of the chaos expansion by $f \circ X_0$.

We first investigate how multiplication by ΔZ_k^α acts on a monomial $\Delta Z_{i_1}^{\alpha_1} \ldots \Delta Z_{i_p}^{\alpha_p}$. If this monomial does not contain any ΔZ_k^β relative to time k, the multiplication simply adds a factor ΔZ_k^α, and this corresponds to a *creation operator* $\Delta a_0^\alpha(k)$ (discrete Evans notation). If the monomial contains ΔZ_k^β with $\beta \neq \alpha$, the multiplication formula replaces the term ΔZ_k^β by the sum of all terms $C_\gamma^{\alpha\beta}(X_{k-1})\, dZ_k^\gamma$, while the replacement of ΔZ_k^β by ΔZ_k^γ is interpreted as a *number or exchange operator* $\Delta a_\beta^\gamma(k)$. Finally, if the term ΔZ_k^α appears in the monomial, the term $C_0^{\alpha\beta} = \delta^{\alpha\beta}$ in formula (9.1) produces a monomial from which ΔZ_k^α has disappeared, and this corresponds to an *annihilation operator* $\Delta a_\alpha^0(k)$. Taking (8.4) into account, we have

$$(10.1) \qquad \xi_k(f) = \xi_{k-1}(f) + \sum_{\sigma\tau} \xi_{k-1}(\mathbf{L}_\tau^\sigma(f))\, \Delta a_\sigma^\tau(k) \,,$$

with the notation $\mathbf{L}_\tau^\sigma(f) = \sum_\rho L_\rho f \, C_\tau^{\rho\sigma}$. Let us denote by $\Lambda(f)$ the matrix $\mathbf{L}_\tau^\sigma(f)$ with coefficients in $\mathcal{A}$. Performing a short explicit computation using the associativity property (9.2), formula (10.1) becomes an universal formula no longer containing the coefficients $C_\tau^{\rho\sigma}$, which is a discrete structure equation

$$(10.2) \qquad \mathbf{L}_\tau^\sigma(fg) - f \mathbf{L}_\tau^\sigma(g) - \mathbf{L}_\tau^\sigma(f)\,g = \sum_\rho \mathbf{L}_\tau^\rho(f)\, \mathbf{L}_\rho^\sigma(g) \,.$$

---

[2] The fact that i is written first does not make it necessarily a "left" action : the algebra $\mathcal{A}$ is commutative.

Note that on the right side, the summation concerns all indices, 0 included. This relation can also be written (with the same convention on the matrix product as in subsection 8) $\Lambda(fg) - f\Lambda(g) - \Lambda(f)g = \Lambda(f)\Lambda(g)$, and means that $f \longmapsto \Sigma(f) = fI + \Lambda(f)$ is a *homomorphism from* $\mathcal{A}$ *into the algebra of matrices with coefficients in* $\mathcal{A}$. This is the discrete analogue of the Evans–Hudson twisted left multiplication, and since the index 0 plays no particular role, the story ends there. The meaning is very clear : the algebra of functions of two variables $g(i,j)$ admits two multiplications by functions of one variable, multiplication by $f(i)$ and multiplication by $f(j)$, and the z^σ constitute a basis of this algebra considered as an $\mathcal{A}$-module for the first multiplication, while $\Lambda(f)$ is the matrix in this basis of the *second* multiplication by f.

COMMENT. In the theory of classical s.d.e.'s, Brownian motion appears as an universal noise using which all diffusions with reasonable generators can be constructed. In the case of Markov chains, Brownian motion can be replaced in the Ito formula or chaos expansions by the martingales Z^α, but the situation is artificial, since they are are ad hoc constructs from the chain itself. It is only when the setup is extended to include operators ("quantization") that we find again universal objects, the creation, annihilation and number processes. On the other hand, we have no obvious analogue of the *flow of diffeomorphisms* associated with a classical s.d.e..

The continuous time case

11 We consider now a continuous time Markov chain on the finite state space E. The standard terminology from martingale theory (angle and square brackets) and from the theory of Markov processes will be used without explanation. We denote by $p_t(i,j)$ the transition matrix, by P_t the transition semigroup. The generator is called A (later L_0 as in the discrete case), with matrix $a(i,j) = p_0'(i,j)$. We assume for simplicity, as we did in the discrete case, that all $a(i,j)$ are different from 0. One also defines $n(i,j) = a(i,j)$ for $i \neq j$, $n(i,i) = 0$; given a function $g(i,j)$ of two variables we define $Ng(i) = \sum_j n(i,j)g(i,j)$; N is called the *Lévy kernel* of the chain (it allows compensating sums over jumps, as in the case of processes with independent increments).

The martingale additive functionals of the chain are compensated sums over jumps

$$(11.1) \qquad M_z(t) = \sum_{s \leq t} z(X_{s-}, X_s) - \int_0^t Nz(X_s)\,ds \ ,$$

where $z(i,j)$ is a function of two variables, equal to zero on the diagonal. Our hypothesis implies that the mapping $z \longmapsto M_z$ is injective, and we may identify completely the space $\mathcal{B}$ of functions of two variables equal to 0 on the diagonal with the space of martingale additive functionals.

We can map the space $\mathcal{A}$ of functions f of one variable in $\mathcal{B}$, defining $\widetilde{f}(i,j) = f(j) - f(i)$. The corresponding martingale is

$$(11.2) \qquad S_f(t) = \sum_{s \leq t, \Delta X_s \neq 0} (f(X_s) - f(X_{s-})) - \int_0^t Nf(X_s)\,ds \ .$$

This is the same as the usual martingale constructed from the generator

$$(11.3) \qquad S_f = f(X_t) - f(X_0) - \int_0^t Af \circ X_s \, ds \ .$$

The space $\mathcal{B}$ is a ring under pointwise multiplication (on the corresponding additive functionals, this "product" of M_z and M_w is the compensated square bracket $[M_z, M_w] - <M_z, M_w>$). On the other hand, $\mathcal{B}$ is an $\mathcal{A}$–bimodule with left and right actions

$$(11.4) \qquad fz(i,j) = f(j)\, z(i,j) \ , \qquad zf(i,j) = z(i,j)\, f(i) \ .$$

The fact that the variable corresponding to X_{s-} is written as the first one has no special meaning, and the algebra $\mathcal{A}$ being commutative, we are completely free to call either action right or left; our choice will be explained shortly. From the point of view of additive functionals, the right action of f is the previsible stochastic integral of the process $f(X_{t-})$, and the left one is the optional (compensated) stochastic integral of $f(X_t)$. We are going to show that it can be interpreted as the Evans–Hudson twisted left action. Note that N is linear with respect to multiplication by functions $f(i)$, but not by functions $f(j)$.

The angle bracket of two martingales (11.1) is given by

$$(11.5) \qquad <M_z, M_w>_t = \int_0^t N(zw) \circ X_s \, ds \ ,$$

and we put

$$(11.6) \qquad <z, w> = N(\bar{z}w) \ ,$$

which will turn out to be the "scalar product on $\mathcal{A}^d$ with values in $\mathcal{A}$" from the Evans–Hudson theory. Note that for functions of one variable we have $\Gamma(f,g) = N(\bar{f}g)$, where the left hand side is the usual squared-field operator.

As in the discrete case, let us try to construct (real) martingales Z_t^α of the form (11.1), such that

$$<Z^\alpha, Z^\beta>_t = \delta^{\alpha\beta} t \ .$$

On the corresponding functions z^α, this is expressed as $\sum_j n(i,j)\, z^\alpha(i,j)\, z^\beta(i,j) = \delta^{\alpha\beta}$, and the problem is similar to the discrete case, but now concerns functions equal to 0 on the diagonal. Since all $a(i,j)$ are different from 0, it is again possible to find $d = \nu$ such functions, and therefore d orthogonal martingales with respect to which multiple integrals can be defined, with the same properties as those relative to d–dimensional Brownian motion. Biane has proved that the chaotic representation property holds for finite Markov chains — otherwise stated, he has given a new probabilistic interpretation of multiple Fock space. We refer the reader to [Bia2], or to Chapter XXI in [DMM]. For the moment we do not make an explicit choice of the basis.

Since we have a martingale basis, the compensated martingale of the square bracket ([ ] isn't a commutator!) has a representation of the form

$$(11.7) \qquad d[Z^\alpha, Z^\beta]_t - \delta^{\alpha\beta} dt = \sum_\gamma C_\gamma^{\alpha\beta} \circ X_{t-}\, dZ_t^\gamma \ .$$

Then $C_\gamma^{\alpha\beta} \circ X_{t-}$ can be computed as the density of the angle bracket with Z^γ, i.e. as $N(z^\alpha z^\beta z^\gamma)$ (our martingales are real). On the other hand, we have

(11.8)
$$\sum_\gamma C_\gamma^{\alpha\beta}(i)\, z^\gamma(i,j) = z^\alpha(i,j)\, z^\beta(i,j) \ .$$

Since multiplication of functions of two variables is associative (and commutative), we have as in the discrete case

(11.9)
$$\sum_\varepsilon C_\varepsilon^{\alpha\beta} C_\eta^{\varepsilon\gamma} = \sum_\varepsilon C_\eta^{\alpha\varepsilon} C_\varepsilon^{\beta\gamma} \ .$$

We extend this notation to a complete set of coefficients $C_\tau^{\rho\sigma}$, putting $C_0^{\alpha\beta} = \delta^{\alpha\beta}$, and $C_\tau^{\rho\sigma} = 0$ if one index ρ,σ is equal to 0. Then putting $Z_t^0 = t$ as usual, we have in full generality

$$dZ_t^\rho\, dZ_t^\sigma = \sum_\tau C_\tau^{\rho\sigma} \circ X_{t-}\, dZ_t^\tau \ .$$

This can be translated, as for (10.1) in the discrete case, into a representation of the multiplication operator by dZ_t^ρ. However, the multiplication operator by dZ_t^0 is not 0, and we must take into account the relation $dZ_t^0 1 = dt$ by means of a new set of coefficients $\widetilde{C}_\tau^{\rho\sigma}$, differing only from the preceding one by the coefficient $\widetilde{C}_0^{00} = 1$ instead of 0

(11.10)
$$dZ_t^\rho = \sum_{\sigma\tau} \widetilde{C}_\tau^{\rho\sigma} \circ X_t\, da_\sigma^\tau(t) \ .$$

The case $\rho = 0$ is uninteresting, but the formula for $\rho \neq 0$ is worth rewriting

(11.10')
$$dZ_t^\alpha = dQ_t^\alpha + \sum_{\beta\gamma} C_\gamma^{\alpha\beta} \circ X_t\, da_\beta^\gamma(t) \ ,$$

where dQ^α denotes, as usual, the sum of the creator and the annihilator of index α.

The basis property also leads to an "Ito formula"

(11.11)
$$f \circ X_t = f \circ X_0 + \sum_\rho \int_0^t D_\rho(f) \circ X_{s-}\, dZ_s^\rho \ .$$

We first compute the operators D_ρ. We have

(11.12) $D_0(f) = Af = -qf + N\tilde{f}$,   and $D_\alpha(f) = N(\tilde{f} z_\alpha)$,

where q is the positive function $q(x) = -a(x,x)$. Again from Ito's formula, and comparing jumps, we deduce

(11.13)
$$\sum_\alpha D_\alpha f(i)\, z^\alpha(i,j) = f(j) - f(i) \ ,$$

and on the other hand

(11.14) $D_\rho(fg) - f D_\rho(g) - (D_\rho f) g = \sum_{\sigma\tau} C_\rho^{\sigma\tau} D_\sigma f\, D_\tau g$, which splits in two

$$D_\alpha(fg) - fD_\alpha(g) - (D_\alpha f)g = \sum_{\beta\gamma} C_\alpha^{\beta\gamma} D_\beta f D_\gamma g \ ,$$
$$D_0(fg) - fD_0(g) - (D_0 f)g = \sum_\alpha D_\alpha f D_\alpha g \ .$$

The multiplication operator by $f \circ X_t$ can be computed from (11.10) and (11.11), and what we get is a quantum flow equation

$$(11.15) \qquad f \circ X_t = f \circ X_0 + \sum_{\rho\sigma\tau} \int_0^t D_\rho(f) \circ X_s \tilde{C}_\tau^{\rho\sigma} \circ X_s \, da_\sigma^\tau(s) \ .$$

The structure maps are given by

$$(11.16) \qquad \mathbf{L}_\sigma^\tau(f) = \sum_\rho D_\rho(f) \tilde{C}_\sigma^{\rho\tau} \ .$$

In particular, $L_0^0 = D_0$, and for all the other coefficients we may replace $\tilde{C}$ by C. We now compute the matrix $\Sigma_\alpha^\beta(f) = \delta_\alpha^\beta f + \mathbf{L}_\alpha^\beta(f)$ Using (11.16), then (11.13), we have

$$\sum_\alpha \mathbf{L}_\alpha^\beta(f) z^\alpha(i,j) = \sum_\varepsilon D_\varepsilon f(i) C_\alpha^{\varepsilon\beta}(i) z^\alpha(i,j)$$
$$= \sum_\varepsilon D_\varepsilon f(i) z^\varepsilon(i,j) z^\beta(i,j) = (f(j) - f(i)) z^\beta(i,j) \ .$$

Adding $\sum_\alpha \delta_\alpha^\beta f(i) z^\beta(i,j)$ we get the identification of $\Sigma(f)$ with multiplication by $f(j)$.

As in the discrete case, there is a simple explicit choice of the martingale basis. We denote by $e_u(i)$ the indicator function of the point u, and by F^{uv} the martingale (11.1) corresponding to $g(i,j) = e_u(i) e_v(j)$ with $u \neq v$. The martingales relative to different pairs (u,v) have no common jumps, and therefore a square bracket equal to 0, while we have

$$d[F^{uv}, F^{uv}]_t = n(u,v) e_u \circ X_t \, dt + dF_t^{uv} \quad ; \quad d<F^{uv}, F^{uv}>_t = n(u,v) e_u \circ X_t \, dt \ .$$

Then we put $M^{uv} = F^{uv}/\sqrt{n(u,v)}$, so that the angle bracket of M^{uv} with itself gets simplified as $e_u \circ X_t \, dt$, and its square bracket is

$$d[M^{uv}, M^{uv}]_t = e_u \circ X_t \, dt + dM_t^{uv}/\sqrt{n(u,v)} \ .$$

Finally, we identify E with the group of integers mod $(\nu+1)$ and put

$$(11.17) \qquad Z^\alpha = \sum_i M^{i,i+\alpha} \ .$$

This constitutes the required basis of the space of martingales. In this case, the only non–zero coefficients are $C_\alpha^{\alpha\alpha}(i) = 1/\sqrt{n(i,i+\alpha)}$.

The case of finite Markov chains is simple, but as soon as one tries to extend the results to infinite Markov chains, non–explosion conditions appear. The subject is beginning now to be well understood, thanks to the work of Chebotarev, Mohari, Fagnola (references [Che], [ChF], [Fag], [Moh], [MoP], [MoS]).

Classical s.d.e.'s as quantum flows

12   We introduced quantum flows by analogy with classical s.d.e.'s in subs. 5. We
return to Ito equations in a more detailed way, following Applebaum [App3]. It should
be noted that Applebaum's articles concern a more general definition of flows than
Evans–Hudson's, as the coefficients L_σ^ρ do not map the algebra into itself. This feature
does not occur in the simple example below.

We consider a C^∞ manifold E of dimension n, compact for simplicity, and a classical
s.d.e. in Stratonovich form

$$(12.1) \qquad X_t = X_0 + \sum_\rho \int_0^t \xi_\rho(X_s) \star dB_s^\rho \ ,$$

where $\star$ denotes a Stratonovich differential, ξ_ρ are vector fields on the manifold, dB_t^α
($\alpha > 0$) are independent standard Brownian motions and $dB_t^0 = dt$. Its Ito form is

$$(12.2) \qquad f \circ X_t = f \circ X_0 + \int_0^t \sum_\alpha \xi_\alpha f \circ X_s \, dB_s^\alpha + Lf \circ X_s ds \ ,$$

where the generator L is $\xi_0 + \frac{1}{2}\sum_\alpha \xi_\alpha \xi_\alpha$. Our purpose is to interpret this classical
flow as a quantum flow $U_t^* f U_t$ (Applebaum considers $U_t f U_t^*$) associated with a left
unitary evolution. We take as initial space $\mathcal{J} = L^2(\mu)$, where μ is a "Lebesgue measure"
of E, i.e. has a smooth, strictly positive density in every local chart — if the manifold
is orientable this amounts to choosing a smooth n–form that vanishes nowhere, but n–
forms are not natural in this problem. We take as a convenient domain $\mathcal{D}$ the space of
complex C^∞ functions, and we also denote by $\mathcal{A}$ the algebra of multiplication operators
associated with C^∞ functions. We denote by $\mathcal{L}$ the family of operators of the form
$K = \xi + k$, where ξ is a smooth *real* vector field on E and k is a smooth real function
(considered as a multiplication operator). These operators preserve $\mathcal{D}$. We note the
following facts :

1) $\mathcal{L}$ is a real Lie algebra, with $[\xi + k, \eta + \ell] = [\xi, \eta] + (\xi\ell - \eta k)$.

2) $\mathcal{L}$ is stable under $*$, since for a real field ξ we have $\xi + \xi^* = -\mathrm{div}_\mu \xi$, where
div_μ is the mapping from vector fields to functions given in local coordinates x^i by
$\mathrm{div}_\mu \zeta = \sum_i D_i \zeta^i + \zeta(\log \mu)$, μ being identified here with its local density w.r.t. $\prod_i dx^i$.

3) Therefore, if ξ is a smooth vector field, $K = \xi + \frac{1}{2}\mathrm{div}_\mu \xi$ defines a (formally) skewadjoint operator in $L^2(\mu)$, and in fact iK can be shown to be essentially self-adjoint
on $\mathcal{D}$.

We consider a left s.d.e. satisfying a formal unitarity condition

$$(12.3) \qquad dU_t = U_t \left(\sum_\alpha K_\alpha \, da_0^\alpha(t) - K^\alpha \, da_\alpha^0(t) + (iH - \frac{1}{2}\sum_\alpha K^\alpha K_\alpha) \, dt \right)$$

where the operators K_α belong to $\mathcal{L}$, and $K^\alpha = K_\alpha^*$, while H is a self-adjoint operator
mapping $\mathcal{D}$ to $\mathcal{D}$. Note that the coefficients of the equation do not belong to $\mathcal{A}$, but
we do have $[K_\alpha, \mathcal{A}] \subset \mathcal{A}$ (and the same with K^α), and we demand from H the same
property. We assume that the equation (12.3) has a unique unitary solution — at the
time this report is being written, the theory of s.d.e.'s with unbounded coefficients is
beginning to offer concrete conditions implying this assumption.

Let us compute the coefficients of the flow $F \circ X_t = U_t^* F U_t$ on bounded operators. According to (4.3), they are given by

$$\mathbf{L}_\sigma^\rho (F) = \overset{*}{L}_\sigma^\rho F + F L_\sigma^\rho + \sum_\alpha \overset{*}{L}_\sigma^\alpha F L_\alpha^\rho .$$

The only non–zero coefficients are

(12.4) $$\mathbf{L}_\alpha^0(F) = -[K_\alpha, F] , \quad \mathbf{L}_0^\alpha(F) = [K^\alpha, F] ,$$

$$\mathbf{L}_0^0(F) = -i\,[H, F] + \frac{1}{2} \sum_\alpha (2K^\alpha F K_\alpha - K^\alpha K_\alpha F - F K^\alpha K_\alpha)$$

$$= -i\,[H, F] + \frac{1}{2} \sum_\alpha (K^\alpha [F, K_\alpha] + [K^\alpha, F] K_\alpha) .$$

These mappings preserve $\mathcal{A}$. Indeed, it suffices to check that for $f \in \mathcal{A}$ and $M = \xi + k \in \mathcal{L}$ we have $M^*[f, M] + [M^*, f] M \in \mathcal{A}$. Now we have a relation of the form $M^* = -M + m$ with $m \in \mathcal{A}$, and then the above operator is equal to $[M, [M, f]] - m[M, f]$, which belongs to $\mathcal{A}$.

Assuming the general theorems are applicable, the flow itself will preserve the commutative algebra $\mathcal{A}$, and we try to identify the corresponding flow on $\mathcal{A}$ with the classical flow (12.2). Of course dB_t^α corresponds to dQ_t^α, $\xi_\alpha f$ should correspond to $[K_\alpha, f]$ (where f is interpreted as a multiplication operator), which according to property 1) above implies $K_\alpha = \xi_\alpha$ and leaves k_α unrestricted. The operator $\mathbf{L}_0^0(f)$ is explicitly given by

$$-i\,[H, f] + \frac{1}{2} \sum_\alpha \xi_\alpha \xi_\alpha f + \sum_\alpha (\tfrac{1}{2} \operatorname{div} \xi_\alpha - k_\alpha) \xi_\alpha f ,$$

where the last two terms are interpreted as multiplication operators. It seems natural to choose $k_\alpha = \frac{1}{2} \operatorname{div} \xi_\alpha$, so that the last function vanishes, and this choice has an intrinsic meaning, since it is precisely that which turns K_α into a skew–symmetric operator on $L^2(\mu)$. Then we do the same with the additional vector field ξ_0 : we associate with it the operator $K_0 = \xi_0 + \frac{1}{2} \operatorname{div} \xi_0$ and take for H the s.a. operator $H = iK_0$; then $i\,[H, f]$ is the multiplication operator by $\xi_0 f$, and $\mathbf{L}_0^0$ can be identified with L. The exponential equation becomes

$$dU_t = U_t \Big(\sum_\alpha K_\alpha dQ_t^\alpha - (K_0 + \frac{1}{2} \sum_\alpha K^\alpha K_\alpha) dt \Big) ,$$

which is very similar to (12.2). Thus the coefficients K_α depend (in a simple, intrinsic way) on the arbitrary "Lebesgue measure" μ.

13   One can avoid this arbitrary "choice of gauge" if one takes as initial space the *intrinsic Hilbert space of the manifold* E, also called the space of (square integrable) half–densities, or wave functions on E. For the significance of half–densities in quantum mechanics, see Accardi [Acc2].

A half–density ψ is a geometric object on a manifold which at every point x and in a local chart V around x has one single complex component $\psi_V(x)$; in a change of charts this component gets transformed according to the formula

$$\psi_W = \psi_V\,(dx_V / dx_W)^{1/2} ,$$

this ratio denoting the absolute value of the jacobian determinant. Otherwise stated, $\psi_V \sqrt{d x_V}$ is invariant, and one may define globally, for a measurable half–density, using a partition of unity

$$\| \psi \|^2 = \int_M |\psi(x)|^2 \, dx \ .$$

The half–densities for which this integral is finite are called *wave functions* and constitute the *intrinsic Hilbert space* $\mathcal{J}$ of the manifold.

Wave functions can be defined globally : given a Lebesgue measure μ with density m_V in a local chart V (with respect to dx_V), the reading of a wave function ψ w.r.t. μ at x is defined to be

$$\psi_\mu(x) = \psi_V(x) / \sqrt{m_V(x)}$$

which does not depend on V. Then ψ_μ becomes an ordinary element of $L^2(\mu)$, while the wave function itself may be denoted symbolically as $\psi_\mu \sqrt{\mu}$. Otherwise stated, the intrinsic Hilbert space is $L^2_{\mathbb{C}}(\mu)$ with a rule saying how the description changes when μ is changed

$$(13.1) \qquad\qquad \psi_\lambda = \psi_\mu \, (\mu/\lambda)^{1/2} \ .$$

Wave functions can be multiplied by Borel bounded functions, and the initial algebra $\mathcal{C}^\infty$ can thus be realized as an algebra of multiplication operators on the intrinsic Hilbert space $\mathcal{J}$.

Let S be a diffeomorphism of the manifold E, let φ be a half–density with reading φ_μ with respect to μ, and let ν be the image measure μS. We define the half–density $\psi = \varphi \circ S$ by its reading w.r.t. μ

$$(13.2) \qquad\qquad \psi_\mu = \varphi_\mu \circ S \, \sqrt{\frac{\nu}{\mu}} \ .$$

This is a unitary mapping U, since

$$< \psi, \psi > = \int |\varphi \circ S|^2 \frac{\nu}{\mu} \mu = \int |\varphi \circ S|^2 \nu = \int |\varphi|^2 \mu = <\varphi, \varphi> \ .$$

It is easy to see that, if $M(f)$ denotes the multiplication operator by a bounded function f, then $U M(f) U^* = M(f \circ S)$.

A complete vector field ξ defines a flow of diffeomorphisms $S_t = e^{t\xi}$, which acts on functions by $S_t f = f \circ S_t$, on measures λ taking images λS_t, and the corresponding derivatives at 0 are given by $DS_t f|_0 = \xi f$, and for a measure α with smooth density a in local charts, $D(\alpha S_t)|_0$ has the smooth density $\xi a + \sum_i a D_i \xi^i$. In particular, if α is a Lebesgue measure, the density of $D(\alpha S_t)|_0$ with respect to α is $\operatorname{div}_\alpha \xi$. Then if we let the group act on half–densities, we find that its generator is read on $L^2(\mu)$ as the skew-adjoint operator

$$(13.3) \qquad\qquad (K\psi)_\mu = \xi \psi_\mu + \tfrac{1}{2} \operatorname{div}_\mu \xi \psi_\mu \ .$$

This explains out the additional term in the preceding subsection.

Introduction to Azéma martingales and their flows

14   As a last example in this introductory section, we describe a recently discovered class of classical processes, which have non–commutative interpretations on Fock space.

The *Azéma martingale* appeared first in [Azé] as a particular case of martingales naturally associated with random sets. Let (B_t) be a Brownian motion starting at 0, and let H be the Cantor like set $\{B = 0\}$. For any t, we put

(14.1) $$g_t = \sup\{s \leq t : B_s = 0\}, \quad X_t = \operatorname{sgn}(B_t)\sqrt{2(t - g_t)}.$$

If we make the convention that $\operatorname{sgn}(0) = 0$, this process is right continuous, with sample functions which in each zero-free interval have a deterministic absolute value, and a random sign equal to that of the Brownian excursion. Thus the only information needed to reconstruct the process is the knowledge of H, and of a sign for each component of H^c (interval contiguous to H). Otherwise stated, X may also describes a one dimensional spin field, with parallel spins in each interval contiguous to H. Azéma proved :

THEOREM. *The processes X_t and $X_t^2 - t$ are martingales (in the filtration $(\mathcal{F}_t)$ generated by (X_t), not in the Brownian filtration).*

> In probabilistic language, the angle bracket $< X, X >_t$ is equal to t. The same holds for the martingale $|X_t| - L_t$ (where L is local time), which has the predictable representation property — but whether it has the chaotic expansion property is an open problem.

This allows to define multiple integrals and chaos spaces with respect to X, and Emery proved :

THEOREM. *The Hilbert sum of the chaos spaces of X is the whole of $L^2(\mathcal{F}_\infty)$.*

For a proof of these results, we refer to [Eme1], [Eme2], or to Protter [Pro]. For the point of view of quantum probability, see Parthasarathy [Par3].

Otherwise stated, Emery discovered a *new probabilistic interpretation of simple Fock space*. To prove chaotic representation, Emery made use of the following fact, which can be directly verified on (14.1) : the martingale $[X, X]_t - t$ (involving the *square bracket* from martingale theory, not a commutator!) has a simple stochastic integral representation with respect to X_t

(14.2) $$d[X, X]_t = dt - X_{t-}\,dX_t, \quad X_0 = 0.$$

Such a relation, expressing the square bracket of a martingale in a "predictable" way, is called a (martingale) *structure equation*. The simplest structure equations are those of Brownian motion and compensated Poisson processes,

$$d[X, X]_t = dt, \qquad\qquad X_0 = 0,$$
$$d[X, X]_t = dt + c\,dX_t, \qquad X_0 = 0.$$

and Emery defined a whole family of "Azéma martingales" containing all martingales with structure equations of the next degree of complexity

(14.3) $$d[X, X]_t = dt + (\alpha + \beta X_{t-})\,dX_t, \quad X_0 = x.$$

It has become standard to put $\beta = c - 1$, but we do not use this notation here. These equations are themselves particular cases of structure equations of "homogeneous Markov type"

$$(14.4) \qquad d[X,X]_t = dt + \varphi(X_{t-})\,dX_t, \qquad X_0 = x.$$

which can be shown to have solutions if the function φ is continuous. If the solution starting at x is unique in law, the process (X_t) is a homogeneous Markov process.

Let us return to the Azéma family with $\alpha = 0$, $x = 0$. The interval $\beta \in [-2,0]$ is the "good" interval : the case $\beta = 0$ is that of Brownian motion, and for $[-2 \le \beta < 0$ the r.v. X_t can be shown to be bounded. The standard Azéma martingale $(\beta = -1)$ belongs to the good interval, and the limiting case $\beta = -2$ also has a simple description, ([Eme1], p.79). The chaotic representation property holds in the "good interval", with the same proof for all values of β. On the other hand, it rests on the use of moments (or equivalently of analytic vectors), and for β outside the good interval, the moments of X_t are too large and very little is known.

15   For a moment, let us consider the case of martingale structure equations (14..4), with a continuous function $\varphi(x)$, assuming the solution constitutes a homogeneous Markov process. Then the processes $Z_t = X_t - X_0$ is an additive functional of this process. Given a smooth function f, we have an "Ito formula" as follows

$$(15.1) \qquad f(X_t) - f(X_0) - \int_0^t Af \circ X_s\,ds = \int_0^t Hf \circ X_{s-}\,dZ_s$$

where the operators A (the generator) and H are defined as follows

$$(15.2) \qquad Af(x) = \begin{cases} \dfrac{f(x+\varphi(x)) - f(x) - f'(x)\,\varphi(x)}{\varphi^2(x)} & \text{if } \varphi(x) \ne 0, \\[2mm] \frac{1}{2}f''(x) & \text{if } \varphi(x) = 0. \end{cases}$$

$$(15.3) \quad Hf(x) = \dfrac{f(x+\varphi(x)) - f(x)}{\varphi(x)} \quad \text{if } \varphi(x) \ne 0, \quad = f'(x) \quad \text{if } \varphi(x) = 0.$$

Let us also assume the chaotic representation property holds with the martingale (Z_t). Then the structure equation allows us to compute the operator of multiplication by X_t : the formula $dZ_t^2 = dt + \varphi(X_0 + Z_{t-})\,dZ_t$ can be read as

$$(15.4) \qquad dZ_t = dQ_t + \varphi(X_t)\,dN_t.$$

And then we may compute the multiplication operator by $f \circ X_t$, in the notation of quantum flows

$$(15.5) \quad f \circ X_t = f \circ X_0 + \int_0^t Af \circ X_s\,ds + \int_0^t Hf \circ X_s\,dQ_s + \int_0^t Hf \circ X_s\,\varphi \circ (X_s)\,dN_s.$$

In the case of the Azéma martingales, the multiplication operator by X_t on simple Fock space with initial space $\mathbb{R}$ satisfies a linear s.d.e.

$$(15.6) \qquad X_t = X_0 + Q_t + \int_0^t \beta X_s\,dN_s.$$

§4. S.D.E.'S AND FLOWS : SOME RIGOROUS RESULTS

The existence and uniqueness theorem for classical (*i.e.* commutative) stochastic differential equations in $\mathbb{R}^\nu$ under a Lipschitz condition includes the standard existence theorem for ordinary differential equations. One can dream of a "stochastic Stone theorem" which would similarly include the standard "deterministic" theorem that e^{itH} is well defined and unitary, provided H is symmetric and has a dense set of analytic vectors, and on the other hand would include the existence and uniqueness theorem for stochastic differential equations with C^∞ coefficients on a compact manifold. The theory of quantum s.d.e.'s and flows is reaching this stage after decisive contributions of Frigerio, Chebotarev, Mohari and Fagnola, though the stochastic case depends more than expected on the deterministic case. It would not be reasonable to include here results from just distributed preprints, but we may at least sketch the main ideas in the last subsections of this chapter.

The development of the theory has gone through three stages. The first one, that of s.d.e.'s or flows with bounded coefficients on a Fock space of finite multiplicity, was settled by Hudson–Parthasarathy [HuP1] and by Evans [Eva1]. These fundamental results have been generalized in two directions : unbounded coefficients and finite multiplicity ([HuP1], Applebaum [App2], Vincent-Smith [ViS], Fagnola [Fag1] for s.d.e.'s, Fagnola–Sinha [FaS] for flows), and bounded coefficients with infinite multiplicity (Mohari–Sinha [MoS]). This last setup (not much studied in classical probability) will be our main subject here. Then the last turn in the theory reaches the case of unbounded coefficients and infinite multiplicity by an approximation argument.

The right exponential

1   We deal here with a Fock space of possibly infinite multiplicity. Thus our exponential domain $\mathcal{E}$ is generated by vectors $\mathcal{E}(j\mathbf{u})$ such that only finitely many components of $\mathbf{u}$ are different from 0 (we recall the notation $S(\mathbf{u})$ for the set of indexes of the non-trivial components of $\mathbf{u}$, including 0). *We also assume that the components u_α are locally bounded.*

We begin with the study of the *right exponential equation* with (adapted) coefficients $L_\sigma^\rho(t)$ and "initial condition" (H_t) (usually $H_t \equiv I$)

$$(1.1) \qquad V_t = H_t + \sum\nolimits_{\rho\sigma} \int_0^t L_\sigma^\rho(s)\, V_s\, da_\rho^\sigma(s) = H_t + K_t \;.$$

It may be worthwhile to deal once with the general time dependent case. From subsection 2 on, we restrict ourselves to coefficients L_σ^ρ which act only on the initial space, with an independent proof — therefore, the reader may skip subs. 1 if he wishes to.

We assume that the coefficients $L_\sigma^\rho(s)$ are bounded operators, and that $L_\sigma^\rho(\cdot)f$ is measurable for given $f \in \Psi$ (and similar conditions for H_t). Our first step consists in describing the natural integrability conditions under which the equation (1.1) has a solution. To this end, we will reduce (1.1) to an ordinary s.d.e. driven by (infinitely many) semimartingales.

Given an exponential vector $\mathcal{E}(j\mathbf{u})$, we put $v_t = V_t\mathcal{E}_t(j\mathbf{u})$, $h_t = H_t\mathcal{E}_t(j\mathbf{u})$, $k_t = K_t\mathcal{E}_t(j\mathbf{u})$. By the definition of the operator stochastic integral, the vector curve

k_t satisfies the equation

$$k_t = \int_0^t k_s \, \mathbf{u}(s) \cdot d\mathbf{X}_s + \int_0^t \sum_{\rho\sigma} u_\rho(s) \, L_\sigma^\rho(s) \, V_s \, \mathcal{E}_s(j\mathbf{u}) \, dX_s^\sigma \ .$$

Therefore we have

$$v_t = \left[h_t - \int_0^t h_s \, \mathbf{u}(s) \cdot d\mathbf{X}_s \right] + \int_0^t v_s \, \mathbf{u}(s) \cdot d\mathbf{X}_s + \int_0^t \sum_{\rho\sigma} L_\sigma^\rho(s) \, v(s) \, u_\rho(s) \, dX_s^\sigma$$

$$(1.2) \qquad\qquad = \eta_t + \sum_\sigma \int_0^t \widetilde{L}_\sigma(\mathbf{u}, s) \, v(s) \, dX_s^\sigma \ ,$$

where $\widetilde{L}_\sigma(\mathbf{u}, \cdot) = \sum_\rho u_\rho(\delta_\sigma^\rho I + L_\sigma^\rho)$ (Evans symbol) and η_t is the bracket on the left. We omit the mention of $\mathbf{u}$, which now remains fixed.

We begin with a discussion of the case $H_t \equiv I$, in which we also have $\eta_t \equiv I$. We first solve the deterministic operator differential equations

$$(1.3) \qquad A_t = I + \int_0^t \widetilde{L}_0(s) \, A_s \, ds \ , \qquad B_t = I - \int_0^t B_s \, \widetilde{L}_0(s) \, ds \ ,$$

under the natural assumption that, for finite t

$$(1.4) \qquad\qquad \int_0^t \| \widetilde{L}_0(s) \| \, ds < \infty \ .$$

The solution satisfies the properties

$$B_t A_t = I \ , \qquad \| A_t \|, \| B_t \| \le e^{\int_0^t \| \widetilde{L}_0(s) \| \, ds} \ .$$

Then, putting $v_t = A_t z_t$ (hence $z_t = B_t v_t$) and $\widehat{L}_\alpha(t) = B_t \widetilde{L}_\alpha(t) A_t$ we have

$$(1.5) \qquad\qquad z_t = j + \sum_\alpha \int_0^t \widehat{L}_\alpha(s) \, z_s \, dX_s^\alpha \ ,$$

a familiar Ito equation, solvable under the assumptions

$$(1.6) \qquad \text{for } f \in \Psi, \quad \sum_\alpha \| \widehat{L}_\alpha(s) f \|^2 \le C^2(s) \| f \|^2 \ , \qquad \int_0^t C^2(s) \, ds < \infty \ ,$$

which are in fact equivalent to the same conditions without hats. Then the solution satisfies the inequalities

$$(1.7) \qquad \| z_t \|^2 \le \| j \|^2 \, e^{\int_0^t C^2(s) \, ds} \ , \quad \| z_t - j \|^2 \le \| j \|^2 \, (e^{\int_0^t C^2(s) \, ds} - 1) \ ,$$

from which similar inequalities follow for v_t. Returning to our initial problem, let us prove that conditions (1.4) and (1.6) are implied by the following ones (for all ρ)

$$(1.8) \qquad \sum_\sigma \| L_\sigma^\rho(s) f \|^2 \le C_\rho^2(s) \| f \|^2 \ , \qquad \text{with} \quad \int_0^t C_\rho^2(s) \, ds < \infty \ .$$

Indeed, we have

$$\sum_\sigma \| \sum_{\rho \in S(\mathbf{u})} u_\rho(s) L_\sigma^\rho(s) f \|^2 \le \sum_\sigma (\sum_{\rho \in S(\mathbf{u})} |u_\rho(s)|^2)(\sum_{\rho \in S(\mathbf{u})} \| L_\sigma^\rho(s) f \|^2)$$

(1.9)
$$\le (\sum_{\rho \in S(\mathbf{u})} |u_\rho(s)|^2)(\sum_{\rho \in S(\mathbf{u})} C_\rho^2(s)) \| f \|^2 ,$$

and we may conclude since the sum is finite and the first factor is a locally bounded function.

The case of general "initial data" (η_t) is not more difficult. We use a coarser method, assuming $t \le T$ and using the Schwarz inequality

$$\| v_t \|^2 \le 3 (\| \eta_t \|^2 + \| \int_0^t \tilde{L}_0(s) v_s \, ds \|^2 + \| \sum_\alpha \int_0^t \tilde{L}_\alpha(s) v_s \, dX_s^\alpha \|^2)$$

$$\le 3 (\| \eta_t \|^2 + T \int_0^t \| \tilde{L}_0(s) v_s \|^2 \, ds + \sum_\alpha \int_0^t \| \tilde{L}_\alpha(s) v_s \|^2 \, ds) ,$$

from which we may deduce an inequality

(1.10)
$$\| v(t) \|^2 \le 3 \| \eta(t) \|^2 + \int_0^t C^2(s) \| v(s) \|^2 \, ds ,$$

where the function $C^2(s)$ is locally integrable. Applying Gronwall's lemma we see that, whenever $\eta(t)$ tends to 0 uniformly on bounded intervals (for instance), the same is true for $\| v(t) \|$.

REMARK. The seminal paper [HuP1] on non–commutative s.d.e.'s dealt with the *left* equation. In the time dependent situation it would raise domain problems, because U_s is generally defined on the exponential domain $\mathcal{E}$ while $L_\sigma^\rho(s)$ does not preserve $\mathcal{E}$.

2 The most important particular case is that of a s.d.e. whose coefficients L_σ^ρ are time independent, and are ampliations to Ψ of bounded operators on $\mathcal{J}$. In this case, the above conditions for the existence of the solution of the equation

(2.1)
$$V_t = H_t + \sum_{\rho\sigma} \int_0^t L_\sigma^\rho V_s \, da_\rho^\sigma(s)$$

take the following form :

1) The adapted operator process (H_t) is defined on $\mathcal{E}$, and for arbitrary $j, \mathbf{u}$ the mapping $H_t \mathcal{E}(j\mathbf{u})$ is locally square integrable w.r.t. the measure $\nu(\mathbf{u}, dt)$.

2) For every ρ we have the Mohari–Sinha conditions

(2.2)
$$\boxed{ \forall j \quad \sum_\sigma \| L_\sigma^\rho j \|^2 \le C_\rho^2 \| j \|^2 } .$$

Such a condition is trivially satisfied in the case of finite multiplicity and bounded coefficients. We shall give it another interpretation after (2.8) at the end of this subsection, and see in subs. 5 that it gives a meaning to the infinite sums which occur in the expression for the generator.

Property (2.2) may be considered as an estimate of the norm of a mapping L_σ from $\mathcal{J}$ to $\mathcal{J} \otimes \mathcal{K}$ (the multiplicity space). Since tensoring with I_Φ does not increase the norm, we also have for $f \in \Psi$

$$(2.2') \qquad\qquad \sum_\sigma \| (L_\sigma^\rho \otimes I) f \|^2 \leq C_\rho^2 \| f \|^2 \,,$$

which implies the main assumption (1.8) of the preceding subsection. Instead of relying upon subs. 1, let us give a quick, but complete proof in this case. We start the Picard method with $V_0(t) = H_t$, and put

$$(2.3) \qquad\qquad V_{n+1}(t) = H_t + \sum_{\rho\sigma} \int_0^t L_\sigma^\rho V_n(s) \, da_\rho^\sigma(s) \,.$$

Therefore $V_1(t) - V_0(t) = \int_0^t \sum_{\rho\sigma} L_\sigma^\rho H_s \, da_\rho^\sigma(s)$, and then

$$V_{n+1}(t) - V_n(t) = \sum_{\rho\sigma} \int_0^t L_\sigma^\rho (V_n(s) - V_{n-1}(s)) \, da_\rho^\sigma(s) \,.$$

Denoting $V_n(t)\,\mathcal{E}(j\mathbf{u})$ as $v_n(t)$ and writing ν for $\nu(\mathbf{u},\cdot)$ we have for $t \leq T$, using the domination inequalities (9.7) in §1

$$\| v_1(t) - v_0(t) \|^2 \leq 2e^{\nu(T)} \int_0^t \sum_{\rho \in S(\mathbf{u}),\sigma} \| L_\sigma^\rho H_s \mathcal{E}(j\mathbf{u}) \|^2 \nu(ds)$$

$$\leq 2e^{\nu(T)} \Big(\sum_{\rho \in S(\mathbf{u})} C_\rho^2 \Big) \int_0^t \| H_s \mathcal{E}(j\mathbf{u}) \|^2 \nu(ds) \,.$$

Let us put $K = 2e^{\nu(T)} \sum_{\rho \in S(\mathbf{u})} C_\rho^2$, and $c = \int_0^T \| H_s \mathcal{E}(j\mathbf{u}) \|^2 \nu(ds)$, two quantities depending on $\mathbf{u}$ and T. Then the right side is dominated by cK, and an easy induction proof following the same lines gives the result

$$\| v_{n+1}(t) - v_n(t) \|^2 \leq cK \frac{K^n \nu^n(t)}{n!} \,.$$

Therefore the left side of (2.3) converges strongly on the exponential domain to a limit $V_t \mathcal{E}(j\mathbf{u})$. To prove that this process solves the right equation, we must check that

$$\Big(\sum_{\rho\sigma} \int_0^t L_\sigma^\rho V_n(s) \, da_\rho^\sigma(s) \Big) \mathcal{E}(j\mathbf{u}) \to \Big(\sum_{\rho\sigma} \int_0^t L_\sigma^\rho V_s \, da_\rho^\sigma(s) \Big) \mathcal{E}(j\mathbf{u}) \,,$$

or that

$$\sum_{\rho \in S(\mathbf{u}),\sigma} \int_0^t \| L_\sigma^\rho (V_n(s) - V_s) \mathcal{E}(j\mathbf{u}) \|^2 \nu(ds) \to 0 \,,$$

which follows from the inequality

$$\sum_{\rho \in S(\mathbf{u}),\sigma} \| L_\sigma^\rho (V_n(s) - V_s) \mathcal{E}(j\mathbf{u}) \|^2 \leq \sum_{\rho \in S(\mathbf{u})} C_\rho^2 \| (V_n(s) - V_s) \mathcal{E}(j\mathbf{u}) \|^2 \,.$$

The way uniqueness on the exponential domain is proved is standard (consider the difference W of two solutions, and iterate to prove that $\| W(t) \|^2 \leq K^n/n!$ for some K and all n). This uniqueness implies that, if $H_t \equiv I$ and the coefficients L_σ^ρ are time independent, $V(t)$ is a *right cocycle,* the idea being that for fixed s, the process

$$V_t' = V_t \quad \text{for } t \leq s, \qquad V_t' = \Theta_s V_{t-s} V_s \quad \text{for } t > s,$$

also is a solution of (2.1). Otherwise stated, to construct V_{s+t}, we perform the same construction as for V_t, taking Ψ_s as the new initial space and V_s as the new initial operator. One easily sees that $V_t \mathcal{E}(j\mathbf{u})$ is continuous in t.

The Picard approximation of order n, in this particular case, has an explicit form as a sum on all multi-indexes $\mu = (\varepsilon_1,\dots,\varepsilon_m)$ of order $m \leq n$, where $\varepsilon_1,\dots,\varepsilon_m$ are Evans indexes including $\binom{0}{0}$ (see §2, formula (2.2))

$$(2.4) \qquad\qquad V_n(t) = \sum_{|\mu| \leq n} L_\mu I^\mu(t)$$

where I^μ is an iterated integral, and the order of indices in L^μ thus corresponds to the higher times standing on the left

$$I_t^\mu = \int_{s_1 < \dots < s_m < t} da^{\varepsilon_m}(s_m)\dots da^{\varepsilon_1}(s_1) \quad ; \quad L_\mu = L_{\varepsilon_m}\dots L_{\varepsilon_1} .$$

REMARKS. 1) Assume that $H_t \equiv I$ and the coefficients are time independent. There is no difficulty in proving the existence and uniqueness results for solutions U_t of the left equation, under the Mohari–Sinha condition (2.2). Indeed, introducing the Picard approximation scheme, we have

$$(2.5) \qquad\qquad U_n(t) = \sum_{|\mu| \leq n} L_\mu' I^\mu(t) ,$$

the $'$ in L_μ' indicating that operators are composed in the reverse ordering. Then we have exactly the same domination properties for $U_n(t)$ as for $V_n(t)$, and the proof that U_t satisfies the left equation is even a little easier, since L_σ^ρ preserves the exponential domain.

2) Consider the case of a finite–dimensional initial space with o.n. basis (e_i), and equations in matrix form

$$(2.6) \quad U_i^j(t) = \delta_i^j + \sum_k \int_0^t U_k^j(s)\,(L_i^k ds + da_s^+(|\lambda_i^k>) + da_s^\circ(\Lambda_i^k) + da_s^-(<\widetilde{\lambda}_i^k |)) ,$$

with the following assumptions : the multiplicity space $\mathcal{K}$ is the completion of a prehilbert space $\mathcal{K}_0$; λ_i^k and $\widetilde{\lambda}_i^k$ are elements of $\mathcal{K}$, while the operators Λ_i^k map $\mathcal{K}_0$ to $\mathcal{K}$, as do their adjoints. We take an orthonormal basis in $\mathcal{K}_0$, denoted by e^α because of our perverse notation system, denote by $<e_\alpha|$ the corresponding "bra", and rewrite the s.d.e. in standard form, with coefficients L_σ^ρ given by

$$(2.7) \qquad L_0^0 e_i = \sum_k L_i^k e_k \quad , \quad L_\alpha^0 e_i = \sum_k <e_\alpha, \lambda_i^k > e_k ,$$

$$L_\alpha^\beta e_i = \sum_k <e_\alpha, \Lambda_i^k e^\beta > e_k \quad , \quad L_0^\alpha e_i = \sum_k <\widetilde{\lambda}_i^k , e^\alpha > e_k .$$

Since $\mathcal{J}$ is finite dimensional, L_ρ^σ is always bounded and the M–S conditions are satisfied : they reduce to the fact that λ_i^k and $\Lambda_i^k e^\beta$ belong to $\mathcal{K}$, while the dual M–S conditions are also satisfied as $\widetilde{\lambda}_i^k$ and $\Lambda_i^{k*} e^\alpha$ belong to $\mathcal{K}$.

We may choose an arbitrary o.n.b. in $\mathcal{K}_0$, its first element being any normalized vector. Therefore *the domain of U_t includes all exponential vectors $\mathcal{E}(j\mathbf{u})$ with $\mathbf{u} \in \mathcal{K}_0$*.

If $\mathcal{J}$ is arbitrary, we also deduce from the M–S theorem that a s.d.e. in the basis free form, for instance the left equation

$$(2.8) \qquad dU_t = U_t(\, Ldt + da_t^+(\lambda) + da_t^\circ(\Lambda) + da_t^-(\widetilde{\lambda})\,)$$

is solvable whenever its coefficients are bounded (L from $\mathcal{J}$ to $\mathcal{J}$, λ from $\mathcal{J}$ to $\mathcal{J} \otimes \mathcal{K}$, etc.), the exponential domain including all vectors $\mathcal{E}(j\mathbf{u})$ with $\mathbf{u}$ locally bounded (no "finite support" restriction).

Introduction to the unbounded case

3   We give now an idea of the hypotheses that lead to the simplest results on unbounded coefficients. In view of the recent results of Fagnola, it is not clear whether this approach remains interesting or not.

Instead of assuming norm boundedness conditions on the operators L_σ^ρ, we assume they are defined on a dense *stable* domain $\mathcal{D} \subset \mathcal{J}$, and satisfy for $j \in \mathcal{D}$ the following inequalities. Given a finite set U of indexes, we write $\mu \in U$ to mean that all the *upper* indexes ρ_i of μ belong to U. We choose $K_n(j, U)$, for $j \in \mathcal{D}$, such that

$$(3.1) \qquad \sum\nolimits_{|\mu|=n,\, \mu \in U} \| L_\mu j \|^2 \le K_n^2(j, U)$$

and we assume that for all $j \in \mathcal{D}$ and finite U

$$(3.2) \qquad \sum_n K_n(j, U) \frac{r^n}{\sqrt{n!}} < \infty \quad \text{for all } r > 0.$$

We also assume that *the operators L_σ^ρ are closable.* If the multiplicity is $d < \infty$, the number of multi-indices of order n is $(d+1)^n$, which gets absorbed into the coefficient r^n, and assumption (3.1) may be replaced by a similar estimate on each coefficient $L_\mu j$, instead of on their sum.

The same estimates as in the bounded case on the squared norm of

$$(V_{n+1}(t) - V_n(t))\,\mathcal{E}(j\mathbf{u}) = \sum_{|\nu|=n+1,\, \mu \in S(\mathbf{u})} L_\mu I_t^\mu \mathcal{E}(j\mathbf{u})$$

then prove that the operators $V_n(t) = \sum_{|\mu| \le n} L_\mu(j)\, I_t^\mu$ have a strong limit V_t on exponential vectors. We have simple boundedness properties for $V_t \mathcal{E}(j\mathbf{u})$ as in the bounded case.

It is also easy to prove that the relation

$$V_{n+1}(t) = I + \sum_{\rho\sigma} \int_0^t L_\sigma^\rho V_n(s)\, da_\rho^\sigma(s)$$

becomes in the limit the right equation. One remarks that $L_\sigma^\rho(V_{n+1} - V_n)$ is also a sum on multi-indices of order $n + 2$ and the series is normally convergent on the exponential domain, following which one can define $L_\sigma^\rho V_t$ using the assumption that L_σ^ρ is closable. There is no serious difficulty and details are left to the reader.

Similar results hold for the left equation, even without a closability assumption.

REMARK. It is intuitive that the result could be improved using the following rough idea : since we are defining an "exponential", it should be sufficient to construct it on some fixed interval $[0, a]$, and to extend it using its algebraic properties. Thus instead of demanding convergence for all $r > 0$, we could assume it converges for $r < r_0$ independent of j. Such an idea is made rigorous in Fagnola [Fag1] and Fagnola–Sinha [FaS] for flows.

On the other hand, if the only non–zero coefficient is L_0^0 (skew–adjoint under the unitarity condition), we would like to recover Nelson's analytic vectors theorem, i.e. in this case the coefficient $\sqrt{n!}$ in (3.2) should be replaced by $n!$. This may be partially realized using a finer estimate for $\| I_t^\mu \mathcal{E}(j\mathbf{u}) \|$, taking into account the number of times $\binom{0}{0}$ appears in μ, and a corresponding finer assumption in (3.2). Such a domination property has been given in §1, subs.9. In fact one can do slightly better, taking also into account the number of annihilation indexes (see [Fag1]). Such inequalities have also been used by Fagnola–Sinha for flows on multiple Fock space.

The contractivity condition

4   In this subsection we give (following Mohari) a natural condition implying that the solution V_t to the right equation is contractive on the exponential domain.

Instead of one single exponential vector, we must consider a finite linear combination, $f = \sum_k \mathcal{E}(j_k \mathbf{u}_k)$, $1 \leq k \leq p$, the components of each $\mathbf{u}_k$ being denoted $u_{k\rho}$. We recall the notation in §3 subs.4 : we define an operator $\mathbf{L}$ on $\mathcal{B} = \mathcal{J} \otimes (\mathbb{C} \oplus \mathcal{K})$ (algebraic tensor product) with matrix $\mathbf{L}_\sigma^\rho$ given by

$$(4.1) \qquad \mathbf{L}_\sigma^\rho = \overset{*}{L}{}_\sigma^\rho + L_\sigma^\rho + \sum_\alpha \overset{*}{L}{}_\sigma^\alpha L_\sigma^\rho \, ,$$

Since only the individual coefficients L_σ^ρ are assumed to be bounded, the infinite sum requires a discussion; see the next subsection. Then the *contractivity condition* is

$$(4.2) \qquad < \mathbf{b}, \mathbf{L}\mathbf{b} > \, \leq 0 \quad \text{for } b \in \mathcal{B}.$$

To prove it implies contractivity, we rewrite Ito's formula as

$$\frac{d}{dt} < V_t \mathcal{E}(\ell \mathbf{v}), V_t \mathcal{E}(j\mathbf{u}) > \, = \, < V_t \mathcal{E}(\ell \mathbf{v}), \sum_{\rho\sigma} v^\sigma(t) u_\rho(t) L_\sigma^\rho V_t \mathcal{E}(j\mathbf{u}) >$$

from which we deduce (f being defined as above)

$$\frac{d}{dt} \| V_t f \|^2 = \sum_{km} < V_t \mathcal{E}(j_m \mathbf{u}_m), \sum_{\rho\sigma} u_m^\sigma(t) u_{k\rho}(t) L_\sigma^\rho V_t \mathcal{E}(j_k \mathbf{u}_k) > \, .$$

We put $b_\rho(t) = \sum_k u_{k\rho}(t) \mathcal{E}(j_k \mathbf{u}_k)$, and consider these vectors as the components (all but finitely many of which are $= 0$) of a vector $\mathbf{b}$. Then the above derivative can be written as $< \mathbf{b}(t), \mathbf{L}\mathbf{b}(t) >$, and contractivity follows from (4.2).

If the contractivity condition is satisfied, the fact that $V_t \to V_0$ strongly on the exponential domain implies the same result on all vectors. Then using the right cocycle property we find that *the mapping $V_\bullet$ is strongly continuous*. Note also that the contractivity condition implies that $I + \Lambda$ is a contraction; therefore Λ can be extended as a bounded operator.

The same reasoning applies to the case of unbounded coefficients (subs. 3).

Under the isometry condition $\mathbf{L}$ vanishes, and the reasoning proves V_t is isometric.

Let us anticipate a little the time reversal principle (subsections 6–9) to prove that under (2.2) *the formal unitarity condition on the coefficients does imply V_t and U_t are unitary*. If the coefficients L^ρ_σ and $\overset{*}{L}{}^\rho_\sigma$ satisfy the Mohari–Sinha condition (2.2) and the isometry condition, then the solutions V_t and $\overset{*}{V}_t$ to the corresponding right equations are isometric. By time reversal, U_t is isometric too. On the other hand, it is easy to see that the solution U_t of the left equation with coefficients L^ρ_σ is adjoint to $(\overset{*}{V}_t)$ on the exponential domain, and therefore its bounded extension is coisometric. Therefore U_t is unitary, and so is V_t by time reversal.

5 Let us return to the justification of the infinite sum in (4.1) — it is precisely this point that makes (2.2) such a natural condition. We use a lemma from Mohari–Sinha [MoS].

LEMMA. *Let (A_n), (B_n) be two sequences of bounded operators on a Hilbert space $\mathcal{J}$, such that $\sum_n A_n^* A_n$, $\sum_n B_n^* B_n$ are strongly convergent. Then $\sum_n A_n^* B_n$ is also strongly convergent.*

PROOF. The sequence of positive operators $P_n = \sum_{k \leq n} A_k^* A_k$ is increasing, and to say that it converges strongly is equivalent to saying that $\| P_n \|$ is bounded, or that $\| \sqrt{P_n} \|$ is bounded, or (thanks to the uniform boundedness principle) that for every j

$$\| \sqrt{P_n} j \|^2 = <j, P_n j> = \sum_{k \leq n} \| A_k j \|^2 \quad \text{is bounded.}$$

Assuming these properties are satisfied, let us prove that $\sum_{k=1}^n A_k^* B_k j$ is a Cauchy sequence in norm. Defining $C_{mn} = \sum_m^n A_k^* B_k$, we have

$$| <h, C_{mn} j> | \leq \sum_m^n \| A_k h \| \, \| B_k j \| \leq (\sum_m^n \| A_k h \|^2)^{1/2} (\sum_m^n \| B_k j \|^2)^{1/2}$$

$$\leq C \| h \| (\sum_m^n \| B_k j \|^2)^{1/2} .$$

It only remains to take a supremum over h on the unit ball. The same lemma gives a meaning to infinite sums with a bounded operator inserted between A_k^* and B_k (§3, (4.3)).

The right equation as a multiplicative integral

6 Our aim in the next two subsections is to express the solution to the right equation as a limit of "Riemann products". Our inspiration in the proof comes from a paper by

Holevo [Hol2], which however deals with a much more difficult problem. Then in a third subsection we apply this result to "Journé's time reversal principle", following Mohari. For simplicity, we deal only with the case of time independent coefficients.

Our first result will concern the boundedness of "Riemann products". We work on a fixed interval $(0, T)$, and its finite subdivisions $0 = t_0 < \ldots < t_n = T$. We put

$$\Delta_i = \sum_{\rho\sigma} M_\sigma^\rho(i) \left(a_\rho^\sigma(t_{i+1}) - a_\rho^\sigma(t_i) \right)$$

where the operators $M_\sigma^\rho(i)$ are assumed to be bounded on Ψ_i (abbreviated notation for $\Psi_{t_i]}$). We are going to estimate

$$\| Q_n \ldots Q_1 \mathcal{E}(j\mathbf{u}) \|^2 \quad \text{with } Q_i = I + \Delta_i .$$

We begin with the first interval, omitting the index 1 in t_1 etc. Using the iterated integral inequality (§1, (9.8)) we have

$$\| (I + \sum_{\rho\sigma} \int_0^t M_\sigma^\rho \, da_\rho^\sigma(s)) \mathcal{E}(j\mathbf{u}) \|^2 \leq e^{\nu_\mathbf{u}(t)} \| \mathcal{E}(\mathbf{u}) \|^2 (\| j \|^2 + 2 \sum_{\rho\sigma} \| M_\sigma^\rho j \|^2 \nu_\mathbf{u}(t))$$

(the summation over ρ is restricted to the finite set $S(\mathbf{u})$). Let us assume the operators M_σ^ρ satisfy the Mohari–Sinha hypothesis (2.2)

$$\sum_{\rho\sigma} \| M_\sigma^\rho j \|^2 \leq C \| j \|^2$$

where C depends only on $S(\mathbf{u})$. If $\mathbf{u} = 0$ outside the interval $(0, t)$, we first have $\| \mathcal{E}(\mathbf{u}) \|^2 \leq e^{\nu_\mathbf{u}(t)}$, and the right side is bounded by

$$\| j \|^2 e^{2\nu_\mathbf{u}(t)} (1 + 2C\nu_\mathbf{u}(t)) \leq \| j \|^2 e^{C\nu_\mathbf{u}(t)}$$

(the meaning of C has changed). We extend this result by induction on i as follows. The part of $\mathbf{u}$ outside the large interval $[0, T]$ is uninteresting because of adaptation, and we may assume it is equal to 0. Then we decompose $\mathbf{u}$ in pieces u_i corresponding to the subdivision intervals (t_i, t_{i+1}). Assume we have computed the vector

$$j_i = Q_{i-1} \ldots Q_1 \mathcal{E}(j(\mathbf{u}_1 + \ldots + \mathbf{u}_{i-1}))$$

which belongs to Ψ_i. We take this space as our new initial space (note that the space of exponential vectors changes too), and we compute on (t_i, t_{i+1})

$$\| (I + \Delta_i) \mathcal{E}(j_i \mathbf{u}_i) \|^2 .$$

To this end, we apply the above reasoning, which gives us a bound of the form $\| j_i \|^2 e^{C(\nu(t_{i+1}) - \nu(t_i))}$, with the same C as above if the Mohari–Sinha condition holds uniformly in i. These bounds multiply well, and we get a global estimate by $\| j \|^2 e^{C\nu_\mathbf{u}(T)}$.

7 We return to the right exponential equation, with constant coefficients for simplicity, and under the Mohari–Sinha hypotheses,

$$V_t = I + \sum_{\rho\sigma} \int_0^t L_\sigma^\rho V_s \, da_\rho^\sigma(s) \ .$$

We take as before a subdivision of $[0,T]$ and put

$$Q_i = I + \sum_{\rho\sigma} L_\sigma^\rho (a_\rho^\sigma(t_i) - a_\rho^\sigma(t_{i-1})) \ .$$

We are going to show that *the Riemann products* $Q_n \dots Q_1 \mathcal{E}(j\mathbf{u})$ *tend weakly to* $V_T \mathcal{E}(j\mathbf{u})$, along the sequence of dyadic subdivisions for instance. Since we have just shown these vectors are uniformly norm bounded, it is sufficient to test weak convergence on exponential vectors $\mathcal{E}(\ell\mathbf{v})$. The arguments $\mathbf{u},\mathbf{v}$ of both exponential vectors are assumed to be bounded on $[0,T]$.

We denote by V_{st} the value at time t of the solution to the right equation, on the interval (s,t) with initial value I at time s. For the interval (t_i, t_{i+1}) we simplify this to V_i. Then we have $V_T = V_n \dots V_1$, and the difference $V_T - Q_n \dots Q_1$ can be written as follows (the abbreviated notation on the last line is clear enough)

$$V_n V_{n-1} \dots V_1 - Q_n Q_{n-1} \dots Q_1 =$$
$$(V_n - Q_n)Q_{n-1} \dots Q_1 + V_n(V_{n-1} - Q_{n-1})Q_{n-2} \dots Q_1 + \dots V_n \dots V_2(V_1 - Q_1)$$
(7.1)
$$= D_n Q'_{n-1} + V'_n D_{n-1} Q'_{n-2} + \dots + V'_2 D_1 \ .$$

We are going to show that

(7.2)
$$< \mathcal{E}(\ell\mathbf{v}), V'_{i+1} D_i Q'_{i-1} \mathcal{E}(j\mathbf{u}) > = o(t_{i+1} - t_i) \ ,$$

with sufficient uniformity in i to sum these inequalities. A similar argument (see **8** below) is used by Holevo to prove *strong* convergence of much more complicated Riemann products, in a finite dimensional situation. Since now we are working on a fixed subdivision interval, let us simplify our notation, putting $t_i = a$, $t_{i+1} = a+s$, $T = a+t$. We take Ψ_a as our new initial space, with new initial vectors

$$j' = Q'_{i-1}(j \otimes \mathcal{E}(\mathbf{u}I_{]0,a]})) \ , \quad \ell' = \ell \otimes \mathcal{E}(\mathbf{v}I_{]0,a]}) \ .$$

From the preceding subsection we shall use the norm boundedness of j', independently of the interval we work on (within $[0,T]$). Then our scalar product can be written as follows

$$< \ell' \otimes \mathcal{E}(\mathbf{v}'_1) \otimes \mathcal{E}(\mathbf{v}'_2), V_{st}(D\mathcal{E}(j'\mathbf{u}'_1) \otimes \mathcal{E}(\mathbf{u}'_2)) >$$

where the arguments $\mathbf{u}'_1, \mathbf{v}'_1$ are carried by $(0,s)$, $\mathbf{u}'_2, \mathbf{v}'_2$ by (s,t), and the notation V_{st} is the solution on (s,t) described above.

We expand everything using multi-index notation :

$$D\mathcal{E}(j'\mathbf{u}'_1) = \sum_{|\lambda|\geq 2} L_\lambda j' \otimes I_s^\lambda \mathcal{E}(\mathbf{u}'_1)$$

$$V_{st}(D\mathcal{E}(j'\mathbf{u}'_1)) = \sum_{|\lambda|\geq 2, |\mu|\geq 0} L_\mu L_\lambda j' \otimes I_s^\lambda \mathcal{E}(\mathbf{u}'_1) \otimes I_{t-s}^\mu \mathcal{E}(\mathbf{u}'_2)$$

and the scalar product we have to compute appears as

$$\sum_{|\lambda|\geq 2, |\mu|\geq 0} <\ell', L_\mu L_\lambda j'> <\mathcal{E}(v_1'), I_s^\lambda \mathcal{E}(u_1')> <\mathcal{E}(v_2'), I_{t-s}^\mu \mathcal{E}(u_2')> .$$

This sum can be estimated along the following lines

$$\sum_{m\geq 0, n\geq 2} \sum_{|\mu|=n, |\lambda|=m} \|\ell'\| \, \| L_\mu L_\lambda j'\| \, a_\mu b_\lambda \leq$$

$$\|\ell'\| \sum_{m\geq 0, n\geq 2} \Big(\sum_{|\mu|=n, |\lambda|=m} \| L_\mu L_\lambda j'\|^2\Big)^{1/2} \Big(\sum_{|\mu|=n} a_\mu\Big) \Big(\sum_{|\lambda|=m} b_\lambda\Big) .$$

The first inner sum is of order C^{m+n} using the Mohari–Sinha assumptions. Next, if the lower and upper indices of μ are respectively $\sigma_1,\dots,\sigma_n$ and $\rho_1,\dots,\rho_n$, we have (omitting indexes appended to $\mathbf{u}, \mathbf{v}$)

$$a_\mu = |<\mathcal{E}(\mathbf{v}), I_{t-s}^\mu \mathcal{E}(\mathbf{u})>| \leq \frac{1}{n!} <|v_{\sigma_1}|\otimes \dots \otimes |v_{\sigma_n}|, |u_{\rho_1}|\otimes \dots |u_{\rho_n}|>$$

and therefore, the finitely many non–zero components u_ρ, v_σ being bounded on $[0,T]$

$$\sum_{|\mu|=n} a_\mu \leq \frac{C^n (t-s)^n}{n!} .$$

For the sum on λ, we assume we work with subdivisions of step smaller than 1. Since the sum is only over $|\lambda|\geq 2$ we then bound s^m by s^2

$$\sum_{|\lambda|=m} b_\lambda \leq s^2 \frac{C^m}{m!} .$$

The coefficient of s^2 being uniformly bounded, (7.2) is proved.

8 ADDITIONAL RESULTS. 1) We are going to sketch a proof, due to Holevo, that leads to strong convergence in many cases. This discussion will not be used in the sequel.

We return to formula (7.1), and compute

$$(8.1) \quad \|(V_T - Q_n \dots Q_1)\mathcal{E}(j\mathbf{u})\|^2 = \sum_{ik} <V_{i+1}' D_i Q_{i-1}' \mathcal{E}(j\mathbf{u}), V_{k+1}' D_k Q_{k-1}' \mathcal{E}(j\mathbf{u})> .$$

It is sufficient to prove the terms with $i = k$ are $o(t_{i+1}-t_i)$, while those with $i \neq k$ are $o((t_{i+1} - t_i)(t_{k+1} - t_k))$. One first considers the case of a multiple Fock space (possibly infinite dimensional), without initial space — therefore the coefficients L_σ^ρ are scalars, and the initial vector j disappears from (8.1). Then $V_{i+1}' D_i Q_{i-1}'$ is a tensor product of three operators acting on different pieces of Fock space, and the scalar products on the right side decompose. The scalar product with $i = k$ appears as the product of $\| D_i \mathcal{E}(\mathbf{u}_i)\|^2$ by a bounded factor, and we leave it to the reader, using Ito's formula, to prove this is $O((t_{i+1} - t_i)^2)$. On the other hand, the scalar product with $i < k$ is the product of a bounded factor, and *two* factors

$$<D_i \mathcal{E}(\mathbf{u}_i), Q_i \mathcal{E}(\mathbf{u}_i)> , \quad <V_k \mathcal{E}(\mathbf{u}_k), D_k \mathcal{E}(\mathbf{u}_k)> .$$

Since $V_k = Q_k + D_k$ and the case of equal indexes has been considered above, only the first scalar product needs to be studied, which we again leave to the reader.

The case of a non-trivial initial space then is reduced to the preceding one using Holevo's domination principle for kernels (Chap.V, §2, subs. 5), as follows. We associate with the operators L_σ^ρ on $\mathcal{J}$ the scalars $\widehat{L}_\sigma^\rho = \| L_\sigma^\rho \|$, and solve the corresponding s.d.e. with trivial initial Fock space, marking with hats $\widehat{V}_i$, $\widehat{D}_i$ ... the corresponding operators. We also denote by $\widehat{\mathbf{u}}$ the element of $L^2(\mathbb{R}_+, \mathcal{K})$ with components $\widehat{u}_\alpha = |u_\alpha|$. Then the absolute value of any scalar product

$$< V'_{i+1} D_i Q'_{i-1} \mathcal{E}(j\mathbf{u}) , V'_{k+1} D_k Q'_{k-1} \mathcal{E}(j\mathbf{u}) >$$

is bounded by the corresponding scalar product

$$\| j \|^2 < \widehat{V}'_{i+1} \widehat{D}_i \widehat{Q}'_{i-1} \mathcal{E}(\widehat{\mathbf{u}}) , \widehat{V}'_{k+1} \widehat{D}_k \widehat{Q}'_{k-1} \mathcal{E}(\widehat{\mathbf{u}}) >$$

and therefore, strong convergence for the scalar equation implies the same result for the vector one — however, if the multiplicity is infinite, the Mohari–Sinha conditions for the scalar equation are stronger than for the vector equation, as they demand that $\sum_\sigma \| L_\sigma^\rho \|^2 < \infty$.

This method can be extended to s.d.e.'s whose coefficients depend on time, but act only on the initial space, while the method we have used for weak convergence can be used for s.d.e.'s with coefficients $L_\sigma^\rho(s)$ acting on Ψ_s.

2) Formally, what we did was to define a right multiplicative integral

$$V_t = \prod_{s<t} (I + \sum_{\rho\sigma} L_\rho^\sigma(s) \, da_\rho^\sigma(s))$$

(the higher times standing to the left of the product), and to identify V_t as the solution of a right s.d.e.. Holevo does the same thing with the product

$$W_t = \prod_{s<t} \exp(\sum_{\rho\sigma} L_\rho^\sigma(s) \, da_\rho^\sigma(s)) .$$

Both the convergence problem and the identification of the limit are much more difficult, and we can only refer to the article [Hol2].

Time reversal

9 We fix an interval (a, b) and consider the following unitary operator on $L^2(\mathbb{R}_+, \mathcal{K})$

(9.1) $\widehat{\mathbf{u}}(s) = \mathbf{u}(s)$ for $s < a$ or $s > b$, $= \mathbf{u}(a+b-s)$ for $a < s < b$.

We extend it as an unitary operator on Φ by second quantization, and then to Ψ tensoring with $I_{\mathcal{J}}$. Since the "hat" notation may be ambiguous, we also introduce a notation R or R_{ab} (simply R_b if $a = 0$) for this unitary operator, called the *time reversal operator on* (a, b). Note that $R^2 = I$. As usual, we have it act on operators on Ψ by $R(A) = RAR$. This preserves b–adapted operators.

We have $\int_0^\infty \widehat{\mathbf{u}}_s \cdot d\mathbf{X}_s = \int_0^\infty \mathbf{u}_s \cdot d\widehat{\mathbf{X}}_s$, where

(9.2) $\widehat{\mathbf{X}}_t = \mathbf{X}_t$ for $t \le a$ or $t \ge b$, $= \mathbf{X}_a + \mathbf{X}_b - \mathbf{X}_{a+b-t}$ for $a < t < b$, is a new Brownian motion, which generates the same Fock space (but not the same filtration on Fock space). Note that the same transformation applied to $X_t^0 = t$ preserves this function.

It is then very simple to describe how R acts on vectors and operators. First it transforms any multiple integral w.r.t. $\mathbf{X}$ into the same integral w.r.t. $\widehat{\mathbf{X}}$. It preserves the exponential domain, and transforms the Evans operators $a_\rho^\sigma(t)$ into the corresponding operators $\widehat{a}_\rho^\sigma(t)$ w.r.t. $\widehat{\mathbf{X}}$. Finally, any operator given by a kernel w.r.t. the family a_ρ^σ is transformed into the operator with the same kernel w.r.t. the family $\widehat{a}_\rho^\sigma$.

We will need the following elementary lemma : *if the operator V is a–adapted, then*

$$(9.3) \qquad R_{a+b}(R_a(V)) = \Theta_b(V) \ .$$

PROOF. We assume for simplicity the initial space is trivial, and denote by W an a–adapted operator. It will be sufficient to test equality on vectors $k = f \otimes \theta_b g \, \theta_{a+b} h$, with $f \in \Phi_b$, $g \in \Phi_a$, $h \in \Phi$. Then we have $R_{a+b}k = \widehat{g} \otimes \theta_a \widehat{f} \otimes \theta_{a+b} h$, where $\widehat{f}, \widehat{g}$ denote f and g reversed respectively at b and a. Applying W gives $W(\widehat{g}) \otimes \theta_a \widehat{f} \otimes \theta_{a+b} h$, and applying R_{a+b} gives

$$R_{a+b} W R_{a+b} \, k = f \otimes \theta_a (R_a W R_a g) \otimes \theta_{a+b} h \ .$$

If we now take $W = R_a(V)$, what we find is $R_{a+b}(R_a(V)) \, k = \Theta_a(V) \, k$.

Applying R_{a+b} to (9.3) gives, with the same notation

$$(9.4) \qquad R_{a+b}(\Theta_b V) = R_a(V) \ .$$

Consider now a right cocycle (V_t) :

$$V_{s+t} = \Theta_t(V_s) \, V_t \ ,$$

and define $U_t = R_t V_t$. Applying R_{s+t} to this relation and using (9.4) and (9.3) we have

$$U_{s+t} = (\, R_{s+t}(\Theta_t R_s(U_s))\,)(\, R_{s+t}(R_t(U_t))\,) = U_s \Theta_s U_t \ ,$$

and we see that *a right cocycle has been transformed into a left cocycle by time reversal.*

If there is a non–trivial initial space, properties (9.3) and (9.4) extend immediately to operators of the form $A \otimes V$, where A acts on the initial space and V is a–adapted on Φ. Then they are extended to general a–adapted operators by taking linear combinations and weak closure. The proof then is finished in the same way.

10 We take for our right cocycle the solution (V_t) to a right exponential equation and try to identify the corresponding left cocycle. We know that V_t is a weak limit (on the exponential domain) of products

$$(10.1) \qquad V_t = \lim \prod_i \Big(I + \sum_{\rho\sigma} \mathbf{L}_\sigma^\rho (a_\rho^\sigma(t_{i+1}) - a_\rho^\sigma(t_i)) \Big) \ .$$

Applying R_t preserves weak convergence on the exponential domain, and therefore we have

$$(10.2) \qquad U_t = \lim \prod_i \Big(I + \sum_{\rho\sigma} \mathbf{L}_\sigma^\rho (\widehat{a}_\rho^\sigma(t_{i+1}) - \widehat{a}_\rho^\sigma(t_i)) \Big) \ .$$

On the other hand, these increments of the process $\widehat{a}_\rho^\sigma$ are also increments of a_ρ^σ relative to a reversed subdivision, *i.e.* the order of factors is now reversed in time. Therefore U_t is a good candidate to be a solution to the *left* equation. That it is indeed so is "Journé's time reversal principle".

We are able to prove the "principle" (following Mohari) whenever the Mohari–Sinha regularity condition is satisfied by the adjoint coefficients. Indeed, in this case we may apply the Riemann product approximation to the right equation with adjoint coefficients, then take adjoints, and find a Riemann product approximation for the left equation, which is the same as (10.2).

> If the adjoint regularity condition is not satisfied, it is not clear in which sense (U_t) constructed by time reversal can be considered a solution to the left equation.

REMARK. Our proof of unitarity in subs.4 used time reversal, which has been established only in the case of bounded coefficients. For a proof which applies to unbounded coefficients (and finite multiplicity), see Applebaum [App2], Theorem 4.2.

Existence of quantum flows

11 It will be useful for applications to generalize the setup of quantum flows, replacing the algebra $\mathcal{A}$ of operators on the initial space $\mathcal{J}$ by a mere linear space, the coefficients $\mathbf{L}_\sigma^\rho$ being linear mappings from $\mathcal{A}$ to $\mathcal{A}$, and the role of "random variables" X_s being played by linear mappings from $\mathcal{A}$ to s-adapted operators on Ψ. It is only assumed that a linear mapping X_0 is given, from $\mathcal{A}$ to operators on $\mathcal{J}$ (or equivalently 0-adapted operators on Ψ). Then we are going to discuss an equation

$$(11.1) \qquad F \circ X_t = F \circ X_0 + \int_0^t \sum_{\rho\sigma} \mathbf{L}_\sigma^\rho(F) \circ X_s \, da_\rho^\sigma(s) \,,$$

where we have kept the notation $F \circ X_t$ for $X_t(F)$.

In many cases, $\mathcal{A}$ has an involution, and the mappings $\mathbf{L}_\sigma^\rho$ satisfy the condition $\mathbf{L}_\sigma^\rho(F^*) = (\mathbf{L}_\rho^\sigma(F))^*$, and X_0 the condition $F^* \circ X_0 = (F \circ X_0)^*$. Then we expect also to have $F^* \circ X_t = (F \circ X_t)^*$. However, this will not play an important role in the discussion.

Equation (11.1) includes the standard equation of quantum flows, and also the usual right equation (1.1), taking $\mathcal{A}=\mathcal{L}(\mathcal{J})$, $\mathbf{L}_\sigma^\rho(F) = L_\sigma^\rho F$ and $F \circ X_t = V_t F$, while the left equation corresponds to $\mathbf{L}_\sigma^\rho(F) = F L_\sigma^\rho$, $F \circ X_t = F U_t$. In these cases, the cumbersome $F \circ X_0$ is simply F. The reader may wish to consider only this situation at a first reading, though we will mention later an application of (11.1) where $\mathcal{J} = \mathbb{C}$ and $F \circ X_0$ is of the form $\delta(F) I$, δ being a linear functional on $\mathcal{A}$.

To avoid lengthy expressions, we use multi-index notation : μ is a multi-index of length $|\mu| = n$, with components $\varepsilon_i = \binom{\sigma_i}{\rho_i}$, and I_t^μ is the corresponding multiple integral $\int_{s_1<\ldots<s_n<t} da_{\rho_n}^{\sigma_n}(s_n) \ldots da_{\rho_1}^{\sigma_1}(s_1)$; $\mathbf{L}_\mu(F)$ is $\mathbf{L}_{\sigma_n}^{\rho_n}(\ldots(\mathbf{L}_{\sigma_1}^{\rho_1}(F))\ldots)$ (upper and lower indices have been exchanged). *Finally, we denote by* $\mathbf{u}_n$ *the set of multi-indexes* μ *of length* n *whose indexes* ρ_i *belong to the "support"* $S(\mathbf{u})$.

Given any fixed $F \in \mathcal{A}$, $j \in \mathcal{J}$ and $\mathbf{u} \in L^2(\mathbb{R}_+, \mathcal{K})$ with finite "support", we put

$$(11.2) \qquad \boxed{S_n^2(F, \mathbf{u}, j) = \sum_{\mu \in \mathbf{u}_n} \| \mathbf{L}_\mu(F) \circ X_0 j \|^2} \,.$$

Explicitly, the sum bears on multi-indexes of length n with all ρ_i belonging to $S(\mathbf{u})$, but σ_i arbitrary. In a somewhat inaccurate way, we will call *Mohari–Sinha hypothesis* the following assumption (see remark 3 below) : *there exists a constant K (depending on $F, j, \mathbf{u}$) such that*

(11.3)
$$S_n^2(F, j, \mathbf{u}) \le K^n .$$

We will call a *weak Mohari–Sinha hypothesis* the following assumption

(11.3')
$$S_n^2(F, j, \mathbf{u}) \le K \, n! \, R^{-n} \quad \text{for all } R > 0 .$$

The constants K appearing in these formulas are examples of "uninteresting constants", and we use for them dummy notations $K, C, a \ldots$, without caring to change the letter from place to place.

REMARKS 1) In the case of flows, the finiteness of $S_1(F, j, \mathbf{u})$ for all j means that for given ρ and F, the sum $\sum_\sigma \mathbf{L}_\sigma^\rho(F)^* \mathbf{L}_\sigma^\rho(F)$ is strongly convergent; using the lemma from subsection 5, this gives a meaning to the infinite sums $\sum_\alpha \mathbf{L}_\alpha^\rho(G)\mathbf{L}_\sigma^\alpha(F)$ appearing in the structure equations.

2) If $\mathcal{A}$ has an involution satisfying the hypotheses mentioned above, replacing F by F^* and taking adjoints we may reverse the roles of upper and lower indexes.

3) A sufficient condition for (11.3), involving the mappings $\mathbf{L}_\sigma^\rho$ and not their products, is taken by Mohari–Sinha as their main assumption :

(11.4) $\sum_\sigma \| \mathbf{L}_\sigma^\rho(F) \circ X_0 \, j \|^2 \le \sum_\sigma \| F \circ X_0 D_\sigma^\rho j \|^2$ with $\sum_\sigma \| D_\sigma^\rho j \|^2 \le C_\rho^2 \| j \|^2$

(in particular, the operators D_σ^ρ on $\mathcal{J}$ are bounded). However, the operators D_σ^ρ themselves are never used, and only (11.3) appears in the proof given in [MoS]. We prefer to emphasize the really basic property, at the cost of historical accuracy.

4) In the case of flows induced by unitary solutions of s.d.e.'s, §3 subs.4, the mappings $\mathbf{L}_\sigma^\rho$ being given by §3, (4.3), it is not trivial to relate the two systems of "Mohari–Sinha conditions".

12 We are going to prove that equation (11.1) has a unique solution under (11.3), using the Picard method.

The standard iteration gives an approximation of X_s by a sequence of mappings X_s^n from $\mathcal{A}$ to s–adapted operators on Ψ. We use again the notation $F \circ X_s^n$ for $X_t^n(F)$ (though in the case of flows it should not suggest that X_t^n is a random variable, *i.e.* an algebra homomorphism). The value of $F \circ X_0$ is known, while we have

(12.1)
$$F \circ X_t^{n+1} = F \circ X_0 + \int_0^t \sum_{\rho\sigma} \mathbf{L}_\sigma^\rho(F) \circ X_s^n \, da_\rho^\sigma(s) .$$

In multi-index notation, we have an explicit form for $F \circ X_t$,

(12.2)
$$F \circ X_t = \sum_\mu \mathbf{L}_\mu(F) \circ X_0 \otimes I_t^\mu ,$$

this series being summed by blocks according to the multi-indexes' lengths, and $F \circ X_t^n$ being its partial sum corresponding to $|\mu| \leq n$. We apply this series to an exponential vector $\mathcal{E}(j\mathbf{u})$ and prove it is strongly convergent. We have

$$(F \circ X_t^n - F \circ X_t^{n-1}) \mathcal{E}(j\mathbf{u}) = \sum_{\mu \in \mathbf{u}_n} \mathbf{L}_\mu(F) \circ X_0 j \otimes I_t^\mu \mathcal{E}(\mathbf{u})$$

whose squared norm (according to §1, formula (9.8)) is bounded for $t \leq T$ by the following quantity

$$(12.3) \qquad \boxed{a \frac{(C\nu_\mathbf{u}(t))^n}{n!} \sum_{\mu \in \mathbf{u}_n} \| \mathbf{L}_\mu(F) \circ X_0 j \|^2} \quad \text{with} \quad a = \| \mathcal{E}(\mathbf{u}) \|^2 , \, C = 2e^{\nu_\mathbf{u}(T)} .$$

The coefficient $\nu_\mathbf{u}(t)$ is significant, since it tends to 0 with t. However, our discussion here is not refined, and we pack it with the other ones. Using the notation (11.2) we get the inequality

$$(12.4) \qquad \| (F \circ X_t^n - F \circ X_t^{n-1}) \mathcal{E}(j\mathbf{u}) \| \leq \frac{C^n}{\sqrt{n!}} S_n(F, j, \mathbf{u}) ,$$

which under (11.3) or (11.3$'$) has a finite sum over n. Therefore we may pass to the limit and define $F \circ X_t \mathcal{E}(j\mathbf{u})$. We also get a bound for $\| F \circ X_t^n \mathcal{E}(j\mathbf{u}) \|$, uniform in n and $t \leq T$

$$(12.5) \qquad \boxed{\| F \circ X_t^n \mathcal{E}(j\mathbf{u}) \| \leq a \sum_n \frac{C^n}{\sqrt{n!}} S_n(F, j, \mathbf{u})} ,$$

with a different constant C and $a = \mathcal{E}(\mathbf{u})$. *We denote by $T_0(F, j, \mathbf{u})$ the right side of this equation.*

The weak Mohari–Sinha assumption (11.4) leads to an estimate that will be useful for the sequel. Applying the Schwarz inequality to

$$T_0(F, j, \mathbf{u}) = a \sum_n (C/R)^n \, S_n(F, j, \mathbf{u}) \, R^n / \sqrt{n!}$$

we get that

$$(12.6) \qquad \boxed{T_0^2(F, j, \mathbf{u}) \leq a \sum_n \frac{C^n}{\sqrt{n!}} S_n^2(F, j, \mathbf{u})} .$$

Let us return to the mappings X_t, and prove they satisfy the equation (11.1) — and first, that the stochastic integral $\int_0^t \sum_{\rho\sigma} \mathbf{L}_\sigma^\rho(F) \circ X_s \, da_\rho^\sigma(s)$ has a meaning on the exponential domain. According to §1, (9.8) this amounts to saying that

$$\sum_{\rho\sigma} \int_0^t \| (\mathbf{L}_\sigma^\rho(F) \circ X_s \mathcal{E}(j\mathbf{u}) \|^2 \, \nu_\mathbf{u}(ds)$$

is finite. On the other hand, we have by definition $\mathbf{L}_\sigma^\rho(F) \circ X_t = \lim_n \mathbf{L}_\sigma^\rho(F) \circ X_t^n$ which can be explicitly computed and is easy to dominate. Then one checks that

$$\int_0^t \sum_{\rho\sigma} \mathbf{L}_\sigma^\rho(F) \circ X_s \, da_\rho^\sigma(s) = \lim_n \int_0^t \sum_{\rho\sigma} \mathbf{L}_\sigma^\rho(F) \circ X_s^n \, da_\rho^\sigma(s) \;,$$

which follows from the fact that

$$\sum_{\rho\sigma} \int_0^t \| (\mathbf{L}_\sigma^\rho(F) \circ X_s - \mathbf{L}_\sigma^\rho(F) \circ X_s^n) \, \mathcal{E}(j\mathbf{u}) \|^2 \, \nu_{\mathbf{u}}(ds)$$

tends to 0 (similar reasoning). This being done, the equation is deduced from the induction relation defining $F \circ X_t^n$ by a passage to the limit.

We leave to the reader a few easy details, like the strong continuity in t of $F \circ X_t$ on the exponential domain, or the relation $F^* \circ X_t = (F \circ X_t)^*$.

It is interesting to note that instead of (11.3) we may assume that, for all $R > 0$ we have an inequality

(12.6) $$S_n(F, j, \mathbf{u}) \le C \sqrt{n!} \, R^{-n} \;,$$

similar to that we used for s.d.e.'s with unbounded coefficients.

The homomorphism property

13 Here we deal with the specific case of Evans–Hudson flows, and discuss the multiplicative property of the flow, $(GF) \circ X_t = (G \circ X_t)(F \circ X_t)$. This is the main algebraic difficulty of the theory. The first proof (due to Evans) was transformed by Mohari–Sinha in a very remarkable way to adapt to infinite dimensional situations. Since now $\mathcal{A} \subset \mathcal{L}(\mathcal{J})$ and $F \circ X_0 = F \otimes I$, we can write Fj, $F\mathcal{E}(j\mathbf{u})$ without mentioning X_0.

In view of later applications, the reader should read this proof carefully and convince himself that, in the final estimate, *the product GF disappears altogether,* and the result amounts to proving that some complicated sums of scalar products tend to 0, involving only vectors of the form $\mathbf{L}_\mu(F) \mathcal{E}(j\mathbf{u})$, $\mathbf{L}_\mu(G) \mathcal{E}(\ell\mathbf{v})$.

We want to prove that

$$B_t(G, F) = \; <\mathcal{E}(\ell\mathbf{v}), \, (G \circ X_t \, F \circ X_t) \mathcal{E}(j\mathbf{u}) > \; - \; <\mathcal{E}(\ell\mathbf{v}), \, (GF) \circ X_t \mathcal{E}(j\mathbf{u}) > \; = 0$$

The first scalar product can be written $<G^* \circ X_t \mathcal{E}(\ell\mathbf{v}), \, F \circ X_t \, \mathcal{E}(j\mathbf{u}) >$. Since strong convergence holds on the exponential domain, it suffices to show that

(13.1) $B_t^n(G, F) = \; <G^* \circ X_t^n \mathcal{E}(\ell\mathbf{v}), \, F \circ X_t^n \, \mathcal{E}(j\mathbf{u}) > \; - \; <\mathcal{E}(\ell\mathbf{v}), \, (GF) \circ X_t^n \mathcal{E}(j\mathbf{u}) >$

tends to 0 as $n \to \infty$.

Using Ito's formula, we transform this expression into $\sum_{\rho\sigma} \int_0^t v^\sigma(s) \, u_\rho(s) \, C_\sigma^\rho(s) \, ds$, putting

$C_\sigma^\rho(s) = \; <G^* \circ X_s^n \mathcal{E}(\ell\mathbf{v}), \, \mathbf{L}_\sigma^\rho F \circ X_s^{n-1} \, \mathcal{E}(j\mathbf{u}) > + <\mathbf{L}_\rho^\sigma G^* \circ X_s^{n-1} \mathcal{E}(\ell\mathbf{v}), \, F \circ X_s^n \, \mathcal{E}(j\mathbf{u}) >$

$+ \sum_\alpha <\mathbf{L}_\alpha^\sigma G^* \circ X_s^{n-1} \mathcal{E}(\ell\mathbf{v}), \mathbf{L}_\alpha^\rho F \circ X_s^{n-1} \, \mathcal{E}(j\mathbf{u}) > \; - \; <\mathcal{E}(\ell\mathbf{v}), \, \mathbf{L}_\sigma^\rho(GF) \circ X_s^{n-1} \, \mathcal{E}(j\mathbf{u}) >$

Note that indexes are not balanced! If X_s^{n-1} were a homomorphism, we could use the structure equation to transform $B_t^n(G, F)$ into the following expression (the sum is over $\sigma \in S(\mathbf{v})$ and $\rho \in S(\mathbf{u})$)

$$R_t^n(G, F) = \sum_{\rho\sigma} \int_0^t v^\sigma(s)\, u_\rho(s) \qquad \times$$

$$(13.2) \quad (< (G^* \circ X_s^n - G^* \circ X_s^{n-1}) \mathcal{E}(\ell \mathbf{v}),\, \mathbf{L}_\rho^\rho F \circ X_s^{n-1}\, \mathcal{E}(j\mathbf{u}) >$$
$$+ < \mathbf{L}_\rho^\sigma(G^*) \circ X_s^{n-1} \mathcal{E}(\ell \mathbf{v}),\, (F \circ X_s^n - F \circ X_s^{n-1}) \mathcal{E}(j\mathbf{u}) >)\, ds \quad ;$$

but since X_s^{n-1} is not multiplicative, a second term appears

$$Z_t^n(G, F) = \sum_{\rho\sigma} \int_0^t v^\sigma(s)\, u_\rho(s) \qquad \times$$

$$(13.3) \quad (B_s^{n-1}(G, \mathbf{L}_\sigma^\rho F) + B_s^{n-1}(\mathbf{L}_\sigma^\rho G, F) + \sum_\alpha B_s^{n-1}(\mathbf{L}_\sigma^\alpha G, \mathbf{L}_\alpha^\rho F))\, ds \ .$$

For possible extensions, it may be useful to note that the structure equation is used only within a scalar product, *i.e.* the infinite sum appearing in it need not be defined as an operator, only as a form.

We start with the domination of $R_t^n(G, F)$, which consists of two similar terms with G, F interchanged, of which we study the first. We recall the notation $\nu_{\mathbf{v}}(ds) = (1 + \|\mathbf{v}(s)\|^2)\, ds$. Several measures of this kind occur in the proof, and for simplicity we use only the largest one $\nu(ds)$ with density $(1 + \|\mathbf{v}(s)\|^2)(1 + \|\mathbf{u}(s)\|^2) = \sum_{\rho\sigma} |v^\sigma(s)\, u_\rho(s)|^2$.

We dominate the scalar product

$$< (G^* \circ X_s^n - G^* \circ X_s^{n-1}) \mathcal{E}(\ell \mathbf{v}),\, \sum_{\rho\sigma} v^\sigma(s)\, u_\rho(s)\, \mathbf{L}_\sigma^\rho F \circ X_s^{n-1}\, \mathcal{E}(j\mathbf{u}) > \ = \ < A(s),\, B(s) >$$

by $\|A(s)\| \, \|B(s)\|$, and the integral $R_t^n(G, F)$ then is bounded using the Schwarz inequality. We first apply the Mohari–Sinha inequality extended to multi-indexes (§1, (9.8)) : since $A(s) = \sum_{\mu \in \mathbf{u}_n} \mathbf{L}_\mu(G^*)\, I^\mu(s)\, \mathcal{E}(\ell \mathbf{v})$ we have

$$\|A(s)\|^2 \leq \int_{s_1 < \ldots < s_n < s} \|\mathbf{L}_\mu(G^*)\, \ell \otimes I_s^\mu \mathcal{E}(\mathbf{v})\|^2\, \nu(ds_1) \ldots \nu(ds_n)$$

$$= \sum_\mu \|\mathbf{L}_\mu(G^*)\, \ell\|^2\, \frac{\nu(s)^n}{n!}\, \|\mathcal{E}(\mathbf{v})\|^2 \ ,$$

still to be integrated from 0 to t. Otherwise stated

$$(13.4) \qquad \int_0^t \|A(s)\|^2\, ds \leq a\, \frac{\nu(t)^{n+1}}{(n+1)!}\, S_n^2(G^*) \ .$$

Again a denotes an uninteresting quantity, and $S_n(G)$ abbreviates $S_n(G, \ell, \mathbf{v})$, since the last two data always go hand in hand with G and no ambiguity occurs. Similarly

$$(13.5) \qquad \|B(s)\|^2\, ds \leq$$
$$(\sum_{\rho\sigma} \|\mathbf{L}_\sigma^\rho(F) \circ X_s^{n-1}\, \mathcal{E}(j\mathbf{u})\|^2)(\sum_{\rho\sigma} |v^\sigma(s)\, u_\rho(s)|^2)\, ds \leq a\, T_1^2(F)\, \nu(ds) \ ,$$

where we put generally

(13.6) $$T_n^2(F) = T_n^2(F, j, \mathbf{u}) = \sum_{\mu \in \mathbf{u}_n} T_0^2(\mathbf{L}_\mu(F), j, \mathbf{u}) \ .$$

Integrating from 0 to t has no interesting effect, and contributes only to the constant a. Therefore, the first term in $R_t^n(G, F)$ is bounded by

(13.7) $$\boxed{a \frac{\nu(t)^{(n+1)/2}}{\sqrt{(n+1)!}} \, S_n(G^*) \, T_1(F)} \ ,$$

and the second term is similar, with $T_1(G^*)$ and $S_n(F)$.

We decompose each one of the three B_s^{n-1} which appear in the expression of $Z_t^n(G, F)$ into the corresponding $R_s^{n-1} + Z_s^{n-1}$ and apply the preceding method to each R_s^{n-1}. Repeating this reasoning, each Z_s^{n-1} generates iterated integrals of order 2 acting on R_s^{n-2}, etc. Finally, we will have to evaluate a large number of iterated integrals of order $k \leq n$, bearing on functions $R_s^{n-k}(\cdot, \cdot)$ where the operators involved are suitable $\mathbf{L}_\lambda(G)$ on the left and $\mathbf{L}_\lambda(F)$ on the right.

Putting everything together, the quantity $B_t^n(G, F)$ that should tend to 0 is estimated as the product of a factor $C^n / \sqrt{(n+1)!}$ by the following sum

$$\sum_{k \leq n} \sum_\theta \frac{C^k}{\sqrt{k!}} \Big(\sum_{\mu \in \mathbf{v}_{q+r}} S_{n-k}^2(\mathbf{L}_\mu G^*) \Big)^{1/2} \Big(\sum_{\mu \in \mathbf{u}_{p+r}} T_1^2(\mathbf{L}_\mu F) \Big)^{1/2}$$

and a similar sum we have forgotten. The internal sum depends on θ only through p, q, r. We are going to use the weak M–S estimate $(11.3')$. In the first sum on the right, $S_{n-p}^2(G^*)^{1/2}$ can be bounded by $a\sqrt{(n-p)!} \, S^{-(n-p)}$ with S arbitrarily large.

In the second sum, we have from (12.6) that $T_0^2(F) \leq a \sum_m (C^m / m!) \, S_m^2(F)$. Therefore we have an estimate

$$\sum_m \frac{C^m}{m!} S_{m+1+p+r}^2(F) \leq a \sum_m \frac{C^m}{m!} \sqrt{(m+1+p+r)!} \, R^{-(m+1+p+r)} \ .$$

We bound the ratio $(m+1+p+r)! / (m+1)!$ by $2^{m+1+p+r} (p+r)!$ and we include 2^m in C^m and 2^{p+r} in $R^{-(p+r)}$. There remains something of order $a R^{-(p+r)} (p+r)!$. Multiplying the two sums and taking a square root, we get

$$a \sum_{k \leq n} \sum_\theta \frac{C^k}{\sqrt{k!}} \sqrt{(n-p)!(p+r)!} \, S^{-(n-p)/2} R^{-(p+r)/2} \ .$$

The ratio $(p+r)!/k!$ can be replaced by 1, we take $R = S > 1$, we include the C^k in the C^n that comes in front, and get finally

$$\frac{C^n}{\sqrt{(n+1)!}} \sum_{k \leq n} \sum_\theta a\sqrt{(n-p)!} \, S^{-n} \ .$$

We again replace $(n-p)!/(n+1)!$ by 1, and since the total number of mappings from a subset of $\{1,\dots,n\}$ to $\{1,2,3\}$ is 4^n, our sum is bounded by $a(4C/S)^n$, which is kind enough to tend to 0 if we have taken S large enough.

REMARK. In the case of a finite dimensional initial space, the solution $F \circ X_t$ can be expressed as an explicit kernel $K(F)$ taking values in the initial algebra

$$F \circ X_t = \sum \int_{s_1 < \dots < s_n < t} K_{\varepsilon_1 \dots \varepsilon_n}(s_1, \dots s_n\,;\, F)\, da_{s_n}^{\varepsilon_n} \dots da_{s_1}^{\varepsilon_1}$$
$$K_{\varepsilon_1 \dots \varepsilon_n}(s_1, \dots, s_n\,;\, F) = L_{\varepsilon_n} \dots L_{\varepsilon_1}(F)$$

the index ε_n corresponding to the larger time stands to the left in the product. One may wonder whether the formula for composition of kernels can be combined with the structure equation to give a purely combinatorial proof of the homomorphism property.

14   Let us prove $\xi = X_t$ is a *contractive* homomorphism. The reasoning comes from C^*-algebraic folklore, arranged by Evans. We consider a vector $h = \sum_i j_i \otimes \mathcal{E}(\mathbf{u}_i)$ in the exponential domain and compute

$$\|\,\xi(F)\,h\,\|^2 = |<h\,,\,\xi(F^*F)\,h>| \;\leq\; \|\,h\,\|\;\|\,\xi(F^*F)\,h\,\|\;.$$

Since now $F^*F = G$ is selfadjoint, we have $G^*G = G^2$ and we iterate

$$\|\,\xi(F)\,h\,\|^2 \leq \|\,h\,\|^{1+\frac{1}{2}+\dots+\frac{1}{2^n}}\;\|\,\xi(G)^{2^n}\,h\,\|^{1/2^n}\;.$$

This last norm is bounded by

$$\sum_i \|\,\xi(G^{2^n})\,(j_i \otimes \mathcal{E}(\mathbf{u}_i))\,\| \leq \sum_i C(\mathbf{u}_i)\,\|\,G^{2^n}\,\|\;\|\,j_i\,\|$$

and since $\|\,G^{2^n}\,\| = \|\,G\,\|^{2^n}$, we get, after raising the inequality to the power $1/2^n$, the product of $\|\,G\,\| = \|\,F^2\,\|$ by a factor which tends to 1. Thus $\|\,\xi(F)\,h\,\|^2 \leq \|\,h\,\|^2\,\|\,F\,\|^2$ and contractivity is proved.

We mention a useful result of Mohari, according to which $F \circ X_t$ belongs to the von Neumann algebra generated by $\mathcal{A} \otimes \mathcal{F}_t$ (the algebra of operators on Φ adapted at time t).

15   We sketch, still following Mohari, the proof of a simple uniqueness result a much deeper one will be given later). Consider two *contractive* versions of the flow — here the notation $F \circ X_t$ becomes very misleading, and we call $\xi_t(F)$ and $\eta_t(F)$ the two versions, and $\zeta_t(F)$ their difference. We have

$$\zeta_t(F) = \sum_{\rho\sigma} \int_0^t \zeta_s(\mathbf{L}_\sigma^\rho F)\, da_\rho^\sigma(s)\;,\qquad \|\,\zeta_t(F)\,\| \leq C\,\|\,F\,\|\;,$$

this last property (with $C = 2$) being a consequence of contractivity. Then we consider the scalar product $<\mathcal{E}(\ell\mathbf{v})\,,\,\zeta_t(F)\,\mathcal{E}(j\mathbf{u})>$ relative to two exponential vectors with bounded arguments $\mathbf{u}, \mathbf{v}$, and introduce the operator (depending on s)

$$L_s(F) = \sum_{\rho\sigma} v^\sigma(s)\, u_\rho(s)\, \mathbf{L}_\sigma^\rho(F)\;.$$

Then we apply Ito's formula

$$< \mathcal{E}(\ell\mathbf{v}),\, \zeta_t(F)\,\mathcal{E}(j\mathbf{u}) > = \int_0^t < \mathcal{E}(\ell\mathbf{v}),\, \zeta_{s_1}(L_{s_1}F)\,\mathcal{E}(j\mathbf{u}) > ds_1$$

We replace inside the integral

$$< \mathcal{E}(\ell\mathbf{v}),\, \zeta_{s_1}(L_{s_1}F)\,\mathcal{E}(j\mathbf{u}) > = \int_0^{s_1} < \mathcal{E}(\ell\mathbf{v}),\, \zeta_{s_2}(L_{s_2}L_{s_1}F)\,\mathcal{E}(j\mathbf{u}) > ds_2$$

and iterate n times. Then using contractivity and $(11.3')$ it is easy to show that the integral tends to 0.

16   Let us prove a beautiful theorem of Parthasarathy–Sinha, using which one may construct classical stochastic processes by means of non commutative flows : *if the initial algebra $\mathcal{A}$ is commutative, then $F \circ X_s$ and $G \circ X_{s+t}$ commute for $t \geq 0$*, thus leading to a commutative process of operators. To see this, we take Ψ_s as initial space, considering $\mathcal{A}$ as an operator algebra on Ψ_s. We also put

$$\zeta_t(G) = \xi_{s+t}(G)\,\xi_s(F) - \xi_s(F)\,\xi_{s+t}(G)$$

which for $t = 0$ is equal to $\xi_s(GF - FG) = 0$. Then it is easy to check that

$$\zeta_t(G) = \sum_{\rho\sigma} \int_0^t \zeta_r(L_\sigma^\rho G)\, da_\rho^\sigma(r)$$

and the same reasoning as above proves that $\zeta_t(G) \equiv 0$.

Mohari's uniqueness theorem

17   The following result of Mohari is so elegant and simple that it may represent the definitive result on uniqueness. We will present in the next subsection the very recent results of Fagnola which provide the corresponding existence theorem. Thus the dream of a "stochastic Stone theorem" is becoming a reality.

Consider a left exponential equation with coefficients L_σ^ρ (possibly unbounded, admitting a common domain $\mathcal{D}$ dense in $\mathcal{J}$), and assume it has a contractive solution, strongly continuous in t on $\mathcal{D} \otimes \mathcal{E}$

$$(17.1) \qquad\qquad U_t = I + \sum_{\rho\sigma} \int_0^t U_s L_\sigma^\rho\, da_\rho^\sigma(s)\ .$$

Then Mohari proves the remarkable uniqueness result :

THEOREM. *If the closure of the operator $L_0^0 = L$ on $\mathcal{D}$ is the generator of a strongly continuous semigroup of contractions then the solution is unique.*

PROOF. Taking a difference of two solutions, we are reduced to proving that a strongly continuous family of uniformly bounded operators W_t satisfying on $\mathcal{D}$ the relation

$$W_t = \sum_{\rho\sigma} \int_0^t W_s L_\sigma^\rho\, da_\rho^\sigma(s)$$

must be identically 0. For fixed $k, \mathbf{u}, \mathbf{v}$ we define linear functionals on $\mathcal{J}$ by

$$h_t^{m,p}(j) = <k\mathbf{v}^{\otimes p}, W_t(j\mathbf{u}^{\otimes m})> \quad , \quad h_t^m(j) = <k\mathcal{E}(\mathbf{v}), W_t(j\mathbf{u}^{\otimes m})> ,$$

and we will prove that $h_t^m = 0$ for all t and m. It would be misleading to put k or j inside the exponentials here. We introduce a complex parameter z and apply the simple Ito formula 1.(8.4) for exponential vectors

$$\sum_m \frac{z^m}{m!} <k\mathcal{E}(\mathbf{v}), W_t(j\mathbf{u}^{\otimes m})> \; = \; <k\mathcal{E}(\mathbf{v}), W_t(j\mathcal{E}(z\mathbf{u}))> =$$

$$= \sum_{\rho\sigma} \int_0^t z^{(\rho)} v^\sigma(s) u_\rho(s) <k\mathcal{E}(\mathbf{v}), W_s L_\sigma^\rho j\mathcal{E}(z\mathbf{u})> ds$$

where $(\rho)=1$ if $\rho\neq 0$, 0 if $\rho=0$. Expanding the right side gives

$$\sum_{\rho\sigma} \int_0^t \frac{z^{n+(\rho)}}{n!} v^\sigma(s) u_\rho(s) <k\mathcal{E}(\mathbf{v}), W_s L_\sigma^\rho (j\mathbf{u}^{\otimes n})> ds .$$

We identify coefficients and get for $m > 0$

$$<k\mathcal{E}(\mathbf{v}), W_t(j\mathbf{u}^{\otimes m})> \; = \sum_\sigma \int_0^t v^\sigma(s) <k\mathcal{E}(\mathbf{v}), W_s L_\sigma^0 (j\mathbf{u}^{\otimes m}> ds$$

$$+ m \sum_{\sigma,\alpha} \int_0^t v^\sigma(s) u_\alpha(s) <k\mathcal{E}(\mathbf{v}), W_s L_\sigma^\alpha (j\mathbf{u}^{\otimes(m-1)})> ds .$$

For $m = 0$ the term on the right disappears. We start with $m = 0$

$$<k\mathcal{E}(\mathbf{v}), W_t j> \; = \sum_\sigma \int_0^t v^\sigma(s) <k\mathcal{E}(\mathbf{v}), W_s L_\sigma^0 j> ds .$$

We again replace $\mathbf{v}$ by $\zeta\mathbf{v}$ and expand in powers ζ^n. For $n = 0$ we have

$$<k, W_t j> \; = \int_0^t <k, W_s L_0^0 j> ds .$$

or $h_t^{00} j = \int_0^t h_s^{00} Lj \, ds$. We put $k_p = \int_0^\infty e^{-ps} h_s^{00} \, ds$, a bounded linear functional, and we have $pk_p = k_p L$, or $k_p(pI - L) = 0$ on $\mathcal{D}$. Now the image $(pI - L)\mathcal{D}$ is dense, hence $k_p = 0$, and then inverting the Laplace transform $h_t^{00} = 0$. Then we proceed to $n = 1$,

$$<k\mathbf{v}, W_t(j)> \; = \sum_\alpha \int_0^t <k\mathbf{v}, W_s L_\alpha^0 j> ds = 0 ,$$

from the preceding result. An easy induction shows that $h_t^{0n} = 0$ for all n. Then we pass to $m = 1$, etc.

Fagnola's existence theorem

18 We give now a sketch of the results contained in very recent preprints of Fagnola [Fag4] [Fag5] [Fag6], one of which is the existence result corresponding to Mohari's uniqueness theorem.

Knowing Mohari's result, it is natural to make the following assumption on the operator $L = L_0^0$ with domain $\mathcal{D}$: *the closure of L is the generator of a strongly continuous contraction semigroup* (P_t). We denote by $W_p = \int_0^\infty e^{-ps} P_s \, ds$ its resolvent, whose range $\mathcal{G}$ is (according to the Hille–Yosida theory) the generator's domain, and we put $R_p = pW_p$, a strong approximation of identity as $p \to \infty$.

Secondly, we assume that *the mappings λ and $\widetilde{\lambda}^*$ are defined on $\mathcal{D}$ and controlled by L*

$$(18.1) \qquad \|\lambda j\|, \|\widetilde{\lambda}^* j\| \leq C(\|j\| + \|Lj\|)$$

(the norm on the left hand side is that of $\mathcal{J} \otimes \mathcal{K}$), with the consequence that their closures are defined on $\mathcal{G}$ and λR_p, $\widetilde{\lambda}^* R_p$ are bounded mappings which converge to λ, $\widetilde{\lambda}^*$ as $p \to \infty$.

Finally, we assume that *the contractivity condition*

$$(18.2) \qquad < \binom{j}{h}, \mathbf{L}\binom{j}{h} > \, \leq 0$$

is satisfied for $j \in \mathcal{D}$ and $h \in \mathcal{D} \otimes \mathcal{K}$ (uncompleted), defining as in §3, subs. 3

$$\mathbf{L} = \begin{pmatrix} L^* + L + \lambda^* \lambda & \lambda^* + \widetilde{\lambda} + \lambda^* \Lambda \\ \widetilde{\lambda}^* + \lambda + \Lambda^* \lambda & \Lambda^* + \Lambda + \Lambda^* \Lambda \end{pmatrix} .$$

Property (18.2) as we have written it must be properly interpreted, since the operators and their products are not everywhere defined. First of all, Λ must be a bounded operator on $\mathcal{J} \otimes \mathcal{K}$. For the other operators, we use an interpretation in the form sense :

$$< j, (L^* + L + \lambda^* \lambda) j > \quad \text{means} \quad 2 \, \Re e < j, Lj > + < \lambda j, \lambda j > ,$$
$$< j, (\lambda^* + \widetilde{\lambda} + \lambda^* \Lambda) h > \quad \text{means} \quad < (\lambda + \widetilde{\lambda}^*) j, h > + < \lambda j, \Lambda h > .$$

The meaning of the two other terms is clear. The main remark of Fagnola [Fag5] is the following : if we define

$$(18.3) \qquad L_p = R_p^* L R_p , \quad \lambda_p = \lambda R_p , \quad \widetilde{\lambda}_p = R_p^* \widetilde{\lambda} , \quad \Lambda_p = \Lambda ,$$

then *the corresponding matrix $\mathbf{L}_p$ defines a bounded operator, and still satisfies the contractivity condition* — simply because, for $j \in \mathcal{J}$ and $h \in \mathcal{J} \otimes \mathcal{K}$

$$< \binom{j}{h}, \mathbf{L}_p \binom{j}{h} > \, = \, < \binom{R_p j}{h}, \mathbf{L} \binom{R_p j}{h} > .$$

Using then the results of Mohari, one solves the left and right equations with these bounded coefficients, thus constructing a family of contractive left cocycles $U_t^{(p)}$ and right cocycles $V_t^{(p)}$ (transformed into each other by time reversal). And now, a weak compactness argument, adapted from a paper of Frigerio, allows us to take weak limits as $p \to \infty$, which are contractive cocycles U_t and V_t solving the original s.d.e.'s. Mohari's uniqueness theorem in subs. 17 plays an essential role in proving convergence. There are of course many technical details to check, but the proof after (18.3) is very natural.

It should be mentioned in this connection that Fagnola [Fag6], completing the work of Journé [Jou] and Accardi–Journé–Lindsay [AJL], proves that all cocycles satisfying a weak differentiability property are given by stochastic differential equations with unbounded coefficients of the preceding form.

There remain several questions to be treated.

1) Assuming the coefficients satisfy, not only the contractivity condition, but the formal unitarity condition, can one prove that the solution is unitary? An abstract necessary and sufficient condition is given, and a more practical one deriving from results of Chebotarev.

2) An application of this last result is given to the construction of solutions of Ito stochastic differential equations on $\mathbb{R}^n$ as quantum flows on the algebra C_c^∞, which applies to a large class of elliptic diffusions. This is much better than all previous results, though not yet completely satisfactory since ellipticity plays no role in Ito's theory.

Chapter VII
Independent Increments

We present here an introduction to some recent work of the Heidelberg school, due to W. von Waldenfels, P. Glockner, and specially M. Schürmann, on processes with independent increments in a non–commutative setup. This chapter is reduced to the bare essentials, and its purpose is only to lead the reader to the much richer original articles. Some very recent and interesting remarks from Belavkin [6] could be inserted into the chapter just before departure to the publisher's office. I am in great debt to lectures by Sauvageot in Paris, and Parthasarathy in Strasbourg, which made clear for me how natural coalgebras are in this setup.

Another definition of conditionally positive functions in a non–commutative setup was developed by Holevo [Hol3], and seems disjoint from Schürmann's : it concerns maps from a group to the space of non–commutative "kernels" (bounded linear maps from a C^*–algebra into itself).

§1 COALGEBRAS AND BIALGEBRAS

1 We have been long familiar with the idea that a "non–commutative state space E" is represented by a unital *–algebra $\mathcal{A}$, whose elements are "functions $F(x)$" on E. Unless it is a C^*–algebra, $\mathcal{A}$ is meant to represent smooth test functions rather than continuous functions, and therefore linear functionals on $\mathcal{A}$ represent distributions rather than measures — though *positive* functionals play the role of measures.

A non–commutative probability space Ω is represented as usual by a concrete operator *–algebra $\mathcal{B}$ on some Hilbert space, with a chosen state (we use the probabilistic notation $\mathbb{E}$ for the corresponding expectation). Then a random variable X on Ω with values in E is defined as a *–homomorphism from $\mathcal{A}$ to $\mathcal{B}$. Instead of the algebraists' notation $X(F)$ we like to use the probabilists' $F \circ X$ or even $F(X)$. In most cases $\mathcal{B}$ will be realized on Fock space, and the state will be the vacuum.

From now on, we say algebra instead of *–algebra with unit, homomorphism instead of *–homomorphism, etc.

An algebraic structure on E, like an internal multiplication $(x, y) \longmapsto xy$, is described through its reflection on functions, namely, the mapping (homomorphism) $F(x) \longmapsto F(xy)$ from the algebra of "functions of one variable" to the algebra of "functions of two variables". This is called the *coproduct,* and the function $F(xy)$ is denoted by ΔF. For the algebra of functions of two variables, the only reasonable choice when we have no topological structure on $\mathcal{A}$ is the algebraic tensor product $\mathcal{A} \otimes \mathcal{A}$, though it represents only finite sums of functions $F(x) G(y)$. Besides that, we would like to have a unit element in E, meaning that we have a mapping (homomorphism) $F(x) \longmapsto F(e)$ from $\mathcal{A}$ to $\mathbb{C}$, called the *co-unit* δ. The mappings Δ and δ should be homomorphisms, the product and involution on $\mathcal{A} \otimes \mathcal{A}$ being the standard ones [1]

$$(x \otimes y)(x' \otimes y') = (xx') \otimes (yy') , \qquad (x \otimes y)^* = y^* \otimes x^* .$$

We demand that

$$F((xy)z) = F(x(yz)) \quad \text{(coassociativity)} \quad , \quad F(xe) = F(ex) = F(x) .$$

The first property reads in algebraic notation $(Id \otimes \Delta)\Delta = (\Delta \otimes Id)\Delta$ as mappings from $\mathcal{A}$ to $\mathcal{A} \otimes \mathcal{A} \otimes \mathcal{A}$, and the second one means that $(Id \otimes \delta)\Delta = (\delta \otimes Id)\Delta = Id$ as mappings from $\mathcal{A}$ to $\mathcal{A}$. The structure we have thus described is called a *bialgebra*, more precisely a *–bialgebra. Note that the co-unit, as a *–homomorphism from $\mathcal{A}$ to $\mathbb{C}$, is also a *state* on the algebra $\mathcal{A}$.

It turns out, particularly when dealing with stochastic calculus, that the bialgebra axioms are very conveniently split into the standard algebra axioms and new *coalgebra axioms* involving only the coproduct and co-unit. If the axioms are dissociated in this way, a "random variable" taking values in a coalgebra is simply a linear mapping X from $\mathcal{A}$ to the operator algebra $\mathcal{B}$ such that $F^* \circ X = (F \circ X)^*$. We will keep using the functional notation with variables $F(x), F(xy)$, which is easily translatable into tensor notation, suggestive, and rarely misleading. The main idea from these axioms is that of a representation as a finite sum

$$F(xy) = \sum_i F_i'(x) F_i''(y) \quad \text{translating} \quad F = \sum_i F_i' \otimes F_i'' .$$

To represent a group, we would also need a mapping $x \longmapsto x^{-1}$, and therefore a linear mapping (homomorphism) $F \longmapsto SF$ representing $F(x^{-1})$. It is called the *antipode*, and a bialgebra with antipode is called a *Hopf algebra*. The antipode axiom expresses $F(xx^{-1}) = F(x^{-1}x) = F(e)$ and reads

$$M(Id \otimes S)\Delta F = M(S \otimes Id)F = \delta(F)1 .$$

Here M is the multiplication in $\mathcal{A}$, interpreted as a mapping from $\mathcal{A} \otimes \mathcal{A}$ to $\mathcal{A}$. One doesn't demand (and it is not always true) that $S^2 = Id$ — thus the functional notation $F(x^{-1})$ *is* misleading in the case of Hopf algebras. We never use an antipode in what follows.

We return to probabilistic ideas : the coalgebra structure allows one to form products of random variables, as we multiply (or add) group–valued random variables in classical probability. Given two random variables X, Y (mappings from $\mathcal{A}$ to $\mathcal{B}$) we define $XY = Z$ as follows : taking a representation $\Delta F = \sum_i F_i' \otimes F_i''$, we put

$$F \circ Z = \sum_i (F_i' \circ X)(F_i'' \circ Y) \quad \text{(product in } \mathcal{B}).$$

The algebraic theory of tensor products implies that the left side is independent of the choice of the representation. The involution axiom is trivially satisfied, but

---

[1] There are other possibilities for graded algebras.

in the bialgebra setup, X and Y are homomorphisms, and we should see whether Z is also. Taking a second element of $\mathcal{A}$, G with $\Delta G = \sum_k G'_k \otimes G''_k$, we have $\Delta(FG) = \sum_{ik} F'_i G'_k \otimes F''_i G''_k$, and therefore, putting $F'_i \circ X = f'_i \in \mathcal{B}$, etc

$$FG \circ Z = \sum_{ik} (f'_i g'_k)(f''_i g''_k) \quad , \quad (F \circ Z)(G \circ Z) = \sum_{ik} (f'_i f''_i)(g'_k g''_k)$$

and there is no reason why these products should be equal. They are equal in two important cases : 1) if the operators $F \circ X$ and $G \circ Y$ take their values in two commuting von Neumann subalgebras (this will be often the case with our processes on Fock space). 2) If Δ takes its values in the *symmetric* subspace of $\mathcal{A} \otimes \mathcal{A}$ (verification left to the reader), in which case $\mathcal{A}$ is said to be *co-commutative*.

2   EXAMPLES. 1) The addition on $\mathbb{R}^n$ is reflected by bialgebra structures on two useful sets of functions. The problem is to interpret $f(x+y)$ as a sum of products of functions of the variables x, y. This is clear for the coordinate mappings themselves : $X_i + Y_i$ is read as $X_i \otimes 1 + 1 \otimes X_i$, and we have a bialgebra structure on polynomials arising from the rules $\Delta 1 = 1 \otimes 1$ (simply denoted 1) and

$$(2.1) \qquad \qquad \Delta X_i = X_i \otimes 1 + 1 \otimes X_i \ ,$$

and extended by multiplication. Since the coordinates take the value 0 at the 0, the co-unit is given by extension of the rules $\delta(1) = 1$, $\delta(X_i) = 0$. Note that the space of affine functions is a coalgebra.

The coproduct is also clear for exponentials $e_u(x) = e^{iu \cdot x}$: the property $e_u(x+y) = e_u(x) e_u(y)$ leads to a bialgebra structure on trigonometric polynomials, arising from the rule

$$(2.2) \qquad \qquad \Delta e_u = e_u \otimes e_u \ .$$

Since $e_u(0) = 1$, the co-unit is given by extension of $\delta(e_u) = 1$. The involution is given by $e_u^* = e_{-u}$.

2) Let $\mathcal{G}$ be a semi-group (the multiplication is written xy without a product sign) with unit e. The matrix elements $m_i^j(x)$ of a finite dimensional representation of $\mathcal{G}$ and their complex conjugates generate a commutative $*$–algebra of functions on $\mathcal{G}$, and the relation $m_i^j(xy) = \sum_{ij} m_k^j(x) m_i^k(y)$ — extended naturally to products — turns it into a bialgebra. Of course, $\mathcal{G}$ may be itself a semi-group of matrices, and the representation the identity mapping. To take a simple example related to the interpretation of Azéma martingales, consider the matrix group (affine group)

$$(2.3) \qquad \qquad \begin{pmatrix} a & 0 \\ b & 1 \end{pmatrix} \begin{pmatrix} a' & 0 \\ b' & 1 \end{pmatrix} = \begin{pmatrix} aa' & 0 \\ ba' + b' & 1 \end{pmatrix} \ ,$$

We take as the basic algebra of functions on the affine group the complex polynomials in the two variables a, b, the involution being complex conjugation (otherwise stated, $a^* = a$, $b^* = b$). We have a coalgebra structure on the linear space generated by $1, a, b$

$$(2.4) \qquad \Delta 1 = 1 \otimes 1 \ , \qquad \Delta a = a \otimes a \ , \qquad \Delta b = b \otimes a + 1 \otimes b \ ,$$

and the co-unit corresponds to the matrix of the unit element : $\delta(1) = 1 = \delta(a)$, $\delta(b) = 0$. We extend these rules to polynomials by multiplication, to get a bialgebra structure.

The simplest and most useful example is the coalgebra structure on a space of dimension n^2 with basis E_i^j, given by the matrix product itself

$$(2.5) \qquad \Delta E_i^j = \sum_k E_k^j \otimes E_i^k , \qquad \delta(E_i^j) = \delta_i^j .$$

Interesting bialgebras, related to the theory of *quantum groups*, arise when we take the matrix bialgebras above, and consider the coefficients m_i^j not as numbers, but as non–commutative indeterminates. We refer the reader to Schürmann [11].

3) We return to a semi–group $\mathcal{G}$ with unit and involution $*$ (which reverses products), and call $\mathcal{M}$ the space of measures with finite support $\sum_i c_i \varepsilon_{x_i}$ on $\mathcal{G}$, which becomes a bialgebra as follows

$$(2.6) \qquad \begin{array}{c} \varepsilon_x \varepsilon_y = \varepsilon_{xy} \quad \text{(convolution)} \quad , \quad (\varepsilon_x)^* = \varepsilon_{x^*} , \\ \delta(\varepsilon_x) = 1 \quad , \quad \Delta(\varepsilon_x) = \varepsilon_x \otimes \varepsilon_x . \end{array}$$

The co-unit is the total mass functional. The linear functionals on $\mathcal{M}$ are arbitrary functions on $\mathcal{G}$, states being *functions of positive type* normalized to have the value 1 at e. The choice of the coproduct means that "convolution" of states is ordinary multiplication of functions.

3 The (algebraic) dual space $\mathcal{A}'$ of a coalgebra is an algebra, under *convolution*, defined as follows

$$(3.1) \qquad (F, \lambda \star \mu) = (\Delta F, \lambda \otimes \mu) .$$

The coalgebra is *co-commutative* if and only if convolution is commutative. This requires only the coalgebra structure, but a useful lemma says that *the convolution* $\lambda \star \mu$ *of two positive linear functionals on a bialgebra is positive*. Indeed, if $\Delta a = \sum_i b_i \otimes c_i$ we have $\lambda \star \mu(a) = \sum_i \lambda(b_i) \mu(c_i)$. Applying this to $\Delta(a^*a) = \sum_{ij}(b_i^* b_j) \otimes (c_i^* c_j)$, we have to show that

$$\sum_{ij} \lambda(b_i^* b_j) \mu(c_i^* c_j) \geq 0 .$$

The quadratic form $\sum_{ij} \bar{u}_i u_j \lambda(b_i^* b_j)$ is positive, as is the similar one with μ, and it is classical that the product coefficient by coefficient of two positive quadratic forms is positive, and the sum of all the coefficients of a positive quadratic form is a positive number.

4 We conclude this section on algebraic preliminaries with the main algebraic tool for Schürmann's theory, the so called *fundamental theorem on coalgebras* :

THEOREM. *In a coalgebra* $\mathcal{A}$, *every finite subset is contained in a finite dimensional sub–coalgebra.*

A sub–coalgebra of $\mathcal{A}$ is a linear subspace $\mathcal{B}$ such that for $b \in \mathcal{B}$, Δb lies in $\mathcal{B} \otimes \mathcal{B}$. It is clear that sums of sub–coalgebras are sub–coalgebras, but the case of intersections is not obvious (though the result is true). This is why we do not mention the coalgebra generated by finitely many elements, and give a weaker statement. The involution plays no role : to make a coalgebra stable, we add its adjoint to it.

PROOF. Instead of using the tensor notation, we reason as if elements of $\mathcal{A}$ were functions of one variable, elements of $\mathcal{A} \otimes \mathcal{A}$ functions of two variables, the coproduct and co-unit came from a true product with neutral element e, and linear functionals were measures. Then the proof (which we owe to J.–C. Sauvageot) becomes very transparent, and translating it back into tensor notation is a good exercise.

We consider an element c of $\mathcal{A}$, and the representation of Δc

$$(4.1) \qquad \Delta c = \sum_i a_i \otimes b_i \quad i.e. \quad c(xy) = \sum_i a_i(x)\, b_i(y) \ .$$

This is of course not unique. Taking $y = e$ (i.e. applying $I \otimes \delta$) we see that c is a linear combination of the a_i, and also of the b_i. If the a_i are not linearly free, we express them as linear combinations of a free subsystem and get a representation with fewer indices. Therefore, if one choses a representation such that the number of indices is minimal, both the a_i and the b_i are linearly free. Then one can find linear functionals λ_i, μ_i on $\mathcal{A}$, such that $\lambda_i(a_j) = \delta_{ij}$ and $\mu_i(b_j) = \delta_{ij}$.

We write the coassociativity property in $\mathcal{A} \otimes \mathcal{A} \otimes \mathcal{A}$

$$(4.2) \qquad c(xyz) = \sum_i a_i(xy)\, b_i(z) = \sum_i a_i(x)\, b_i(yz) \ ,$$

and we get down to $\mathcal{A} \otimes \mathcal{A}$ by means of $\lambda_k \otimes I \otimes I$, i.e.

$$(4.3) \qquad b_k(yz) = \int \lambda_k(dx)\, c(xyz) = \sum_i \Big(\int \lambda_k(dx)\, a_i(xy) \Big)\, b_i(z) \ .$$

Similarly

$$(4.3') \qquad a_j(xy) = \int c(xyz)\, \mu_j(dz) = \sum_i a_i(x) \Big(\int b_i(yz)\, \mu_j(dz) \Big) \ .$$

"Integrating" (4.2), we introduce the elements

$$(4.4) \qquad c_{ij}(y) = \int \lambda_i(dx)\, c(xyz)\, \mu_j(dz) = \int b_i(yz)\, \mu_j(dz) = \int \lambda_i(dx)\, a_j(xy) \ .$$

The relations (4.3) (4.3$'$) then give us

$$(4.5) \qquad b_k(yz) = \sum_i c_{ki}(y)\, b_i(z) \ , \qquad a_j(xy) = \sum_i a_i(x)\, c_{ij}(y) \ .$$

In the first formula, we replace y by xy and use coassociativity

$$\sum_i c_{ki}(xy)\, b_i(z) = b_k((xy)z) = b_k(x(yz)) =$$

$$\sum_i c_{ki}(x)\, b_i(yz) = \sum_{ij} c_{ki}(x)\, c_{ij}(y)\, b_j(z) \ ,$$

and "integrating" with respect to $\mu_j(dz)$ we get

$$(4.6) \qquad c_{kj}(xy) = \sum_i c_{ki}(x)\, c_{ij}(y) \ .$$

Therefore the linear space $\mathcal{B}$ generated by the c_{ij} is a sub–coalgebra. According to (4.5) it contains the $b_k = \sum_i c_{ki}\delta(b_i)$ (as well as the a_k), and finally c itself.

REMARK. Relation (4.6) is the coproduct arising from the law of composition of matrices. Here it appears as linked with one particular proof, but it comes naturally whenever we have a finite dimensional coalgebra $\mathcal{A}$, as follows. Let e_i be a basis of $\mathcal{A}$. Then we have a coproduct representation

$$(4.7) \qquad \Delta e_i = \sum_{jk} c_i^{jk} e_j \otimes e_k .$$

Let us put $e_i^j = \sum_k c_i^{jk} e_k$ so that we have, putting $\delta_i = \delta(e_i)$,

$$(4.8) \qquad \Delta e_i^j = \sum_k e_k^j \otimes e_i^k , \qquad \delta(e_i^j) = \delta_i^j , \qquad e_i = \sum_j e_i^j \delta_j .$$

The first two properties show that the coalgebra formulas are the same as for the coalgebra of matrices. Of course the e_i^j are not linearly free in $\mathcal{A}$ since there are too many of them, but they generate $\mathcal{A}$ by the last formula. Then computations on matrices become universal tools. This will be a good method to handle stochastic differential equations in coalgebras without needing a new general theory.

Processes with independent increments

5 We give the axioms for a process with (multiplicative) independent increments, separating the structure necessary for each property.

We consider a linear space $\mathcal{A}$ and a family of "random variables" X_{rs} ($0 \leq r \leq s$), i.e. mere mappings from $\mathcal{A}$ to an operator algebra $\mathcal{B}$ provided with a state. They represent the (multiplicative) increments of the process $X_t = X_{0t}$.

Independence means that for $s_1 < t_1 \ldots < s_n < t_n$ and $F_1 \ldots F_n \in \mathcal{A}$, the operators $F_1 \circ X_{s_1 t_1} \ldots F_n \circ X_{s_n t_n}$ commute in $\mathcal{B}$ and

$$(5.1) \qquad \mathbb{E}\left[F_1 \circ X_{s_1 t_1} \ldots F_n \circ X_{s_n t_n} \right] = \mathbb{E}\left[F_1 \circ X_{s_1 t_1} \right] \ldots \mathbb{E}\left[F_n \circ X_{s_n t_n} \right] .$$

This involves no structure on $\mathcal{A}$, and will be almost automatically realized in our constructions, as the operators X_{st} will act between s and t on a Fock space, and $\mathbb{E}$ will be a vacuum expectation.

Another property that does not require a structure on $\mathcal{A}$ is the *stationarity* of increments : for arbitrary $F \in \mathcal{A}$, $h \geq 0$

$$(5.2) \qquad \mathbb{E}\left[F \circ X_{rs} \right] = \mathbb{E}\left[F \circ X_{r+h,t+h} \right] .$$

Multiplicativity of increments is the following property — a *coalgebra* structure on $\mathcal{A}$ is needed to define the "product" involved in the formula

$$(5.3) \qquad X_{rt} = X_{rs} X_{st} \quad \text{for } r \leq s \leq t .$$

Using (5.2) we define φ_{t-s} to be the law of X_{st}, i.e. the linear functional on $F \longmapsto \mathbb{E}\left[F \circ X_{st} \right]$ on $\mathcal{A}$. Then the three preceding axioms imply that $\varphi_s \star \varphi_t = \varphi_{s+t}$, *i.e.* we have a convolution semigroup. We add a *continuity condition* involving the co-unit δ

(5.4) $$\lim_{t \downarrow s} \varphi_{st} = \varphi_{ss} = \delta \ .$$

Finally, the last axiom is the only one that involves the algebra structure of $\mathcal{A}$: it consists in requiring that the X_{st} be true random variables, *i.e.* *algebra homomorphisms*.

Schürmann triples

6 We consider a bialgebra $\mathcal{A}$ and a convolution semigroup of states φ_t on $\mathcal{A}$, tending to the co-unit δ as $t \to 0$. This is the first part of Schürmann's main theorem.

THEOREM *For every* $F \in \mathcal{A}$, *the generator*

(6.1) $$\psi(F) = \lim_{t \to 0} (\varphi_t(F) - \delta(F))/t$$

exists, and φ_t *is the convolution exponential of* $t\psi$

(6.2) $$\varphi_t = \delta + t\psi + \frac{t^2}{2} \psi \star \psi + \dots$$

We have $\psi(1) = 0$, *and* ψ *is conditionally positive : on the subspace* $K = \operatorname{Ker} \delta$ *(which is an ideal in* $\mathcal{A}$*) it satisfies the properties*

(6.3) $$\psi(F^*) = \overline{\psi}(F) , \quad \psi(F^*F) \geq 0 \ .$$

PROOF. The main point is (6.1) and follows from 4 : let $\mathcal{A}_0$ be a finite dimensional coalgebra containing F. Then the restrictions to $\mathcal{A}_0$ of the functionals φ_t constitute a semigroup in the finite dimensional convolution algebra $\mathcal{A}'_0$, continuous at 0, therefore it is differentiable at 0 and is the convolution exponential of its generator. The remainder is almost obvious. Note that (6.3) depends on the co-unit, but not on the coproduct.

The next step is a construction for conditionally positive functionals which parallels the GNS construction for positive ones (Appendix 4, §2, 2–3). We mix with Schürmann's results some remarks due to Belavkin [Bel6].

We consider a *–algebra $\mathcal{A}$ with unit 1 and a "co-unit" or "mass functional" δ, *i.e.* a homomorphism from $\mathcal{A}$ to $\mathbb{C}$. Let $\mathcal{A}_0$ be the kernel of δ. On $\mathcal{A}$, we consider a linear functional ψ such that $\psi(1) = 0$, which is conditionally of positive type. We provide $\mathcal{A}_0$ with the scalar product

(6.4) $$< G, F > = \psi(G^*F) ,$$

denote by $\mathcal{N}_0$ the null subspace of this positive hermitian form, and by $\mathcal{K}$ the prehilbert space $\mathcal{A}_0/\mathcal{N}_0$. *We do not complete* $\mathcal{K}$ *unless we say so explicitly.*

If F belongs to $\mathcal{A}_0$ so does GF for $G \in \mathcal{A}$, and this defines a left action of $\mathcal{A}$ on $\mathcal{A}_0$. If $\psi(F^*F) = 0$ we have $\psi(G^*F) = 0$ for $G \in \mathcal{A}_0$ by the Schwarz inequality, and since the case of $G = 1$ is trivial $\mathcal{N}_0$ is stable under the action of $\mathcal{A}$ on $\mathcal{A}_0$. Therefore $\mathcal{A}$ *acts on the left on* $\mathcal{K}$. We denote by $\rho(G)$ the corresponding operator. It may be unbounded, but it preserves the prehilbert space $\mathcal{K}$, as does its adjoint $\rho(G^*)$.

For $F \in \mathcal{A}$ we define $F_0 = F - \delta(F) 1 \in \mathcal{A}_0$, and denote by $\eta(F) \in \mathcal{K}$ as the class of F_0. Then we have by an immediate computation

$$(6.5) \qquad \eta(GF) = \rho(G)\eta(F) + \eta(G)\delta(F) .$$

In algebraic language, $\mathcal{K}$ is considered as an $\mathcal{A}$–bimodule with the left action ρ and the right action δ, and η is a *cocycle*. Note that the range of η is the whole of $\mathcal{K}$. Finally, a simple computation gives the identity

$$(6.6) \qquad \psi(G^*F) - \delta(G^*)\psi(F) - \psi(G^*)\delta(F) = <\eta(G), \eta(F)> .$$

A system (ρ, η, ψ) consisting of a *–representation ρ of $\mathcal{A}$ in a prehilbert space $\mathcal{K}$, of a cocycle η with values in $\mathcal{K}$, and a scalar valued mapping ψ satisfying (6.5-6) will be called a *Schürmann triple* in this chapter — maybe δ should be asked to sit for the picture too, and we then have a *Schürmann quadruple*. One may reduce the prehilbert space $\mathcal{K}$ to the range of η, but this plays no important role.

7 EXAMPLE. Which is the Schürmann triple corresponding to a convolution semigroup (π_t) of probability measures on $\mathbb{R}^n$? We use the bialgebra $\mathcal{A}$ of trigonometric polynomials, and denote by φ_t the restriction of π_t to $\mathcal{A}$ — otherwise stated, we take Fourier transforms. Another interpretation uses the dual group $\mathcal{G}$ of $\mathbb{R}^n$, so that a trigonometric polynomial appears as a measure of finite support on $\mathcal{G}$ as described in subsection 2, example 3. Then $\varphi_t(u)$ is a normalized function of positive type on $\mathcal{G}$, and "convolution" of states on $\mathcal{M}$ is plain multiplication. Introducing the generator $\psi(u)$ — the opposite of the standard Lévy function — we may equivalently write $\psi(\lambda)$ (a measure of finite support on $\mathcal{G}$) or $\psi(F)$ (a trigonometric polynomial). However, writing the Lévy-Khinchin formula requires the second notation :

$$(7.1) \qquad \psi(F) = i \sum_k m^k D_k F(0) - \tfrac{1}{2} \sum_{ij} \sigma^{ij} D_{ij} F(0) + \int \left(F(x) - F(0) - h(x) x^i D_i F(0) \right) \nu(dx) ,$$

where ν is the Lévy measure, h is a bounded function with compact support, equal to 1 in a neighbourhood of 0, and σ^{ij} has a representation as $\sum_k a_k^i a_k^j$. Then the *completed* space $\overline{\mathcal{K}}$ is

$$(7.2) \qquad \overline{\mathcal{K}} = \mathbb{C}^n \oplus L^2(\nu) .$$

The mapping η is given by

$$(7.3) \qquad \eta(F) = AF'(0) \oplus (F - F(0)) ,$$

the notation $AF'(0)$ denoting the vector with components $\sum_k a_i^k D_k F(0)$. Then the incomplete space $\mathcal{K}$ is the range of η. The representation ρ is given by

$$(7.4) \qquad \rho(F) = F(0) I \oplus F ,$$

where F is interpreted as a multiplication operator on $L^2(\nu)$. Then it becomes an easy computation to check that

$$\psi(GF) - G(0)\psi(F) - \psi(G) F(0) =$$

$$\sum_{ij} \sigma^{ij} D_i G(0) D_j F(0) + \int (G(x) - G(0))(F(x) - F(0)) \nu(dx) = (\eta(G), \eta(F)) .$$

where the right side is a bilinear inner product, since we used G instead of G^*.

8 We return to the general case, keeping the notation from 6. Belavkin had the idea of introducing on $\mathcal{A}$ instead of $\mathcal{A}_0$ a non-positive definite hermitian scalar product

(8.1) $$[G \mid F] = \psi(G^* F) .$$

Note that $[G^* \mid F^*] = \overline{[G \mid F]}$, hence the kernel $\mathcal{N}$ of this hermitian form is stable under *. Note the relation

(8.2) $$[G \mid F] = \; <G_0 , \, F_0> + \, \delta(G^*)\psi(F) + \psi(G^*) \, \delta(F) .$$

Using this remark, it is easy to prove that *the mapping* $F \longmapsto (\delta(F), \eta(F), \psi(F))$ *from* $\mathcal{A}$ *to* $\widehat{\mathcal{K}} = \mathbb{C} \oplus \mathcal{K} \oplus \mathbb{C}$ *is an isometric isomorphism from* $\mathcal{A}/\mathcal{N}$ *into* $\widehat{\mathcal{K}}$ *provided with the non-positive definite scalar product*

(8.3) $$[u + k + v \mid u' + k' + v'] = \overline{u}v' + \; <k, k' > + \; \overline{v}u' .$$

Next, we define a representation on $\mathcal{A}$ on $\widehat{\mathcal{K}}$, which on the image of $\mathcal{A}$ by the above mapping corresponds to $\mathcal{A}$ acting on itself by left multiplication. An element of $\widehat{\mathcal{K}}$ being a column (!) (u, k, v), we consider the following $(3, 3)$ matrix acting on $\mathcal{K}$

(8.4) $$R(H) = \begin{pmatrix} \delta(H) & 0 & 0 \\ |\eta(H)> & \rho(H) & 0 \\ \psi(H) & <\eta(H^*)| & \delta(H) \end{pmatrix} .$$

We then have

(8.5) $$\begin{pmatrix} \delta(HF) \\ \eta(HF) \\ \psi(HF) \end{pmatrix} = R(H) \begin{pmatrix} \delta(F) \\ \eta(F) \\ \psi(F) \end{pmatrix} .$$

Then it is very easy to see that $R(H) R(K) = R(HK)$ (this is the definition of a Schürmann triple), while $R(H^*)$ *is Belavkin's "twisted adjoint"*, which involves a symmetry w.r.t. the upward diagonal. If $\delta(H) = 0$, (8.4) corresponds to Belavkin's notation for coefficients of a quantum differential : otherwise stated, *whenever conditionally positive functions appear, the way is open to quantum stochastic calculus*.

A striking application arises when we consider a *–algebra $\mathcal{D}$ without unit, and a function *of positive type* ψ on it. Then we take for $\mathcal{A}$ the unital extension of $\mathcal{D}$, and put for $c \in \mathbb{C}, d \in \mathcal{D}$

$$\delta(c\mathbf{1} + d) = c \quad , \quad \psi(c\mathbf{1} + d) = \psi(d) .$$

Then ψ has become conditionally positive, the preceding theory applies and leads to a Schürmann triple. For example, $\mathcal{D}$ may have the four generators dt, da_t^ε $(\varepsilon = -, \circ, +))$ with multiplication given by the Ito table and the usual involution, ψ taking the value 1 on dt, 0 on the other generators. This is why Belavkin uses the name of *Ito algebras* to describe the general situation.

§2. CONSTRUCTION OF THE PROCESS

In this section we use stochastic calculus to construct a process with independent increments from a Schürmann triple on a bialgebra.

Construction on a coalgebra

1 The construction becomes very clear if we separate the coalgebraic and the algebraic structure. We consider here a coalgebra $\mathcal{A}$ with an involution, and a "Schürmann triple" consisting simply of a prehilbert space $\mathcal{K}$, of a linear mapping ρ from $\mathcal{A}$ to operators on $\mathcal{K}$ satisfying only the property $\rho(F^*) = \rho(F)^*$, of a linear mapping η from $\mathcal{A}$ to $\mathcal{K}$, and a linear mapping ψ from $\mathcal{A}$ to $\mathbb{C}$. Since we have no algebraic structure on $\mathcal{A}$, we cannot write the other conditions for a Schürmann triple.

We consider a Fock space Φ of multiplicity $\mathcal{K}$, with initial space $\mathbb{C}$. We are going to give a meaning to Schürmann's stochastic differential equation for mappings X_s from $\mathcal{A}$ to s–adapted operators on Φ (writing as usual $F \circ X_s$ for $X_s(F)$)

$$(1.1) \qquad F \circ X_t = \delta(F) I \;+$$
$$\int_0^t F \circ X_s \star (da_s^+(|\eta(F)>) + da_s^\circ(\rho(F) - \delta(F)I) + da_s^-(<\eta(F^*)|) + \psi(F)\, ds)\,,$$

and more generally, for $0 \le s \le t$

$$(1.2) \qquad F \circ X_{st} = \delta(F) I +$$
$$\int_s^t F \circ X_r \star (da_r^+(|\eta(F)>) + da_r^\circ(\rho(F) - \delta(F)I) + da_r^-(<\eta(F^*)|) + \psi(F)\, dr)\,.$$

We generally deal with (1.1), leaving (1.2) to the reader. The explicit meaning of the convolution symbol in (1.1) is the following : if $\Delta F = \sum_i F_i' \otimes F_i''$, we have

$$F \circ X_t = \delta(F) I +$$
$$\sum_i \int_0^t F_i' \circ X_s (da_s^+(|\eta(F_i'')>) + da_s^\circ(\rho(F_i'') - \delta(F_i'')I) + da_s^-(<\eta(F_i''^*)|) + \psi(F_i'')\, ds)\,.$$

This is still very abstract : choose a finite–dimensional coalgebra $\mathcal{A}_0$ containing all the elements of $\mathcal{A}$ involved, take a basis e_i of it, and define the coproduct and co-unit in this basis as $\Delta e_i = \sum_{jk} c_i^{jk} e_j \otimes e_k$, $\delta(e_i) = \delta_i$. Then we are reduced to a linear equation in a finite dimensional space

$$(1.3) \qquad X_i(t) = \delta_i I +$$
$$\sum_{jk} \int_0^t c_i^{jk} X_j(s)(da_s^+(|\eta_k>) + da_s^\circ(\rho_k - \delta_k I) + da_s^-(<\tilde{\eta}_k)|) + \psi_k\, ds)\,.$$

where $\rho_k = \rho(e_k)$, $\tilde{\eta}_k = \eta(e_k^*)$, etc.. We take an o.n.b. e^α of the prehilbert space $\mathcal{K}$ (the upper index corresponds to the perverse notation system of Chapter V), and we are reduced to a standard linear s.d.e. with finite dimensional initial space — and therefore bounded coefficients, given by

$$(1.5) \qquad L^0_\alpha(e_i) = \sum_{jk} < e_\alpha, \eta_k > c_i^{jk} e_j \;, \; L^\alpha_0(e_i) = \sum_{jk} < \tilde{\eta}_k, e^\alpha > c_i^{jk} e_j \;, \text{ etc.}$$

This system with bounded coefficients and infinite multiplicity *satisfies the Mohari–Sinha conditions* VI.4.(11.3), as well as the dual Mohari–Sinha conditions for the adjoint equation. Indeed, the sums $\sum_\sigma \| L^\rho_\sigma(e_i) \|^2$ are finite as all the vectors $\eta_k, \tilde{\eta}_k$ and $\rho_k(e^\alpha)$ belong to $\mathcal{K}$, and this is sufficient to imply the conditions in a finite dimensional space. Therefore the equation has a unique solution on the exponential domain. This exponential domain, as described by the Mohari–Sinha theorem, depends on the choice of the basis e^α, and consists of vectors which have finitely many non–zero components in this basis, but changing the basis one reaches general exponential vectors with argument in $\mathcal{K}$ (uncompleted).

Once uniqueness is known, we may prove easily the relation $X_t(F^*) = (X_t(F))^*$. Let us also put

$$X_i^j(t) = \sum_k c_i^{jk} X_k(t) \quad, \quad \eta_i^j(t) = \sum_k c_i^{jk} \eta_k(t) \;, \text{ etc.}$$

from which $X_i(t)$ can be recovered as $\sum_j \delta_j X_i^j(t)$. Then the operator process $(X_i^j(t))$ is the unique solution of the standard matrix equation

$$X_i^j(t) = \delta_i^j + \sum_k \int_0^t X_k^j(s) \, (da_s^+(|\eta_i^k>) + da_s^\circ(\rho_i^k - \delta_i^k I) + da_s^-(<\tilde{\eta}_i^k|) + \psi_i^k \, ds) \;.$$

To prove that for $r < s < t$ the increments are multiplicative,

$$X_i(r,t) = \sum_{kl} c_i^{kl} X_k(r,s) \, X_l(s,t)$$

it suffices to prove the easy cocycle property

$$(1.6) \qquad X_i^j(r,t) = \sum_k X_k^j(r,s) \, X_i^k(s,t) = \sum_{kl} X_k^j(r,s) \, c_i^{kl} X_l(s,t) \;,$$

to multiply by δ_j and sum over j. Though the operators in (1.6) are unbounded, and do not preserve the exponential domain, there is no serious domain problem, since they act on different parts on Fock space : we may consider that the domain for $X_i^j(s,t)$ consists of linear combinations of vectors $a \otimes \mathcal{E}(u) \otimes b$, where a belongs to the past of s, b to the future of t, and u vanishes outside $[s,t]$.

2 Now comes the main result :

THEOREM. *If $\mathcal{A}$ is a bialgebra, and if the hypotheses of Schürmann triples are satisfied by ρ, η, ψ, then the mappings X_{st} are algebra homomorphisms, i.e. random variables.*

This statement is proved as follows, considering $X_t = X_{0t}$ for simplicity : First of all, the properties of Schürmann triples are expressed on the coefficients $L^\rho_\sigma(F)$ as the Evans–Hudson structure equations. Then a careful analysis of the Mohari–Sinha proof (prepared in the preceding chapter) shows that we may prove the weak homomorphism property

$$(2.1) \qquad < \mathcal{E}(\ell \mathbf{v}), \, X_t(GF) \, \mathcal{E}(j\mathbf{u}) > \; = \; < X_t(G^*) \, \mathcal{E}(\ell \mathbf{v}), \, X_t(F) \, \mathcal{E}(j\mathbf{u}) >$$

in a finite–dimensional coalgebra containing F and G, without need of finite-dimensional *bialgebras*. However, in the theory of Evans–Hudson flows we dealt with an algebra $\mathcal{A}$ of bounded operators, and we had no difficulty in defining $X_t(G) \, X_t(F)$ and in deducing the homomorphism property from (2.1). Here the situation is more delicate, and the way Schürmann solves this problem consists in showing that $X_t(F)$ has a Maassen kernel, operating on a stable domain of "test-functions." This requires a new definition of Maassen kernels since we deal with a Fock space of infinite multiplicity. We give no details.

Unboundedness also creates a difficulty in the probabilistic interpretation of the results. Indeed, to construct classical stochastic processes from quantum probabilistic methods one would need a theorem allowing to extend commuting families of symmetric operators into commuting families of selfadjoint operators.

Example : Azéma martingales

3   We consider first a general situation : let $\mathcal{A}$ be the (non-commutative) free algebra generated by n indeterminates $X_1 \ldots X_n$; we define the involution by $X_i^* = X_i$ and trivial extension to products, and we give it a coalgebra structure by a formula

$$\Delta X_i = \sum_{jk} c_i^{jk} X_j \otimes X_k \,, \qquad \delta(X_i) = \delta_i \,,$$

with coefficients subject to a few conditions expressing coassociativity, etc; Then we extend these operations uniquely as algebra homomorphisms (in particular, we have $\Delta 1 = 1 \otimes 1$).

Then we choose a finite dimensional Hilbert space $\mathcal{K}$, arbitrary hermitian operators ρ_i on it, arbitrary vectors η_i, and real scalars c_i. There exists a unique extension of these mappings as a Schürmann triple on the bialgebra $\mathcal{A}$.

The case of the Azéma martingale arises when one takes two generators, and deduce the coalgebra structure from example (2.4) in section 1 :

$$(6.2) \qquad \Delta X_1 = X_1 \otimes X_1 \,, \;\; \Delta X_2 = X_2 \otimes X_1 + 1 \otimes X_2 \,, \;\; \delta(X_1) = 1 \,, \;\; \delta(X_2) = 0 \,.$$

The only non-zero coefficients are $c_1^{11}, c_2^{21}, c_2^{02}$. The space $\mathcal{K}$ is taken one-dimensional so that everything is scalar. Then the Schürmann equations can be written as

$$dX_1(t) = X_1(t)(m_1 dt + dQ_t(\eta_1) + da_t^\circ (\rho_1 - I))$$
$$dX_2(t) = X_2(t)(m_1 dt + dQ_t(\eta_1) + da_t^\circ (\rho_1 - I)) + X_0(t)(m_2 dt + dQ_t(\eta_2) + da_s^\circ (\rho_2 - I))$$

If we take $m_1 = m_2 = 0$, $\eta_1 = 0$, $\eta_2 = 1$, and finally $\rho_2 = I$, $\rho_1 = cI$, the system separates into a pair of independent equations

$$X_1(t) = 1 + (c-1) \int_0^t X_1(s) \, da_s^\circ \,, \qquad X_2(t) = Q_t + (c-1) \int_0^t X_2(s) \, da_s^\circ \,.$$

The first one represents a (commutative) process, which can be explicitly computed and is a.s. equal to 1 in the vacuum state. *The second equation is the quantum s.d.e. which defines the family of Azéma martingales.*

Example : Conditionally positive functions on semi-groups

4 We consider the case of a unital *– algebra $\mathcal{A}$ on which are defined a "mass functional" δ and a conditionally positive function ψ vanishing at 1, from which we construct as before the representation ρ operating on $\mathcal{K}$, and the cocycle η. On the other hand, $e^{t\psi} = \varphi_t$ is a state for every $t > 0$, and we are going to show how the GNS representation of φ_t can be constructed by "exponentiation" (Fock second quantization). An outstanding example is the algebra $\mathcal{M}$ of measures of finite support on a semi-group $\mathcal{G}$ with unit and involution — if $\mathcal{G}$ is a locally compact abelian group, this corresponds to processes with independent increments on the dual of $\mathcal{G}$.

We will make use of two Fock spaces : $\Gamma(\mathcal{K})$ over $\mathcal{K}$, and $\Gamma'(\widehat{\mathcal{K}})$ over $\widehat{\mathcal{K}} = \mathbb{C} \oplus \mathcal{K} \oplus \mathbb{C}$, the $'$ indicating that the Fock space is the usual one, but provided with an unusual scalar product corresponding to the non-positive scalar product (8.3) on $\widehat{\mathcal{K}}$. We begin by explaining this.

We identify Fock space over $\widehat{\mathcal{K}}$ with the space of sequences $(f) = f_{km}$ of elements of $\Gamma(\mathcal{K})$ such that $\sum_{km} \| f_{km} \|^2 / k! m! < \infty$, the scalar product being $\sum_{km} < f_{km}, f'_{km} > / k! m!$. Then the exponential vector $\mathcal{E}(u + k + v)$ is read as the sequence $u^k v^m \mathcal{E}(k)$. We then twist the scalar product on this space, putting

$$[(f)|(g)] = \sum_{km} \frac{< f_{km}, g_{mk} >}{k! m!}.$$

The twisted scalar product of two (standard) exponential vectors $\mathcal{E}(u + k + w)$ and $\mathcal{E}(u' + k' + w')$ then is $\exp[u + k + v \,|\, u' + k' + v']$ as it should be.

The operators $R(H)$ (§1 subs. 8) on $\widehat{\mathcal{K}}$ are then extended by second quantization to the exponential domain of $\Gamma(\widehat{\mathcal{K}})$, with the same notation. In this way we get a representation of $\mathcal{A}$, which is a *–representation w.r.t. the twisted adjoint. We want to deduce from it a true *–representation on $\Gamma(\mathcal{K})$.

To accomplish that, Belavkin defines for every real p an isometry from $\Gamma(\mathcal{K})$ to $\Gamma'(\widehat{\mathcal{K}})$ as follows : given $f \in \Gamma(\mathcal{K})$, he puts

$$(Jf)_{km} = 0 \quad \text{if } m \neq 0 \qquad , \qquad (Jf)_{k0} = p^k f .$$

Thus for $k \in \mathcal{K}$ we have $J\mathcal{E}(k) = \mathcal{E}(p + k + 0)$. The (twisted) adjoint $J^\star$ of J maps the sequence $(g) = g_{km}$ to

$$J^\star(g) = \sum_m \frac{p^m g_{0m}}{m!} .$$

Indeed, we have

$$[Jf \,|\, (g)] = \sum_n p^k < f, g_{0k} > / k! = < f, J^\star(g) > .$$

Hence $J^\star$ maps $\mathcal{E}(u+k+v)$ to $e^{pv}\mathcal{E}(k)$. It is clear that $J^\star J = I$ since J is an isometry, while $JJ^\star\mathcal{E}(u+k+v) = e^{pv}\mathcal{E}(p+k+0)$. In particular, $JJ^\star = I$ on exponential vectors $\mathcal{E}(p+k+0)$. We then let $\mathcal{A}$ operate on $\Gamma(\mathcal{K})$ by

$$(4.1) \qquad\qquad S(G) = J^\star R(G)\, J\ .$$

It is clear that $S(G^*) = S(G)^*$ (standard adjoint), and a small computation gives, for F of mass 0 and arbitrary G

$$(4.2) \qquad S(G)\,\mathcal{E}(F) = e^{p^2\psi(G)+p<\eta(G^*)\,,\,F>}\mathcal{E}(p\eta(G)+\rho(G)\,F)\ .$$

If G has mass 1, we then have $S(H)\,S(G) = S(HG)$, and we get a $*$–representation of the multiplicative semi-group of elements of mass 1 in $\mathcal{A}$ in the Fock space $\Gamma(\mathcal{K})$, which is a generalized Weyl representation. In particular, if $F = 0$ we have

$$< S(H)\,\mathbf{1}\,,\,S(G)\,\mathbf{1}> = <e^{p^2\psi(H)}\mathcal{E}(p\eta(H))\,,\,e^{p^2\psi(G)}\mathcal{E}(p\eta(G))>$$
$$= e^{p^2(\psi(H^*)+\psi(G)+<\eta(H)\,,\,\eta(G)>} = e^{p^2\,[\,H\,|\,G\,]}\ .$$

Thus what we constructed is the GNS representation associated with the function of positive type $e^{\psi(H^*G)}$, on the space $\mathcal{A}_p$ of all elements G, H of mass p.

EXAMPLE. Take for semi-group $\mathcal{G}$ the additive group of a prehilbert space, put $x^* = -x$, $x^*y = y - x$, and $\psi(x) = -\|x\|^2/2$. Then $\mathcal{M}_0$ is generated by the measures $\varepsilon_x - \varepsilon_0$, the corresponding scalar product on $\mathcal{K}$ being equal to $<x,y>$. The semi-group of measures of mass 1 is generated by the measures ε_x, operating on $\mathcal{G}$ by translation. The representation we get is exactly Weyl's.

5   Let us assume $\mathcal{A} = \mathcal{M}$, which has a simple bialgebra structure, and compare the preceding construction with Schürmann's general theorem. The prehilbert space $\mathcal{K}$ is spanned by the vectors $\varepsilon_x - \varepsilon_e$ which we write simply as $x - e$. Since the coproduct is $\Delta\varepsilon_x = \varepsilon_x \otimes \varepsilon_x$, the stochastic differential equations relative to different points can be handled separately, and become

$$X_t(x) = I + \int_0^t X_s(x)\,\big(\,da_s^+(x-e) + da_s^\circ(\rho(x)-I) + da_s^-(x^*-e) + \psi(x)\,ds\,\big)\ .$$

This equation may be solved explicitly. Denoting by 1_t the indicator of $[\,0,t\,[\,$, we have

$$X_t(x) = e^{t\psi(x)}\exp(a^+((x-e)\otimes 1_t))\,\Gamma(\rho(x)1_t + I(1-1_t))\exp(a^-((x^*-e)\otimes 1_t))\ .$$

Indeed, to check that these operators satisfy the appropriate s.d.e. we remark that the three terms can be differentiated without Ito correction since this product is normally ordered. To compute the differential of the middle term we use the second example of Chapter VI, §1 subs. 13 : for every operator U the process $Y_t = \Gamma(U1_t + I(1-1_t))$ solves the s.d.e. $Y_t = I + \int_0^t Y_s(U-I)\,dN_s$. Finally, one sees that X_t on the Fock space up to time t is the same as that of the above representation, with $p = \sqrt{t}$.

Appendix 1

Functional Analysis

Hilbert space functional analysis plays for quantum probability the same role measure theory plays for classical probability. Many papers in quantum probability are unreadable by a non–specialist, because of their heavy load of references to advanced functional analysis, and in particular to von Neumann algebras. However, one can do a lot (not everything, but still a great deal) with a few simple tools. Such tools will be summarily presented in these Appendices, essentially in three parts : here, elementary results of functional analysis in Hilbert space : later, the basic theory of C^*–algebras, and finally, the essentials of von Neumann algebras. For most of the results quoted, a proof will be presented, sometimes in a sketchy way.

These sections on functional analysis have nothing original : my main contribution has consisted in choosing the *omitted* material — this is almost as important as choosing what one includes! The material itself comes from two books which I have found excellent in their different ways : Bratteli–Robinson [BrR1] *Operator Algebras and Quantum Statistical Mechanics I*, and Pedersen [Ped] *C*-algebras and their Automorphism Groups*. The second book includes very beautiful mathematics, but may be *too* complete. The first one has greatly helped us by its useful selection of topics. For the results in this section, we also recommend Reed and Simon [ReS], and specially Parthasarathy's recent book [Par1], which is specially intended for quantum probability,.

Hilbert–Schmidt operators

1 We denote by $\mathcal{H}$ the basic Hilbert space. Given an orthonormal basis (e_n) and an operator $\mathbf{a}$, we put

$$(1.1) \qquad \| \mathbf{a} \|_2 = \left(\sum_n \| \mathbf{a} e_n \|^2 \right)^{1/2} \leq +\infty .$$

Apparently this depends on the basis, but let (e'_n) be a second o.n.b. ; we have

$$\| \mathbf{a} \|_2^2 = \sum_{nm} |<\mathbf{a} e_n, e'_m>|^2 = \sum_{nm} |<e_n, \mathbf{a}^* e'_m>|^2 = \| \mathbf{a}^* \|_2'^2$$

where the $'$ on the extreme right indicates that the "norm" is computed in the new basis. Taking first the two bases to be the same we get that $\| \mathbf{a} \|_2 = \| \mathbf{a}^* \|_2$, and then one sees that $\| \mathbf{a} \|_2$ does not depend on the choice of the basis. It is called the *Hilbert–Schmidt norm* of $\mathbf{a}$, and the space of all operators of finite HS norm is denoted by *HS* or by $\mathcal{L}^2$. In contrast to this, the space $\mathcal{L}(\mathcal{H})$ of all bounded operators on $\mathcal{H}$ will sometimes be denoted by $\mathcal{L}^\infty$ and its norm by $\| \ \|_\infty$. Since every unit vector can be included in some o.n.b., we have $\| \mathbf{a} \|_\infty \leq \| \mathbf{a} \|_2$. According to (1.1) we have for every bounded operator $\mathbf{b}$

$$(1.2) \qquad \| \mathbf{ba} \|_2^2 = \sum_n \| \mathbf{ba} e_n \|^2 \leq \| \mathbf{b} \|_\infty^2 \| \mathbf{a} \|_2^2$$

and taking adjoints

(1.3) $\| \mathbf{ab} \|_2^2 \leq \| \mathbf{b} \|_\infty^2 \| \mathbf{a} \|_2^2$.

Note in particular that $\| \mathbf{ab} \|_2 \leq \| \mathbf{a} \|_2 \| \mathbf{b} \|_2$.

It is clear from (1.1) that a (hermitian) scalar product between HS operators can be defined by $< \mathbf{a}, \mathbf{b} >_{HS} \, = \sum_n < \mathbf{a} e_n, \mathbf{b} e_n >$. It is easily proved that the space HS is complete.

Trace class operators

2 Let first $\mathbf{a}$ be bounded and *positive*, and let $\mathbf{b} = \sqrt{\mathbf{a}}$ be its positive square root (if $\mathbf{a}$ has the spectral representation $\int_0^\infty t \, dE_t$, its square root is given by $\mathbf{b} = \int \sqrt{t} \, dE_t$). We have in any o.n. basis (e_n)

(2.1) $\sum_n < e_n, \mathbf{a} e_n > \, = \sum_n < \mathbf{b} e_n, \, \mathbf{b} e_n > \, = \| \mathbf{b} \|_2^2 \leq +\infty$.

Thus the left hand side does not depend on the basis : it is called the *trace* of $\mathbf{a}$ and denoted by $\mathrm{Tr}(\mathbf{a})$. For a positive operator, the trace (finite or not) is always defined. Since it does not depend on the basis, it is unitarily invariant ($\mathrm{Tr}(\mathbf{u}^* \mathbf{a} \mathbf{u}) = \mathrm{Tr}(\mathbf{a})$ if $\mathbf{u}$ is unitary).

Consider now a product $\mathbf{a} = \mathbf{bc}$ of two HS operators. We have

(2.2) $\sum_n |{<}e_n, \mathbf{a} e_n{>}| = \sum_n |{<}\mathbf{b}^* e_n, \mathbf{c} e_n{>}| \leq \| \mathbf{b} \|_2 \| \mathbf{c} \|_2 < \infty$

and

(2.3) $\sum_n < e_n, \mathbf{a} e_n > \, = \sum_n < \mathbf{b}^* e_n, \mathbf{c} e_n > \, = \, < \mathbf{b}^*, \mathbf{c} >_{HS}$.

Since the right hand side does not depend on the basis, the same is true of the left hand side. On the other hand, the left hand side does not depend on the decomposition $\mathbf{a} = \mathbf{bc}$, so the right side does not depend on it either. Operators $\mathbf{a}$ which can be represented in this way as a product of two HS operators are called *trace class operators* (sometimes also *nuclear* operators), and the complex number (2.1) is denoted by $\mathrm{Tr}(\mathbf{a})$ and called the *trace* of a. One sees easily that, for positive operators, this definition of the trace is compatible with the preceding one. Note also that, given two arbitrary HS operators $\mathbf{b}$ and $\mathbf{c}$, their HS scalar product $<\mathbf{b}, \mathbf{c}>_{HS}$ is equal to $\mathrm{Tr}(\mathbf{b}^* \mathbf{c})$.

Intuitively speaking, HS operators correspond to "square integrable functions", and a product of two square integrable function is just an "integrable function", the trace corresponding to the integral. This is why the space of trace class operators is sometimes denoted by $\mathcal{L}^1$. In other contexts it may be denoted by $\mathcal{M}(\mathcal{H})$, a notation suggesting a space of bounded measures. The same situation occurs in classical probability theory on a discrete countable space like $\mathbb{N}$, on which all measures are absolutely continuous w.r.t. the counting measure. Indeed, non–commutative probability on $\mathcal{L}(\mathcal{H})$ is an extension of such a situation, the trace playing the role of the counting integral. More general σ–fields are represented in non–commutative probability by arbitrary von Neumann algebras, which offer much more variety than their commutative counterparts.

Given a bounded operator $\mathbf{a}$, we denote by $|\mathbf{a}|$ the (positive) square root of $\mathbf{a}^*\mathbf{a}$ — this is usually not the same as $\sqrt{\mathbf{a}\mathbf{a}^*}$, and this "absolute value" mapping has some pathological properties : for instance it is not subadditive. An elementary result called *the polar decomposition of bounded operators* asserts that $\mathbf{a} = \mathbf{u}|\mathbf{a}|$, $|\mathbf{a}| = \mathbf{u}^*\mathbf{a}$ where $\mathbf{u}$ is a unique *partial isometry*, i.e. is an isometry when restricted to $(\mathrm{Ker}\ \mathbf{u})^{\perp}$. We will not need the details, only the fact that $\mathbf{u}$ always has a norm ≤ 1, and is unitary if $\mathbf{a}$ is invertible. For all this, see Reed–Simon, theorem VI.10.

THEOREM. *The operator $\mathbf{a}$ is a product of two HS operators (i.e. belongs to the trace class) if and only if $\mathrm{Tr}(|\mathbf{a}|)$ is finite.*

PROOF. Let us assume $\mathrm{Tr}(|\mathbf{a}|) < \infty$, and put $\mathbf{b} = \sqrt{|\mathbf{a}|}$, a HS operator. Then $\mathbf{a} = \mathbf{u}|\mathbf{a}| = (\mathbf{u}\mathbf{b})\mathbf{b}$ is a product of two HS operators. Conversely, let $\mathbf{a} = \mathbf{h}\mathbf{k}$ be a product of two HS operators. Then $|\mathbf{a}| = \mathbf{u}^*\mathbf{a} = (\mathbf{u}^*\mathbf{h})\mathbf{k}$ and the same reasoning as (2.1) gives (the trace being meaningful since $|\mathbf{a}|$ is positive)

$$\mathrm{Tr}(|\mathbf{a}|) \leq \|\mathbf{u}^*\mathbf{h}\|_2 \|\mathbf{k}\|_2 \leq \|\mathbf{h}\|_2 \|\mathbf{k}\|_2$$

since $\|\mathbf{u}\|_\infty \leq 1$.

The same kind of proof leads to other useful consequences. Before we state them, we define the *trace norm* $\|\mathbf{a}\|_1$ of the operator $\mathbf{a}$ as $\mathrm{Tr}(|\mathbf{a}|)$. We shall see later that this is indeed a norm, under which $\mathcal{L}^1$ is complete.

a) *If $\mathbf{a}$ is a trace class operator, we have $|\mathrm{Tr}(\mathbf{a})| \leq \|\mathbf{a}\|_1$.*

Indeed, putting $\mathbf{b} = \sqrt{|\mathbf{a}|}$ as above, we have from the polar decomposition $\mathbf{a} = (\mathbf{u}\mathbf{b})\mathbf{b}$

$$|\mathrm{Tr}(\mathbf{a})| = |<\mathbf{b}^*\mathbf{u}, \mathbf{b}>_{HS}| \leq \|\mathbf{b}^*\mathbf{u}\|_2 \|\mathbf{b}\|_2 \leq \|\mathbf{b}\|_2^2 = \|\mathbf{a}\|_1 .$$

b) *For $\mathbf{a} \in \mathcal{L}^1$, $\mathbf{h} \in \mathcal{L}^\infty$, we have $\|\mathbf{a}\mathbf{h}\|_1 \leq \|\mathbf{a}\|_1 \|\mathbf{h}\|_\infty$.*

Indeed, we have with the same notation $\mathbf{a} = \mathbf{u}\mathbf{b}\mathbf{b}$, $\mathbf{a}\mathbf{h} = (\mathbf{u}\mathbf{b})(\mathbf{b}\mathbf{h})$, then

$$\|\mathbf{a}\mathbf{h}\|_1 \leq \|\mathbf{u}\mathbf{b}\|_2 \|\mathbf{u}\mathbf{h}\|_2 \leq \|\mathbf{b}\|_2 \|\mathbf{b}\|_2 \|\mathbf{h}\|_\infty .$$

c) *If $\mathbf{a} \in \mathcal{L}^1$ we have $\mathbf{a}^* \in \mathcal{L}^1$, $\|\mathbf{a}\|_1 = \|\mathbf{a}^*\|_1$.*

Indeed $\mathbf{a} = \mathbf{u}|\mathbf{a}|$ gives $\mathbf{a}^* = |\mathbf{a}|\mathbf{u}^*$ and the preceding property implies $\|\mathbf{a}^*\|_1 \leq \|\mathbf{a}\|_1$, from which equality follows. Knowing this, one may take adjoints in b) and get the same property with $\mathbf{h}$ to the left of $\mathbf{a}$.

d) *For $\mathbf{a} \in \mathcal{L}^1$, $\mathbf{h} \in \mathcal{L}^\infty$, we have $\mathrm{Tr}(\mathbf{a}\mathbf{h}) = \mathrm{Tr}(\mathbf{h}\mathbf{a})$.*

This fundamental property has little to do with the above method of proof : if $\mathbf{h}$ is unitary, it reduces to the unitary invariance of the trace class and of the trace itself. The result extends to all bounded operators, since they are linear combinations of (four) unitaries.

The last property is a little less easy :

e) *If $\mathbf{a} \in \mathcal{L}^1$, $\mathbf{b} \in \mathcal{L}^1$, we have $\mathbf{a} + \mathbf{b} \in \mathcal{L}^1$ and $\|\mathbf{a} + \mathbf{b}\|_1 \leq \|\mathbf{a}\|_1 + \|\mathbf{b}\|_1$.*

To see this, write the three polar decompositions $\mathbf{a} = \mathbf{u}|\mathbf{a}|$, $\mathbf{b} = \mathbf{v}|\mathbf{b}|$, $\mathbf{a}+\mathbf{b} = \mathbf{w}|\mathbf{a}+\mathbf{b}|$. Then $|\mathbf{a} + \mathbf{b}| = \mathbf{w}^*(\mathbf{a} + \mathbf{b}) = \mathbf{w}^*\mathbf{u}^*|\mathbf{a}| + \mathbf{w}^*\mathbf{v}^*|\mathbf{b}|$. Let (e_n) be a finite o.n. system; we have

$$\sum_n <e_n, |\mathbf{a}+\mathbf{b}|e_n> = \sum_n <e_n, \mathbf{w}^*\mathbf{u}^*|\mathbf{a}|e_n> + \sum_n <e_n, \mathbf{w}^*\mathbf{v}^*|\mathbf{b}|e_n> \leq \|\mathbf{a}\|_1 + \|\mathbf{b}\|_1 .$$

Then we let (e_n) increase to an orthonormal basis, etc.

EXAMPLE. Let $\mathbf{a}$ be a selfadjoint operator. Then it is easy to prove that $\mathbf{a}$ belongs to the trace class if and only if $\mathbf{a}$ has a discrete spectrum (λ_i), and $\sum_i |\lambda_i|$ is finite; then this sum is $\|\mathbf{a}\|_1$, and $\mathrm{Tr}(\mathbf{a}) = \sum_i \lambda_i$. It follows easily that the two operators $\mathbf{a}^+$, $\mathbf{a}^-$ belong to $\mathcal{L}^1$ if $\mathbf{a}$ does, and that $\|\mathbf{a}\|_1 = \|\mathbf{a}^+\|_1 + \|\mathbf{a}^-\|_1$, a result which corresponds to the Jordan decomposition of bounded measures in classical measure theory.

Duality properties

3 The results of this subsection are essential for the theory of von Neumann algebras, and have some pleasant probabilistic interpretations.

Let us denote by E_{xy} the operator of rank one

$$E_{xy}z = <y,z>x \qquad (\ |x><y|\ \text{ in Dirac's notation.})$$

Then we have $(E_{xy})^* = E_{yx}$, $(E_{xy})^* E_{xy} = \|x\|^2 E_{yy}$, $\|E_{xy}\|_1 = \|x\|\|y\|$. The space $\mathcal{F}$ generated by these operators consists of all *operators of finite rank*. One can show that its closure in operator norm consists of all *compact operators*. Our aim is to prove

THEOREM. *The dual space of $\mathcal{F}$ is $\mathcal{L}^1$, and the dual of $\mathcal{L}^1$ is the space $\mathcal{L}^\infty$ of all bounded operators.*

The proof also makes explicit the duality functional between these spaces, in both cases the bilinear functional $(\mathbf{a},\mathbf{b}) \longmapsto \mathrm{Tr}(\mathbf{ab})$). We leave it to the reader to check that the theorem remains true if all three spaces are restricted to their selfadjoint elements.

Note that the theorem implies that $\mathcal{L}^1$ is complete.

PROOF. 1) We remark first that for $\mathbf{a}$ positive we have $\mathrm{Tr}(\mathbf{a}) = \sup_{\mathbf{h}\in\mathcal{F}_1}\mathrm{Tr}(\mathbf{ah})$, where $\mathcal{F}_1$ denotes the unit ball of $\mathcal{F}$. To see this, if $\mathbf{a}$ has discrete spectrum, diagonalize it and choose for $\mathbf{h}$ a diagonal matrix whose diagonal coefficients are zeroes and finitely many ones, tending to the identity matrix. If $\mathbf{a}$ has some continuous spectrum, then the right side of the relation is $+\infty$, and on the other hand $\mathbf{a}$ dominates $c\mathbf{k}$ where $c > 0$ is a constant and $\mathbf{k}$ is the projector on some infinite dimensional subspace. Then taking $\mathbf{h}$ to be almost the identity matrix of this subspace we see that the right side is also equal to $+\infty$.

We extend this relation to arbitrary $\mathbf{a}$ using the polar decomposition $\mathbf{a} = \mathbf{u}|\mathbf{a}|$, $|\mathbf{a}| = \mathbf{u}^*\mathbf{a}$. Then we have from the above discussion

$$\|\mathbf{a}\|_1 = \mathrm{Tr}(|\mathbf{a}|) = \sup_{\mathbf{k}\in\mathcal{F}_1} \mathrm{Tr}(|\mathbf{a}|\mathbf{k})$$
$$= \sup_{\mathbf{k}\in\mathcal{F}_1} \mathrm{Tr}(\mathbf{u}^*\mathbf{ak}) = \sup_{\mathbf{k}\in\mathcal{F}_1} \mathrm{Tr}(\mathbf{aku}^*)$$

We now put $\mathbf{ku}^* = \mathbf{h}$, which belongs to $\mathcal{F}_1$, and we get $\mathrm{Tr}(\mathbf{a}) \le \sup_{\mathbf{h}\in\mathcal{F}_1}\mathrm{Tr}(\mathbf{ah})$; the reverse inequality is obvious.

2) Let φ be a continuous linear functional on $\mathcal{F}$. Then it is continuous for the topology induced by HS, which is *stronger* than the operator norm topology. Hence it can be written $\varphi(\cdot) = <\mathbf{a},\cdot>_{HS}$ for some HS operator $\mathbf{a}$. Using the preceding computation, one sees that $\|\mathbf{a}\|_1 = \|\varphi\|$, and in particular that $\mathbf{a}$ belongs to $\mathcal{L}^1$.

3) Let x, y be two normalized vectors. and let (e_n) be an o.n. basis whose first element is y. Then for every operator $\mathbf{h}$ we have $\mathbf{h}E_{xy}(e_n) = \mathbf{h}(x)$ for $n = 1$, 0 otherwise, hence $\mathrm{Tr}(\mathbf{h}E_{xy}) = <y, \mathbf{h}(x)>$. Since E_{xy} belongs to the unit ball of $\mathcal{L}^1$, we have

$$\| \mathbf{h} \|_\infty \leq \sup_{\|\mathbf{a}\|_1 \leq 1} \mathrm{Tr}(\mathbf{h}\mathbf{a}) \, ,$$

whence the equality. If $\mathbf{h}$ is selfadjoint one can take $y = x$ in this argument, and therefore $\mathbf{a}$ can be taken to be selfadjoint.

4) Finally, let φ be a continuous linear functional on $\mathcal{L}^\infty$. Then $(y, x) \longmapsto \varphi(E_{xy})$ is a continuous hermitian bilinear functional, which can be written as $<y, \mathbf{h}x>$ for some bounded operator $\mathbf{h}$. The computation above shows that the operator norm of $\mathbf{h}$ is equal to the norm of φ. The two functionals $\varphi(\cdot)$ and $\mathrm{Tr}(\mathbf{h}\cdot)$ are equal on operators of finite rank, hence on all of $\mathcal{L}^1$. The proof is concluded.

Weak convergence properties

4 We are going now to study weak convergence properties for "measures" and "functions". The results concerning measures are pleasant, but not very important, and we do not prove them in detail.

We stop using boldface letters for operators.

We start with measures, recalling first Bourbaki's terminology for weak convergence on a locally compact space E. One says that a sequence (we leave it to the reader to rewrite the statements for filters, if he cares to) of probability measures (μ_n) converges *vaguely* to μ if $\mu_n(f)$ converges to $\mu(f)$ for every continuous function f with compact support. Then the limit is a positive measure with total mass ≤ 1. If μ is a true probability measure, then the sequence is said to converge *narrowly*, and then $\mu_n(f)$ converges to $\mu(f)$ for every continuous bounded function f. Convergence of sequences in the narrow topology implies *Prohorov's condition* : for every $\varepsilon > 0$ there exists a compact set K whose complement has measure $\leq \varepsilon$ for every measure μ_n. Conversely, every sequence μ_n which satisfies Prohorov's condition is relatively compact in the narrow topology. In fact, we do not need any sophisticated theory : quantum probability in the situation of this chapter is similar to measure theory on $\mathbb{N}$.

What corresponds to *vague* convergence for a sequence ρ_n of density matrices is the convergence of $\mathrm{Tr}(\rho_n a)$ to $\mathrm{Tr}(\rho a)$ for every operator a *of finite rank,* while *narrow* convergence holds if in addition $\mathrm{Tr}(\rho) = 1$. Since the closure in norm of the space of finite rank operators is the space of all compact operators, whose dual is $\mathcal{L}^1$, the unit ball of $\mathcal{L}^1$ is compact in the vague topology, just as in classical probability. If (ρ_n) converges narrowly to ρ, then $\rho_n(a)$ converges to $\rho(a)$ for every bounded operator a, just as in classical probability. *Sketch of proof* : we reduce to a self adjoint, then we approximate a in norm by operators with discrete spectrum : using the spectral representation of a this amounts to approximating uniformly the function t by a step function $f(t)$ on a compact interval. Then a can be diagonalized in some o.n.b. (e_n) and we have $\mathrm{Tr}(\rho_n a) = \sum_n \rho_{nn} \lambda_n$, where λ_n is the eigenvalue of a corresponding to the eigenvector e_n and ρ_{nn} is the n–th diagonal matrix element of ρ. Then we are reduced to a trivial problem of narrow convergence on $\mathbb{N}$.

"Prohorov's condition" takes the following form : for every $\varepsilon > 0$ there exists a projector P of finite rank such that $\|\rho_n - P\rho_n P\|_1 \leq \varepsilon$ for all n. If (ρ_n) converges narrowly to ρ, then Prohorov's condition in this sense holds ([Dav], p. 291, lemma 4.3), and in fact ρ_n tends to ρ *in trace norm*. This is entirely similar to the fact that weak and strong convergence are the same *for sequences* in ℓ^1 (Dunford–Schwartz , *Linear Operators I*, p. 296, Cor. 14). This theorem is wrong for states on a general von Neumann algebra (see Dell'Antonio, [D'A]).

The "normal" topology for operators

5 The results in this subsection are fundamental for the theory of von Neumann algebras. They describe several weak topologies on the space $\mathcal{L}^\infty$ of all bounded operators on $\mathcal{H}$.

The best known among them are the *weak* and the *strong* topologies. As usual with locally convex topologies, they are defined by a family of seminorms $(p_\lambda)_{\lambda \in \Lambda}$, operators $(a_i)_{i \in I}$ converging to 0 (along some filter on I) if and only if for every λ the numbers $p_\lambda(a_i)$ converge to 0. In the case of the weak topology, Λ is the set of all pairs (x,y) of vectors, and $p_{x,y}(a) = |< y, ax >|$. In the case of the strong topology, Λ is the set of all vectors, and $p_x(a) = \|ax\|$.

There is a third basic topology, which arises from the fact that $\mathcal{L}^\infty$ is the dual of $\mathcal{L}^1$: in this case, Λ is the set of all trace class operators (or, what amounts to the same, the set of all density matrices ρ), and $p_\rho(a) = |\operatorname{Tr}(\rho a)|$. It turns out that it is the most important one, and that it has no satisfactory name. Its old name "ultraweak topology" is bad, since it seems to mean "still weaker than the weak topology", while it is *stronger* than the weak topology! Many people call it σ-*weak topology* (σ has the same meaning here as in σ-*finite*, the explanation being given below). My own tendency, given its importance, and the very frequent use of the adjective *normal* in the context of von Neumann algebras, to mean *continuous* in this topology, consists in calling it the *normal topology*.

The usual description of the normal topology is as follows : note that every trace class operator can be written as a product bc of two Hilbert–Schmidt operators, and that in an o.n.b. (e_n) one can write

$$\operatorname{Tr}(bca) = \operatorname{Tr}(cab) = \sum_n <c^* e_n \, , \, ab(e_n)>$$

with $\sum_n \|x_n\|^2 < \infty$, $\sum_n \|y_n\|^2 < \infty$. Consider now a direct sum $\mathcal{K}$ of countably many copies of $\mathcal{H}$; an element of this space is a sequence $\mathbf{x} = (x_n)$ of elements of $\mathcal{H}$ such that $\|\mathbf{x}\|^2 = \sum_n \|x_n\|^2 < \infty$. On the other hand, associate with every bounded operator a on $\mathcal{H}$ the operator $\mathbf{a}$ on $\mathcal{K}$ which maps (x_n) into (ax_n). Then note that for every sequence $\mathbf{y}$ there is exactly one operator b which transforms (e_n) into (y_n). So the above displayed expression can be written $<\mathbf{x}, \mathbf{a}\mathbf{y}>$, and we see that *convergence of a in the normal topology of $\mathcal{H}$ amounts to the convergence of $\mathbf{a}$ in the usual weak topology of $\mathcal{K}$*.

> One can define in the same way the "σ-strong topology", which is interesting but will not be considered here.

Since $\mathcal{L}^\infty$ is the dual of $\mathcal{L}^1$, its unit ball is compact in the normal topology, hence on the unit ball there cannot be a different weaker Hausdorff topology. Otherwise stated, as far as one restricts oneself to some norm bounded set of operators, there is no distinction between the normal and the weak topologies.

We end this section with a simple and well known result.

THEOREM. Let φ be a linear functional on $\mathcal{L}^\infty(\mathcal{H})$, continuous for the strong topology of operators. Then φ is a finite linear combination of functionals of the type $a \longmapsto < y_k, ax_k >$ (in particular, φ is also continuous for the weak topology).

PROOF. By definition of the strong topology, there exist finitely many vectors x_i such that $\sup_i \| ax_i \| \leq 1 \Rightarrow |\varphi(a)| \leq 1$. Let p be the projector on the subspace $\mathcal{K}$ generated by these vectors. The relation $ap = 0$ implies $ax_i = 0$ for all i, hence $\varphi(a) = 0$, so we must have $\varphi(a) = \varphi(ap)$ for all a. Choose an o.n.b. (e_n) whose first k elements generate $\mathcal{K}$, and express $\varphi(a)$ as a linear combination of matrix elements of a in this basis, etc.

Since the dual space of $\mathcal{L}^\infty$ in the normal topology is $\mathcal{L}^1$, we see that the two topologies are rather different. On the other hand, a classical theorem of Banach (of which a very simple proof has been given by Mokobodzki : see for instance [DeM3], chap. X, n° 52), asserts that a linear functional on a dual Banach space is continuous in the weak* topology if and only if its restriction to the unit ball is continuous. Here, a linear functional on $\mathcal{L}^\infty$ is normally continuous (or, as stated briefly, is *normal*) if and only if its restriction to the unit ball is normally continuous. But on the unit ball the weak and normal topologies coincide. This will be useful as a convenient characterization of normal functionals without looking at the normal topology itself.

Tensor products of Hilbert spaces

6 The tensor product of two Hilbert spaces plays in quantum probability the role of the ordinary product of measure spaces in classical probability. This subsection is very important, and nearly trivial.

Given two Hilbert spaces $\mathcal{A}$ and $\mathcal{B}$, we define their Hilbert space tensor product to consist of 1) a Hilbert space $\mathcal{C}$ and 2) a bilinear mapping $(f,g) \longmapsto f \otimes g$ from $\mathcal{A} \times \mathcal{B}$ to $\mathcal{C}$ such that

$$(6.1) \qquad\qquad <f \otimes g, h \otimes k>_\mathcal{C} \;=\; < f, h >_\mathcal{A} < g, k >_\mathcal{B} \;.$$

On the other hand, the set of all vectors $f \otimes g$ should span $\mathcal{C}$.

From these two properties it is easy to deduce that, given two o.n. bases (f_α) of $\mathcal{A}$ and (g_β) of $\mathcal{B}$, the family $(f_\alpha \otimes g_\beta)$ is an o.n.b. of $\mathcal{C}$. It follows immediately that the space $\mathcal{C}$ and the bilinear mapping $\otimes$ are defined up to isomorphism. It is possible also to define uniquely a canonical tensor product, *i.e.* $\mathcal{A} \otimes \mathcal{B}$ and all the mappings involved are constructed from $\mathcal{A}$ and $\mathcal{B}$ in the language of set theory, but the need for such a precise definition is never felt.

What about existence? Assume $\mathcal{A}$ and $\mathcal{B}$ are given as concrete Hilbert spaces $L^2(E,\mathcal{E},\lambda)$ and $L^2(F,\mathcal{F},\mu)$ (for instance, o.n.b.'s (e_α) and (f_β) have been chosen, and λ,μ are the counting measures on the index sets). Then $\mathcal{C}$ is nothing but $L^2(E \times F, \lambda \otimes \mu)$, the product measure, $f \otimes g$ being interpreted as the function of two variables $f(x)\,g(y)$ on $E \times F$ (which is quite often denoted $f \otimes g$ in a more general context). The verification of the two axioms is very easy. Besides that, the concrete case of L^2 spaces covers all practical situations.

Returning to abstract Hilbert spaces, it is convenient to have a name for the linear span (no closure operation) of all vectors $f \otimes g$, f and g ranging respectively over two subspaces $\mathcal{A}_0$ and $\mathcal{B}_0$ of $\mathcal{A}$ and $\mathcal{B}$: we call it the *algebraic tensor product of $\mathcal{A}_0$ and $\mathcal{B}_0$*. We introduce no specific notation for it (usually, one puts some mark like $\overline{\otimes}$ on the tensor sign).

Let $\mathcal{H}$ and $\mathcal{K}$ be two Hilbert spaces, U and V be two continuous linear mappings from $\mathcal{A}$ to $\mathcal{H}$, $\mathcal{B}$ to $\mathcal{K}$. There exists a unique continuous mapping $W = U \otimes V$ from the algebraic tensor product of $\mathcal{A}$ and $\mathcal{B}$ into $\mathcal{H} \otimes \mathcal{K}$ such that

(6.2) $$W(f \otimes g) = (Uf) \otimes (Vg) \,,$$

and we are going to prove that

(6.3) $$\| W \| \leq \| U \| \, \| V \| \,.$$

Then it will extend by continuity to the full tensor product. To prove (6.3) we may proceed in two steps, assuming one of the two mappings to be identity — for instance $\mathcal{B} = \mathcal{K}$, $V = I$. Let (f_α), (g_β) be o.n. bases for $\mathcal{A}$ and $\mathcal{B}$, and let $\mathcal{A}_0$, $\mathcal{B}_0$ be their linear spans (without completion). Since the vectors $f_\alpha \otimes g_\beta$ are linearly independent, there is a unique operator W with domain $\mathcal{A}_0 \otimes \mathcal{B}_0$ which maps $f_\alpha \otimes g_\beta$ to $U(f_\alpha) \otimes g_\beta$. Every vector in the domain can be written as a finite sum $z = \sum_\beta x_\beta \otimes g_\beta$, and its image is a sum $z' = Wz = \sum_\beta U(x_\beta) \otimes g_\beta$ of *orthogonal* vectors. Then we have

$$\| z' \|^2 = \sum \| U(x_\beta) \otimes g_\beta \|^2 \leq \| U \|^2 \sum \| x_\beta \|^2 \| g_\beta \|^2 = \| U \|^2 \| z \|^2 \,.$$

Relation (6.2) is proved on the domain, which is dense, and then the extension of W to an everywhere defined operator satisfying (6.2) and (6.3) is straightforward.

In classical probability, when we want to consider simultaneously two probabilistic objects, *i.e.* spaces $(E, \mathcal{E}, \mathbb{P})$ and $(F, \mathcal{F}, \mathbb{Q})$, we consider the product $(E \times F, \mathcal{E} \times \mathcal{F})$, and then a joint law on it which, if the two objects are physically unrelated, is the product law $\mathbb{P} \otimes \mathbb{Q}$ (classical probabilistic independence), and is some other law if there is a non-trivial correlation between them. The quantum probabilistic analogue consists in forming the tensor product $C = \mathcal{A} \otimes \mathcal{B}$ of the Hilbert spaces describing the two quantum objects, the state of the pair being the tensor product $\rho \otimes \sigma$ of the individual states if there is no interaction. This explains the basic character of the tensor product operation. On the other hand, we met in Chapter IV a new feature of quantum probability : systems of *indistinguishable objects* are not adequately described by the ordinary tensor product, but rather by subspaces of the tensor product subject to symmetry rules.

7   There are some relations between tensor products and HS operators. We describe them briefly.

First of all, we consider a Hilbert space $\mathcal{H}$ and its dual space $\mathcal{H}'$, *i.e.* the space of complex linear functionals on $\mathcal{H}$. Mapping a *ket* $x = |x>$ to the corresponding *bra*

$x^* = <x|$ provides an *antilinear* 1-1 mapping from $\mathcal{H}$ onto $\mathcal{H}'$. Since $\mathcal{H}'$ is a space of complex functions on $\mathcal{H}$, the tensor product $u \otimes v$ of two elements of $\mathcal{H}'$ has a natural interpretation as the function $u(x)v(y)$ on $\mathcal{H} \times \mathcal{H}$, which is bilinear. If we denote by (e_α) an o.n.b. for $\mathcal{H}$, by e^α its dual basis, an o.n.b. for $\mathcal{H}' \otimes \mathcal{H}'$ is provided by the bilinear functionals $e^{\alpha\beta} = e^\alpha \otimes e^\beta$, and the elements of $\mathcal{H}' \otimes \mathcal{H}'$ are bilinear forms which may be expanded as $\sum c_{\alpha\beta} e^{\alpha\beta}$ with a square summable family of complex coefficients $c_{\alpha\beta}$. This is much smaller than the space of all continuous bilinear forms on $\mathcal{H} \times \mathcal{H}$, and is called the space of *Hilbert-Schmidt forms*. One may define similarly multiple tensor products, and Hilbert–Schmidt n–linear forms.

Why Hilbert–Schmidt? Instead of considering $\mathcal{H}' \otimes \mathcal{H}'$ let us consider $\mathcal{H}' \otimes \mathcal{H}$, a space of bilinear functionals on $\mathcal{H} \times \mathcal{H}'$. Any operator A on $\mathcal{H}$ provides such a bilinear form, namely $(x, y') \longmapsto (y', Ax)$ — everything here is bilinear and the hermitian scalar product is not used — and the basis elements $e^\alpha \otimes e_\beta$ are read in this way as the operators $|e_\beta > < e_\alpha|$ (Dirac's notation) which constitute an o.n.b. for the space of Hilbert–Schmidt operators.

More generally, the tensor product $x^* \otimes y$ can be interpreted as the rank one operator $|y><x|$, and the algebraic tensor product $\mathcal{H}' \bar{\otimes} \mathcal{H}$ as the space of all operators of finite rank.

Appendix 2

Conditioning and Kernels

Conditioning is one of the basic ideas of classical probability, and the greatest success of the Kolmogorov system of axioms was its inclusion of conditioning as a derived notion, without need of special axioms. Therefore classical probabilists will expect a discussion of conditioning in quantum probability. On the other hand, the subject is very delicate : in quantum physics, "knowing" a random variable does not concern only the mind of the physicist. It may be impossible to "know" without destroying some features of the system. Thus conditional expectations do not exist in general, and the simplest constructions of classical probability (like elementary Markov chains with given transition probabilities) run into serious difficulties. The end of this Appendix has been modified to include an important idea of Bhat and Parthasarathy, which extends a classical construction of Stinespring to give a clear (though restricted) meaning to the quantum Markov property.

This Appendix is not necessary to read the main body of these notes. It can also be read independently, except that its last section uses a little of the language of C^*-algebras and von Neumann algebras (Appendix 4).

Conditioning : discrete case

1 The discussion in this and the following subsection is inspired from Davies' book [Dav], *Quantum Theory of Open Systems*, p. 15–17.

Consider first a classical probability space $(\Omega, \mathcal{F}, \mathbb{P})$ and a random variable X taking values in some nice measurable space E. Then the space Ω is decomposed into the "slices" $\Omega_x = X^{-1}(x)$, $x \in E$, and the measure $\mathbb{P}$ can be disintegrated according to the observed value of X. Namely, for every $x \in E$, there exists a law $\mathbb{P}_x$ on Ω, carried by Ω_x, such that for $A \in \mathcal{F}$ (μ denoting the law of X under $\mathbb{P}$)

$$\mathbb{P}(A) = \int \mathbb{P}_x(A) \, \mu(dx) \, .$$

Intuitively speaking, if we observe that $X = x$, then the absolute law $\mathbb{P}$ is reduced to the conditional law $\mathbb{P}_x$. What about getting a similar disintegration in quantum probability ?

We now denote by Ω, as we did previously, the basic Hilbert space, and by X a r.v. on Ω, taking values in E (a spectral measure over E). We begin with the case of a countable space E : then for each $i \in E$ we have an "event" (subspace) A_i and its "indicator" (spectral projector) P_i. Let Z denote a real valued random variable (selfadjoint operator), bounded for simplicity.

The easiest thing to describe is the decomposition of Ω in slices : these are simply the subspaces $A_i = \{X = i\}$. For continuous random variables, this will be non–trivial, and we deal with this problem in subsection 4.

In quantum physics, a measurement is a model for the concrete physical process of installing a macroscopic apparatus which filters the population of particles according to some property, here the value of X. When this is done (physically : when the power is turned on) the state of the system changes. If it was described by the density operator ρ, it is now described by

$$(1.1) \qquad\qquad \widetilde{\rho} = \sum_i P_i \rho P_i \ .$$

Note that $\widetilde{\rho}$ is a density operator, which commutes with all the spectral projections P_i of X (we simply say it commutes with X). Also note that, if ρ originally did commute with X, no change has occurred. This change of state has nothing to do with our looking at the result of the experiment. Indeed, *if we do*, that is if we filter the population which goes out of the apparatus and select those particles for which X takes the value i, a further change takes place, which this time is the same familiar one as in classical probability : $\widetilde{\rho}$ is replaced by the conditional density operator

$$(1.2) \qquad\qquad \widetilde{\rho}_i = \frac{P_i \rho P_i}{\mathrm{Tr}(\rho P_i)} \ .$$

Starting from these assumptions about the effect of a measurement of X on the state of the system, we are going to compute the expectation of Z and "joint distributions".

First of all, the expectation of Z under the new law $\widetilde{\rho}$ is equal to

$$\widetilde{\mathbb{E}}\,[\,Z\,] = \mathrm{Tr}(\widetilde{\rho}Z) = \sum_i \mathrm{Tr}(P_i \rho P_i Z)$$
$$= \sum_i \mathrm{Tr}(\rho P_i Z P_i)$$

(the property that $\mathrm{Tr}(AB) = \mathrm{Tr}(BA)$ has been used). This is not the same as the original expectation of Z — otherwise stated, contrary to the classical probability case, *conditioning changes expectation values.* The difference is

$$\mathbb{E}\,[\,Z\,] - \widetilde{\mathbb{E}}\,[\,Z\,] = \sum_{ik} \mathrm{Tr}(\rho P_i Z P_k) - \sum_i \mathrm{Tr}(\rho P_i Z P_i)$$
$$= \sum_{i \neq k} \mathrm{Tr}(\rho P_i Z P_k)$$

(note that the difference vanishes if either Z or ρ commutes with X). Instead of deciding that the state has changed (Schrödinger's picture) we might have decided that the random variable Z has been replaced by $\widetilde{Z} = \sum_i P_i Z P_i$, the state remaining unchanged (Heisenberg's picture). Though both points of view are equivalent for Hamiltonian evolutions, here the first point of view seems more satisfactory. Indeed, basic properties of Z like being integer valued (having a spectrum contained in $\mathbb{N}$, with the possible physical meaning of being the output of a counter) is not respected by the above operation on r.v.'s, while one has no objection to a shift from discrete spectrum to continuous spectrum in a change of density matrix.

Let us consider a second r.v. Y taking values in a countable space F (points denoted j, events B_j, spectral projections Q_j), and let us compute a "joint distribution" for X and Y according to the preceding rules. If it has been observed that $X = i$, then the probability that $Y = j$ is $\mathrm{Tr}(\tilde{\rho}_i Q_j) = \mathrm{Tr}(\rho P_i Q_j P_i)/\mathrm{Tr}(\rho P_i)$, and the probability that *first* the measure of X yields i and *then* the measure of Y yields j is

$$p_{ij} = \mathrm{Tr}(\rho P_i Q_j P_i)$$

These coefficients define a probability law, but they depend on the order in which the measurements are performed. Note also that the mapping $(i, j) \longmapsto P_i Q_j P_i$ does not define an observable on $E \times F$, since the selfadjoint operators $P_i Q_j P_i$ generally are not projectors (unless X and Y commute). This obviously calls for a generalization of the notion of observable (see subs. 4.1).

Conditioning : continuous case

2 We are going now to deal with the case of a continuous r.v. X, taking their values in a measurable space $(E, \mathcal{E})$ with a countably generated σ–field. Consider an increasing family of finite σ–fields $\mathcal{E}_n$ whose union generates $\mathcal{E}$. For notational simplicity, let us assume that the first σ–field $\mathcal{E}_0$ is trivial, and that every atom H_i of $\mathcal{E}_n$ is divided in two atoms of $\mathcal{E}_{n+1}$, so that the atoms of $\mathcal{E}_n$ are indexed by the set $W(n)$ of dyadic words with n letters. Let X_n be the random variable X considered as a spectral measure on $(E, \mathcal{E}_n)$. Since X_n is a discrete random variable, the conditional expectation of Z given X_n is the following function η_n on E (I_{H_i} here is an indicator function in the usual sense, not a projector, and our conditional expectation is a r.v. in the usual sense)

$$\eta_n = \sum_{i \in W(n)} \frac{\mathrm{Tr}(\rho P_i Z P_i)}{\mathrm{Tr}(\rho P_i)} I_{H_i} .$$

The analogy with the theory of derivation leads us to compute $\eta_n - \mathbb{E}\left[\eta_{n+1} | \mathcal{E}_n\right]$ (ordinary conditional expectation of the r.v. X, with respect to the law $\mathbb{P}$). To help intuition, let us set for every "event" (subspace) A with associated projector P, $\mathbb{E}[A ; Z] = \mathrm{Tr}(\rho P Z P)$. Then the function we want to compute is equal on H_i to

$$\frac{1}{\mathbb{P}(H_i)} \left(\mathbb{E}\left[\{X \in H_i\} ; Z\right] - \mathbb{E}[\{X \in H_{i0}\} ; Z] - \mathbb{E}[\{X \in H_{i1}\} ; Z] \right)$$

and we have

$$\mathbb{E}\left[|\eta_n - \mathbb{E}\left[\eta_{n+1} | \mathcal{E}_n\right]|\right] = \sum_{i \in W(n)} |\mathrm{Tr}(\rho P_{i0} Z P_{i1}) + \mathrm{Tr}(\rho P_{i1} Z P_{i0})|$$

Let H belong to $\mathcal{E}$ and P_H denote the projection on the subspace $\{X \in H\}$. Then the function

$$(2.1) \qquad\qquad (H, K) \longmapsto \mathrm{Tr}(\rho P_H Z P_K)$$

is a *complex bimeasure* ν, and we will assume it has bounded variation, *i.e.* can be extended into a bounded complex measure on $E \times E$, still denoted by ν (we return to the discussion of ν at the end of this subsection). The above expectation then is bounded by

$$\sum_{i \in W(n)} \left(|\nu|(H_{i0} \times H_{i1}) + |\nu|(H_{i1} \times H_{i0}) \right)$$

and finally, summing over n, we get that

$$\sum_n \mathbb{E}\,[\,|\eta_n - \mathbb{E}\,[\,\eta_{n+1}|\mathcal{E}_n\,]\,|\,] \;\le\; |\nu|(E \times E \setminus \Delta)$$

Δ denoting the diagonal. Thus the (ordinary) random variables η_n constitute a *quasimartingale.* Since we have $\mathbb{E}\,[\,|\eta_n|\,] = \sum_{i \in \omega(n)} |Tr(\rho P_i Z P_i)| \le \sum_i |\nu|(H_i \times H_i) \le |\nu|(E \times E)$, this quasimartingale is bounded in L^1, and finally η_n *converges* $\mathbb{P}$-*a.s.*.

Let us end with a remark on the bimeasure $\nu(H, K) = \mathrm{Tr}(P_H Z P_K)$. We have used it to estimate the change in expectation due to conditioning, that is an expression of the following form, where (H_i) is some partition of E

$$\mathbb{E}\,[\sum_i P_{H_i} Z P_{H_i} - Z] = \mathbb{E}\,[\frac{1}{2}\sum_i (2P_{H_i} Z P_{H_i} - P_{H_i} Z - Z P_{H_i})]$$

$$= \mathbb{E}\,[\frac{1}{2}\sum_i [P_{H_i}, [Z, P_{H_i}]]]\;.$$

Hence instead of studying the complex bimeasure ν, we may as well study the real bimeasure $\theta(H, K) = \mathrm{Tr}\,\big(\rho\,[P_H, [Z, P_K]]\big)\,.$

3 We illustrate the preceding computations, by the basic example of the *canonical pair* (studied in detail in Chapter 3). Explicitly, Ω is the space $L^2(\mathbb{R})$ (Lebesgue measure); E is the line, the initial partition is given by the dyadic integers, and then we proceed by cutting each interval in two equal halves; X is the identity mapping from Ω to $\mathbb{R}$, *i.e.* for a Borel set A of the line P_A is the operator of multiplication by I_A. For ρ we choose the pure state ε_ω corresponding to the "wave function" ω. Finally, the operator we choose for Z will not be a bounded selfadjoint operator, but rather the *unitary* operator $Z f(x) = f(x-u)$: since Z is a complex linear combination of two (commuting) selfadjoint operators, the preceding theory can be applied without problem. It is clear that Z is the worst possible kind of operator from the point of view of the "slicing" : instead of operating along the slices it interchanges them. We shall see later that $Z = e^{-iuY}$, where $Y = -iD$ is the momentum operator of the canonical pair.

The bimeasure we have to consider in this case is

$$\nu(H, K) = \mathrm{Tr}\,(\rho P_H Z P_K) = \;<\omega, P_H Z P_K \omega>$$

$$= \int \bar{\omega}(s)\, I_H(s)\, I_K(s - u)\, \omega(s - u)\, ds\;.$$

Let $\lambda(ds)$ be the complex measure on the line with density $\bar{\omega}(s)\omega(s-u)$; since ω is normalized, the total mass of λ is at most 1. Let n be the image of λ under the mapping $s \longmapsto (s, s+u)$; then $\nu(H, K) = n(H \times K)$, and therefore the basic assumption for the convergence of the conditional expectations is satisfied ; we may forget the notation n and use ν for both objects, bimeasure and measure.

In the case we are considering, *the limit of* $\mathbb{E}\,[Z|\mathcal{E}_n]$ *is a.s. equal to* 0. Indeed, $\mathrm{Tr}(\rho P_i Z P_i) = 0$ when the partition is fine enough, simply because then the square $H_i \times H_i$ does not meet the line $y = x + u$ which carries ν. Otherwise stated, we have

$$\mathbb{E}\,[\,e^{-iuY}\,|\,X\,] = 0 \quad \text{if} \quad u \ne 0\,, \;= 1 \quad \text{if} \quad u = 0\,.$$

In weak convergence problems, this degenerate characteristic function is typical of cases where all the mass escapes to infinity.

Multiplicity theory

4 We are going now to extend the idea of "slicing" or "combing" the basic Hilbert space Ω to the case of a continuous r.v. X. This is also called "multiplicity theory", and applies just as well to *real* Hilbert spaces (for an application of the real case, see *Sém. Prob IX*, LN. 465, p. 73–88, which also contains a proof of the theorem itself). A complete proof is also given in [Par1].

Note first that, a nice measurable space being isomorphic to a Borel subset of $\mathbb{R}$, we lose no generality by assuming that X is real valued. Hence X is associated with an orthogonal resolution of identity $(\mathcal{H}_t)$ with spectral projectors E_t. Let us call *martingale* any curve $x = x(\cdot)$ such that $x(s) = E_s x(t)$ for $s < t$, and denote by η the "bracket" of the martingale, *i.e.* the measure on $\mathbb{R}$ such that $\eta(\,]\,s,t\,]\,) = \|\,x(t) - x(s)\,\|^2$. Let us call *stable subspace of* Ω any closed subspace which is stable under all projectors E_t. An example of such a subspace is given by $S(x)$, the set of all "stochastic integrals" $\int f(s)\,dx(s)$ with $f \in L^2(\eta)$; its orthogonal space $S(x)^\perp$ is a stable subspace too.

The mapping $f \longmapsto \int f(s)\,dx(s)$ is an isomorphism from the Hilbert space $\mathcal{M}(\eta) = L^2(\mathbb{R}, \eta)$, which we call the *model space*, onto $S(x)$; it is slightly more than that : the model space carries a natural spectral family $(\mathcal{I}_t)$, corresponding to functions supported by the half–line $]-\infty, t]$ (the selfadjoint operator it generates is multiplication by the function x), and $\mathcal{I}_t$ is carried by the isomorphism into the resolution of identity $\mathcal{H}_t \cap S(x)$, induced on $S(x)$ by our original spectral family. *Thus* Ω *contains a stable subspace which is a copy of the model.*

We now replace Ω by $\Omega_1 = S(x)^\perp$ and (if it is not reduced to 0) extract from it a second copy of the model, possibly with a different measure η_1. Iterating transfinitely this procedure, it is very intuitive that Ω *can be decomposed into a direct sum of copies of model spaces.* This intuition can be made rigorous using Zorn's lemma. Since Ω is always assumed to be separable, this direct sum decomposition is necessarily countable. Rearranging the indexes into a single sequence, denote by $S(x_n)$ the spaces and by η_n the corresponding measures, choose a measure θ such that every η_n is absolutely continuous w.r. to θ (a probability measure if you wish), and denote by h_n a density of η_n w.r. to θ.

Every $\omega \in \Omega$ can be represented uniquely in the form

$$\omega = \sum_n \int f_n(s)\,dx_n(s) \quad \text{with} \quad \sum \int |f_n(s)|^2 \eta_n(ds) < \infty \ .$$

Associate with every $s \in \mathbb{R}$ the Hilbert space $\mathcal{F}_s$ consisting of the sequences (x_n) of complex numbers such that $\sum_n |x_n|^2 h_n(s) < \infty$ (for simplicity we assume $h_n(s) > 0$ for all n ; if this condition is not fulfilled, consider only those n such that $h_n(s) > 0$). Then for θ–a.e. s the sequence $(f_n(s))$ belongs to $\mathcal{F}_s$, and the above isomorphism realizes the "slicing" of Ω we were looking for. Such a "slicing" has an official name : in Hilbert space language, one says that Ω *is isomorphic to the continuous sum of the measurable family of Hilbert spaces* $\mathcal{F}_s$ *over the measure space* $(\mathbb{R}, \theta)$. We do not need to make here an axiomatic theory of continuous sums of Hilbert spaces, since the above description was entirely explicit. For a general discussion, see Dixmier [Dix], part II, Chapter 1.

Up to now, we have not risen much above the level of triviality. Things become more interesting when we try to get some kind of uniqueness. Here we shall explain the results, without even sketching a proof.

First of all, the "models" we use are not uniquely determined. If γ and η are two equivalent measures on $\mathbb{R}$, and j is a density of η w.r. to γ, the mapping $f \longmapsto f\sqrt{\gamma}$ is an isomorphism of $\mathcal{M}(\eta)$ onto $\mathcal{M}(\gamma)$ which preserves the given resolutions of identity. Thus it is the *equivalence class* of η (also called the *spectral type* of the model) which matters. Next, if η is decomposed into a sum of two mutually singular measures λ and μ, the model space $\mathcal{M}(\eta)$ gets decomposed into a direct sum $\mathcal{M}(\lambda) \oplus \mathcal{M}(\mu)$, compatible with the given resolutions of identity. Hence if we want to add as many "models" as we can in a single operation, our interest is to choose a spectral type as strong as we can. Then the above construction can be refined : one starts with a measure η which has *maximal spectral type*. At the following step, when one restricts oneself to the orthogonal Ω_1 of the first model, one again chooses the maximal spectral type allowed in Ω_1, and so on. Then one can show that transfinite induction is unnecessary, and (more important) *the spectral types of $\eta_1, \eta_2, \ldots$, which become weaker and weaker, are uniquely determined.* This is the well known *spectral multiplicity theorem* (Hellinger-Hahn theorem). For a detailed proof, see [ReS], Chap. VII. On the other hand, the "slicing" itself (the decomposition of Ω into a continuous sum of Hilbert spaces) can be shown to be unique once the measure θ is chosen.

Finally, let us show briefly the relation of the above slicing with the definition of a complete observable, as given above in this chapter (§2, subs. 3) : X is complete if and only if one single copy of the "model" is sufficient to exhaust Ω, *i.e.* if the slices are one-dimensional. Given that operators which operate slice by slice commute with X, it is easy to prove that the observable X is complete if and only if all operators which commute to X are of the form J_f^X.

Transition kernels and completely positive mappings[1]

5   Let us first recall the definition of a *(transition) kernel* in classical probability.

Consider two measurable spaces $(E_0, \mathcal{E}_0)$ and $(E_0, \mathcal{E}_1)$. Then a kernel P from E_0 to E_1 is a mapping which associates with every point x of E_0 a probability measure $P(x, dy)$ on E_1, depending measurably on x. Thus for every bounded measurable function f on E_1 (note the reverse order), the function $Pf : x \longmapsto \int P(x, dy) f(y)$ is measurable on E_0, and P may be described as a mapping from (measurable) functions on E_1 to (measurable) functions on E_0, which is positive, transforms 1 into 1, and satisfies a countable additivity property.

On the other hand, let $\mathcal{P}_0$ be the convex set of probability measures on E_0, and similarly $\mathcal{P}_1$. The kernel P defines an affine mapping from $\mathcal{P}_0$ to $\mathcal{P}_1$

$$\lambda \longmapsto \lambda P = \int \lambda(dx) P(x, \cdot)$$

---

[1]   This subsection has been entirely rewritten.

However, one cannot characterize simply, by a property like countable additivity, those affine mappings which arise in this way from transition kernels.

Every measurable mapping h from E_0 to E_1 defines a kernel H, such that $H(x, dy) = \varepsilon_{h(x)}(dy)$. Such a kernel is deterministic, in the following sense : if one imagines a kernel $P(x, dy)$ as the probability distribution of bullets fired on E_1 from a gun located at $x \in E_0$, then the gun corresponding to the kernel H has no spreading : all bullets from x fall at the same point $h(x)$. This is reflected into the algebraic property that $H(fg) = H(f)H(g)$ for any two (bounded measurable) functions f, g on F.

Let P be a kernel from E_0 to E_1, and let μ be a probability law (a state) on E_0. Then we may build a probability space $(\Omega, \mathcal{F}, \mathbb{P})$, a pair X_0, X_1 of random variables on Ω, taking values in E_0, E_1, such that the law of X_0 is μ and the conditional law of X_1 given that $X_0 = x$ is $P(x, dy)$. The construction is well known : we take Ω to be $E_0 \times E_1$ with the product σ–field, we define X_0 and X_1 to be the two coordinate mappings on $E_0 \times E_1$, and we define the law $\mathbb{P}$ as the only measure on Ω such that — as usual, the tensor product $a_0 \otimes a_1$ of two bounded measurable functions $a_0(x), a_1(y)$ on E_0, E_1 respectively denotes the function $a_0(x)\, a_1(y)$ on $E_0 \times E_1$

$$(5.1) \qquad \mathbb{E}\left[a_0 \otimes a_1\right] = \int_{E_0} \mu(dx)\, a_0(x)\, P a_1(x) \,.$$

We will now translate this construction into non-commutative language. What we get is a well known construction due to Stinespring, which will lead us to the proper definition of non-commutative kernels.

6   It is a fundamental principle of non-commutative geometry that a "non-commutative space" E *is* an abstract algebra $\mathcal{A}$, whose elements intuitively represent "functions" on E — additional structure on $\mathcal{A}$ will tell whether they must be undertood as measurable, continuous, differentiable... functions.

Thus the two spaces E_0, E_1 become now two algebras $\mathcal{A}_0, \mathcal{A}_1$, our "kernel from E_0 to E_1" will be a linear mapping P from $\mathcal{A}_1$ to $\mathcal{A}_0$, and our measure μ will be a state on $\mathcal{A}_0$ [1]. As a minor technical assumption, we will assume $\mathcal{A}_0, \mathcal{A}_1$ are C^*–*algebras*. The intuitive idea is that in this case we are dealing with the non–commutative analogue of two *compact* spaces and their algebras of continuous functions, and of *Feller kernels* which preserve continuity. Then we need not worry about the countable additivity of measures and kernels. If the reader does not know yet what a C^*–algebra is, let him simply omit the corresponding details.

It is clear that we must assume P maps I_1 to I_0, and is a positive mapping. However, and this will be our main point, *positivity is not sufficient*.

We are going to construct a pair or "random variables X_0, X_1 with values in E_0, E_1" such that the law of X_0 is μ, and P somehow describes the conditioning of X_1 by X_0. The idea is to extract, from the commutative case, the way the Hilbert space $L^2(\mathbb{P})$ is constructed from $L^2(\mu)$ and the transition kernel P. To achieve this, we first construct the analogue of $L^2(\mu)$. That is, we give ourselves a Hilbert space $\mathcal{K}_0$, a representation X_0 from $\mathcal{A}_0$ on $\mathcal{K}_0$, and a unit vector ψ implementing the state μ

$$\mu(a_0) = \,<\psi, (a_0 \circ X_0)\psi> \qquad (a_0 \in \mathcal{A}_0) \,.$$

---

[1]   Since E_0 and E_1 do not exist as spaces, one generally says P is a kernel from $\mathcal{A}_1$ to $\mathcal{A}_0$.

We recall that $I_0 \circ X_0$ is assumed to be the identity operator on $\mathcal{K}_0$. We may for instance construct $\mathcal{K}_0$ using the GNS construction (Appendix 4, §2), but it is better to keep some freedom here : changing the initial vector will then give different states μ — though a simultaneaous construction for *all* possible initial states (as in the classical case) could not be achieved with a separable $\mathcal{K}_0$.

Our purpose is to construct a Hilbert space $\mathcal{K}_1$ containing $\mathcal{K}_0$, a representation X_1 from $\mathcal{A}_1$ on $\mathcal{K}_1$ ($I_1 \circ X_1$ being the identity of $\mathcal{K}_1$), such that the following extension of (5.1) holds (the notation $a_0 \circ X_0$ means $X_0(a_0)$, in order to stress the intuitive meaning of a_0 as a function on E_0 ; similarly for $a_1 \circ X_1$).

(6.1) $< \psi , a_1 \circ X_1 a_0 \circ X_0 \psi > \, = \, < \psi , (Pa_1 a_0) \circ X_0 \psi > .$

Though $a_0 \circ X_0$ is an operator on $\mathcal{K}_0$, not on $\mathcal{K}_1$, still it can be applied to $\psi \in \mathcal{K}_0$ [1]. On the other hand $a_0 \circ X_0 a_1 \circ X_1 \psi$ would be meaningless. Thus we are constructing something coarser than a "joint law" for X_0 and X_1 — intuitively speaking, P predicts the value of X_1 while X_0 is under observation.

A necessary condition for the existence of such a Hilbert space $\mathcal{K}_1$ is that, given any finite families of elements a_i^0 of $\mathcal{A}_0$ and a_i^1 of $\mathcal{A}_1$, and putting $A = \sum_i a_i^1 \circ X_1 a_i^0 \circ X_0$, $< \psi , A^* A \psi >$ should be positive, that is,

$$\mu \big(\sum_{ij} a_j^{0*} P(a_j^{1*} a_i^1) a_i^0 \big) \geq 0 .$$

It should be a property of the mapping P that the preceding assumption is satisfied for an arbitrary state μ on $\mathcal{A}_0$, and thus we are led naturally to the basic definition of *complete positivity*

(6.2) $\sum_{ij} a_j^{0*} P(a_j^{1*} a_i^1) a_i^0 \geq 0$ for arbitrary $a_i^1 \in \mathcal{A}_1$, $a_i^0 \in \mathcal{A}_0$.

A trivial case where this condition is satisfied is that of an algebra homomorphism P from $\mathcal{A}_1$ to $\mathcal{A}_0$ (corresponding to the "no spreading" case in the commutative situation).

On the other hand, if this condition is satisfied, we can construct the Hilbert space $\mathcal{K}_1$ and the representation X_1, using the following procedure, which generalizes the GNS method :

1) Provide the algebraic tensor product $\mathcal{K}_0 \otimes \mathcal{A}_1$ with the only hermitian form such that, for $k, h \in \mathcal{K}_0$, $b_1, a_1 \in \mathcal{A}_1$

(6.3) $< k \otimes b_1 , h \otimes a_1 > \, = \, < k, P(b_1^* a_1) h > .$

Let us deduce from (6.2) that we get in this way a positive (possibly degenerate) scalar product. Otherwise stated, we have for finite families $k_i, h_i \in \mathcal{K}_0$, $b_i, a_i \in \mathcal{A}_1$

$$\sum_{ij} < k_j, P(b_j^* a_i) h_i > \, \geq 0 .$$

This follows from the complete positivity assumption on P if $h_i = (a_i^0 \circ X_0) \psi$, $k_i = (b_i^0 \circ X_0) \psi$ for some vector $\psi \in \mathcal{K}_0$ and elements a_i^0, b_i^0 of $\mathcal{A}_0$. Then we get the

---

[1]   In fact, nothing depends on the particular choice of $\psi \in \mathcal{K}_0$.

result by density for arbitrary $h_i, k_i \in \mathcal{K}_0$ if the representation $\mathcal{K}_0$ is cyclic with generating vector ψ (GNS case). Finally, we decompose $\mathcal{K}_0$ into a direct sum of cyclic representations to reach the general case.

2) Knowing this scalar product is positive, we construct the Hilbert space $\mathcal{K}_1$ by killing the null-space and completing. If we map $h \in \mathcal{K}_0$ to $h \otimes I_1 \in \mathcal{K}_0 \otimes \mathcal{A}_1$, we get an isometry : its image does not intersect the null space, and we may identify h with $h \otimes I_1$, thus imbedding $\mathcal{K}_0$ into $\mathcal{K}_1$. In particular, we identify ψ with $\psi \otimes I_1$.

We define a representation X_1 of $\mathcal{A}_1$ in $\mathcal{K}_1$ as follows. Given $a \in \mathcal{A}_1$, we put

$$(6.4) \qquad (a \circ X_1)(h \otimes b) = h \otimes (ab) .$$

It is clear that this operation behaves well with respect to sums and products, a little less obvious that $(a \circ X)^* = a^* \circ X$, and that the action of a is bounded with norm at most $\| a \|$ — to prove this last point, assume $\| a \| \leq 1$ and remark that $a^* a \leq I$ may be written as $a^* a = I - c^* c$. Then

$$\| \sum_i h_i \otimes ab_i \|^2 = \sum_{ij} < h_j \otimes (ab_j), h_i \otimes ab_i > = \sum_{ij} < h_j, P(b_j^* a_j^* a_i b_i) \, h_i >$$

$$= \| \sum_i h_i \otimes b_i \|^2 - \sum_{ij} < h_j, P(b_j^* c_j^* c_i b_i) \, h_i >$$

and the last term to the right is positive. Then the operators are extended to $\mathcal{K}_1$ by continuity and our representation of $\mathcal{A}_1$ is constructed.

In contrast with this, one generally cannot define a representation of $a \in \mathcal{A}_0$ in $\mathcal{K}_1$, extending its action $a \circ X_0$ on $\mathcal{K}_0$, by the formula

$$(6.5) \qquad (a \circ X_0)(h \otimes b) = ((a \circ X_0) h) \otimes b .$$

This operation behaves well with respect to sums and products, but *not* with respect to adjoints, and generally it is *not* continuous. We may still consider X_0 as a representation of $\mathcal{A}_0$ in $\mathcal{K}_1$, but not quite in the usual sense : $I_0 \circ X_0$ is not the identity operator, but the orthogonal projection on $\mathcal{K}_0$. If we really want to have a representation respecting the unit, then we assume $\mathcal{A}_0$ has a multiplicative state δ and put, for $x = y + z \in \mathcal{K}_0 \oplus \mathcal{K}_0^\perp$

$$(a \circ X_0) x = (a \circ X_0) y + \delta(a) z .$$

Uniqueness. The GNS construction is unique up to isomorphism, under an assumption of cyclicity. Here the assumption that ensures uniqueness is the following : *the (closed) invariant subspace for the representation X_1 generated by $\mathcal{K}_0$ is the whole of $\mathcal{K}_1$.* Otherwise stated, the vectors $(a_1 \circ X_1) h$ $(a_1 \in \mathcal{A}_1, h \in \mathcal{K}_0)$ form a total set in $\mathcal{K}_1$. This property it satisfied here from the construction, and it is easily seen that any two situations for which it is satisfied are unitarily equivalent. This assumption is called *minimality.*

In particular, if $\mathcal{K}_1$ is the GNS space relative to a state μ, with generating vector ψ, the vectors $(a_1 \circ X_1)(a_0 \circ X_0) \psi$ are total in $\mathcal{K}_1$, and the matrix elements between such vectors are known.

COMMENT. This construction describes the structure of completely positive mappings $P : \mathcal{A}_1 \rightarrow \mathcal{A}_0$. Assuming $\mathcal{A}_0$ acts faithfully [1] on $\mathcal{K}_0$, so that we may consider $\mathcal{A}_0$ as a sub-algebra of $\mathcal{L}(\mathcal{K}_0)$, and P as a mapping from $\mathcal{A}_1$ to $\mathcal{L}(\mathcal{K}_0)$, whose image lies in $\mathcal{A}_0$. Then P appears as a composition of two elementary models of completely positive mappings :

1) a homomorphism from $\mathcal{A}_1$ to $\mathcal{L}(\mathcal{K}_1)$,

2) a mapping $A \rightarrow V^*AV$ from $\mathcal{L}(\mathcal{K}_1)$ to $\mathcal{L}(\mathcal{K}_0)$, V denoting a mapping from $\mathcal{K}_0$ to $\mathcal{K}_1$ — here, V is an isometry.

7   REMARKS. 1) Though (6.5) does not waork in general, it does work when a belongs to the *center* $\mathcal{Z}_0$ of $\mathcal{A}_0$, and this representation of $\mathcal{Z}_0$ commutes with the representation of $\mathcal{A}_1$. More generally (anticipating on definitions concerning von Neumann algebras), we may define a representation of the commutant $\mathcal{A}_0'$ of (the image of) $\mathcal{A}_0$ acting on $\mathcal{K}_0$, defined for $\alpha \in \mathcal{A}'$ by $\alpha(h \otimes a_1) = \alpha h \otimes a_1$ as in (6.5), and this extension clearly commutes with all operators $b_1 \circ X_1$. Since $\alpha(h \otimes I) = (\alpha h) \otimes I$ it preserves $\mathcal{K}_0$, hence $\mathcal{K}_0^\perp$. Then it commutes with all the operators $a_0 \circ X_0$ as extended in Remark 3 below to the whole of $\mathcal{K}_1$, and therefore with the whole algebra generated by the "random variables" X_0 and X_1.

2) Let us return to the commutative case of two random variables X_i taking values in compact state spaces E_i (thus $\mathcal{A}_0 = \mathcal{C}(E_i)$). Then $\mathcal{K}_0$ is $L^2(E_0, \mu)$ on which $\mathcal{A}_0$ operates by multiplication, and $\mathcal{K}_1$ is $L^2(E_0 \times E_1, \mathbb{P})$, where $\mathbb{P}$ is the joint law of (X_0, X_1). This situation satisfies the minimality assumption, in the sense that vectors of the form $f_1(x_1)f_0(x_0)1$ are total in $\mathcal{K}_1$. Thus the quantum situation is a true extension of the commutative one. On the other hand, in the classical situation a product $a_0 \circ X_0 a_1 \circ X_1$ is well defined, while it is not defined in the quantum situation.

3) If we insist on dealing with operators defined on the whole of $\mathcal{K}_1$, the simplest way to extend $a_0 \circ X_0$, and more generally operators on $\mathcal{K}_0$, to operators on $\mathcal{K}$, consists in deciding their extension kills $\mathcal{K}_0^\perp$. Thus $\mathcal{L}(\mathcal{K}_0)$ gets identified with the "σ–algebra" $\mathcal{F}_0$ of all operators $E_0 A E_0$ on $\mathcal{K}$ — not a true von Neumann algebra on $\mathcal{H}$ since its unit is E_0 not I. Extending $H \in \mathcal{L}(\mathcal{K}_0)$ in this way, let us compute a matrix element like $< k, b_1 \circ X_1 H a_1 \circ X_1 h >$, with $h, k \in \mathcal{K}_0$. Since $H a_1 \circ X_1 h$ belongs to $\mathcal{K}_0$ we may replace $b_1 \circ X_1$ by $P b_1 \circ X_0$. Then we move $H P b_1 \circ X_0$ to the left side, taking adjoints. Again we may replace $a_1 \circ X_1 h$ by its projection $P a_0 \circ X_0 h$ on $\mathcal{K}_0$, and then move all operators to the right. In probabilistic notation, we have a "Markov property"

$$\mathbb{E}\left[b_1 \circ X_1 H a_1 \circ X_1 \mid \mathcal{F}_0 \right] = P b_1 \circ X_0 H P a_1 \circ X_0 .$$

This computation is easily extended by induction to a product of an arbitrary number of $a_i \circ X_1$ and elements of $\mathcal{F}_0$. This simple extension remains non-commutative even in the classical set-up. More generally, if a belongs to the center $\mathcal{Z}_0$, the extension of $a \circ X_0$ killing $\mathcal{K}_0^\perp$ does not coincide with the action of the center as defined in Remark 1.

The computation above concerns matrix elements only : if we are to compute the *operator* $a_0 \circ X_0 a_1 \circ X_1$ (say), it is not sufficient to tell how it acts on $h \in \mathcal{K}_0$; we must

---

[1]   This is not a serious restriction : any C^*–algebra has a faithful representation, except that the Hilbert space involved may not be separable.

compute it on a vector $c_1 \circ X_1 h$, and what we get is $(a_0 P(a_1 c_1)) \circ X_0 h$. General rules can be found in the original article of Bhat and Parthasarathy.

4) We again anticipate on the theory of von Neumann algebras. Assume $\mathcal{A}_0$ and $\mathcal{A}_1$ are weakly closed (= von Neumann) algebras, X_0 is a normal homomorphism, and P is normal — in both cases this means they behave well with respect to strong limits of bounded increasing families $f_\alpha \uparrow f$ of positive operators in $\mathcal{A}_0$. *Then X_1 is also a normal homomorphism.* Indeed, let $f_\alpha \uparrow f$ be a bounded increasing family of positive elements of $\mathcal{A}_1$, and let $F = \sup_\alpha f_\alpha \circ X_1$. To check that $F = f \circ X_1$, it suffices to check equality of diagonal matrix elements on a total set, here by the minimality assumption, for $a_1 \in \mathcal{A}_1$, $h_0 \in \mathcal{K}_0$

$$\sup_\alpha < a_1 \circ X_1 h_0, f_\alpha a_1 \circ X_1 h_0 > = < a_1 \circ X_1 h_0, f a_1 \circ X_1 h_0 > \ .$$

We apply the Markov property and are reduced to

$$\sup_\alpha P(a_1^* f_\alpha a_1) \circ X_0 = P(a_1^* f a_1) \circ X_0$$

That is, to the normality of X_0 and P.

5) A problem of practical importance is : *how to make sure that $\mathcal{K}_1$ is separable ?* In classical situations, we would assume that $\mathcal{K}_0$ is separable and $\mathcal{A}_1$ is norm-separable, from which the separability of $\mathcal{K}_1$ would follow at once. But (as the example of the Weyl algebra on Boson Fock space shows) norm separability is not a welcome assumption in quantum probability. Instead, let us assume the following properties, taking $\mathcal{A}_0 = \mathcal{A}_1$ for simplicity.

i) $\mathcal{K}_0$ is separable, $\mathcal{A}$ is an operator algebra on $\mathcal{K}_0$ and X_0 is the identity homomorphism; hence $\mathcal{L}(\mathcal{K}_0)$ is separable in all the usual weak operator topologies, and so is $\mathcal{A}$.

ii) $P : \mathcal{A} \to \mathcal{A}$ is strong/strong continuous (or even strong/weak continuous) on the unit ball (this is meant to be easy to check in practical cases).

Then $\mathcal{K}_1$ is separable.

To prove this, we will prove first that $c \longmapsto c \circ X_1$ is continuous from the unit ball of $\mathcal{A}$ with the strong topology, to $\mathcal{L}(\mathcal{K}_1)$ given the weak topology. It suffices to check weak convergence on a total set, and therefore to prove that $< b_1 \circ X_1 k_0 , c \circ X_1 a_1 \circ X_1 h_0 >$ is continuous in c. We rewrite this as $< k_0 , P(b_1^* c a_1) \circ X_0 h_0 >$. Then if $c_\alpha \to c$ strongly $b_1^* c_\alpha a_1 \to b_1^* c a_1$ strongly, thus applying P we get weak convergence, which is preserved under X_0.

This being proved, consider the total set of vectors $a_1 \circ X_1 h_0$ where a_1 may be chosen in the unit ball of $\mathcal{A}$. Choose a countable strongly dense set D in the unit ball of $\mathcal{A}$, a countable norm dense set K in $\mathcal{K}_0$, and take $a_\alpha \in D$ tending to a_1 strongly, $h_n \in K$ tending to h in norm. Then $c_\alpha \circ X_1$ tends weakly to $a_1 \circ X_1$ from the above, with norm bounded by 1, and therefore $(c_\alpha \circ X_1) h_n$ tends weakly to $(a_1 \circ X_1) h$. That's it.

If $\mathcal{A}$ is given as a von Neumann algebra, it will be convenient instead of ii) to assume that P is normal.

Quantum Markov processes

8 As in the commutative case, once the case of two algebras has been fully worked out, the case of a finite number of algebras $\mathcal{A}_0, \ldots, \mathcal{A}_n$ and completely positive "kernels" $P_{i,i+1} : \mathcal{A}_{i+1} \to \mathcal{A}_i$ follows by an easy induction[1] leading to the construction of an increasing family of Hilbert spaces $\mathcal{K}_0 \subset \mathcal{K}_1 \ldots \subset \mathcal{K}_n$ and of homomorphisms $X_i : \mathcal{A}_i \to \mathcal{L}(\mathcal{K}_i)$ such that, for $h_i, k_i \in \mathcal{K}_i$ and $a_{i+1}, b_{i+1} \in \mathcal{A}_{i+1}$ (and writing $a_{i+1} \circ X_{i+1}$ for $X_{i+1}(a_{i+1})$)

$$(8.1) \qquad < (b_{i+1} \circ X_{i+1})\, k_i\,,\, (a_{i+1} \circ X_{i+1})\, h_i > \; = \; < k_i\,,\, (P_{i+1}(b_{i+1}^* a_{i+1} \circ X_i)) > .$$

and in particular, given elements $h_0, k_0 \in \mathcal{K}_0$, we may compute by an easy induction matrix elements of the form

$$(8.2) \qquad < b_n \circ X_n \ldots b_1 \circ X_1 \, k_0\,,\, a_n \circ X_n \ldots a_1 \circ X_1 \, h_0 >$$

by a formula which closely resembles the computation of probabilities relative to a classical Markov process. Taking $h_0 = k_0 = \psi$ we are computing expectations relative to the initial state μ. On the other hand, our "operators" are not everywhere defined, and thus cannot be composed in a different order.

The corresponding minimality assumption, which ensures uniqueness, is the density of vectors of the form $a_n \circ X_n \ldots a_1 \circ X_1 \, h_0$ with $h_0 \in \mathcal{A}_0$, $a_i \in \mathcal{A}_i$.

As in classical probability, the stationary case is the most important : $\mathcal{A}_i = \mathcal{A}$, $P_i = P$ do not depend on i. There is no difficulty in extending the above construction to that of an infinite "quantum Markov chain (X_n) with transition probability P". A very important additional element of structure is the *central process* : as we could define in (6.5) an action of the center of $\mathcal{A}_0$ on $\mathcal{K}_1$, extending the given action of the center on $\mathcal{K}_0$, then in the discrete case we may define homomorphisms Y_n from the center $\mathcal{Z}$ of $\mathcal{A}$ to $\mathcal{L}(\mathcal{K})$, such that for $a \in \mathcal{Z}$ $a \circ Y_n$ extends $a \circ X_n$. *All these homomorphisms commute*, and therefore constitute together a classical process — though not necessarily a classical Markov process, unless the transition kernel also maps $\mathcal{Z}$ into itself.

A similar construction applies in continuous time. Let $\mathcal{A}$ be a C^*–algebra. A *quantum dynamical semigroup* or simply a *Markov semigroup* on $\mathcal{A}$ is a family $(P_t)_{t>0}$ of completely positive mappings, such that $P_t 1 = 1$, possessing the semigroup property $P_{s+t} = P_s P_t$, and continuous in some reasonable topology. One generally assumes it is *strongly* continuous in the sense that for any fixed element of $\mathcal{A}$, $\| P_t a - a \| \to 0$ as $t \to 0$, which implies (by the general theory of semigroups on Banach spaces) that $t \to P_t a$ is norm-continuous. Since there are non-pathological algebras $\mathcal{A}$ which are not norm-separable (this occurs most often with von Neumann algebras), continuity in weak topologies must be considered too. Let us just mention that in the case of von Neumann algebras, the mappings P_t are assumed to be normal, and the natural kind of continuity involves the "ultraweak" topology.

Let us now describe the "quantum Markov process" associated with a quantum dynamical semigroup (P_t) and an initial state μ on $\mathcal{A}$. As usual, we assume that $\mathcal{A}$

---

[1] Here one sees the interest of starting from an arbitrary Hilbert space $\mathcal{K}_0$, since in the induction the space $\mathcal{K}_n$ already constructed takes the place of $\mathcal{K}_0$.

acts on an initial Hilbert space $\mathcal{K}_0$ and μ corresponds to an unit vector ψ in this space. Then it is possible to construct :

— a Hilbert space $\mathcal{K}$ containing $\mathcal{K}_0$, and an increasing family $(\mathcal{K}_t)$ of intermediate Hilbert spaces,

— a family (X_t) of homomorphisms from $\mathcal{A}$ to $\mathcal{L}(\mathcal{K}_t)$ (not to $\mathcal{L}(\mathcal{K})$) such that for $s < t$, for $h, k \in \mathcal{K}_s$, for $a, b \in \mathcal{A}$, we have a "Markov property"

$$< b \circ X_t\, k\, ,\, a \circ X_t\, h > = < k, P_{t-s}(b^* a) \circ X_s\, h > .$$

Besides that, uniqueness is ensured by a minimality property : vectors of the form

$$a_n \circ X_{t_n} \ldots a_1 \circ X_{t_1}\, h_0 \quad , \quad (t_1 < \ldots < t_n,\, a_1, \ldots, a_n \in \mathcal{A},\, h_0 \in \mathcal{K}_0)$$

are dense in $\mathcal{K}$.

We refer to the original article for this construction, though the essential ideas have been given, and it only remains to take the limit of an increasing system of Hilbert spaces. The central process too can be defined in continuous time. These results corresponds in the classical case to the construction of a coarse Markov process, while the non-trivial part of the classical theory (starting with the strong Markov property) requires choosing a better version, usually right continuous. What corresponds to such a choice in quantum probability is not yet known [1].

Quantum dynamical semigroups are interesting objects by themselves, and much work has been devoted to understanding their structure. It is natural to look for an "infinitesimal Stinespring theorem" giving the structure of their generators, but the search for such a theorem has been successful only in the case of a *bounded* generator, or equivalently, of the generator of a *norm-continuous* quantum dynamical semigroup. We will mention a partial result without proof. For a complete account see Parthasarathy's book, (p. 267–268, the Gorini–Kossakowski–Sudarshan theorem) and the original paper by Gorini et *al* [GKS]. For comments and applications, see also the Lecture Notes [AlL] by Alicki and Lendi.

The (bounded) generator G of a uniformly continuous quantum dynamical semi-group on the C^*-algebra $\mathcal{L}(\mathcal{H})$ of a Hilbert space $\mathcal{H}$ has the following form :

$$G(X) = i\,[H,X] - \frac{1}{2}\sum_i (L_i^* L_i X + X L_i^* L_i - 2L_i^* X L_i)$$

where H is a bounded self-adjoint operator on $\mathcal{H}$, and (L_i) is a sequence of bounded operators such that $\sum_i L_i^* L_i$ is strongly convergent to a bounded operator.

This is often called, for historical reasons, the *Lindblad form* ([Lin]) of the generator. It is similar to the Hörmander form $G = X + \sum_i X_i^2$ of the generator of a classical diffusion on a manifold, where the X_i are vector fields on the manifold, and as the Hörmander form, it lends itself to a construction of the semi-group and the process by means of *stochastic differential equations*.

---

[1] A quantum probabilistic interpretation of the strong Markov property has recently (1995) been given by Attal and Parthasarathy.

Appendix 3

Two Events

This Appendix illustrates a strange feature of quantum probability : the "σ–field" generated by two non–commuting events contains uncountably many projectors. The results are borrowed from M.A. Rieffel and A. van Daele, [RvD1] (1977), and through them from Halmos [Hal] (1969). As one may guess from the title of [RvD1], the results are useful for the proof of the main theorem of Tomita–Takesaki, which will be given in Appendix 4.

The operators denoted here with lower case or greek letters $p, q, j, d, \gamma, \sigma \ldots$ are denoted P, Q, J, D, S, C in the section on Tomita's theory.

1 Let $\mathcal{H}$ be a Hilbert space, complex or real (the applications to Tomita–Takesaki essentially use the *real* case) and let A and B be two "events" (= closed subspaces) and I_A, I_B be the corresponding projectors. These projectors leave the four subspaces $A \cap B$, $A \cap B^\perp$, $A^\perp \cap B$, $A^\perp \cap B^\perp$ invariant. On the sum $\mathcal{K}$ of these four subspaces, the projectors I_A and I_B commute, and nothing interesting happens. If $\mathcal{K} = \{0\}$ we say that the two subspaces are *in general position*. This property can be realized by restriction to the invariant space $\mathcal{K}^\perp$. Thus we assume

$$(1.1) \qquad A \cap B = A^\perp \cap B^\perp = \{0\} \, (= A \cap B^\perp = A^\perp \cap B) \, .$$

However, in the application to the Tomita–Takesaki theory only the first two conditions are realized, and we indicate carefully the places where "general position" assumptions come into play.

To simplify notation we set $I_A = p$, $I_B = q$. Then we put

$$(1.2) \qquad \sigma = p - q \quad ; \quad \gamma = p + q - I$$

These operators are bounded and selfadjoint, and we have the following relations (depending only on p and q being projectors : $p^2 = p$, $q^2 = q$: none of the conditions (1.1) is used)

$$(1.3) \qquad \sigma^2 + \gamma^2 = 1 \quad ; \quad \sigma\gamma + \gamma\sigma = 0$$

The notations σ and γ suggest a "sine" and a "cosine". The first relation implies that $< \sigma x, \sigma x > + < \gamma x, \gamma x > \, = \, < x, x >$, so the spectrum of each operator lies in the interval $[-1, 1]$, and we may write the spectral representations (valid also in the real case !)

$$(1.4) \qquad \sigma = \int_{-1}^{1} \lambda dE_\lambda \quad ; \quad \gamma = \int_{-1}^{1} \lambda dF_\lambda$$

The relation $\sigma x = 0$ means $px = qx$, which implies $px = 0 = qx$ since $A \cap B = \{0\}$; then $x \in A^{\perp} \cap B^{\perp}$ which in turn implies $x = \{0\}$. Otherwise stated, σ *is injective*. Similarly, one can prove that $I \pm \gamma$ *is injective*. Using the remaining relations (1.1), one may prove that γ, $I - \sigma$, $I + \sigma$ are injective. In the application to Tomita–Takesaki theory, only the left side of (1.1) is true, so these last three operators will not be injective. Note that the injectivity of σ means that the spectral measure dE_λ has no jump at 0, *i.e.* $E_{0-} = E_0$.

We now define

$$(1.5) \qquad j = \mathrm{sgn}(\sigma) = \int_{-1}^{1} \mathrm{sgn}(\lambda)\, dE_\lambda \quad ; \quad d = |\sigma| = \int_{-1}^{1} |\lambda|\, dE_\lambda$$

Since the spectral measure has no jump at 0, it is not necessary to define the sign of 0 and we have $j^2 = I$: j is a symmetry, which we call the *main symmetry*. On the other hand, d is self-adjoint positive. Since we have $d^2 = \sigma^2$, d is the only positive square root of $1 - \gamma^2$. Then it commutes with γ and, since it already commutes with σ, it commutes with *all the operators* we are considering. Finally, we have

$$djp = \sigma - p = (p - q)p = (I - q)(p - q) = (I - q)dj = d(I - q)j .$$

Since d is injective, this implies

$$(1.6) \qquad jp = (I - q)j \quad \text{whence taking adjoints} \quad jq = (I - p)j .$$

Adding these relations, we get

$$(1.7) \qquad\qquad j\gamma = -\gamma j \quad ; \quad j\sigma = \sigma j$$

(the second equality is not new, we recall it for symmetry). The fact that the spaces are in general position has not been fully used yet. If we do, operating on F_λ as we did on E_λ gives a second symmetry $k = \mathrm{sgn}(\gamma)$ and a second positive operator $e = |\gamma|$ such that

$$\gamma = ke = ek , \quad e = |\gamma| = (I - \sigma^2)^{1/2} , \quad k\sigma = -\sigma k , \quad k\gamma = \gamma k$$

and e commutes with all other operators. We call k the *second symmetry*. It turns out that *the two symmetries k and j anticommute*. Indeed, k anticommutes with every odd function of σ, and in particular with $\mathrm{sgn}(\sigma) = j$.

Let us add a few words on the application of the above results to the Tomita–Takesaki theory : in this situation, $\mathcal{H}$ is a complex Hilbert space (also endowed with the *real* Hilbert space structure given by the real part of its scalar product), and the two subspaces A and B are *real*, and such that $B = iA$. Another way to state this is the relation $ip = qi$. Then the "cosine" operator γ is complex linear, but the "sine" operator σ and the main symmetry j are *conjugate linear*. We shall return to this situation in due time.

The σ–field generated by two events

2 Assuming that the two subspaces are in general position, denote by $\mathcal{M}$ the eigenspace $\{jx = x\}$ of the main symmetry. Then $\mathcal{M}^\perp$ is the eigenspace $\{jx = -x\}$, and it is not difficult to see that k is an isomorphism of $\mathcal{M}$ onto $\mathcal{M}^\perp$. Note that since $\mathcal{H} = \mathcal{M} \oplus \mathcal{M}^\perp$ it must have even or infinite dimension. If we identify $\mathcal{M}$ to $\mathcal{M}^\perp$ by means of k, we identify $\mathcal{H}$ with $\mathcal{M} \oplus \mathcal{M}$ (every vector in $\mathcal{H}$ having a unique representation as $x + ky$, $x, y \in \mathcal{M}$) and every bounded operator on $\mathcal{H}$ is represented by a $(2,2)$ matrix of operators on $\mathcal{M}$. In particular, $j(x+ky) = x-ky$, $k(x+ky) = y+kx$, and therefore

$$j = \begin{pmatrix} I & 0 \\ 0 & -I \end{pmatrix} \quad ; \quad k = \begin{pmatrix} 0 & I \\ I & 0 \end{pmatrix} \quad .$$

It is now clear that $\mathcal{H} \approx \mathcal{M} \otimes \mathbb{C}^2$ with j corresponding to $I \otimes \sigma_z$ and K to $I \otimes \sigma_x$. If we assume our Hilbert space is complex and put $l = -ijk$ (where i denotes the operator of multiplication by the complex scalar i !) this operator is represented by the third Pauli matrix

$$l = \begin{pmatrix} 0 & -iI \\ iI & 0 \end{pmatrix} \quad .$$

The operator d commutes with j and k and leaves $\mathcal{M}$ invariant. Consider now the family $\mathcal{A}$ of all operators

$$tI + xk + yl + zj = \begin{pmatrix} t+z & x-iy \\ x+iy & t-z \end{pmatrix} \quad ,$$

where x, y, z, t are not scalars, but are restrictions to $\mathcal{M}$ of functions of d. Then it is very easy to see that $\mathcal{A}$ is closed under multiplication and passage to the adjoint, and one can show that $\mathcal{A}$ is exactly the *von Neumann algebra generated by p and q*. So this family contains uncountably many projectors, *i.e.* events in the probabilistic sense.

An interesting by–product of the preceding discussion is the fact that two anti-commuting symmetries j and k on $\mathcal{H}$ necessarily look like two of the Pauli matrices, and in fact (taking an o.n.b. of $\mathcal{M}$) the space decomposes into a direct sum of copies of $\mathbb{C}^2$ equipped with the Pauli matrices. On the other hand, consider a Hilbert space $\mathcal{H}$ and two mutually adjoint bounded operators b^+ and b^- such that $b^{+2} = b^{-2} = 0 = b^+b^- + b^-b^+$. Then it is easy to see that $b^+ + b^-$ and $i(b^+ - b^-)$ are anticommuting symmetries . Thus the Hilbert space splits into a countable sum of copies of $\mathbb{C}^2$ equipped with the standard creation and annihilation operators. This is the rather trivial analogue, for the canonical anticommutation relations, of the Stone-von Neumann uniqueness theorem for the canonical commutation relations.

Appendix 4

C^*–Algebras

This appendix contains a minimal course on C^* and von Neumann algebras, with emphasis on probabilistic topics, given independently of the course on Fock spaces. Von Neumann algebras play in non–commutative probability the role of measure theory in classical probability, though in practice (as in the classical case!) deep results on von Neumann algebras are rarely used, and a superficial knowledge of the language is all one needs to read the literature. Our presentation does not claim to be original, except possibly by the choice of the material it excludes.

§1. ELEMENTARY THEORY

Definition of C^*–algebras

The proofs in this section are very classical and beautiful, and can be traced back to Gelfand and Naimark, with several substantial later improvements. For details of presentation we are specially indebted to the (much more complete) books by Bratelli–Robinson and Pedersen. We assume that our reader has some knowledge of the elementary theory of Banach algebras, which is available in many first year functional analysis courses.

1 By a C^*–*algebra* we mean a complex algebra $\mathcal{A}$ with a norm $\|\cdot\|$, an involution $*$ *and also a unit* I (the existence of an unit is not always assumed in standard courses, but we are supposed to do probability), which is complete with respect to its norm, and satisfies the basic axiom

$$(1.1) \qquad\qquad \| a^*a \| = \|a\|^2 \ .$$

> The most familiar $*$–algebra of elementary harmonic analysis, the convolution algebra $L^1(G)$ of a locally compact group, is not a C^*–algebra.

C^*–algebras are non–commutative analogues of the algebras $\mathcal{C}(K)$ of complex, continuous functions over a compact space K (whence the C) with the uniform norm, the complex conjugation as involution, and the function 1 as unit. Thus bounded linear functionals on C^*–algebras are the non–commutative analogues of (bounded) complex measures on a compact space K.

The relation $(ab)^* = b^*a^*$ implies that I^* is a unit, thus $I^* = I$, and (1.1) then implies $\| I \| = 1$. On the other hand, to prove (1.1) is suffices to check that $\| a^*a \| \geq \| a \|^2$. Indeed, this property implies (since the inequality $\| ab \| \leq \| a \| \| b \|$ is an axiom of normed algebras) $\|a^*\|\,\|a\| \geq \|a\|^2$, hence $\|a^*\| \geq \|a\|$, hence equality. Finally $\|a^*a\| \leq \|a^*\|\,\|a\| = \|a\|^2$, and equality obtains.

This allows us to give the fundamental example of a non–commutative C^*–algebra : the algebra $\mathcal{L}(\mathcal{H})$ of all bounded linear operators on a complex Hilbert space, with the operator norm as $\|\cdot\|$, the adjoint mapping as involution, and the operator I as unit. To see this, we remark that for an operator A

$$\|A\|^2 = \sup_{\|x\|\leq 1} <Ax, Ax> = \sup_{\|x\|\leq 1} <A^*Ax, x> \leq \|A^*A\|\ .$$

One can show that every C^*–algebra is isomorphic to a closed subalgebra of some $\mathcal{L}(\mathcal{H})$.

By analogy with the case of $\mathcal{L}(\mathcal{H})$, one says that a in some C^*–algebra $\mathcal{A}$ is *selfadjoint* or *hermitian* if $a = a^*$, is *normal* if a and a^* commute, and is *unitary* if $aa^* = a^*a = I$. The word *selfadjoint* is so often used that we abbreviate it into *s.a.*.

Application of spectral theory

2 Let us recall a few facts from the elementary theory of Banach algebras. First of all, the *spectrum* $\mathrm{Sp}(a)$ of an element a of a Banach algebra $\mathcal{A}$ with unit I is the set of all $\lambda \in \mathbb{C}$ such that $\lambda I - a$ is not invertible in $\mathcal{A}$. It is a non–empty compact set of the complex plane, contained in the disk with radius $\|a\|$, but the *spectral radius* $\rho(a)$ of a, i.e. the radius of the smallest disk containing $\mathrm{Sp}(a)$, may be smaller than $\|a\|$. The complement of $\mathrm{Sp}(a)$ is called the *resolvent set* $\mathrm{Res}(a)$. The spectrum, and hence the spectral radius, are defined without reference to the norm. On the other hand, the spectral radius is given explicitly by

$$(2.1) \qquad\qquad \rho(a) = \limsup_n \|a^n\|^{1/n}\ .$$

We use many times the *spectral mapping theorem*, which in its simplest form asserts that if $F(z)$ is an entire function $\sum_n c_n z^n$ on $\mathbb{C}$, and $F(a)$ denotes the sum $\sum_n c_n a^n \in \mathcal{A}$, then $\mathrm{Sp}(F(a))$ is the image $F(\mathrm{Sp}(a))$. This result is true more generally for functions which are not holomorphic in the whole plane, but only on a neighbourhood of the spectrum. From this theorem we only need here the following particular case

$$(2.2) \qquad\qquad \textit{If } a \textit{ is invertible, then } \mathrm{Sp}(a^{-1}) = (\mathrm{Sp}(a))^{-1}.$$

Another property we will use is the following

$$(2.3) \qquad\qquad \mathrm{Sp}(ab) \textit{ and } \mathrm{Sp}(ba) \textit{ differ at most by } \{0\}.$$

The proof goes as follows : assume $\lambda I - ba$ has an inverse c. Then the easy equality

$$(\lambda I - ab)(I + acb) = (I + acb)(\lambda I - ab) = \lambda I$$

implies, for $\lambda \neq 0$, that $\lambda I - ab$ is invertible too.

3 The following properties are specific to C^* algebras. They are very simple and have important consequences.

a) *If a is normal, we have $\rho(a) = \|a\|$.* This follows from the computation (2.1) of $\rho(a)$, and the following consequence of (1.1) (note that $b = a^*a$ is s a.)

$$\|a^{2^n}\|^2 = \|a^{*\,2^n} a^{2^n}\| = \|(a^*a)^{2^n}\| = \|(a^*a)^{2^{n-1}}\|^2 = \ldots = \|a\|^{2^{n+1}}\,.$$

As a corollary, note that a Banach algebra $\mathcal{A}$ has at most one C^* norm, given by $\|a\| = (\rho(a^*a))^{1/2}$. Also, if $\mathcal{A}$ and $\mathcal{B}$ are two C^* algebras, f a morphism from $\mathcal{A}$ to $\mathcal{B}$ (*i.e.* f preserves the algebraic operations, involutions and units), it is clear that $f(a)$ has a smaller spectral radius than a, hence also a smaller norm.

b) *The spectrum of a unitary element a is contained in the unit circle.* Indeed, the fact that $a^*a = I$ implies $\rho(a^*a) = 1$. Therefore the spectrum of a lies in the unit disk. The same applies to a^{-1}, and using (2.2) we find their spectra are on the unit circle.

c) *A s.a. element a has a real spectrum, and at least one of the points $\pm\|a\|$ belongs to $\mathrm{Sp}(a)$.* One first remarks that e^{ia} is unitary, hence has its spectrum on the unit circle. The first statement then follows from the spectral mapping theorem. On the other hand, $\mathrm{Sp}(a)$ is compact, and therefore contains a point λ such that $|\lambda| = \rho(a)$. Since the spectrum is real and $\rho(a) = \|a\|$, we have $\lambda = \pm\|a\|$.

Positive elements

4 We adopt the following definition of a positive element a in the C^* algebra $\mathcal{A}$: *a is s.a. and its spectrum is contained in $\mathbb{R}_+$.* For instance, since every s.a. element has a real spectrum, its square is positive by the spectral mapping theorem. In subsection 7 we prove that positive elements are exactly those of the form b^*b. We start with a few elementary results which have extremely clever and elegant proofs.

a) *If a finite sum of positive elements is equal to 0, each of them is 0.* It is sufficient to deal with a sum $a + b = 0$. Since $\mathrm{Sp}(a) = -\mathrm{Sp}(b)$ are in $\mathbb{R}_+$ we must have $\mathrm{Sp}(a) = \{0\}$, hence $\rho(a) = 0 = \|a\|$.

b) *Let a be s.a. with $\|a\| \leq 1$. Then $a \geq 0 \iff \|I - a\| \leq 1$.* Indeed, if $a \geq 0$ and $\|a\| \leq 1$ then $\mathrm{Sp}(I - a) \subset [0,1]$, hence $\rho(I - a) = \|I - a\| \leq 1$. Conversely, from $\|a\| \leq 1$ and $\|I - a\| \leq 1$ we deduce that $\mathrm{Sp}(a)$ lies in the intersection of the real axis, and of the two disks of radius 1 with centers at 0 and 1. That is, in the interval $[0,1] \subset \mathbb{R}_+$ and a is positive.

Let $\mathcal{A}_+$ be the set of positive elements. It is clearly a cone in A. The preceding property easily implies that its intersection with the unit ball is convex and closed, from which it easily follows that *$\mathcal{A}_+$ is closed, and closed under sums (i.e. is a closed convex cone)*

The following result is extremely important :

c) *Every positive element a has a unique square root b; it belongs to the closed algebra generated by a* (it is not necessary to add the unit element).

We may assume $\|a\| \leq 1$, and then $\alpha = I - a$ has norm ≤ 1. We recall that $(1 - z)^{1/2} = 1 - \sum_{n \geq 1} c_n z^n$ with positive coefficients c_n of sum 1. We define $b = I - \sum_n c_n \alpha^n = \sum_n c_n(I - (I - a)^n)$. Then it is very easy to see that b is a positive square root of a. It appears as a limit of polynomials in a without constant term, whence the last statement.

Before we prove uniqueness, we note that a *product of two commuting positive elements a, a' is positive*. Indeed, their (just constructed) square roots b, b' commute as limits of polynomials in a, a', hence bb' is s.a. with square aa'.

To prove uniqueness, consider any positive c such that $c^2 = a$. Then $a = c^2$ belongs to the closed algebra generated by c, and then the same is true for the square root b considered above. Thus all three elements commute and we have $0 = (b^2 - c^2)(b - c) = (b - c)^2(b + c)$. Now $(b - c)^2 b$ and $(b - c)^2 c$ are positive from the preceding remark and their sum is 0. By a) above, they are themselves equal to 0, and so is their difference $(b - c)^3$. This implies $\rho(b - c) = 0$, and finally $\| b - c \| = 0$.

Symbolic calculus for s.a. elements

5 Symbolic calculus in an algebra $\mathcal{A}$ consists in giving a reasonable meaning to the symbol $F(u)$, where u ranges over some class of elements of $\mathcal{A}$ and F over some algebra of functions on $\mathbb{C}$. Notations like e^a or $\sqrt{a}$ are examples of symbolic calculus in algebras. It is clear that polynomials operate on any algebra with unit, and entire functions on every Banach algebra; more generally, the symbolic calculus in Banach algebras defines $F(u)$ whenever F is analytic in a neighbourhood of $\mathrm{Sp}(u)$. We are going to show that *continuous functions on $\mathbb{R}$ operate on s.a. elements of C^* algebras*. Later, we will see that in von Neumann algebras this can be extended to Borel bounded functions.

Let u be a s.a. element of a C^*–algebra, with (compact) spectrum $\mathrm{Sp}(u) = K$. Let $\mathcal{P}$ be the *–algebra of all complex polynomials on $\mathbb{R}$, with the usual complex conjugation as involution. Then the mapping $P \longmapsto P(u)$ is a morphism from $\mathcal{P}$ to the subalgebra of $\mathcal{A}$ generated by u and I (a *–subalgebra since u is s.a.). By the spectral mapping theorem, $\rho(P(u))$ (which is equal to $\| P(u) \|$ since this element is normal) is exactly the uniform norm of P on $\mathrm{Sp}(u) = K$. Polynomials being dense in $\mathcal{C}(K)$ by Weierstrass' theorem, we extend the mapping to $\mathcal{C}(K)$ by continuity, to define an *isomorphism* between $\mathcal{C}(K)$, and the closed *–algebra generated by u and I. *This isomorphism is the continuous symbolic calculus*. Essentially the same reasoning can be extended to normal u instead of s.a., starting from polynomials $P(z, \overline{z})$ on $\mathcal{C}$, and more generally one may define continuous functions of several *commuting* s.a. or normal elements.

6 Let us give some applications of the symbolic calculus.

a) The decomposition $f = f^+ - f^-$ of a continuous function gives rise to a decomposition $u = u^+ - u^-$ of any s.a. element of $\mathcal{A}$ into a difference of two (commuting) positive elements. Also we may define $|u|$ and extend to the closed real algebra generated by u the order properties of continuous functions.

b) Given a *state* μ of $\mathcal{A}$ (see §2 below), μ induces a state of the closed subalgebra generated by u, which translates into a positive linear functional of mass 1 on $\mathcal{C}(\mathrm{Sp}(u))$, called the *law of u in the state μ*. More generally, one could define in this way the joint law of a finite number of commuting s.a. elements of $\mathcal{A}$.

c) Let u be s.a. (or more generally, normal) and invertible in $\mathcal{A}$. Then its spectrum K does not contain 0, and the function $1/x$ thus belongs to $\mathcal{C}(K)$. Therefore u^{-1} belongs to the closed algebra generated by u. More generally, for v not necessarily s.a.,

v is invertible if and only if $u = v^*v$ is invertible, and therefore if v is invertible in $\mathcal{A}$, it is invertible in any C^*–subalgebra $\mathcal{B}$ of $\mathcal{A}$ containing v. It follows that $\mathrm{Sp}(v)$ also is the same in $\mathcal{A}$ and $\mathcal{B}$.

d) Every element a in a C^*–algebra is a linear combination of (four) unitaries. We first decompose a into $p + iq$ with $p = (a + a^*)/2$, $q = i(a - a^*)/2$ selfadjoint. Then we remark that if the norms of p, q are ≤ 1, the elements $p \pm i\sqrt{1 - p^2}$, $q \pm i\sqrt{1 - q^2}$ are unitary.

Characterization of positive elements

7 Every positive element in a C^*–algebra $\mathcal{A}$ has a s.a. square root, and therefore can be written in the form a^*a. Conversely, in the operator algebra $\mathcal{L}(\mathcal{H})$ it is trivial that A^*A is positive in the operator sense, hence has a positive spectrum, and finally satisfies the algebraic definition of positivity. So it was suspected from the beginning that in an arbitrary C^*–algebra every element of the form a^*a is positive, but Gelfand and Naimark could not prove it : it was finally proved by Kelley and Vaught.

We put $b = a^*a$. Since b is s.a., we may decompose it into $b^+ \cdot\!\cdot b^-$ and we want to prove that $b^- = 0$. We put $ab^- = s + it$ with s, t s.a. and define $u = (ab^-)^*(ab^-)$, $v = (ab^-)(ab^-)^*$. We make two remarks.

1) $u = (ab^-)^*(ab^-) = (b^-)^*a^*ab^- = b^-bb^- = (-b^-)^3$ is s.a. ≤ 0. On the other hand $u = (s - it)(s + it) = s^2 + t^2 + i(st - ts)$, thus $i(st - ts) \leq 0$.

2) Then $v = (ab^-)(ab^-)^* = s^2 + t^2 - i(st - ts)$ is s.a. ≥ 0 as a sum of two such elements.

According to (2.3), u and v have the same spectrum, except possibly the value 0. Since $u \leq 0$ and $v \geq 0$, this spectrum is $\{0\}$. Since u, v are s.a., this implies $u = v = 0$. Then the relation $u := -(b^-)^3$ implies $\rho(b^-) = 0$, and $b^- = 0$ since b^- is s.a..

Non–commutative inequalities

8 Many inequalities that are used everyday in commutative life become non–commutative traps. For instance, it is not true that $0 \leq a \leq b \Rightarrow a^2 \leq b^2$. If we try to define $|a|$ as $(a^*a)^{1/2}$ it is not true that $|a + b| \leq |a| + |b|$, etc. Thus it is worth making a list of true inequalities.

a) Given one single positive element a, we have $a \leq \|a\|\,\mathrm{I}$, and $a^2 \leq \|a\|\,a$: this amounts to the positivity of the polynomials $\|a\| - x$ and $\|a\|\,x - x^2$ on $\mathrm{Sp}(a)$.

b) $a \geq 0$ implies $c^*ac \geq 0$ for arbitrary c (write a as b^*b). By difference, $a \geq b \Rightarrow c^*ac \geq c^*bc$.

c) $a \geq b \geq 0 \Rightarrow \|a\| \geq \|b\|$. Indeed, $\|a\|\,\mathrm{I} \geq a \geq b$, hence $\mathrm{Sp}(b) \subset [0, \|a\|]$.

d) $a \geq b \geq 0 \Rightarrow (\lambda\mathrm{I} + a)^{-1} \leq (\lambda\mathrm{I} + b)^{-1}$ for $\lambda \geq 0$. This is less trivial : one starts from $0 \leq \lambda\mathrm{I} + b \leq \lambda\mathrm{I} + a$, then applies b) to get

$$\mathrm{I} \leq (\lambda\mathrm{I} + b)^{-1/2}(\lambda\mathrm{I} + a)(\lambda\mathrm{I} + b)^{-1/2} \ .$$

Then the function x^{-1} reverses inequalities on positive s.a. operators, and we deduce

$$\mathrm{I} \geq (\lambda\mathrm{I} + b)^{1/2}(\lambda\mathrm{I} + a)^{-1}(\lambda\mathrm{I} + b)^{1/2} \ ,$$

and we apply again b) to get the final result.

e) The preceding result gives an example of increasing functions $F(x)$ on $\mathbb{R}_+$ such that $0 \leq a \leq b \Longrightarrow F(a) \leq F(b)$, namely the function $1 - (\lambda + x)^{-1} = x/(x + \lambda)$ for $\lambda \geq 0$. We get other examples by integration w.r.t. a positive measure $\mu(d\lambda)$, the most important of which are the powers x^α for $0 < \alpha < 1$, and in particular the function $\sqrt{x}$.

§2. STATES ON C^*-ALGEBRAS

1   We gave in Chapter I the general definition of a probability law, or state, on a complex unital *-algebra $\mathcal{A}$. Let μ be a linear functional on a C^*-algebra are no exception, such that $\mu(a^*a) \geq 0$ for arbitrary a. According to subsection 7, this means exactly that μ is positive on positive elements, hence real on s.a. elements. Since every element of $\mathcal{A}$ can be written as $a = p + iq$, $a^* = p - iq$, we see that $\mu(a^*) = \overline{\mu(a)}$, and one of the general axioms becomes unnecessary.

The same familiar reasoning that leads to the Schwarz inequality gives here

(1.1) $$|\mu(b^*a)|^2 \leq \mu(b^*b)\,\mu(a^*a) .$$

In particular, we have $|\mu(a)|^2 \leq \mu(a^*a)\,\mu(I)$. On the other hand, we have $a^*a \leq \|a\|^2 I$, and since $\mu(I) = 1$ we find that $|\mu(a)| \leq \|a\|$ — otherwise stated, μ is bounded and has norm 1.

Conversely, a bounded linear functional on $\mathcal{A}$ such that $\mu(I) = \|\mu\| = 1$ is a state. Consider indeed a s.a. positive element a with spectrum K; define a linear functional on $\mathcal{C}(K)$ by $m(f) = \mu(f(a))$. Since μ is bounded, m is a complex measure, of norm $\|m\| \leq \|\mu\| = \mu(1) = m(1)$. Then m is positive from standard measure theory, and we have $\mu(a) = m(x) \geq 0$.

C^* algebra have many states : let us prove that for every s.a. $a \in \mathcal{A}$, there exists a state μ such that $|\mu(a)| = \|a\|$. We put $K = \mathrm{Sp}(a)$, $\|a\| = \alpha$, and recall (§1, subs.2, c)) that at least one of the two points $\pm\alpha$ belongs to K; for definiteness we assume α does. The linear functional $f(a) \longmapsto f(\alpha)$ on the C^*-algebra generated by a is bounded with norm 1, and by the Hahn–Banach theorem it can be extended to a linear functional μ of norm 1 on $\mathcal{A}$. The equality $\mu(1) = 1 = \|\mu\|$ implies that μ is a state on $\mathcal{A}$ such that $\mu(a) = \alpha$.

REMARK. The set of all bounded linear functionals on $\mathcal{A}$ is a Banach space, the unit ball of which is compact in its weak* topology. The subset of all positive linear functionals is weakly closed and hence compact $\mathcal{A}$, and finally the set of all states itself is a weakly compact convex set. According to the Krein–Milman theorem, it is the closed convex hull of the set of its extreme points, also called *pure states*. For instance, the state $\mu(A) = <x, Ax>$ of the algebra $\mathcal{L}(\mathcal{H})$, associated with an unit vector x, can be shown to be pure in this sense. It is not difficult to show that the state μ is the preceding proof can be chosen to be pure.

> The space of all states of a C^*-algebra has received considerable attention, specially because of the (unsuccessful) attempts to classify all states of the CCR and CAR algebras.

Representations and the GNS theorem

2 A *representation* of a C^*–algebra $\mathcal{A}$ in a Hilbert space $\mathcal{H}$ is a morphism φ from $\mathcal{A}$ into the C^*–algebra $\mathcal{L}(\mathcal{H})$. Then every unit vector $x \in \mathcal{H}$ gives rise to a state $\mu(a) = \,<x,\varphi(a)\,x>$ of $\mathcal{A}$ (generally not a pure one). The celebrated theorem of Gelfand–Naimark–Segal, which we now prove, asserts that every state of a C^*–algebra $\mathcal{A}$ arises in this way — and pure states arise from *irreducible* representations, an important point we do not need here.

> From an analyst's point of view, the main point of the GNS theorem is to show that a C^*–algebra has many (irreducible) representations. For us, its interest lies in the construction itself, which is a basic step in the integration theory w.r.t. a state.

The proof is very easy. For $a, b \in \mathcal{A}$ one defines $<b,a> = \mu(b^*a)$. Then $\mathcal{A}$ becomes a prehilbert space. Let $\mathcal{H}$ denote the associated Hilbert space, which means that first we neglect the space $\mathcal{N}$ of all $a \in \mathcal{A}$ such that $\mu(a^*a) = 0$, and then complete $\mathcal{A}/\mathcal{N}$. We denote by $\tilde{a}$, for a few lines only, the class $mod\,\mathcal{N}$ of $a \in \mathcal{A}$.

We have $\mu((ba)^*ba) = \mu(a^*(b^*b)a)$, and since $b^*b \leq \|b\|^2 I$ this is bounded by $\|b\|^2\mu(a^*a)$. Thus the operator ℓ_b of left multiplication by b is bounded on $\mathcal{H}$. We have $\ell_u\ell_v = \ell_{uv}$, etc. so that $\ell.$ is a representation of $\mathcal{A}$ in $\mathcal{H}$. Denoting by $\mathbf{1}$ the vector in $\mathcal{H}$ corresponding to I, we have $\mu(a) = \,<\mathbf{1},\ell_a\mathbf{1}>$.

The relation $aI = a$ for all $a \in \mathcal{A}$ implies that $\ell_a\mathbf{1} = \tilde{a}$. The set of all $\ell_a\mathbf{1}$ is thus *dense* in $\mathcal{H}$: $\mathbf{1}$ is said to be a *cyclic vector* for the representation, and the representation itself is said to be *cyclic*.

Assume some other representation φ is given on a Hilbert space $\mathcal{K}$, with a cyclic vector x such that $\mu(a) = \,<x,\varphi(a)\,x>$ for $a \in \mathcal{A}$. Then the mapping $a \longmapsto \varphi(a)x$ is an isometry from the prehilbert space $\mathcal{A}$ into $\mathcal{K}$, with a dense image. Then it can be transferred to $\mathcal{A}/\mathcal{N}$, extended to $\mathcal{H}$ by continuity, and becomes an isomorphism between $\mathcal{H}$ and $\mathcal{K}$. Thus the pair $(\mathcal{H},\mathbf{1})$ is independent, up to isomorphism, of the particular way it was constructed, and we may speak loosely of *the* GNS representation associated with the state μ.

3 COMMENTS. a) Assume we are in the commutative case, and $\mathcal{A}$ is one of the three usual C^*–algebras of measure theory : the algebra $\mathcal{C}(K)$ of continuous functions on a compact set K (with an arbitrary state μ), the algebra $\mathcal{B}(\mathcal{E})$ of all bounded functions for a σ–field $\mathcal{E}$ (in this case, to develop integration theory we need a countable additivity assumption on μ), and the algebra $L^\infty(\mu)$ (this one is a von Neumann algebra, of which μ is a "normal state"). In all three cases, $\mathcal{H}$ is the space $L^2(\mu)$, on which $\mathcal{A}$ acts by multiplication. Thus it seems that $\mathcal{H}$ is a kind of non commutative L^2 space. It is indeed, but the GNS construction leads only to a "left" L^2 space, on which right multiplication generally defines unbounded operators. We return to this subject later.

b) From the GNS construction, one can understand how every C^*–algebra can be realized as a C^*–algebra of operators on some Hilbert space — a very large one, possibly the direct sum of the Hilbert spaces associated with all irreducible representations of the algebra. We do not insist on this point.

c) The state μ is said to be *faithful* if

(3.1) $\forall a \in \mathcal{A} \quad (\mu(a^*a) = 0) \Longrightarrow (a = 0) .$

This is equivalent to saying that for $a \geq 0$, $\mu(a) = 0 \Rightarrow a = 0$. From the point of view of the representation, we have $\mu(a^*a) = \|\ell_a 1\|$, and (2.1) means that $\ell_a 1 = 0 \Rightarrow a = 0$: the vector $\mathbf{1}$ then is said to be *separating*. Note that the state being faithful is a stronger property than the GNS representation being faithful $((\ell_a = 0) \Rightarrow (a = 0))$.

In classical measure theory, μ is a faithful state on $\mathcal{C}(K)$ if the support of μ is the whole of K ; μ is practically never faithful on the algebra of Borel bounded functions, and is always (by construction) faithful on $L^\infty(\mu)$. In the non–commutative case, we cannot generally turn a state into a faithful one by taking a quotient subalgebra, *because $\mathcal{N}$ usually is only a left ideal, not a two sided one*. This difficulty does not occur for *tracial states*, defined by the property $\mu(ab) = \mu(ba)$.

As a conclusion, while in classical measure theory all states on $\mathcal{C}(K)$ give rise to countably additive measures on the Borel field of K with an excellent integration theory, it seems that not all states on non–commutative C^*-algebras $\mathcal{A}$ are "good". Traces certainly are good states, as well as states which are normal and faithful on some von Neumann algebra containing $\mathcal{A}$. Up to now, the most efficient definition of a class of "good" states has been the so called *KMS condition*. We state it in the next section, subsection 10.

4 EXAMPLES. As in Chapter II, consider the probability space Ω generated by N independent Bernoulli random variables x_k, and the corresponding "toy Fock space" $\Gamma = L^2(\Omega)$ with basis x_A (A ranging over the subsets of $\{1, \ldots, N\}$). We turn Γ into a Clifford algebra with the product

$$x_A x_B = (-1)^{n(A,B)} x_{A \triangle B} .$$

and we embed Γ into $\mathcal{L}(\Gamma)$, each element being interpreted as the corresponding left multiplication operator. Then the vacuum state on $\mathcal{L}(\Gamma)$

$$\mu(U) = <1, U1>$$

induces a state on the Clifford algebra, completely defined by

$$\mu(x_A) = 0 \quad \text{if } A \neq \emptyset, \quad \mu(1) = 1 .$$

Then we have $\mu(x_A x_B) = (-1)^{n(A,A)}$ if $A \neq \emptyset$, 0 otherwise, and *this state is tracial*.

The C^*-algebra generated by all creation and annihilation operators $a_i^\pm$ is the whole of $\mathcal{L}(\Gamma)$. The vacuum state μ is read on the basis $a_A^+ a_B^-$ of normally ordered products as $\mu(a_A^+ a_B^-) = 0$ unless $A = B = \emptyset$, $\mu(I) = 1$. It is not tracial, since for $A \neq \emptyset$ $\mu(a_A^+ a_A^-) = 0$, $\mu(a_A^- a_A^+) \neq 0$ — anyhow, $\mathcal{L}(\Gamma)$ has no other tracial state than the standard one $U \longmapsto 2^{-N} \text{Tr}(U)$. It is also not faithful (in particular, annihilation operators kill the vacuum).

5 We mentioned above several times the construction of a quotient algebra $\mathcal{A}/\mathcal{J}$, where $\mathcal{J}$ is a two sided ideal. This raises a few interesting problems. For instance, whether the quotient algebra will be a C^*-algebra. We mention the classical answers, mostly due to Segal. It is not necessary to study this subsection at a first reading.

Let first $\mathcal{J}$ be a left ideal in $\mathcal{A}$, not necessarily closed or stable under the involution. We assume $\mathcal{J} \neq \mathcal{A}$, or equivalently $I \notin \mathcal{J}$. Let $\mathcal{J}_+$ denote the set of all positive elements in $\mathcal{J}$. We have seen in §1, subs. 8, d) that the mapping $j \longmapsto e_j = j(I+j)^{-1}$ on $\mathcal{J}_+$ is increasing. It maps $\mathcal{J}_+$ into the "interval $[0,I]$" in $\mathcal{J}$. The symbol $\lim_j$ is understood as a limit along the directed set $\mathcal{J}_+$.

The following important result explains why e_j is called an *approximate unit for $\mathcal{J}$*.

THEOREM. For $a \in \mathcal{J}$ we have $\lim_j \| a - ae_j \| = 0$.

PROOF. For $n \in \mathbb{N}$ and $j \geq na^*a$, we have (§1, 8 b))

$$(a - e_j)^*(a - ae_j) = (I - e_j) a^*a (I - e_j) \leq \frac{1}{n}(I - e_j) j (1 - e_j) .$$

We must show that the norm of the right side tends to 0, and taking n large it suffices to find a universal bound for the norm of $j(I-e_j)^2$. Using symbolic calculus this reduces to the trivial remark that $x/(1+x)^2$ is bounded for $x \in \mathbb{R}$. □

APPLICATIONS. a) This reasoning does not really use the fact that $a \in \mathcal{J}$: since $\mathcal{J}$ is a left ideal ae_j belongs to $\mathcal{J}$ for $a \in \mathcal{A}$: it applies whenever a^*a is dominated by some element of $\mathcal{J}_+$, with the consequence that a then belongs to the closure of $\mathcal{J}$.

b) If $a = \lim ae_j$, we have $a^* = \lim e_j a^*$. Thus if $\mathcal{J}$ is a closed two sided ideal, $\mathcal{J}$ is also stable under the involution.

c) Let $\mathcal{J}$ be a closed two sided ideal, and $\tilde{a}$ denote the class $mod \mathcal{J}$ of $a \in \mathcal{A}$. The general definition of $\| \tilde{a} \|$ in the Banach space $\mathcal{A}/\mathcal{J}$ is $\inf_{b \in \mathcal{J}} \| a + b \|$, and therefore we have $\| a - ae_j \| \geq \| \tilde{a} \|$. Let us prove that

(5.1) $$\| \tilde{a} \| = \lim_j \| a - ae_j \| .$$

Since $\| I - e_j \| \leq 1$ we have

$$\| a + b \| \geq \| (a + b)(I - e_j) \| = \| (a - ae_j) + (b - be_j) \|$$

For $b \in \mathcal{J}$, $b - be_j$ tends to 0 in norm, and therefore $\| a + b \| \geq \limsup_j \| a - ae_j \|$, implying (5.1).

d) It follows that $\mathcal{A}/\mathcal{J}$ is a C^*-algebra :

$$\| \tilde{a} \|^2 = \lim_j \| a - ae_j \|^2 = \lim_j \| (a^* - e_j a^*)(a - ae_j) \|$$
$$= \lim_j \| (I - e_j)(a^*a + b)(I - e_j) \| \quad \text{for every } b \in \mathcal{J}$$
$$\leq \| a^*a + b \| \quad \text{for every } b \in \mathcal{J}$$

Taking an $\inf_b$ we have that $\| \tilde{a} \|^2 \leq \| \tilde{a}^*\tilde{a} \|^2$, hence equality (§1, subs. 1).

e) An interesting consequence : let f be a morphism from a C^*-algebra $\mathcal{A}$ into a second one $\mathcal{B}$, and let $\mathcal{J}$ be its kernel. We already know that f is continuous (norm decreasing), hence $\mathcal{J}$ is closed. Then let $\tilde{f}$ be the algebraic isomorphism from the quotient C^*-algebra $\mathcal{A}/\mathcal{J}$ into $\mathcal{B}$; from subs. 3 b), $\tilde{f}$ is also a norm isomorphism, *and therefore its image $f(\mathcal{A})$ is closed in $\mathcal{B}$.*

§3. VON NEUMANN ALGEBRAS

This section is a modest one : more than ever its proofs are borrowed from expository books and papers, and we have left aside the harder results (and specially all of the classification theory).

Weak topologies for operators

1 Though von Neumann (vN) algebras can be defined as a class of abstract C^*-algebras, all the vN–algebras we discuss below will be concrete $*$–algebras of operators on some Hilbert space $\mathcal{H}$. As operator algebras, they are characterized by the property of being closed under the weak operator topologies instead of the uniform topology, and our first task consists in making this precise.

> The most striking characterization of vN algebras as abstract C^*–algebras is possibly that of Sakai : a C^*–algebra $\mathcal{A}$ is a vN algebra if and only if the underlying Banach space is a dual. Also, vN algebras are characterized by the fact that bounded increasing families of s.a. elements possess upper bounds. On the other hand, Pedersen has studied natural classes of algebras satisfying weaker properties, like that of admitting a Borel symbolic calculus for individual s.a. elements, or of admitting upper bounds of bounded increasing *sequences* of s.a. elements. He has shown that, in all practical cases, they turn out to be the same as vN algebras.

There are three really useful weak topologies for bounded operators. The first one is the *weak operator topology* : $a_i \to a$ iff $< y, a_i x > \to < y, ax >$ for fixed x, y — here and below, the use of i rather than n means that convergence is meant for filters or directed sets, not just for sequences. The second is the *strong operator topology* : $a_i x \to ax$ in norm for fixed x. However, the most important for us is the third one, usually called the σ–*weak* or *ultraweak* topology. We have seen in Appendix 1 that the space $\mathcal{L} = \mathcal{L}^\infty$ of all bounded operators is the dual of the space $\mathcal{L}^1$ of trace class operators, and the ultraweak topology is the weak topology on $\mathcal{L}$ relative to this duality. Otherwise stated, convergence of a_i to a means that $\text{Tr}(a_i b) \to \text{Tr}(ab)$ for every $b \in \mathcal{L}^1$. To emphasize the importance of this topology, and also because states which are continuous in this topology are called *normal states*, we call it the *normal topology* (see App. 1, subs. 5).

> More precisely, the word *normal* usually refers to a kind of order continuity, described later on, which will be proved to be equivalent to continuity in the "normal" topology (see subs. 6). Thus our language doesn't create any dangerous confusion.

It is good to keep in mind the following facts from Appendix 1 :

a) The normal topology is stronger than the weak topology, and not comparable to the strong one. The strong topology behaves reasonably well w.r.t. products : if $a_i \to a$ and $b_i \to b$ strongly, $\| b_i \|$ (or $\| a_i \|$) remaining bounded, then $a_i b_i \to ab$ and $b_i a_i \to ba$ strongly. On the other hand, it behaves badly w.r.t. adjoints, while if $a_i \to a$ weakly, then $a_i^* \to a^*$ weakly.

b) The dual of $\mathcal{L}(\mathcal{H})$ is the same for the weak and the strong topologies, and consists of all linear functionals $f(a) = \sum_j < y_j, ax_j >$ where x_j, y_j are (finitely many) elements of $\mathcal{H}$. The dual for the normal topology consists of all linear functionals $f(a) = \mathrm{Tr}(aw)$ where w is a trace class operator.

c) One can give a simple characterization of the normal linear functionals on $\mathcal{L}(\mathcal{H})$, *i.e.* those which are continuous in the normal topology : *a linear functional μ on $\mathcal{L}(\mathcal{H})$ is normal iff its restriction to the unit ball is continuous in the weak topology.*

As an application, since every weakly or strongly convergent sequence is norm bounded, it also converges in the normal topology.

Von Neumann algebras

2 Since the weak and strong topologies on $\mathcal{L}(\mathcal{H})$ have the same dual, the Hahn–Banach theorem implies that the weakly and strongly closed subspaces of $\mathcal{L}(\mathcal{H})$ are the same.

DEFINITION *A (concrete) von Neumann algebra (vNa) is a $*$-subalgebra of $\mathcal{L}(\mathcal{H})$, containing I, closed in the strong or weak topology.*

The *commutant* J' of a subset $J \subset \mathcal{L}(\mathcal{H})$ is the set of all bounded operators a such that $ab = ba$ for every $b \in J$. It is clear that J' is an algebra, closed in the strong topology, and that the *bicommutant* J'' of J contains J. If J is stable under the involution, so is J' (which therefore is a vNa), and so is J''.

Here is the first, and most important, result in this theory.

VON NEUMANN'S BICOMMUTANT THEOREM. *Let $\mathcal{A}$ be a sub-$*$-algebra of $\mathcal{L}(\mathcal{H})$ containing I. Then the bicommutant $\mathcal{A}''$ is exactly the strong $(=$ weak$)$ closure of $\mathcal{A}$. In particular, $\mathcal{A}$ is a von Neumann algebra iff it is equal to its bicommutant.*

Note that it is not quite obvious that the strong closure of an algebra is an algebra!

PROOF. It is clear that the strong closure of $\mathcal{A}$ is contained in $\mathcal{A}''$. We must prove conversely that, for every $a \in \mathcal{A}''$, every $\varepsilon > 0$ and every finite family $x_1, \ldots, x_n$ of elements of $\mathcal{H}$, there exists $b \in \mathcal{A}$ such that $\| bx_i - ax_i \| \leq \varepsilon$ for all i.

We begin with the case of one single vector x. Let K the closed subspace in $\mathcal{H}$ generated by all bx, $b \in \mathcal{A}$; K is closed under the action of $\mathcal{A}$, and the same is true for $K^\perp$ (if $<y, bx> = 0$ for every $b \in \mathcal{A}$, we also have for $c \in \mathcal{A}$ $<cy, bx> = <y, c^*bx> = 0$). This means that the orthogonal projection p on K commutes with every $b \in \mathcal{A}$; otherwise stated, $p \in \mathcal{A}'$. Then $a \in \mathcal{A}''$ commutes with p, so that $ax \in K$. This means there exist $a_n \in \mathcal{A}$ such that $a_n x \to ax$.

Extension to n vectors : let $\widehat{\mathcal{H}}$ be the direct sum of n copies of $\mathcal{H}$. A linear operator on $\widehat{\mathcal{H}}$ can be considered as a (n,n) matrix of operators on $\mathcal{H}$, and we call $\widehat{\mathcal{A}}$ the algebra consisting of all operators $\widehat{a}$ repeating $a \in \mathcal{A}$ along the diagonal $(\widehat{a}(y_1 \oplus \ldots \oplus y_n) = ay_1 \oplus \ldots \oplus ay_n)$. It is easy to see that $\widehat{\mathcal{A}}$ is a vNa whose commutant consists of all matrices with coefficients in $\mathcal{A}'$, and whose bicommutant consists of all diagonals $\widehat{a}$ with $a \in \mathcal{A}''$. Then the approximation result for n vectors $(x_1, \ldots, x_n)$ in $\mathcal{A}$ follows from the preceding result applied to $x_1 \oplus \ldots \oplus x_n$.

COROLLARY. *The von Neumann algebra generated by a subset J of $\mathcal{L}(\mathcal{H})$ is $(J \cup J^*)''$.*

3 We give other useful corollaries as comments, rather than formal statements (the reader is referred also to the beautiful exposition of Nelson [Nel2]). The general idea is, that every "intrinsic" construction of operator theory with an uniquely defined result, when performed on elements of a vNa $\mathcal{A}$, also leads to elements of $\mathcal{A}$. We illustrate this by an example.

First of all, let us say that an operator with domain $\mathcal{D}$, possibly unbounded, is *affiliated* to the vNa $\mathcal{A}$ if $\mathcal{D}$ is stable under every $b \in \mathcal{A}'$ and $abx = bax$ for $x \in \mathcal{D}$ (thus a bounded everywhere defined operator is affiliated to $\mathcal{A}$ iff it belongs to $\mathcal{A}$: this is a restatement of von Neumann's theorem).

Consider now an unbounded selfadjoint operator S on $\mathcal{H}$, and its spectral decomposition $S = \int_{\mathbb{R}} \lambda \, dE_\lambda$. This decomposition is unique. More precisely, if $u : \mathcal{H} \to \mathcal{K}$ is an isomorphism between Hilbert spaces, and T is the operator uSu^{-1} on $\mathcal{K}$, then the spectral projections of T are given by $F_t = uE_tu^{-1}$. Using this trivial remark, we prove by "abstract nonsense" that *if S is affiliated to a vNa $\mathcal{A}$, then its spectral projections E_t belong to $\mathcal{A}$.*

Indeed, it suffices to prove that for every $b \in \mathcal{A}'$ we have $E_t b = bE_t$. Since b is a linear combination of unitaries within $\mathcal{A}'$, we may assume $b = u$ is unitary. Then we apply the preceding remark with $\mathcal{K} = \mathcal{H}$, $T = S$, and get that $E_t = F_t = uE_tu^{-1}$, the desired result.

> More generally, this applies to the so called *polar decomposition* of every closed operator affiliated with $\mathcal{A}$.

Kaplansky's density theorem

4 Let $\mathcal{A}$ be a *-algebra of operators. Since strang convergence without boundedness is not very useful, von Neumann's theorem is not powerful enough as an approximation result of elements in $\mathcal{A}''$ by elements of $\mathcal{A}$. The Kaplansky theorem settles completely this problem, and shows that elements in $\mathcal{A}''$ of one given kind can be approximated boundedly by elements of the same kind from $\mathcal{A}$.

THEOREM. *For every element a from the unit ball of $\mathcal{A}''$, there exists a filter a_i on the unit ball of $\mathcal{A}$ (not necessarily a sequence) that converges strongly to a. If a is s.a., positive, unitary, the a_i may be chosen with the same property.*

PROOF (from Pedersen [Ped]). Let f be a real valued, continuous function on $\mathbb{R}$, so that we may define $f(a)$ for every s.a. operator a. We say that f is *strong* if the mapping $f(\cdot)$ is continuous in the strong operator topology. Obvious examples of strong functions are $f(t) = 1$, $f(t) = t$. The set $\mathcal{S}$ of all strong functions is a linear space, closed in the uniform topology. On the other hand, the product of two elements of $\mathcal{S}$, one of which is bounded, belongs to $\mathcal{S}$. The main remark is

LEMMA. *Every continuous and bounded function on $\mathbb{R}$ is strong.*

We first put $h(t) = 2t/(1+t^2)$ and prove that it is strong. Let a, b be s.a. and put $A = (1+a^2)^{-1}$, $B = (1+b^2)^{-1}$. Then we compute as follows the difference $h(b) - h(a)$ (forgetting the factor 2)

$$Bb - Aa = B\left[b(1+a^2) - (1+b^2)a\right]A = B(b-a)A + Bb(a-b)aA .$$

If $b_i \to a$, B_i and B_ib_i remain bounded in norm, and therefore $h(b_i) \to h(a)$.

This bounded function being strong, we find (by multiplication with the strong function t) that $t^2/(1+t^2)$ by t is strong, and so is $1/(1+t^2)$ by difference. Applying the Stone–Weierstrass theorem on $\overline{\mathbb{R}}$ we find that all bounded continuous functions on $\overline{\mathbb{R}}$ are strong. Then, if f is only continuous and bounded on $\mathbb{R}$, $t f(t)/(1+t^2)$ tends to 0 at infinity and hence is strong and bounded. Then writing

$$f(t) = \frac{f(t)}{1+t^2} + t\,\frac{t\,f(t)}{1+t^2}$$

we find that f itself is strong. This applies in particular to $\cos t$, $\sin t$, and we have everything we need to prove the theorem.

S.a. operators. Let $b \in \mathcal{A}''$ be s.a. with norm ≤ 1. Since b belongs to the weak closure of $\mathcal{A}$ and the involution is continuous in the weak topology, it belongs to the weak closure of $\mathcal{A}_{sa}$. This set being convex, b belongs in fact to its strong closure. We choose selfadjoint $b_i \in \mathcal{A}_{sa}$ converging strongly to b, and we deduce from the lemma that $h(b_i) \to h(b)$ strongly. And now $h(b_i)$ belongs to the unit ball of $\mathcal{A}_{sa}$, while every element a in the unit ball of $\mathcal{A}''_{sa}$ can be represented as $h(b)$, since h induces a homeomorphism of $[-1, 1]$ to itself.

The same reasoning gives a little more : if a is positive so is b. Then we also have $h^+(b_i) \to h^+(b) = a$, and these approximating operators are positive too.

Unitaries. If $u \in \mathcal{A}''$ is unitary, it has a spectral representation as $\int_{-\pi}^{\pi} e^{it} dE_t$ with spectral projections $E_t \in \mathcal{A}''$. The s.a. operator $a = \int_{-\pi}^{\pi} t\,dE_t$ can be approximated strongly by s.a. operators $a_j \in \mathcal{A}$, and then the unitaries e^{ia_j} converge strongly to u.

Arbitrary operators of norm ≤ 1. Consider $a \in \mathcal{A}''$ of norm ≤ 1. On $\mathcal{H} \oplus \mathcal{H}$ let the vNa $\widehat{\mathcal{A}}$ consist of all matrices $\begin{pmatrix} j & k \\ l & m \end{pmatrix}$ with $j, k, l, m \in \mathcal{A}$. Then $\widehat{\mathcal{A}}'$ consists of the matrices $\begin{pmatrix} p & 0 \\ 0 & p \end{pmatrix}$ with $p \in \mathcal{A}'$, and the operator $\begin{pmatrix} 0 & a^* \\ a & 0 \end{pmatrix}$ is s.a. with norm ≤ 1 and belongs to the bicommutant $\widehat{\mathcal{A}}''$. From the result above it can be approximated by operators $\begin{pmatrix} r & s^* \\ s & t \end{pmatrix}$ of norm ≤ 1, with $r, s, t \in \mathcal{A}$. Then s has norm ≤ 1 and converges strongly to a, and the proof gives an additional result : the approximants $s \to a$ can be chosen so that $s^* \to a^*$.

REMARK. Assume $\mathcal{A}$ was closed in the "normal" topology. We have just seen that the unit ball B of $\mathcal{A}''$ is the strong or weak closure of the unit ball of $\mathcal{A}$. But on the unit ball of $\mathcal{L}(\mathcal{H})$ the weak and normal topologies coincide, and therefore B is contained in $\mathcal{A}$. Thus $\mathcal{A} = \mathcal{A}''$ and therefore $\mathcal{A}$ is a vNa.

The predual of a von Neumann algebra

5 The Banach space $\mathcal{L}(\mathcal{H})$ is the dual of $\mathcal{T}(\mathcal{H})$, the space of trace class operators. Therefore every subspace $\mathcal{A} \subset \mathcal{L}(\mathcal{H})$ which is closed for the normal topology is itself a dual, that of the quotient Banach space $\mathcal{T}/\mathcal{A}^\perp$. In the case $\mathcal{A}$ is a vNa, this space is called the *predual of* $\mathcal{A}$ and denoted by $\mathcal{A}_*$. The predual can be described as the space of all complex linear functionals on $\mathcal{A}$ which are continuous in the normal topology. We call them *normal linear functionals* below.

According to the Hahn–Banach theorem, every normal functional μ on $\mathcal{A}$ can be extended to a normal functional on $\mathcal{L}(\mathcal{H})$, and therefore can be represented as

$\mu(a) = \mathrm{Tr}(aw)$ where w is a trace class operator. In particular, decomposing w we see that every normal linear functional is a complex linear combination of four normal laws (states). In the case of laws, the following lemma (which is not essential for the sequel) gives a more precise description

LEMMA. *Every normal law* μ *on* $\mathcal{A}$ *can be extended as a normal law on* $\mathcal{L}(\mathcal{H})$, *hence the trace class operator* w *can be taken positive.*

PROOF. For simplicity, we assume that $\mathcal{H}$ is separable. Since $\mu(a) = \overline{\mu}(a^*)$, μ is also associated with the trace class operator w^*, and Replacing w by $(w + w^*)/2$ we may assume w is s.a.. Then we choose an o.n.b. (e_n) such that $we_n = \lambda_n e_n$ with $\sum_n |\lambda_n| < \infty$. Consider the direct sum $\widehat{\mathcal{H}}$ of countably many copies $\mathcal{H}_n$ of $\mathcal{H}$, on which $\mathcal{A}$ operates diagonally $(\widehat{a}(\sum_n x_n) = \sum_n ax_n)$. Let $\mathbf{1} = c \sum_n \sqrt{|\lambda_n|}\, e_n \in \widehat{\mathcal{H}}$, where c is a normalization constant, and ν be the corresponding state $\nu(a) = \sum_n <\mathbf{1}, \widehat{a}\mathbf{1}>$. We have

$$\mu(a) = \sum_n \lambda_n <e_n, ae_n> , \qquad \nu(a) = c^2 \sum_n \sqrt{|\lambda_n|} <e_n, ae_n> .$$

Let $\mathcal{K}$ be the closed (stable) subspace generated by $\mathcal{A}\mathbf{1}$ in $\widehat{\mathcal{H}}$. Since λ_n is bounded μ is dominated by a scalar multiple of ν, and therefore (by a simple lemma proved in subs. 8 below) is of the form $< T\mathbf{1}, \widehat{a}\mathbf{1} >$ where T is a bounded positive operator on $\mathcal{K}$ which *commutes* with the representation. Then putting $j = \sqrt{T}\,\mathbf{1}$ we have $\mu(a) = <j, \widehat{a}j>$. Taking $a = I$ note that j is an unit vector. In the last expression we may replace a by $b \in \mathcal{L}(\mathcal{H})$ operating diagonally on $\widehat{\mathcal{H}}$, and we have found an extension of μ as a law on $\mathcal{L}(\mathcal{H})$.

6   In this subsection, we prove that for a positive linear functional on $\mathcal{A}$, being normal is equivalent to being *order continuous*, an intrinsic property, in the sense that it can be defined within $\mathcal{A}$, independently of the way it acts on the Hilbert space $\mathcal{H}$. As a consequence, the predual $\mathcal{A}_*$ and the normal topology itself are intrinsic.

> In the literature, properties of a concrete vNa $\mathcal{A}$ of operators on $\mathcal{H}$ which depend explicitly on $\mathcal{H}$ are often called *spatial* properties.

Let (a_i) be a family of positive elements of $\mathcal{A}$, which is filtering to the right and norm bounded. Then for $x \in \mathcal{H}$ the limit $\lim_i < x, a_i x >$ exists, and by polarization also $\lim_i < y, a_i x >$. Therefore $\lim_i a_i = a$ exists in the weak topology, and of course belongs to $\mathcal{A}$. Since $\| a_i \|$ is bounded, the convergence also takes place in the normal topology. It is also strong, as shown by the relation

$$\| (a - a_i) x \|^2 \leq \| a - a_i \| < x, (a - a_i) x > \to 0$$

(the inequality $b^2 \leq \| b \| b$ for $b \geq 0$ is used here). On the other hand, a can be described without reference to $\mathcal{H}$ as the l.u.b. of the family (a_i) in $\mathcal{A}$.

Let μ be a positive linear functional on $\mathcal{A}$. If $a = \sup_i a_i$ implies $\mu(a) = \sup_i \mu(a_i)$ we say that μ is *order continuous*. Since order convergence implies normal convergence, every normal functional is also order continous. We are going to prove the converse.

We assume that μ is order continuous. We associate with every positive element $b \in \mathcal{A}$ a (complex!) linear functional $\mu_b(a) = \mu(ab)$ and if μ_b is normal we say that b is *regular*. Thus our aim is to prove that I is regular. We achieve this in three steps.

1) Let (b_i) a norm bounded, increasing family of regular elements. We prove its upper bound b is regular.

Since the predual $\mathcal{A}_*$ is a Banach space, it suffices to prove that μ_{b_i} converges in norm to μ_b. For a in the unit ball of $\mathcal{A}$ we have

$$|\mu(a(b-b_i))| = |\mu(a(b-b_i)^{1/2}(b-b_i)^{1/2})|$$
$$\leq \mu(a(b-b_i)a^*)^{1/2}\mu(b-b_i)^{1/2} \quad \text{(Schwarz)}$$

The first factor is bounded independently of a, the second one tends to 0 since μ is order continuous.

2) Let $m \in \mathcal{A}$ be positive and $\neq 0$. Then there exists a regular element $q \neq 0$ dominated by m.

We choose a vector x such that $<x, mx>\ >\ \mu(m)$ and choose p positive and $\leq m$, such that $<x, px>\ \leq \mu(p)$. Since μ is order continuous, Zorn's lemma allows us to choose a maximal p. Let us prove that $q = m - p$ fulfills our conditions. First of all, it is positive, dominated by m, and the choice of x forbids that $q = 0$. The maximality of p implies that for every positive $u \leq q$ we have $\mu(u) \leq\ <x, ux>$, and this is extended to the case $u \leq Cq$ by homogeneity.

We now prove that μ_q is strongly continuous, hence also weakly continuous, and finally normal. For $a \in \mathcal{A}$ we have $|\mu(aq)| \leq \mu(I)^{1/2}\mu(qa^*aq)^{1/2}$. On the other hand, $qa^*aq \leq \|a\|^2 q^2 \leq \|a\|^2 \|q\| q$, a constant times q. Therefore

$$\mu(qa^*aq) \leq\ <x, qa^*aq\, x>\ =\ \|aqx\|^2 \ .$$

So finally we have $\mu_q(a) \leq \mu(I)^{1/2}\|aqx\|$, whence the strong continuity of μ_q.

3) Zorn's lemma and 1) allow us to choose a maximal regular element b such that $0 \leq b \leq I$. Using 2) the only possibility is $b = I$, and we are finished.

About integration theory

7   The standard theory of non–commutative integration is developed for *faithful normal laws (states)* on a von Neumann algebra. However, as usual with integration theory, the problem is to start from a functional on a small space of functions and to extend it to a larger space. We are going to discuss (rather superficially) this problem.

First of all, we consider a C^*–algebra denoted by $\mathcal{A}^\circ$, with a probability law μ. The GNS construction provides us with a Hilbert space $\mathcal{H}$ on which $\mathcal{A}^\circ$ acts (we simply write ax for the action of $a \in \mathcal{A}^\circ$ on $x \in \mathcal{H}$), and a cyclic vector $\mathbf{1}$ such that $\mu(a) =\ <\mathbf{1}, a\mathbf{1}>$. Since the kernel $\mathcal{N}$ of the GNS representation is a two sided ideal and the quotient $\mathcal{A}^\circ/\mathcal{N}$ is a C^*–algebra, we do not lose generality by assuming the GNS representation is 1–1. Thus $\mathcal{A}^\circ$ appears as a norm closed subalgebra of $\mathcal{L}(\mathcal{H})$. Let us denote by $\mathcal{A}$ its bicommutant, a von Neumann algebra containing $\mathcal{A}^\circ$, and to which μ trivially extends as the (normal) law $\mu(a) =\ <\mathbf{1}, a\mathbf{1}>$.

What have we achieved from the point of view of integration theory? Almost nothing. Think of the commutative case. Let $\mathcal{F}^\circ$ be a boolean algebra of subsets of some set E, and let μ be a finitely additive law on $\mathcal{F}^\circ$. The space of complex elementary functions (linear combinations of finitely many indicator of sets) is a unital *–algebra, and its completion in the uniform norm is a commutative C^*-algebra $\mathcal{A}^\circ$. If we perform all the construction above, we get a normal law on some huge commutative vNa containing $\mathcal{A}^\circ$, and the essential assumption of integration theory, namely the countable additivity of μ, is completely irrelevant!

Besides that, to develop a rich integration theory, it seems necessary to use a faithful law, which cannot be achieved simply (as we pointed out in §2 subs. 3) by a further passage to the quotient. To demand that the state be faithful on $\mathcal{A}^\circ$ is not sufficient to make sure it will be faithful on $\mathcal{A}$. However, the following easy lemma sets a necessary condition :

LEMMA. *Assume the law μ is faithful, and $(a_i)_{i \in I}$ is a norm bounded family such that* $\lim_i \mu(a_i^* a_i) = 0$ *(along some filter on I). Then we also have* $\lim_i \mu(a_i b) = 0$ *for $b \in \mathcal{A}$.*

PROOF. We use the language of the GNS representation. It is sufficient to prove that $\mu(a_i b) \to 0$ along every ultrafilter on the index set I, finer than the given filter. Since the operators a_i are bounded in norm, they have a limit a in the weak topology of operators (even in the normal one), and $a_i \mathbf{1} \to a\mathbf{1}$ weakly. The condition $\mu(a_i^* a_i) \to 0$ means that $\lim_i a_i \mathbf{1} = 0$ in the strong sense, and therefore $a\mathbf{1} = 0$. Since the vector $\mathbf{1}$ is separating, this implies $a = 0$. On the other hand $\mu(a_i b) = <\mathbf{1}, a_i b\mathbf{1}>$ tends to $<\mathbf{1}, ab\mathbf{1}> = 0$.

The same conclusion would be true under the hypothesis that the norm bounded family a_i converges weakly to 0 : this will be useful in subsection 9.

Knowing this, we assume that the probability law μ on the C^*-algebra $\mathcal{A}^\circ$ satisfies the property

(7.1) $\| a_i \|$ bounded, $\lim_i \mu(a_i^* a_i) = 0$ implies $\lim_i \mu(a_i b) = 0$ for every $b \in \mathcal{A}^\circ$.

(It would be sufficient to assume this for sequences : the easy proof is left to the reader). In particular, $(\mu(a^* a) = 0) \Longrightarrow (\mu(ab) = 0)$ for every b, and therefore the set of all "left negligible elements" is the same as the kernel $\mathcal{N}$ of the GNS representation. Passing to the quotient, we may assume that the GNS representation is faithful, and imbed $\mathcal{A}^\circ$ in $\mathcal{L}(\mathcal{H})$. On the other hand, the property (7.1) is preserved when we pass to the quotient.

Let us prove now that *the cyclic vector $\mathbf{1}$ is separating for the von Neumann algebra $\mathcal{A}$, the bicommutant of $\mathcal{A}^\circ$.*

PROOF. Consider $a \in \mathcal{A}$ such that $a\mathbf{1} = 0$. According to Kaplansky's theorem, there exists a norm bounded family of elements a_i of $\mathcal{A}^\circ$ that converges strongly to a, and in particular $\lim_i \| a_i b\mathbf{1} \| = 0$ for every $b \in \mathcal{A}^\circ$. Taking $b = I$ we have $\mu(a_i^* a_i) \to 0$, and the same for $c^* a_i$ ($c \in \mathcal{A}^\circ$ since left multiplications are bounded. Then we deduce from (7.1) that

$$0 = \lim_i \mu(c^* a_i b) = \lim_i <c\mathbf{1}, a_i b\mathbf{1}> = <c\mathbf{1}, ab\mathbf{1}> .$$

Since b, c are arbitrary and $\mathbf{1}$ is cyclic, we have $a = 0$.

From now on, we use the notation $\|a\|$ for the operator norm, and keep the notation $\| a \|$ for the norm $\mu(a^* a)^{1/2}$.

8 In the commutative case, a space $L^\infty(\mu)$ can be interpreted as the space of essentially bounded functions, and as a space of measures possessing bounded densities with respect to μ. In the non–commutative case, if we think of $\mathcal{A}^{\circ}$ as consisting of "continuous functions", the first interpretation of L^{∞} becomes the von Neumann algebra $\mathcal{A}$, and we are going to show that the second interpretation leads, when $\mathcal{A}$ has a cyclic and separating vector $\mathbf{1}$, to the *commutant of* $\mathcal{A}$.

A positive linear functional π on $\mathcal{A}$ may be said to have a bounded density if it is dominated by a constant c times μ. Then we have for $a, b \in \mathcal{A}$

$$(8.1) \qquad |\pi(b^*a)| \le \pi(b^*b)^{1/2}\,\pi(a^*a)^{1/2} \le c\,\|\,b\,\|\,\|\,a\,\|\,.$$

Forgetting the middle, the inequality is meaningful for a *complex* linear functional π. If it is satisfied, we say that π is a *complex measure with bounded density*. Let $\mathcal{D}$ be the dense subspace $\mathcal{A}\mathbf{1}$ of $\mathcal{H}$ (in the GNS construction, this is nothing but another name for $\mathcal{A}$ itself). Since $\mathbf{1}$ is separating we may define on $\mathcal{D}$ $p(b\mathbf{1}, a\mathbf{1}) = \pi(b^*a)$, a bounded sesquilinear form, which can be extended to $\mathcal{H}$. Given any unitary $u \in \mathcal{A}$ we have $p(ub\mathbf{1}, ua\mathbf{1}) = p(b\mathbf{1}, a\mathbf{1})$, and by continuity this can be extended to $p(uy, ux) = p(y, x)$ for $x, y \in \mathcal{H}$. With the bounded form p on $\mathcal{H} \times \mathcal{H}$ we associate the unique bounded operator α on $\mathcal{H}$ such that $p(y, x) = \,<\alpha y, x>$. Then the unitary invariance of p means that α commutes with every unitary $u \in \mathcal{A}$ *and thus belongs to* $\mathcal{A}'$. Otherwise stated, the commutant $\mathcal{A}'$ of $\mathcal{A}$ appears as a space of measures on $\mathcal{A}$, given by the formula

$$(8.2) \qquad \pi(a) = p(\mathbf{1}, a\mathbf{1}) = \,<\alpha\mathbf{1}, a\mathbf{1}>\,.$$

Conversely, any linear functional on $\mathcal{A}$ of the form (8.2) satisfies (8.1).

REMARK. The name of "measures with bounded density" for functionals of the form $\pi(a) = \mu(\alpha a)$ associated with $\alpha \in \mathcal{A}'$ is reasonable. First assume $\alpha \ge 0$. Then π is positive and for $a \in \mathcal{A}_+$ we have

$$0 \le \pi(a) = \|\,\sqrt{\alpha}\,\sqrt{a}\,\mathbf{1}\,\|^2 \le \|\,\sqrt{\alpha}\,\|^2\,\|\,\sqrt{a}\,\mathbf{1}\,\|^2 = \|\,\sqrt{\alpha}\,\|^2\,\mu(a)\,.$$

Thus π is dominated by a constant times μ. On the other hand, every element in $\mathcal{A}'$ can be written as $p + iq$ using s.a. elements, each of which can be decomposed into two positive elements of $\mathcal{A}'$ to which the preceding reasoning applies.

The symmetry between $\mathcal{A}$ and $\mathcal{A}'$ is very clear in the following lemma (which can easily be improved to show that $\mathbf{1}$ is cyclic (separating) for $\mathcal{A}$ iff it is separating (cyclic) for $\mathcal{A}'$).

LEMMA. *If the vector $\mathbf{1}$ is cyclic and separating for $\mathcal{A}$, it has the same properties with respect to $\mathcal{A}'$.*

PROOF. 1) Let $\mathcal{K}$ be the closed subspace generated by $\mathcal{A}'\mathbf{1}$. The projection on $\mathcal{K}^{\perp}$ commutes with $\mathcal{A}'$ (hence belongs to $\mathcal{A}$) and kills $\mathbf{1}$. Therefore $\mathcal{K}^{\perp} = \{0\}$ and $\mathbf{1}$ is cyclic for $\mathcal{A}'$.

2) Consider $\alpha \in \mathcal{A}'$ such that $\alpha\mathbf{1} = 0$. Then for $a \in \mathcal{A}$ we have $\alpha a\mathbf{1} = a\alpha\mathbf{1} = 0$, hence $\alpha = 0$ on the dense subspace $\mathcal{D}$, thus $\alpha = 0$ and we see that $\mathbf{1}$ is separating for $\mathcal{A}'$.

The linear Radon–Nikodym theorem

9 As we have just seen, measures absolutely continuous w.r.t. μ are naturally defined by "densities" belonging to $\mathcal{A}'$. But if we insist to have densities in $\mathcal{A}$, it is unreasonable to define the "measure with (s.a.) density m" by a formula like $\pi(a) = \mu(ma)$, since it would give a complex result for non commuting s.a. operators. We have a choice between non–linear formulas like $\pi(a) = \mu(\sqrt{m}\, a\, \sqrt{m})$ (for $m \geq 0$, leading to a positive linear functional) and linear ones like

$$(9.1) \qquad \pi(a) = \frac{1}{2}\,\mu(ma + am)\ ,$$

which give a real functional on s.a. operators if the "density" m is s.a., but usually not a positive functional for $m \geq 0$. The possibility of representing "measures" by such a formula is an important lemma for the Tomita–Takesaki theory.

However, if we want to construct a s.a. operator from a pair of s.a. operators a, m, we are not restricted to (9.1) : we can use any linear combination of the form $kam + \bar{k}ma$ where k is complex. For this reason, we put

$$(9.2) \qquad \pi_{m,k}(a) = \frac{1}{2}\,\mu(kam + \bar{k}ma)$$

(written simply as π if there is no ambiguity). Then we can state *Sakai's linear Radon–Nikodym theorem* as follows.

THEOREM. *Let φ be a positive linear functional on $\mathcal{A}$ dominated by μ. For every k such that $\Re e(k) > 0$ there exists a unique positive element $m \in \mathcal{A}$ such that $\varphi = \pi_{m,k}$, and we have $m \leq \Re e(k)^{-1}I$.*

PROOF. *Uniqueness.* Let us assume that $\mu(kam + \bar{k}ma) = 0$ for $a \in \mathcal{A}_{sa}$. Since $\Re e(k) \neq 0$ the case $a = m$ gives $\mu(m^2) = 0$. Since m is s.a. and μ is faithful this implies $m = 0$.

Existence. We may assume that $\Re e(k) = 1$. The set M of all $m \in \mathcal{A}_{sa}$ such that $0 \leq m \leq I$ is compact and convex in the normal topology. For $m \in M$ the linear functional π_m from (9.2) on $\mathcal{A}$ is normal; indeed, if a family (a_i) from the unit ball of $\mathcal{A}$ is weakly convergent to 0, the same is true for ma_i and $a_i m$, and therefore $\pi_m(a_i) \to 0$. Also, the 1–1 mapping $m \longmapsto \pi_m$ is continuous from M to $\mathcal{A}_*$: this amounts to saying that if a family (m_i) from the unit ball of $\mathcal{A}$ converges weakly to 0, then $\mu(m_i a)$ and $\mu(am_i)$ tend to 0, and this was mentioned as a remark following the lemma in subsection 7.

Let the real predual consist of those normal functionals on $\mathcal{A}$ which are real valued on $\mathcal{A}_{sa}$. It is easy to see that it is a real Banach space with dual $\mathcal{A}_{sa}$, and we may embed M in it, via the preceding 1–1 mapping, as a convex, weakly compact subset. We want to prove that $\varphi \in M$. According to the Hahn–Banach theorem, this can be reduced to the following property

$$\text{for every } a \in \mathcal{A}_{sa} \text{ such that } <M, a> \leq 1 \text{ we have } <\varphi, a> \leq 1.$$

The condition $<M, a> \leq 1$ means that $\frac{1}{2}\mu(kam + \bar{k}ma) \leq 1$ for every $m \in \mathcal{A}_{sa}$ such that $0 \leq m \leq I$. We decompose a into $a^+ - a^-$ and take for m the projection

$I_{]0,\infty[}(a)$. Then $ma = am = a^+$, and the condition implies that $\frac{1}{2}(k + \overline{k})\mu(a^+) \leq 1$. Since we assumed that $\Re(k) = 1$ we have $\mu(a^+) \leq 1$. On the other hand, φ is a positive linear functional dominated by μ, and therefore $\varphi(a) \leq \varphi(a^+) \leq \mu(a^+) \leq 1$.

The KMS condition

10   Let $\mathcal{H}$ be a Hilbert space, and let H be a "Hamiltonian" : a s.a. operator, generally unbounded and positive (or at least bounded from below) generating a unitary group $U_t = e^{itH}$ on $\mathcal{H}$, and a group of automorphisms of the C^*-algebra $\mathcal{L}(\mathcal{H})$

$$\eta_t(a) = e^{itH}\, a\, e^{-itH}\ .$$

One of the problems in statistical mechanics is the construction and the study of *equilibrium states* for this evolution. Von Neumann suggested that the natural equilibrium state at (absolute) temperature T is given by

$$\mu(a) = \mathrm{Tr}(aw) \quad \text{with } w = Ce^{-H/\kappa T}$$

where κ would be Boltzmann's constant if we had been using the correct physical notations and units. The coefficient $1/\kappa T$ is frequently denoted by β. However, this definition is meaningful only if $e^{-\beta H}$ is a trace class operator, a condition which is not generally satisfied for large quantum systems. For instance, it is satisfied by the number operator of a finite dimensional harmonic oscillator, but not by the number operator on boson Fock space.

The KMS (Kubo–Martin–Schwinger) condition is a far reaching generalization of the preceding idea, which consists in forgetting about traces, and retaining only the fact that $e^{-\beta H}$ is an analytic extension of e^{itH} to purely imaginary values of t which is well behaved w.r.t. μ. From our point of view, the most remarkable feature of the KMS property is the natural way, completely independent of its roots in statistical mechanics, in which it appears in the Tomita–Takesaki theory. It seems to be a very basic idea indeed!

Given an automorphism group (η_t) of a C^*-algebra $\mathcal{A}$, we say that $a \in \mathcal{A}$ is *entire* if the function $\eta_t a$ on the real axis is extendable as an $\mathcal{A}$-valued entire function on $\mathbb{C}$. Every strongly continuous automorphism group of a C^*-algebra has a dense set of entire elements (see subsection 11).

DEFINITION. *A law μ on the C^*-algebra $\mathcal{A}$ satisfies the KMS condition at β w.r.t. the automorphism group (η_t) if there exists a dense set of entire elements a such that we have, for every $b \in \mathcal{A}$ and t real*

$$(10.1) \qquad\qquad \mu((\eta_t\, a)\, b) = \mu(b(\eta_{t+i\beta}\, a))\ .$$

One can always assume that $\beta = 1$, replacing if necessary η_t by $\eta_{t/\beta}$ (this applies also to $\beta < 0$).

> One can prove that, whenever $e^{-\beta H}$ is a trace class operator, the state defined above satisfies the KMS property at β.

There is another version of the KMS condition, which has the advantage of applying to arbitrary elements instead of entire ones. We say that a function is *holomorphic on a closed strip* of the complex plane if it is holomorphic in the open strip and continuous on the closure.

THEOREM. *The KMS condition is equivalent to the following one : for arbitrary* $a, b \in \mathcal{A}$, *there exists a bounded function* f, *holomorphic on the horizontal unit strip* $\{0 \leq \Im m(z) \leq 1\}$, *such that*

$$(10.2) \qquad f(t + i0) = \mu(b(\eta_t a)) \quad , \quad f(t + i1) = \mu((\eta_t a) b) .$$

PROOF. Assume the KMS condition holds. We may approximate a by a sequence (a_n) of entire elements satisfying (10.1), and we set $f_n(z) = \mu(b(\eta_z a_n))$, an entire function. We prove that f_n converges uniformly on the horizontal unit strip, its limit being then the function f. Since the analytic functions $\eta_{t+z} a_n$ and $\eta_t \eta_z a_n$ (for real t) are equal for $z \in \mathbb{R}$, they are equal, and

$$| \mu(b(\eta_{t+is} a_n)) | \leq \| b \| \| \eta_{t+is} a_n \| = \| b \| \| \eta_{is} a_n \|$$

is bounded in every horizontal strip, and therefore satisfies the maximum principle on horizontal strips. On the other hand

$$| \mu(b\eta_t(a_n - a_m)) | \leq \| b \| \| a_n - a_m \|$$

and using KMS

$$| \mu(b\eta_{t+i}(a_n - a_m)) | = | \mu(\eta_t(a_n - a_m) b) | \leq \| b \| \| a_n - a_m \| .$$

Therefore f_n converges uniformly.

Conversely, assume (10.2) holds and consider an entire element a. The function $f(z) - \mu(b(\eta_z a))$ is holomorphic in the open unit strip and its limit on the real axis is 0. By the reflection principle, it is extendable to the open strip $\{-1 < \Im m(z) < 1\}$, equal to 0 on the real axis, hence equal to 0. Therefore we also have $f(t + i) = \mu(b(\eta_{t+i} a))$. On the other hand, by (10.2) we have $f(t + i) = \mu((\eta_t a) b)$, and (10.1) is established. Note that it has been proved for an arbitrary entire vector a.

We draw two consequences from the KMS condition. The first one is the *invariance* of μ. Let a be an entire element. We apply KMS with $b = I$. We have just remarked that the function $\mu_z(a)$ is bounded on horizontal strips. On the other hand, KMS implies that it is periodic with period i. An entire bounded function being constant, μ is invariant.

The second consequence is the regularity property (7.1) :

$$b_j \text{ bounded, } \lim_j \mu(b_j^* b_j) = 0 \text{ implies } \forall a \in \mathcal{A} \ \lim_j \mu(b_j a) = 0 .$$

We may assume a is an entire vector, and pass to the limit. On the other hand by KMS $\mu(b_j a) = \mu((\eta_{-i} a) b_j)$, and we are reduced to the trivial case of left multiplication by a fixed element.

11 Finally, we mention the result on entire vectors we used above. Let (U_t) be any strongly continuous group of contractions on a Banach space B. For every $b \in B$ set $b_n = c_n \int e^{-ns^2} U_s b \, ds$, where c_n is a normalization constant. Then we have

$$U_t b_n = c_n e^{-nt^2} \int_{-\infty}^{\infty} e^{2nst - s^2} U_s b \, ds$$

and it is trivial to replace t by z in this formula. On the other hand, strong continuity of the semigroup implies that for every b we have $b_n \to b$ in norm. We remark that this function is norm bounded in *vertical* strips of finite width.

§4. THE TOMITA–TAKESAKI THEORY

Nearly all papers on non–commutative integration begin with the statement "let $\mathcal{A}$ be a von Neumann algebra with a faithful normal state..." In the language of GNS representations, the statement becomes " Let $\mathcal{A}$ be a von Neumann algebra of operators on a Hilbert space $\mathcal{H}$, with a cyclic and separating vector...". This is also the starting point for the Tomita–Takesaki theorem.

> Though the theory is extremely beautiful, it does not apply to our main example, that of the vacuum state on the CCR algebra acting on simple Fock space, which isn't a separating state. It applies to the vacuum state of non–Fock representations.

We did our best in the following sections to justify this starting point : μ being a law on a C^*-algebra $\mathcal{A}^\circ$, we know from the GNS theorem that it can be interpreted as the law associated with a cyclic vector $\mathbf{1}$ in a Hilbert space $\mathcal{H}$ on which $\mathcal{A}^\circ$ operates. Then we showed that a weak regularity assumption on μ allows to reduce to the case $\mathbf{1}$ is also separating for the vNa $\mathcal{A}$ generated by $\mathcal{A}^\circ$.

We follow the approach of Rieffel–van Daele [RvD1], which avoids almost completely the techniques of unbounded closable operators. The beginning of their paper has already been given in Appendix 3. It should be mentioned that the same methods lead to a simple proof ([RvD2]) of the commutation theorem $(\mathcal{A} \otimes \mathcal{B})' = \mathcal{A}' \otimes \mathcal{B}'$ for tensor products of von Neumann algebras. There is also a Tomita–Takesaki theorem for *weights* on vN algebras, which are the analogue of σ–finite measures. This result does not concern us here.

Before we start , let us describe the aim of this section. In classical integration theory, bounded measures absolutely continuous w.r.t. the law μ on $L^2(\mu)$ are represented as scalar products

$$\pi(f) = \int \overline{p}(x) f(x) \mu(dx) .$$

for $p \in L^1$, and in particular measures with bounded densities are represented by means of elements of L^∞. Thus we have an antilinear 1–1 mapping $p \longmapsto \overline{p}$ from functions to measures with bounded densities. In the non commutative setup, this becomes an antilinear 1–1 mapping between the vNa $\mathcal{A}$ and its commutant $\mathcal{A}'$, called the *modular conjugation* J.

This conjugation then gives a general meaning to the preceding formula, at least in the case of a positive measure π absolutely continuous w.r.t. μ (translation : a positive element of the predual of $\mathcal{A}$). Its positive density in L^1 is interpreted as a product $\overline{q}q$ for some $q \in L^2$, i.e. a scalar product $\pi(f) = <q, fq>$. A similar representation holds in the non commutative case, but we will only state this result, which is too long to prove.

Finally, when one has defined conveniently L^1, L^2 and L^∞, the way to L^p is open by interpolation and duality. We do not include the theory here.

The main operators

We will use the rotations and results of the Appendix on "two events", beginning with the definition of the real subspaces of $\mathcal{H}$ to which it will be applied.

1 We consider $\mathcal{H}$ as a real Hilbert space, with the scalar product $(x, y) = \Re <x\ y>$. We introduce the real subspaces

$$A_0 = \mathcal{A}_{sa}\mathbf{1} \ , \quad A_0' = \mathcal{A}_{sa}'\mathbf{1} \ , \quad B_0 = iA_0 \ , \quad B_0' = iB_0 \ .$$

We denote the corresponding closures with the same letters, omitting the index $_0$. Since $\mathbf{1}$ is cyclic for $\mathcal{A}$, $A_0 + B_0$ (and therefore $A + B$) is dense in $\mathcal{H}$. Similarly, $A' + B'$ is dense.

The following lemma translates a little of the algebra into a geometric property.

LEMMA. *We have $A^\perp = B'$, and similarly $(A')^\perp = B$.*

PROOF. Consider $a \in \mathcal{A}_{sa}$, $\alpha \in \mathcal{A}_{sa}'$. Since the product of two commuting s.a. operators is s.a., $<a\mathbf{1}, i\alpha\mathbf{1}>$ is purely imaginary, and this is translated as the orthogonality of A and B'.

To say that the orthogonal of A is B' and no larger amounts to saying that the only $x \in \mathcal{H}$ orthogonal to A and B' is $x = 0$. As often, this will be proved by a matrix trick.

We have $a \in \mathcal{A}$ act on $\mathcal{H} \oplus \mathcal{H}$ as a matrix $\hat{a} = \begin{pmatrix} a & 0 \\ 0 & a \end{pmatrix}$. We put also $\hat{x} = \begin{pmatrix} \mathbf{1} \\ x \end{pmatrix}$. Let j be the (complex) projection on the subspace K generated by all vectors $\hat{a}\hat{x}$ ($a \in \mathcal{A}$); since it commutes with each $\hat{a}$, it can be written as $j = \begin{pmatrix} \alpha & \gamma \\ \gamma^* & \beta \end{pmatrix}$ with $\alpha, \beta \in \mathcal{A}_{sa}'$ (more precisely, $0 \leq \alpha \leq 1$), and $\gamma \in \mathcal{A}'$. From the definition of x we deduce three simple properties.

1) *for $a \in \mathcal{A}_{sa}$, $<x, a\mathbf{1}>$ is purely imaginary.* Otherwise stated

$$<x, a\mathbf{1}> = \overline{<a\mathbf{1}, x>} = -<a\mathbf{1}, x> = -<\mathbf{1}, ax> \ .$$

The equality between the extreme members is a $\mathbb{C}$–linear property, and therefore extends from $\mathcal{A}_{sa}$ to $\mathcal{A}$. Then it is interpreted as the complex orthogonality of $\begin{pmatrix} a & 0 \\ 0 & a \end{pmatrix}\begin{pmatrix} \mathbf{1} \\ x \end{pmatrix}$ and $\begin{pmatrix} x \\ \mathbf{1} \end{pmatrix}$, or as the property $j\begin{pmatrix} x \\ \mathbf{1} \end{pmatrix} = 0$. In particular

(i) $\qquad\qquad \alpha x + \gamma\mathbf{1} = 0 \ , \quad \text{implying} \ <x, \alpha x> = -<x, \gamma\mathbf{1}> .$

2) *For $\varepsilon \in \mathcal{A}'_{sa}$, $<x, \varepsilon 1>$ is real.* As above, we deduce

$$< x, \varepsilon 1 > = \overline{< \varepsilon 1, x >} = < \varepsilon 1, x > = < 1, \varepsilon x > .$$

The equality between the extreme members is a $\mathbb{C}$–linear property, and therefore extends to $\varepsilon \in \mathcal{A}'$. Taking then $\varepsilon = \gamma$ we have

(*ii*) $-< x, \gamma 1 > = -< 1, \gamma x > .$

3) *Since $I \in \mathcal{A}$, $\hat{x}$ belongs to the subspace K, therefore $j\hat{x} = \hat{x}$.* In particular

(*iii*) $\alpha 1 + \gamma x = 1$,   implying $-< 1, \gamma x > = -< 1, 1 - \alpha 1 >.$

The relations *(i)–(iii)* taken together imply $< x, \alpha x > = -< 1, 1 - \alpha 1 >$, an equality between a positive and a negative number. Therefore $< x, \alpha x > = 0$, implying $\sqrt{\alpha} x = 0$, then $\alpha x = 0$. Similarly the second relation implies $(I - \alpha) 1 = 0$. Since 1 is separating for $\mathcal{A}'$ we have $\alpha = I$, and finally $x = 0$.

2   In this subsection we define the modular symmetry and the modular group. We refer to Appendix 3 for some details (note that we use capital letters here for the main operators)

The real subspaces A, B are not in "general position", since $A \cap B^\perp = A \cap A'$ contains $c1$ for every c belonging to the center of A, and similarly $A^\perp \cap B$ contains $ic1$. We have

(2.1) $A \cap B = \{0\} = A^\perp \cap B^\perp .$

The projections on A, B being denoted by P, Q, we put

$$S = P - Q \;\; ; \;\; C = P + Q - I .$$

These operators are symmetric and bounded, such that $S^2 + C^2 = 1$, $SC + CS = 0$, and their spectra are contained in $[-1, 1]$. The operator S being injective, we may define $J = \operatorname{sgn}(S)$, which will be called the *modular symmetry*. We also put $D = |S|$, which commutes with P, Q, S, C, J. On the other hand, we have

(2.2) $JP = (I - Q)J , \;\;\; JQ = (I - P)J .$

The operator of multiplication by i satisfies $iP = Qi$, so that C and D commute with i, *i.e.* are complex linear) while S and J *anticommute with i* (are conjugate linear).

3   The operators $I \pm C$ are injective too, and we define the *modular operator* Δ by the following formula

(3.1) $$\Delta = \frac{1 - C}{1 + C} = \int_{-1}^{1} \frac{1 - \lambda}{1 + \lambda} \, dF_\lambda .$$

It is selfadjoint, positive, injective, generally unbounded. Its square root occurs often and will be denoted by ε. The domain of Δ contains the range of $I + C$. For instance, we have $P1 = 1$, $Q1 = 0$, hence $(I + C)1 = 1$ and $\Delta 1 = 1$.

The *modular group* is the unitary group

(3.2) $$\Delta^{it} = (I - C)^{it}(I + C)^{-it} .$$

Note that the right side does not involve unbounded operators. These operators commute with C and D, but they have the fundamental property of also *commuting with* J, hence also with $JD = S$, with P and Q. Otherwise stated, they preserve the two subspaces A and B, and define automorphisms of the whole geometric situation. On the other hand, the really deep result of the Tomita theory is the fact that they preserve, not only the closed subspaces A and B, but also A_0, B_0, A_0', B_0'.

PROOF. It suffices to prove that $J(I \pm C)^{it} J = (I \mp C)^{-it}$. Introducing the spectral decompositions U_λ and V_λ of $I - C$ and $I + C$ respectively, this amounts to

$$ J(\int \lambda^{it} dU_\lambda) J = \int \lambda^{-it} dV_\lambda . $$

Since J and C anticommute, we have $J(I-C) = (I+C) J$, and therefore $JU_\lambda J = V_\lambda$. The result then follows from the fact that J is conjugate linear.

If we exchange the vNa $\mathcal{A}$ with its commutant $\mathcal{A}'$, the space A gets exchanged with $A' = B^\perp$ (this is almost the only place where we need the precise result of subsection 1). Thus $P' = I - Q$, $Q' = I - P$, $S' = S$, $J' = J$, $C' = -C$, and $\Delta' = \Delta^{-1}$. This remark will be used at the very end of the proof.

Interpretation of the adjoint

4 To simplify notation, we often identify $a \in \mathcal{A}$ with the vector $a1 \in \mathcal{H}$. Then the space $A_0 + B_0$ is identified with the vNa $\mathcal{A}$ itself. Every $x \in \mathcal{H}$ can be interpreted as a (generally unbounded) operator $Op_x : a1 \longmapsto ax$ with domain $\mathcal{D} = \mathcal{A}'1$.

The following result is very important, because it sets a bridge between the preceding (i.e. Rieffel and van Daele's) definition of J and ε, and the classical presentation of the Tomita–Takesaki theory, in which $J\varepsilon$ would appear as the polar decomposition of the (unbounded, closable) operator $a + ib \longmapsto a - ib$ for $a, b \in \mathcal{A}_{sa}$, which describes the operation $*$.

THEOREM For $a, b \in \mathcal{A}_{sa}$, the vector $a + ib$ belongs to $\mathrm{Dom}(\varepsilon)$ and we have

(4.1) $$ J\varepsilon(a + ib) = a - ib \qquad (\varepsilon = \Delta^{1/2}) . $$

PROOF. We remark that, for $a \in \mathcal{A}_{sa}$

$$ 2a = 2Pa = (I + C + S)a = (I + C + (I - C^2)^{1/2} J) a = (I + C)^{1/2} x $$

with $x = (I + C)^{1/2} a + (I - C)^{1/2} Ja$, so that a belongs to the range of $(I + C)^{1/2}$, and therefore to the domain of $\varepsilon = (I - C)^{1/2}/(I + C)^{1/2}$. Then we have

$$ 2\varepsilon a = (I - C^2)^{1/2} a + (I - C) Ja = Da + (I - S) Ja $$

and applying J we have $2J\varepsilon a = JDa + (I + C) a = 2a$.

Since C, ε are complex linear and J conjugate linear, this implies $J\varepsilon ib = -b$ for $b \in \mathcal{A}_{sa}$, and (4.1) follows. ∎

The modular property

5   The second impertant result is the so called "modular property" of the unitary group Δ^{it} relative to the real subspace A. It will be interpreted later as a KMS condition. Recall that a function holomorphic in a closed strip of the complex plane is a function holomorphic in the open strip and continuous in the closure. For clarity, we do not identify a and $a1$, etc. in the statement of the theorem (we do in the proof).

THEOREM *Given arbitrary $a, b \in A_{sa}$, there exists a bounded holomorphic function $f(z)$ in the closed strip $0 \leq \Im m\, z \leq 1$, and such that for t real*

$$f(t + i0) = <b1, \Delta^{-it}a1>, \qquad f(t + i1) = <\Delta^{-it}a1, b1>$$

PROOF. We try to extend the relation

$$\Delta^{-it} = (I - C)^{-it}(I + C)^{it}$$

to complex values of t. Since $I \pm C$ is a positive operator, $(I - C)^{-iz}$ can be defined as a bounded operator for $\Im m\, z \geq 0$ and $(I + C)^{iz}$ for $\Im m\, z \leq 0$ there is no global extension outside the real axis. However, we saw at the beginning of the preceding proof that $a \in A$ can be expressed as $(I + C)^{1/2}u$. Then we put

$$\Delta^{-iz}a = F(z) = (I - C)^{-iz}(I + C)^{iz+1/2}u\,,$$

which is well defined, holomorphic and bounded in the closed strip $0 \leq \Im m\, z \leq 1/2$. For $z = t + i0$ we have $<b, F(t)> = <b, \Delta^{-it}a>$. We are going to prove that $<b, F(t + 1/2)>$ is real. By the Schwarz reflection principle, this function will then be extendable to the closed strip $0 \leq \Im m\, z \leq 1$, and assume conjugate values at corresponding points $t + i0, t + i1$ of the boundary.

We start from the relation

$$F(t + i/2) = (I - C)^{-it+1/2}(I + C)^{it}u = \Delta^{-it}(I - C)^{1/2}u\,.$$

On the other hand, $a = (I + C)^{1/2}u$, hence $(I - C)^{1/2}u = Ja$, and

$$<b, F(t + i/2)> = <b, \Delta^{-it}Ja> = <b, Jc> \quad \text{with } c = \Delta^{-it}a \in A.$$

Now Jc belongs to $(iA)^{\perp}$ (the real orthogonal space of iA), meaning that $<ib, Jc>$ is purely imaginary, or $<b, Jc>$ is real. □

Rieffel and van Daele prove that the modular group is the only strongly continuous unitary group which leaves A invariant and satisfies the above property.

Using the linear Radon–Nikodym theorem

6   All the preceding discussion was about the closed real subspaces A, A'. We now discuss the von Neumann algabras $\mathcal{A}$, $\mathcal{A}'$ themselves, and proceed to the deeper results, translating first the Sakai linear R–N theorem, whose statement we recall. Given any k such that $\Re e\, k > 0$, and any $\beta \in \mathcal{A}'_{sa}$ (β is a "measure", whence the greek letter) there exists a unique $b \in \mathcal{A}_{sa}$ (a "function") suc   that for $a \in \mathcal{A}$

(6.1) $<\beta1, a1> = <1, (kab + \bar{k}ba)1>$

(we keep writing explicitly 1 here and below, for clarity). For each k, this extends to a conjugate linear mapping from $\mathcal{A}'$ to $\mathcal{A}$, which for the moment has no name except $\beta \longmapsto b$. We are going to express (6.1) with the help of the operators C and S, as follows

LEMMA. For $k = 1/2$, we have

$$(6.2) \qquad P\beta 1 = b1 , \qquad Q\beta 1 = 0 \quad \text{hence } S\beta 1 = b1.$$

and in the general case

$$(6.3) \qquad S\beta S = k(I + C)\,b\,(I - C) + \overline{k}(I - C)\,b\,(I + C)$$

PROOF. The case $k = 1/2$ goes as follows :

$$< \beta 1 , a1 > = \frac{1}{2}< 1 , (ab + ba)\,1 > =$$
$$\frac{1}{2}(< a1 , b1 > + < b1 , a1 >) = \Re e < a1 , b1 > = (a1, b1) .$$

Since a is arbitrary, this means $b1 = P\beta 1$. On the other hand, $\mathcal{A}'_{sa}1$ is orthogonal to $B = iA$, hence $Q\beta 1 = 0$.

We now consider a general k. We replace in (6.1) a by ca ($a, b, c \in \mathcal{A}_{sa}$, but $ca \in \mathcal{A}$ only), so that

$$< c\beta 1 , a1 > = k< 1 , cab1 > + \overline{k}< 1 , bca1 > = k< c1 , ab1 > + \overline{k}< cb1 , a1 > .$$

Let λ, μ be arbitrary in $\mathcal{A}'_{sa}$, and ℓ, m be the corresponding elements of $\mathcal{A}$ through the preceding construction with $k = 1/2$. In the preceding formula we replace c by ℓ and a by m

$$< \ell\beta 1 , m1 > = k< \ell 1 , mb1 > + \overline{k}< \ell b1 , m1 > .$$

Since $S\lambda 1 = \ell 1$, $S\mu 1 = m1$, and ℓ commutes with β, this is rewritten

$$< \beta S\lambda 1 , S\mu 1 > = k< S\lambda 1 , mb1 > + \overline{k}< \ell b1 , S\mu 1 > .$$

Since S is real self adjoint and complex conjugate linear, it satisfies the relation $< u, Sv > = < Su, v >^{-}$. Therefore we can move the second S in the left member to the "bra" side, the result being $< S\beta S\lambda 1 , \mu 1 >^{-}$.

In the right member, we use the result from subsection 4 :

$$mb1 = J\epsilon(mb)^{*}1 = J\epsilon bm1 = J\epsilon bS\mu 1 , \quad \text{similarly} \quad \ell b1 = J\epsilon bS\lambda 1.$$

Then $< S\lambda 1 , mb1 > = < S\lambda 1 , J\epsilon bS\mu 1 > = < J\epsilon S\lambda 1 , bS\mu 1 >^{-}$, $J\epsilon = \epsilon J$ being real self-adjoint and complex conjugate linear. On the other hand, $J\epsilon S = I - C$. Doing the same on the second term we get

$$< S\beta S\lambda 1 , \mu 1 >^{-} = k< (I - C)\,\lambda 1 , bS\mu 1 >^{-} + \overline{k}< bS\lambda 1 , (I - C)\,\mu 1 >^{-} .$$

On the other hand, $\lambda 1$, $\mu 1$ belong to the kernel of Q, and the relations $S = P - Q$, $I + C = P + Q$ allow us to replace S by $I + C$. Taking out the conjugation sign, and using the fact that C and $S\beta S$ are complex selfadjoint, we get

$$< \lambda 1, \, S\beta S\mu 1 > \, = \overline{k} < \lambda 1, \, (I - C)\, b(I + C)\, \mu 1 > + \, k < \lambda 1, \, (I + C)\, b(I - C)\, \mu 1 > .$$

This is complex linear in μ, and therefore can be extended from $\mathcal{A}'_{sa}$ to $\mathcal{A}'$, and similarly in λ. Since 1 is cyclic for $\mathcal{A}'$, we can write this as the operator relation (6.1).

7 We now make explicit the dependence on k, putting $k = e^{i\theta/2}$ — the condition $\mathfrak{Re}(k) > 0$ then becomes $|\theta| < \pi$. We keep $\beta \in \mathcal{A}'_{sa}$ fixed, and denote by b_θ the element of $\mathcal{A}_{sa}$ corresponding to β for the given value of θ. It is really amazing that b_θ can be explicitly computed by the integral formula

$$(7.1) \qquad\qquad b_\theta = \int_{\mathbb{R}} \Delta^{it} J\beta J \Delta^{-it}\, h_\theta(t)\, dt \, , \qquad h_\theta(t) = \frac{e^{-\theta t}}{2\mathrm{Ch}(\pi t)} \; .$$

PROOF. We consider two analytic vectors $x, y \in \mathcal{H}$ for the unitary group Δ^{it}, as constructed in §2, subs.11. Then $\Delta^z x$, $\Delta^z y$ are entire functions of z, bounded in every vertical strip of finite width, and the same is true for the scalar function

$$f(z) = < \Delta^{\overline{z}} x, \, DbD\Delta^{-z} y > .$$

We recall that $D = |S|$ commutes with C and S, and $S = JD$ (see Appendix 3). We have $\quad \Delta^{1/2} D = I - C$, $\Delta^{-1/2} D = I + C$. Then

$$f(\tfrac{1}{2} + it) = < x, \, \Delta^{-it}\Delta^{1/2} x, \, DbD\Delta^{-1/2}\Delta^{-it} y > \, = \, < x, \, \Delta^{it}(I - C)\, b(I + C)\, \Delta^{-it} y > .$$

Similarly

$$f(-\tfrac{1}{2} + it) = < x, \, \Delta^{it}(I + C)\, b(I - C)\, \Delta^{-it} y > .$$

Then using (6.3) we get the basic formula ($k = e^{i\theta/2}$)

$$k f(\tfrac{1}{2} + it) + \overline{k} f(-\tfrac{1}{2} + it) = < x, \, \Delta^{it} S\beta S \Delta^{-it} y > .$$

We now apply the following Cauchy integral formula for a bounded holomorphic function $g(z)$ in the closed vertical strip $|\mathfrak{Re}(z)| \leq 1/2$

$$g(0) = \int \left(g(\tfrac{1}{2} + it) + g(-\tfrac{1}{2} + it) \right) \frac{dt}{2\,\mathrm{Ch}(\pi t)} \; .$$

We replace $g(z)$ by $e^{iz\theta} f(z)$ with $\theta < \pi$, and get

$$f(0) = \int \left(k f(\tfrac{1}{2} + it) + \overline{k} f(-\tfrac{1}{2} + it) \right) h_\theta(t)\, dt \, ,$$

or explicitly

$$< x. \, DbDy > \, = \, < x, \, \left(\int \Delta^{it} S\beta S \Delta^{-it}\, h_\theta(t)\, dt \right) y > .$$

This equality can be extended by density to arbitrary x, y. Since D is selfadjoint and commutes with J, Δ we may write

$$< Dx, bDy > \; = \; < Dx , \Big(\int \Delta^{it} J\beta J\Delta^{-it} \, h_\theta(t) \, dt \Big) Dy > ,$$

and (7.1) follows, the range of D being dense.

Before we proceed to the main results, we need one more remark :

(7.2)   For $\beta \in \mathcal{A}'$ we have $\Delta^{it} J\beta J\Delta^{-it} \in \mathcal{A}$.

It suffices to prove this operator commutes with every $\alpha \in \mathcal{A}'$. We put $g(t) = \; < x, \; [\Delta^{it} J\beta J\Delta^{-it}, \alpha] y >$. From (7.1) we deduce $\int g(t) \, h_\theta(t) \, dt = \; < x, \; [b_\theta, \alpha] y > \; = 0$. Since θ is arbitrary one may deduce that $g(t) = 0$ a.e., and then everywhere.

The main theorems

8   We first introduce some notation. The modular symmetry J operates on vectors on $\mathcal{H}$; given a bounded operator $a \in \mathcal{L}(\mathcal{H})$ we define $ja = JaJ$. This is a conjugation on $\mathcal{L}(\mathcal{H})$: it is conjugate linear, but doesn't reverse the order of products. Similarly, we extend the modular group as a group of automorphisms on $\mathcal{L}(\mathcal{H})$, putting $\delta_t a = \Delta^{it} a \Delta^{-it}$. The fact that Δ^{it} leaves $\mathbf{1}$ fixed implies that the law μ is invariant (this is made more precise by the KMS condition in theorem 3).

THEOREM 1. *The conjugation j exchanges $\mathcal{A}$ and $\mathcal{A}'$.*

PROOF. If in (7.2) we take $t = 0$, we find that $j(\mathcal{A}') \subset \mathcal{A}$. Exchanging the roles of $\mathcal{A}$ and $\mathcal{A}'$ doesn't change the modular symmetry J, and therefore $j(\mathcal{A}) \subset \mathcal{A}'$. Then we have $\mathcal{A} = j(j(\mathcal{A})) \subset j(\mathcal{A}')$ and equality follows.

COMMENT. We had seen previously that J maps the closed space A onto A'. Now we have the much more precise result that it maps A_0 onto A_0', without taking closures. Also, ja is the unique $\alpha \in \mathcal{A}'$ such that $\alpha\mathbf{1} = Ja\mathbf{1}$.

THEOREM 2. *The group (δ_t) preserves $\mathcal{A}$ and $\mathcal{A}'$.*

PROOF. The fact that $\mathcal{A}$ is preserved follows from (7.2) and theorem 1, since $J\beta J$ is an arbitrary element of $\mathcal{A}$. Then we exchange the roles of $\mathcal{A}$ and $\mathcal{A}'$.

The statement in the next theorem is not exactly the same as that of the KMS condition (10.2) in the preceding section : the boundary lines of the strip are exchanged.

THEOREM 3. *Given arbitrary $a, b \in \mathcal{A}$, there exists a bounded bolomorphic function $f(z)$ in the closed strip $0 \leq \Im m(z) \leq 1$ such that*

(8.1) $f(t + i0) = \mu((\delta_t a) \, b) \, , \quad f(t + i1) = \mu(b\delta_t a) \, .$

PROOF. It is sufficient to prove this result when a, b are selfadjoint. Then the result follows from the modular property on vectors, if we remark that ($\mathbf{1}$ being invariant by $\Delta^{\pm it}$)

$$\mu(b\delta_t a) = \; < \mathbf{1}, \; b\Delta^{it}a\Delta^{-it}\mathbf{1} > \; = \; < b\mathbf{1}, \Delta^{it}a\mathbf{1} >$$
$$\mu((\delta_t a) \, b) = \mu(a\delta_{-t}b) = \; < \mathbf{1}, \; a\Delta^{-it}b\Delta^{it}\mathbf{1} > \; = \; < \Delta^{it}a\mathbf{1}, b\mathbf{1} > .$$

One can prove that this property characterizes the automorphism group (δ_t).

Additional results

9   The Tomita–Takesaki theorem is not an end in itself, but the gateway to new developments. The end of the academic year also put an end to the author's efforts to give examples and self–contained proofs, and the work could not be resumed. We only give a sketch without proofs of important results due to Araki, Connes, Haagerup. They are available in book form in [BrR1].

Everything we are doing depends on the choice of the faithful normal state μ (or the separating cyclic vector **1** in the GNS Hilbert space $\mathcal{H}$). It turns out at the end, as in the classical case, that replacing μ by an equivalent state produces leaves the situation unchanged up to well controlled isomorphisms.

We take the bold step of representing by $\overline{x}$ the vector Jx for $x \in \mathcal{H}$, and by $\overline{a}$ the operator $j(c) = JcJ$ for $c \in \mathcal{L}(\mathcal{H})$, returning to the "$J$" notation whenever clarity demands it. Note that $\overline{ab} = \overline{b}\,\overline{a}$ for $a, b \in \mathcal{A}$. For $x \in \mathcal{H}$, $c \in \mathcal{L}(\mathcal{H})$, we have $\overline{(cx)} = \overline{c}\,\overline{x}$; indeed, $(JcJ)Jx = J(cx)$. Similarly, for $c, d \in \mathcal{L}(\mathcal{H})$, $\overline{(cd)} = \overline{c}\,\overline{d}$ and $\overline{\overline{cd}} = \overline{c}\,d$.

The *positive cone* $\mathcal{P}$ in the Hilbert space $\mathcal{H}$ is by definition the closure of the set of vectors $a\overline{a}\,\mathbf{1}$ for $a \in \mathcal{A}$. Since **1** is invariant by J, the same is true for every element of $\mathcal{P}$ (all elements of $\mathcal{P}$ are "real"). Also, for $a, b \in \mathcal{A}$ we have $b\overline{b}\,a\overline{a} = (ba)\overline{(ba)}$, therefore $\mathcal{P}$ is stable under $b\overline{b}$.

We now give a list of properties of the positive cone

1) $\mathcal{P}$ is the closure of $\Delta^{1/4}(\mathcal{A}_+\mathbf{1})$, and therefore is a convex cone.

2) $\mathcal{P}$ is stable under the modular group Δ^{it}.

3) $\mathcal{P}$ is pointed, i.e. $(-\mathcal{P}) \cap \mathcal{P} = \{0\}$, and is self–dual (i.e. a form $<x, \cdot>$ is positive on $\mathcal{P}$ if and only if $x \in \mathcal{P}$).

4) Every "real" vector x (i.e. $Jx = x$) is a difference of two elements of $\mathcal{P}$, which can be uniquely chosen so as to be orthogonal.

Now, we describe what happens when we change states, assuming first the new state is a pure state $\nu(a) = <\omega, a\omega>$ with $\omega \in \mathcal{P}$ (we will see later that this is not a restriction).

5) The state ν is faithful if and only if ω is a cyclic vector for $\mathcal{A}$ operating on $\mathcal{H}$.

6) Assuming this is true, we have $J_\omega = J$, $\mathcal{P}_\omega = \mathcal{P}$.

Finally, we give the main result, which describes all states on $\mathcal{A}$ which are absolutely continuous with respect to μ.

THEOREM. *Every positive normal linear functional φ on $\mathcal{A}$ can be uniquely represented as*

$$\varphi(a) = <\omega, a\omega> \quad \text{with } \omega \in \mathcal{P}.$$

Formally, this says that every positive element of L^1 is something like the square of a positive element of L^2.

10   EXAMPLES. 1) The trivial case from the point of view of the T–T theory is that of a tracial state μ. The tracial property can be written

$$< a\mathbf{1}, b\mathbf{1} > = \mu(a^*b) = \mu(ba^*) = \overline{< b\mathbf{1}, a\mathbf{1} >}.$$

Then property (8.1) holds with $\delta_t = I$, and taking for granted the uniqueness property mentioned after Theorem 3, we see that the modular group is trivial. According to (4.1), since $\varepsilon = I$ the modular symmetry J must be given by

$$Ja1 = a^*1 \quad \text{for } a \in \mathcal{A}.$$

Indeed, this is a well defined operator on $\mathcal{A}1$, and we have $< Ja1, Jb1 > \,=\, < b1, a1 >$ because the state is tracial, so that J extends to a well defined conjugation on $\mathcal{H}$. To check that $\bar{a} = JaJ$ belongs to $\mathcal{A}'$, it is sufficient to prove that for arbitrary $b, c, d \in \mathcal{A}$,

$$< c1 \,, \bar{a}\,b\,d1 > \,=\, < c1 \,, b\bar{a}\,d1 > .$$

On the left hand side we replace $\bar{a}\,bd1 = JaJbd1$ by Jad^*b^*1 and then by bda^*1. Similarly on the right hand side we replace $b\bar{a}d1$ by bda^*1, and the result is the same.

2) The second case which can be more or less explicitly handled is that of $\mathcal{A} = \mathcal{L}(\mathcal{K})$ with a state of the form $\mu(a) = \text{Tr}(aw)$. Then w is faithful if and only if w is injective. The modular group is given by $\delta_t(a) = w^{it}aw^{-it}$, because this group of automorphisms satisfies the KMS condition of Theorem 3. Indeed, if we put for $a, b \in \mathcal{A}$

$$f(z) = \text{Tr}(w\,w^{iz}aw^{-iz}b)$$

it is easy to see that $f(z)$ is holomorphic and bounded in the unit horizontal strip, with $f(t + i0) = \mu(\delta_t(a)\,b)$ and $f(t + i1) = \mu(b\delta_t(a))$.

3) Let us mention a striking application of the T–T theory, due to Takesaki : given a von Neumann algebra $\mathcal{A}$ with a faithful normal law μ and a von Neumann subalgebra $\mathcal{B}$, a conditional expectation relative to $\mathcal{B}$ exists if and only if $\mathcal{B}$ is stable by the modular automorphism group of $\mathcal{A}$ w.r.t. μ. On the subject of conditional expectations, we advise the reader to consult the two papers [AcC] by Accardi and Cecchini given in the references.

These lectures on von Neumann algebras stopped here, just when things started to be really interesting, and no chance was given to resume work and discuss L^p spaces, etc.. Unfortunately, most results are still scattered in journals, and no exposition for probabilists seems to exist.

Appendix 5

Local Times and Fock Space

This Appendix is devoted to applications of Fock space to the theory of local times. We first present *Dynkin's formula*, ([Dyn2]) which shows how expectations relative to the stochastic process of local times L_∞^x of a symmetric Markov process, indexed by the points of the state space, can be computed by means of an auxiliary Gaussian process whose covariance is the potential density of the Markov process. Though this theorem does not mention explicitly Fock space, it was suggested to Dynkin by Symanzik's program for the construction of interacting quantum fields. Then we mention without proof a remarkable application to the continuity of local times, discovered by Marcus and Rosen. We continue with the "supersymmetric" version of Dynkin's theorem, due to Le Jan, and mention (again without proof) its application to the self-intersection theory of two-dimensional Brownian motion. Thus local times and self-intersection form the probabilistic background to present a variety of results on Fock space (symmetric, antisymmetric, and mixed) which might belong as well to Chapter V. There is an incredible amount of literature on the antisymmetric case, of which I have read very little : so please consider this chapter as an mere introduction to the subject.

Some parts of this Appendix depend on the theory of Markov processes. As they concern mostly the probabilistic motivations, and we presume our reader is interested principally in the non-commutative methods, we have chosen to omit technical details altogether. Everything can be found in the original papers, and the beautiful Marcus-Rosen article also contains information on Markov processes for non-specialists.

§1. Dynkin's formula

Symmetric Markov processes

1 Let E be a state space with a σ–finite measure η. We consider on this space a transient (sub)Markov semigroup (P_t) whose potential kernel G has a potential density $g(x, y)$ with respect to η. Explicitly, for $f \geq 0$ on E

$$(1.1) \qquad \int_0^\infty \int_E P_t(x, dy) f(y) dt = \int G(x, dy) f(y) = \int g(x, y) f(y) \eta(dy) .$$

Then $g(\cdot, y)$ may be chosen to be excessive (= positive superharmonic) for every y, and we may define potentials of positive measures, $G\mu(x) = \int g(x, y) \mu(dy)$. We are mostly interested in the case of a *symmetric* density g. The case of non-transient processes like low-dimensional Brownian motion can be reduced to the transient case, replacing the semigroup (P_t) by $e^{-ct}P_t$ (which amounts to killing the process at an independent exponential time).

On the sample space Ω of all right continuous mappings from $\mathbb{R}_+$ to E with lifetime ζ, and left limits in E up to the lifetime, we define the measures $\mathbb{P}^{\theta}$, under which (X_t) (the co-ordinate process) is Markov with semigroup (P_t) and initial measure θ. This requires some assumptions on E and (P_t), which we do not care to describe. When $\theta = \varepsilon_x$, one writes simply $\mathbb{P}^x, \mathbb{E}^x$, and for every positive r.v. h on Ω, $\mathbb{E}^{\bullet}[h]$ is a function on E.

We will also need to use "conditioned" measures, $\mathbb{P}^{\theta/k}$ where k is an excessive function; the corresponding initial measure is θ, and the semigroup is replaced by

$$(1.2) \qquad P_t^{/k}(x, dy) = \frac{P_t(x, dy)\, k(y)}{k(x)} \quad .$$

The corresponding potential $G^{/k}(x, dy)$ is $G(x, dy)\, k(y)/k(x)$, and the potential density w.r.t. η has the same form[1]. It would take us too far to explain how conditioning establishes a symmetry between initial and final specifications on the process, but this will be visible on the last formula (1.9).

A *random field* is a linear mapping $f \longmapsto \Phi(f)$ from some class of functions to random variables. We shall be interested here in the *occupation field*, which maps f to $A(f)_{\infty} = \int_0^{\infty} f(X_s)\, ds$ — this is always meaningful for positive functions, and transience means that it is meaningful for bounded signed functions of suitably restricted support.

Let f be a function such that $|f|$ has a bounded potential, let us put $A = A(f)$ and define, for n a positive integer,

$$(1.3) \qquad h^{(n)} = \mathbb{E}^{\bullet}[A_{\infty}^n] \quad .$$

In particular, $h = h^{(1)}$ is the potential $\mathbb{E}^{\bullet}[A_{\infty}] = Gf$. Then we have

$$h^{(n)} = \mathbb{E}^{\bullet}\Big[\int_0^{\infty} -d(A_{\infty} - A_s)^n\Big] = \mathbb{E}^{\bullet}\Big[\int_0^{\infty} n(A_{\infty} - A_s)^{n-1}\, dA_s\Big]$$

$$= \mathbb{E}^{\bullet}\Big[\int_0^{\infty} n\,\mathbb{E}\big[(A_{\infty} - A_s)^{n-1}\,|\,\mathcal{F}_s\big]\, dA_s\Big] = \mathbb{E}^{\bullet}\Big[\int_0^{\infty} n\,h^{(n-1)}(X_s)\, dA_s\Big]$$

$$= G(h^{(n-1)}f) \quad .$$

An easy induction then gives the formula

$$(1.4) \qquad h^{(n)}(x) = n! \int g(x, y_1)\, f(y_1)\, \eta(dy_1)\, g(y_1, y_2) \ldots g(y_{n-1}, y_n)\, f(y_n)\, \eta(dy_n) \quad .$$

We put $\mu = f \cdot \eta$ and integrate w.r.t. an initial measure θ, adding an integration to the left

$$(1.5) \qquad \mathbb{E}^{\theta}[(A_{\infty})^n] = n! \int g(x, y_1)\, g(y_1, y_2) \ldots g(y_{n-1}, y_n)\, \theta(dx)\, \mu(dy_1) \ldots \mu(dy_n) \quad .$$

---

[1] Symmetry is lost, but could be recovered using $k^2\eta$ as reference measure.

To add similarly an integration on the right, we apply the preceding formula to the conditioned semigroup $P_t^{/k}$ with $k = G\chi$, assumed for simplicity to be finite and strictly positive. Then we have, still denoting by μ the measure $f \cdot \eta$

$$(1.6) \qquad \mathbb{E}^{k\theta/k}\left[(A_\infty)^n\right] =$$
$$n! \int g(x,y_1)\,g(y_1,y_2)\ldots g(y_{n-1},y_n)\,g(y_n,z)\ \theta(dx)\,\mu(dy_1)\ldots\mu(dy_n)\,\chi(dz)\ .$$

In the computation, the first $1/k(x)$ of $g^{/k}$ simplifies with the k of $k\theta$, the intermediate $k(y_i)$'s collapse to 1, and the last remaining $k(y_n)$ is made explicit as the integral w.r.t. χ on the right.

Now the symmetry comes into play. We define the symmetric bilinear form on measures (first positive ones)

$$(1.7) \qquad\qquad e(\mu,\nu) = \int g(x,y)\,\mu(dx)\,\nu(dy)\ .$$

We say that μ has *finite energy* if $e(|\mu|,|\mu|) < \infty$. For instance, all positive bounded measures with bounded potential have finite energy. One can prove that the space of all measures of finite energy with the bilinear form (1.7) is a prehilbert space $\mathcal{H}$ (usually it is not complete : there exist "distributions of finite energy" which are not measures). On the other hand, with every measure μ of finite energy one can associate a continuous additive functional $A_t(\mu)$ (or A_t^μ for the sake of typesetting) reducing to $A_t(f)$ when $\mu = f \cdot \eta$, and we still have

$$(1.8) \quad \mathbb{E}^{k\theta/k}\left[(A_\infty(\mu))^n\right] =$$
$$n! \int g(x,y_1)\,g(y_1,y_2)\ldots g(y_{n-1},y_n)\,g(y_n,z)\ \theta(dx)\,\mu(dy_1)\ldots\mu(dy_n)\,\chi(dz)\ .$$

Some dust has been swept under the rugs here : there was no problem about defining the same $A(f)$ for the two semigroups (P_t) and $(P_t^{/k})$, because we had an explicit expression for it, but as far as $A(\mu)$ is concerned it requires some work. Take us on faith that it can be done.

Finally, we "polarize" this relation to compute the expectation $\mathbb{E}^{k\theta/k}\left[A_\infty^{\mu_1}\ldots A_\infty^{\mu_n}\right]$ (called by Dynkin "the n-point function of the occupation field") as a sum over all permutations σ of $\{1,\ldots,n\}$

$$(1.9) \qquad \sum_\sigma \int g(x,y_1)\,g(y_1,y_2)\ldots g(y_n,z)\ \theta(dx)\,\mu_{\sigma(1)}(dy_1)\ldots\mu_{\sigma(n)}(dy_n)\,\chi(dz)\ .$$

This is the main probabilistic formula, which will be interpreted using the combinatorics of Gaussian random variables. Note that it is symmetric in the measures μ_i, even if the potential kernel is not symmetric.

A Gaussian process

2   For simplicity, we reduce the prehilbert space $\mathcal{H}$ to the space of all signed measures which can be written as differences of two positive bounded measures with bounded potentials — such measures clearly are of finite energy. Though we will need it only much later, let us give an interpretation of formula (1.9) on this space : we denote by α_i the operator on $\mathcal{H}$ which maps a measure λ to $(G\lambda)\mu_i$ (a bounded measure with bounded potential since $G\lambda$ is bounded), and rewrite (1.9) as

$$(2.1) \qquad \sum_\sigma (\theta, \alpha_{\sigma(1)}\alpha_{\sigma(2)}\cdots\alpha_{\sigma(n)}\chi).$$

On the same probability space as (X_t) and independently of it, we construct a Gaussian (real, centered) linear process Y_μ, indexed by $\mu \in \mathcal{H}$, with covariance

$$(2.2) \qquad <Y_\mu, Y_\nu> = \int \mu(dx)\, g(x,y)\, \nu(dy).$$

We claim that, *if θ and χ are point masses ε_x and ε_y* [1], given a function $F \geq 0$ on $\mathbb{R}^n$, we have *Dynkin's formula*

$$(2.3) \quad \mathbb{E}^{k\theta/k}\left[F(A_1 + (Y_1^2/2),\ldots, A_n + (Y_n^2/2))\right] = \mathbb{E}\left[F(Y_1^2/2,\ldots,Y_n^2/2)Y_0 Y_{n+1}\right].$$

Note that θ and χ have been moved inside the Gaussian field expectation on the right hand side.

We follow the exposition of Marcus and Rosen, which is Dynkin's proof without Feynman diagrams. The first step consists in considering the case of a function of n variables which is a product of coordinates, $F(x_1,\ldots,x_n) = x_1\ldots x_n$. Since the measures μ_i are not assumed to be different, (2.3) will follow for polynomials. Then a routine path will lead us to the general case via complex exponentials, Fourier transforms and a monotone class theorem — this extension will not be detailed, see Dynkin [Dyn2] or Marcus–Rosen [MaR].

Given a set E with an even number $2p$ of elements, a *pairing* π of E is a partition of E into p sets with two elements — one of which is arbitrarily chosen so that we call them $\{\alpha, \beta\}$. We may also consider π as the mapping $\pi(\alpha) = \beta$, $\pi(\beta) = \alpha$, an involution without fixed point. The expectation of a product of an even number $m = 2p$ of (centered) gaussian random variables ξ_i is given by a sum over all pairings π of the set $\{1,\ldots,2p\}$ (see formula (4.3) in Chapter IV, subs. 3)

$$(2.4) \qquad \sum_\pi \prod_\alpha \mathbb{E}[\xi_\alpha \xi_\beta].$$

We apply (2.4) to the following family of $2n+2$ gaussian r.v.'s : for $i = 1,\ldots,n$ we define $x_i = Y_i$, $y_i = Y_i$, and the r.v.'s Y_θ, Y_χ which play a special role are denoted by x_0 and x_{n+1}. Given a pairing π, we separate the indexes i in two classes : "closed" indexes such that x_i is paired to y_i, and "open" indexes i, such that x_i is paired to some x_j or y_j with $j \neq i$, necessarily open. Note that $i = 0, n+1$ is open since there is

---

[1]   An assumption wrongly omitted from the earlier edition, as Prof. Marcus kindly pointed to us.

no y_i, and if $i \neq 0, n+1$ y_i is paired to some x_k or y_k with $k \neq i$. Then we consider the corresponding product in (2.5) : first we have factors corresponding to "closed" indexes, $E[x_i y_i] = \mathbb{E}[Y_i^2]$. Next we start from the open index 0 ; $\pi(x_0)$ is some z_j, where j is open and z_j is either x_j or y_j (unless $j = n+1$ in which case the only choice is $z_j = x_j$). Then $\pi(z_j)$ is some z_k with k open, *etc.* and we iterate until we end at x_{n+1}. This open chain of random variables contributes to the product a factor

$$\mathbb{E}[Y_0 Y_{j_1}] \, \mathbb{E}[Y_{j_1} Y_{j_2}] \ldots \mathbb{E}[Y_{j_k} Y_{n+1}] \ .$$

On the other hand, this chain does not necessarily exhaust the "open" indexes (think of the case x_0 is paired directly with x_{n+1}). Then starting with an "open" index that does not belong to the chain, we may isolate a new chain, this time a closed one, or cycle

$$\mathbb{E}[Y_{k_1} Y_{k_2}] \, \mathbb{E}[Y_{k_2} Y_{k_3}] \ldots \mathbb{E}[Y_{k_n} Y_{k_1}] \quad ;$$

one or more such closed chains may be necessary to exhaust the product. Calling the number of expectation signs the *length* of the chain (open or closed), we see that the case of "closed indexes" appears as that of cycles of minimal length 1.

Thus we are reorganizing the right hand side of (2.3) as a sum of products of expectations over one open and several closed chains of indexes, and we must take into account the numbers of summands and the powers of 2. If we exchange in all possible ways the names x_i, y_i given to each "open" r.v. Y_i, $i = 1, \ldots, n$, we get all the pairings leading to the same decomposition into chains. Assuming the lengths of the chains are m_0 for the open chain, $m_1, \ldots, m_k$ for the cycles, with $m_0 + \ldots + m_k = n + 2$, the number of such pairings corresponding to the given decomposition in chains is $2^{m_0-1} \ldots 2^{m_k-1} = 2^{n+1-k}$. On the other hand, the right side of (4) has a factor of 2^{-n}. Therefore, we end with a factor of 2^{k-1}, k being *the number of cycles*.

We would have a similar computation for a product containing only squares, *i.e.* without the factor Y_0, Y_{n+1} : the open chain linking Y_0 to Y_{n+1} would disappear, and we would get a sum of products over cycles only.

We now turn to the left hand side of (2.3) : we expand the product into a sum of terms $2^{-j} \mathbb{E}[Y_{m_1}^2 \ldots Y_{m_j}^2] \, \mathbb{E}[A_{i_1} \ldots A_{i_{n-j}}]$ where the two sets of indexes partition $\{1, \ldots, n\}$. We have just computed the first expectation as a sum of products over cycles, and formula (1.9) tells exactly that the second expectation contributes the open chain term.

3 Marcus and Rosen use Dynkin's formula to attack the problem of continuity of local times : assuming the potential density is finite and continuous (as in the case of one dimensional Brownian motion killed at an exponential time), point masses are measures of finite energy, and the corresponding additive functionals are the *local times* (L_t^x). On the other hand, we have a Gaussian process Y_x indexed by points of the state space. Then Marcus and Rosen prove that the local times can be chosen to be jointly continuous in (t, x) *if and only if* the Gaussian process has a continuous version. The relationship can be made more precise, local continuity or boundedness properties being equivalent for the two processes. This is proved directly, without referring to the powerful theory of regularity of Gaussian processes — but as soon as the result is known, the knowledge of Gaussian processes can be transferred to local times. We can only refer the reader to the article [MaR], which is beautifully written for non-specialists.

§2 Le Jan's "supersymmetric" approach

Taking his inspiration from Dynkin's theorem and from a paper by Luttinger, Le Jan developed a method leading to an isomorphism theorem of a simpler algebraic structure than Dynkin's. This theorem is "supersymmetric" in the general sense that it mixes commuting and anticommuting variables. That is, functions as the main object of analysis are replaced by something like non homogeneous differential forms of arbitrary degrees. Starting from a Hilbert space $\mathcal{H}$, one works on the tensor product of the symmetric Fock space over $\mathcal{H}$, providing the coefficients of the differential forms, and the antisymmetric Fock space providing the differential elements (the example of standard differential forms on a manifold, in which the coefficients are smooth functions while the differential elements belong to a finite dimensional exterior algebra, suggests that two different Hilbert spaces should be involved here). However, this is not quite sufficient : Le Jan's theorem requires analogues of a *complex* Brownian motion Fock space, *i.e.* , the symmetric and antisymmetric components are built over a Hilbert space which has already been "doubled" We are going first to investigate this structure.

The construction will be rather long, as we will first construct leisurely the required Fock spaces, adding several interesting results (many of which are borrowed from papers by J. Kupsch), supplementing Chapter V on complex Brownian motion.

Computations on complex Brownian motion (symmetric case)

1 We first recall a few facts about complex Brownian motion, mentioned in Chapter V, §1. The corresponding L^2 space, generated by multiple integrals of all orders (m, n) w.r. t. two real Brownian motions X, Y, is isomorphic to the Fock space over $\mathcal{G} = L^2(\mathbb{R}_+) \oplus L^2(\mathbb{R}_+)$. It is more elegant to generate it by multiple integrals w.r.t. two conjugate complex Brownian motions $Z, \overline{Z}$. The first chaos of the ordinary Fock space over $\mathcal{G}$ contains two kinds of stochastic integrals associated with a given $u \in L^2(\mathbb{R}_+)$

$$(1.1) \qquad Z_u = \int u_s \, dZ_s \quad , \quad \overline{Z}_u = \int u_s \, d\overline{Z}_s \ .$$

both of them being linear in u. The conjugation maps Z_u into $\overline{Z}_{\overline{u}}$, exchanging the two kinds of integrals. The Wiener product of random variables is related to the Wick product by

$$(1.2) \qquad Z_u Z_v = Z_u : Z_v \quad , \quad \overline{Z}_u \overline{Z}_v = \overline{Z}_u : \overline{Z}_v \quad , \quad Z_u \overline{Z}_v = Z_u : \overline{Z}_v + (u, v)\mathbf{1} \ .$$

It is convenient to formulate things in a more algebraic way : we have a complex Hilbert space $\mathcal{G}$ with a conjugation, and two complex subspaces $\mathcal{H}$ and $\mathcal{H}'$ exchanged by the conjugation, such that $\mathcal{G} = \mathcal{H} \oplus \mathcal{H}'$. On $\mathcal{G}$ we have a bilinear scalar product $(u, v) = <\overline{u}, v>$, and this bilinear form is equal to 0 on $\mathcal{H}$ and $\mathcal{H}'$ — they are *(maximal) isotropic subspaces*. Thus the Fock space structure over $\mathcal{G}$ is enriched in two ways : by the conjugation, which leads to the bilinear form (u,v) (and to the notion of a Wiener product), and by the choice of the pair $\mathcal{H}, \mathcal{H}'$.

Here is an important example of such a situation : we take any complex Hilbert space $\mathcal{H}$, denote by $\mathcal{H}'$ its dual space — the pairing is written $v'(u) = (v', u)$ — and define $\mathcal{G} = \mathcal{H} \oplus \mathcal{H}'$. There is a canonical antilinear mapping $u \longmapsto u^*$ from $\mathcal{H}$ to $\mathcal{H}'$ (if u is a *ket*, u^* is the corresponding *bra*), and it is very easy to extend it to a conjugation of $\mathcal{G}$. Then every element of $\mathcal{G}$ can be written as $u + v^*$ with $u, v \in \mathcal{H}$, and we have

$$(1.3) \qquad (u + v^*, w + z^*) = \ <v, w> + <z, u> .$$

An element of the incomplete (boson) Fock space over $\mathcal{G}$ is a finite sum of homogeneous elements of order (m, n), themselves linear combinations of elements of the form

$$(1.4) \qquad u_1 \circ \ldots \circ u_m \circ v_1' \circ \ldots \circ v_n'$$

with $u_i \in \mathcal{H}$, $v_i' \in \mathcal{H}'$. Another way of writing this is

$$(1.4') \qquad u_1 \circ \ldots \circ u_m \circ v_1^* \circ \ldots \circ v_n^*$$

where this time all u_i, v_i belong to $\mathcal{H}$, and * is the standard map from $\mathcal{H}$ to $\mathcal{H}'$.

If a conjugation is given on $\mathcal{H}$, we also have a canonical scalar product (bilinear) $(v, u) = \ <\bar{v}, u>$, and a canonical linear mapping from $\mathcal{H}$ to $\mathcal{H}'$, denoted $v \longmapsto v'$. Then we may split each v' in (1.4) into a $v \in \mathcal{H}$ and a $'$. This is essentially what we do when we use the probabilistic notation, a third way of writing the generic vectors :

$$(1.4'') \qquad Z_{u_1} : \ldots : Z_{u_m} : \overline{Z}_{v_1} : \ldots : \overline{Z}_{v_n}$$

again with $u_i, v_i \in \mathcal{H} = L^2(\mathbb{R}_+)$, with its standard conjugation. The symbols $\circ$ and $:$ have the same meaning : generally, we use the first when dealing with abstract vectors, and the second one for random variables — a mere matter of taste.

The form $(1.4'')$ can be rewritten as a multiple Ito integral

$$(1.5) \qquad \int u_1(s_1) \ldots u_m(s_m) v_1(t_1) \ldots v_n(t_n) \, dZ_{s_1} \ldots dZ_{s_m} \, d\overline{Z}_{t_1} \ldots d\overline{Z}_{t_n} \ ,$$

which is incorrectly written : through a suitable symmetrization of the integrand, (1.5) should be given the form

$$(1.6) \qquad \frac{1}{m!n!} \int f(s_1, \ldots, s_m ; t_1, \ldots, t_n) \, dZ_{s_1} \ldots dZ_{s_m} \, d\overline{Z}_{t_1} \ldots d\overline{Z}_{t_n} .$$

extended to $\mathbb{R}_+^m \times \mathbb{R}_+^n$, f being a symmetric function in the variables s_i, t_i separately. In our case, because of the factorials in front of the integral, f is equal to a sum over permutations

$$(1.7) \qquad \sum_{\sigma, \tau} u_{\sigma(1)}(s_1) \ldots u_{\sigma(m)}(s_m) v_{\tau(1)}(t_1) \ldots v_{\tau(n)}(t_n) .$$

2 Instead of the Wick product $(1.4'')$, let us compute the *Wiener* product (usually written without a product sign; if necessary we use a dot)

$$(2.1) \qquad Z_{u_1} \ldots Z_{u_m} \overline{Z}_{v_1} \ldots \overline{Z}_{v_n}$$

According to the complex Brownian motion form of the multiplication formula (Chapter V, §1, formula (2.4)) the Wiener product (2.1) has a component in every chaos of order $(m - p, n - p)$. For $p = 0$ we have the Wick product

$$Z_{u_1} : \ldots : Z_{u_m} : \overline{Z}_{v_1} : \ldots : \overline{Z}_{v_n} ,$$

for $p = 1$, we have a sum of Wick products

(2.2) $$(v_j, u_i) Z_{u_1} : \ldots \widehat{Z_{u_i}} \ldots : Z_{u_m} : \overline{Z}_{v_1} : \ldots \widehat{\overline{Z}_{v_j}} \ldots : \overline{Z}_{v_n} ,$$

where the terms with a hat are omitted. Such a term is called the *contraction* of the indices s_i and t_j in the Wick product. Similarly, the term in the chaos of order $(m - 2, n - 2)$ involves all contractions of two pairs of indexes, etc.

If we had computed a Wiener product in the form (1.4′), in algebraic notation

$$u_1 \cdot \ldots \cdot u_m \cdot v_1^* \cdot \ldots \cdot v_n^* ,$$

then the contractions would have involved the *hermitian* scalar product $< v_j, u_i >$.

The expectation of (2.1) is equal to 0 for $m \neq n$, and for $m = n$ it is given by

(2.3) $$\mathbb{E}\left[Z_{u_1} \ldots Z_{u_m} \overline{Z}_{v_1} \ldots \overline{Z}_{v_n} \right] = \operatorname{per}(v_j, u_i)$$

where *per* denotes a *permanent*. We may translate this into algebraic language, as a linear functional on the incomplete Fock space, given in the representation (1.4) by

(2.4) $$\mathbf{Ex}\left[u_1 \circ \ldots \circ u_m \circ v_1' \circ \ldots \circ v_m' \right] = \operatorname{per}(v_j', u_i) .$$

(the notation **Ex** should recall the word "expectation" without suggesting an "exponential"). In the representation (1.4′) we have

(2.5) $$\mathbf{Ex}\left[u_1 \circ \ldots \circ u_m \circ v_1^* \circ \ldots \circ v_m^* \right] = \operatorname{per} < v_j, u_i > .$$

3 The following two subsections are devoted to comments on the preceding definitions. They are very close to considerations in Slowikowski [Slo1] and Nielsen [Nie].

The Wiener and Wick products can be described in the same manner : consider a (pre)Hilbert space $\mathcal{H}$ with a bilinear functional $\xi(u, v)$. We are going to define on the incomplete Fock space $\Gamma_0(\mathcal{H})$ an associative product admitting **1** as unit element, such that on the first chaos

(3.1) $$u \cdot v = u \circ v + \xi(u, v) \mathbf{1} .$$

Note the analogy with Clifford algebra. This product is commutative if ξ is symmetric, and satisfies in general an analogue of the CCR

$$u \cdot v - v \cdot u = (\xi(u, v) - \xi(v, u)) \mathbf{1} .$$

We must supplement the rule (3.1) by a prescription for higher order products of vectors belonging to the first chaos

(3.2)
$$(u_1 \circ \ldots \circ u_m) \cdot v = u_1 \circ \ldots \circ u_m \circ v + \sum_i \xi(u_i, v) u_1 \circ \ldots \widehat{u_i} \ldots \circ u_m$$

$$u \cdot (v_1 \circ \ldots \circ v_n) = u \circ v_1 \circ \ldots \circ v_n + \sum_i \xi(u, v_i) v_1 \circ \ldots \widehat{v_i} \ldots \circ v_n$$

the hat on a vector indicating that it is omitted. Then an easy induction shows that the product $(u_1 \circ \ldots \circ u_m) \cdot (v_1 \circ \ldots \circ v_n)$ is well defined, with a term in each chaos of order $m + n - 2p$ involving all possible p-fold contractions. The Wick (symmetric) product corresponds to $\xi = 0$; the Wiener product needs a conjugation and corresponds to $\xi(u, v) = \overline{< u, v >}$.

We have, u_o^m and v_o^n denoting symmetric (Wick) powers,

$$(3.3) \qquad u_o^m \cdot v_o^n = \sum_{k \leq m,n} k! \binom{m}{k} \binom{n}{k} \xi^k(u, v) \, u_o^{m-k} \circ v_o^{n-k} \; .$$

Note the analogy with the multiplication formula for Hermite polynomials. We deduce a product formula for exponential vectors

$$(3.4) \qquad \mathcal{E}(u) \cdot \mathcal{E}(v) = e^{\xi(u,v)} \mathcal{E}(u + v) \; ,$$

on which associativity is seen to depend on the identity

$$(3.5) \qquad \xi(u, v + w) + \xi(v, w) = \xi(u, v) + \xi(u + v, w) \; ,$$

trivially satisfied in our case. Another relation with Hermite polynomial occurs if we express the ξ-power u_ξ^m using Wick powers, and conversely (the index u_ξ or u_o indicates in which sense the polynomial operates on u)

$$(3.5) \qquad u_o^m = H_m(u_\xi, \xi(u, u)) \quad , \quad u_\xi^m = H_m(u_o, -\xi(u, u)) \; ,$$

where the generalized Hermite polynomials are defined by

$$(3.5') \qquad e^{tu - at^2/2} = \sum_m \frac{t^m}{m!} H_m(u, a) \; ,$$

and we have in particular, as in the Wiener product case

$$(3.6) \qquad e_\xi^u = e^{\xi(u,u)/2} \mathcal{E}(u) \; .$$

from which a generalized Weyl relation follows

$$\exp_\xi^u \exp_\xi^v = e^{(\xi(u,v) - \xi(v,u))/2} \exp_\xi(u + v) \; .$$

Let us give at least one interesting non-commutative example. We return to the case of $\mathcal{G} = \mathcal{H} \oplus \mathcal{H}'$, with elements represented as $u + v'$, and we identify u with the creation operator a_u^+, and v' with the annihilation operator $a_{v'}^-$ (indexed by an element of $\mathcal{H}'$: see VI.1.4). Then, the *operator multiplication* can be described as an ξ-product involving the non-symmetric bilinear form

$$(3.7) \qquad \xi(u + v^*, w + z^*) = < v, w > \; ,$$

"half" the symmetric bilinear form (1.3) that leads to the Wiener product. For instance

$$(3.8) \qquad (a_u^+ + a_{v'}^-)(a_w^+ + a_{z'}^-) = (a_u^+ + a_{v'}^-) : (a_w^+ + a_{z'}^-) + (v', w) \mathbf{1} \; .$$

One easily recovers the well known fact that the (commutative) Wick product of a string of creation and annihilation operators is equal to its normally ordered operator product. On the other hand, for non normally ordered strings the operator product is reduced to the Wick product using the CCR.

For instance, the operator exponential e^{u+v^*} is given by (3.6), and can be seen to act on coherent vectors as follows, (interpreting the Wick product as a normally ordered product, and then performing the computation)

$$e^{u+v^*}\mathcal{E}(h) = e^{<v,u>/2}e^u e^{v^*}\mathcal{E}(h)$$
$$= e^{<v,h>+<v,u>/2}\mathcal{E}(u+h)$$

which is the formula we gave in Chapter IV for the action of Weyl operators on exponential vectors.

4   We now discuss a topic of probabilistic interest, that of *Stratonovich integrals*. This is close to what [Slo1] and [Nie] call "ultracoherence", with a different language. We again proceed leisurely, including topics which are not necessary for our main purpose. On the other hand, many details have been omitted : see the papers of Hu–Meyer [HuM], Johnson–Kallianpur [JoK], and a partial exposition in [DMM].

First of all, we consider for a while a *real* Hilbert space $\mathcal{H}$, to forget the distinction between the hermitian and bilinear scalar products. As usual, a generic element of Fock space is a sum $f = \sum_n f_n/n!$, f_n being a symmetric element of $\mathcal{H}^{\otimes n}$ and the sum $\sum_n \| f_n \|^2/n!$ being convergent. We call $F = (f_n)$ the *representing sequence* of f, and use the notation $f = I(F)$, I for "Ito" since when $\mathcal{H} = L^2(\mathbb{R}_+)$, F is the sequence of coefficients in the Wiener–Ito expansion of the r.v. f. Many interesting spaces of sequences (with varying assumptions of growth or convergence) can be interpreted as spaces of "test-functions" or of "generalized random variables" (see the references [Kré], [KrR]).

Given a (continuous) bilinear functional ξ on $\mathcal{H}$, we wish to define a mapping Tr_ξ from $\mathcal{H}^{\otimes m}$ to $\mathcal{H}^{\otimes(m-2)}$, which in the case of $\mathcal{H} = L^2(\mathbb{R}_+)$, $\xi(u,v) = (u,v)$ (the bilinear scalar product), should be the trace

(4.1) $$\mathrm{Tr}\, f_m(s_1,\ldots,s_{m-2}) = \int f_m(s_1,\ldots,s_{m-2},s,s)\,ds$$

(for $m = 0,1$ we define $\mathrm{Tr}\, f_m = 0$). If ξ is given by a kernel $a(s,t)$, we will have

(4.2) $$\mathrm{Tr}\, f_m(s_1,\ldots,s_{m-2}) = \int f_m(s_1,\ldots,s_{m-2},s,t)\,a(s,t)\,ds dt .$$

To give an algebraic definition, we begin with the incomplete tensor power $\mathcal{H}^m$ and define

(4.3) $$\mathrm{Tr}_\xi(u_1 \otimes \ldots \otimes u_m) = \xi(u_{m-1},u_m)\,u_1 \otimes \ldots \otimes u_{m-2} .$$

Since we will apply this definition to symmetric tensors, which indexes are contracted is irrelevant, and we made an arbitrary choice. This definition is meaningful for any bounded bilinear functional ξ, and it can be extended to the completed m-th tensor power if ξ belongs to the Hilbert-Schmidt class.

We now consider Tr_ξ as a mapping from sequences $F = (f_n)$ to sequences $\mathrm{Tr}_\xi F = ((\mathrm{Tr}(f_{n+2}))$. For instance, given an element u from the first chaos, the sequence $F = (u^{\otimes n})$ is the representing sequence for $\mathcal{E}(u)$, and we have $\mathrm{Tr}_\xi(F) = \xi(u, u) F$.

At least on the space of finite representing sequences, we may define the iterates Tr_ξ^k and put $T_\xi = e^{(1/2)\,\mathrm{Tr}_\xi} = \sum_n \mathrm{Tr}_\xi^n / 2^n\, n!$. On the example above, we have $T_\xi(F) = e^{(1/2)\xi(u,u)} F$. It is clear on this example that, if we carry the operator T_ξ to Fock space without changing notation, what we get is

$$(4.4) \qquad T_\xi \mathcal{E}(u) = e^{(1/2)\xi(u,u)} \mathcal{E}(u) = \exp_\xi(u) ,$$

the exponential of u for the ξ–product. If ξ is symmetric, we have $\exp_\xi(u + v) = \exp_\xi(u)\exp_\xi(v)$, and this suggests T_ξ is a homomorphism from the incomplete Fock space with the Wick (=symmetric) product onto itself with the ξ–product. It turns out that this result is true (see [HuM] for the case of the Wiener product) — and it can't be true in the non-symmetric case, since the Wick product is commutative.

Given a sequence F, and assuming $I(T_\xi(F))$ is meaningful, it is denoted by $S_\xi(F)$. In the case of $L^2(\mathbb{R}_+)$ and the usual scalar product, $S(F)$ is called a *Stratonovich* chaos expansion. An intuitive description can be given as follows : an Ito multiple integral and a Stratonovich integral of the same symmetric function $f(s_1, \ldots, s_m)$ differ by the contribution of the diagonals, which is 0 in the case of the Ito integral, and computed in the case of the Stratonovich integral according to the rule $dX_s^2 = ds$.

Let us now discuss the most important case for the theory of local times : that of complex Browniam motion. The chaos coefficients are separately symmetric functions, and the "diagonals" which contribute to the Stratonovich integral are the diagonals $\{s_i = t_j\}$, the trace being computed according to the rules

$$dZ_s^2 = d\overline{Z}_s^2 = 0 \quad , \quad dZ_s d\overline{Z}_s = d\overline{Z}_s dZ_s = ds .$$

More generally, there exist contracted integrals using a function $\xi(s, t)$, not necessarily symmetric. The trace in the complex case involves the contraction of one single s with one single t, and the Tr_ξ operator applied to the representing sequence $F = (u^{\otimes m} v^{\otimes n})$ of an exponential vector multiplies it by $\xi(u, v)$. The operator T_ξ (Stratonovich integral) has no coefficient $1/2$: $S_\xi(F) = I(e^{\mathrm{Tr}_\xi} F)$.

The linear functional **Ex** on the space of sequences, *i.e.* the ordinary (vacuum) expectation of the Stratonovich integral, is generalized as follows (we write the formula in the complex case)

$$(4.5) \qquad \mathbf{Ex}_\xi\,[F] = \sum_m \frac{1}{m!}\, \mathrm{Tr}_\xi^m\, f_{mm} .$$

5 The following computation is an essential step in the proof of Le Jan's result. Let G be an element of the chaos of order $(1, 1)$

$$(5.1) \qquad G = Z_{u_1} \circ \overline{Z}_{v_1} + \ldots + Z_{u_n} \circ \overline{Z}_{v_n}$$

In algebraic language, the vectors v_i would rather be written as $v_i' \in \mathcal{H}'$, and G would be $u_1 \circ v_1' + \ldots + u_n \circ v_n'$. We also associate with G the corresponding Wiener product

$$(5.2) \qquad \widehat{G} = Z_{u_1} \overline{Z}_{v_1} + \ldots + Z_{u_n} \overline{Z}_{v_n}$$

which is the "Stratonovich" version of (5.1). Similarly, the Wiener exponential $e^{-\varepsilon \widehat{G}}$ is the "Stratonovich" version of the Wick exponential $e_0^{-\varepsilon G}$, and therefore we have

$$(5.3) \qquad \mathbb{E}\left[e^{-\varepsilon \widehat{G}}\right] = \mathbf{Ex}\left[e_0^{-\varepsilon G}\right],$$

which will be seen to exist for ε close to 0. We are going to compute this quantity.

LEMMA 1. *Let A be the square matrix (v_j, u_i). Then we have*

$$(5.4) \qquad \mathbb{E}\left[e^{-\varepsilon \widehat{G}}\right] = \frac{1}{\det(I + \varepsilon A)}.$$

Here is a useful extension : *for ε small enough, we have*

$$(5.5) \qquad \mathbb{E}\left[Z_{a_1} \overline{Z}_{b_1} \ldots Z_{a_m} \overline{Z}_{b_m} e^{-\varepsilon \widehat{G}}\right] = \frac{\mathrm{per}(b_j, (I + \varepsilon C)^{-1} a_i)}{\det(I + \varepsilon A)}.$$

where C is the matrix (b_j, a_i). We give two proofs of this formula, a combinatorial one and a more probabilistic one (only sketched) — indeed, this is nothing but a complex Gaussian computation.

We work with the Wiener product and $\mathbb{E}$ instead of the Wick product and $\mathbf{Ex}$. We expand the exponential series (5.2)

$$(5.6) \qquad \mathbb{E}\left[e^{-\varepsilon \widehat{G}}\right] = 1 + \sum_k \frac{(-\varepsilon)^k}{k} \mathbb{E}\left[(Z_{u_1} \overline{Z}_{v_1} + \ldots + Z_{u_n} \overline{Z}_{v_n})^k\right]$$

$$(5.7) \qquad = 1 + \sum_k \frac{(-\varepsilon)^k}{k} \sum_{\tau \in F_k} \mathbb{E}\left[Z_{u_{\tau(1)}} \overline{Z}_{v_{\tau(1)}} + \ldots + Z_{u_{\tau(k)}} \overline{Z}_{v_{\tau(k)}}\right]$$

where F_k is the set of all mappings from $\{1, \ldots, k\}$ to $\{1, \ldots, n\}$. Before we compute (5.7), we make a remark : using the basic formula (2.3) for the expectation of a product, we know it can be developed into a permanent of scalar products (v_j, u_i). Therefore, the expectation does not change if we replace each v_i by its projection on the subspace K generated by the vectors u_j. On the other hand, $\widehat{G}$ depends only on the finite rank operator $\alpha = v_1' \otimes u_1 + \ldots v_n' \otimes u_n$ with range K, and from the above we may assume the v_j's also belong to K. Then we may assume that the u_i constitute an orthogonal basis of K, in which the matrix of α is triangular, and the determinant we are interested in is the product of the diagonal elements $(1 + \varepsilon \lambda_i)$ of $I + \varepsilon \alpha$.

We now start computing (5.7) : let $i_1 < \ldots < i_p$ be the range of the mapping τ ; since τ is not necessarily injective, each point i_j has a multiplicity k_j with $k_1 + \ldots + k_p = k$. All the mappings τ with the same range and multiplicities contribute the same term to the sum, and their number is $k!/k_1! \ldots k_p!$. Therefore we rewrite (5.7) as

$$(5.8) \qquad 1 + \sum_k (-\varepsilon)^k \sum_{\substack{k_1 + \ldots + k_p = k \\ i_1 < \ldots < i_p}} \mathbb{E}\left[(Z_{u_{i_1}} \overline{Z}_{v_{i_1}})^{k_1} \ldots (Z_{u_{i_p}} \overline{Z}_{v_{i_1}})^{k_p}\right]/k_1! \ldots k_p!$$

We rewrite the expectation as

$$\mathbb{E}\,[\,Z_{f_1}\overline{Z}_{h_1}\dots Z_{f_k}\overline{Z}_{h_k}]$$

with $f_j = u_{i_1}$, $h_j = v_{i_1}$ for $1 \le j \le k_1$, u_{i_2}, v_{i_2} for $k_1 < j \le k_1 + k_2$, etc., and formula (2.3) expands it as a sum, over all permutations of $\{1, \dots, k\}$, of products $\prod_j (h_j, f_{\sigma(j)})$. The matrix A being triangular, σ does not contribute unless it preserves the intervals $(1, k_1), (k_1 + 1, k_1 + k_2)\dots$; the contribution of such a permutation is $(v_{i_1}, u_{i_1})^{k_1} \dots (v_{i_p}, u_{i_p})^{k_p}$, and the number of such permutations is $k_1! \dots k_p!$. Therefore the expectation (5.6) has the value

$$1 + \sum_k (-\varepsilon)^k \sum_{\substack{k_1 + \dots + k_p = k \\ i_1 < \dots < i_p}} \lambda_{i_1}^{k_1} \dots \lambda_{i_p}^{k_p} = (1 + \varepsilon\lambda_1)^{-1} \dots (1 + \varepsilon\lambda_n)^{-1}$$

for ε small enough. The theorem is proved. The extension is not difficult : assuming the t_i's and ε are small enough, we have from the preceding result

$$\mathbb{E}\,[\exp(-\varepsilon\widehat{G} + t_1 Z_{a_1}\overline{Z}_{b_1} + \dots + t_m Z_{a_m}\overline{Z}_{b_m})] = 1/\det(I + \varepsilon A - U)\,,$$

where $U = \sum_i t_i b_i' \otimes a_i$. Putting $C = I + \varepsilon A$ and assuming C is invertible, this determinant can be written $\det(C)\det(I - C^{-1}U)$. On the other hand, $C^{-1}U$ is the finite rank operator $\sum_i t_i b_i' \otimes c_i$, with $c_i = C^{-1}a_i$, and if the t_i are small enough, we may use again the main result to compute $1/\det(I - C^{-1}U)$. Therefore

$$\mathbb{E}\,[\exp(-\varepsilon\widehat{G} + \sum_i t_i Z_{a_i}\overline{Z}_{b_i})] = \mathbb{E}\,[\exp(\sum_i t_i Z_{a_i}\overline{Z}_{c_i})]/\det(C)$$

We identify the coefficients of $t_1 \dots t_m$ on both sides and get the required formula.

There is a similar computation for the "ξ-product" defined above, if ξ is symmetric : A is now the matrix $\xi(v_j, u_i)$ and C the matrix $\xi(b_j, a_i)$.

Let us now sketch a proof using Gaussian integrals. It has the advantage of expressing the Wick exponential itself as a standard (Wiener) exponential, instead of merely computing its expectation. We start with the elementary Fock space over $\mathbb{C}$ or $\mathbb{C}^2$, i.e. the L^2 space generated by a standard Gaussian variable X (real) or Z (complex). We recall the classical formula (for $c > 0$)

$$(5.9) \qquad\qquad e^{cx^2/2} = \frac{1}{\sqrt{c}} \int e^{xu}\,e^{-u^2/2c}\,du$$

where du is the "Plancherel measure". From the interpretation of $\exp_{:}(uX)$ as $e^{uX - u^2/2}$ we deduce that

$$(5.10) \quad \exp_{:}(cX^2/2) = \frac{1}{\sqrt{1+c}}\,e^{\frac{c}{1+c}X^2/2}\,, \quad e^{aX^2/2} = \sqrt{1-a}\,\exp_{:}(\frac{a}{1-a}X^2/2)\,.$$

What about the ξ-product ? Since the basic Hilbert space $\mathcal{H}$ is simply $\mathbb{C}$, let us put $\xi(1,1) = q$ and assume for the moment that $q > 0$. Then the ξ-product is the Wiener product w.r.t. a Gaussian law of variance q, and the Wick exponential $\exp_{:}(ux)$ takes the form $e^{ux-qu^2/2}$. This has a combinatorial meaning which can be extended to all values of q. The same reasoning as above then gives us

$$(5.11) \qquad \exp_:(cX^2/2) = \frac{1}{\sqrt{1+qc}} \exp_\xi(\frac{c}{1+qc} X^2/2) \ .$$

The computation is similar, but simpler, in the complex case : we have $\exp_:(uZ+v\overline{Z}) = e^{uZ+v\overline{Z}-uv}$, and (returning to the representation $Z = (X+iY)/\sqrt{2}$ as an intermediate step) we deduce

$$(5.12) \qquad \exp_:(cZ\overline{Z}) = \frac{1}{1+c} e^{\frac{c}{1+c} Z\overline{Z}} \ .$$

This elementary formula can now be used to compute Wick exponentials on the complex Brownian motion Fock space, for elements of the second chaos of the form

$$G = Z_{u_1} : \overline{Z}_{v_1'} + \cdots + Z_{u_n} : \overline{Z}_{v_n'}$$

(vectors and linear functionals are identified via the bilinear scalar product). To perform this, one remarks that G depends only on the operator $A = \sum_i u_i \otimes v_i'$, and then one reduces A to a diagonal or at least triangular form : we give no details.

Antisymmetric computations

For a complete exposition of the rich algebra underlying the informal presentation we give here, we refer to the papers by Helmstetter in the references.

6 We start with a translation of the elementary Clifford algebra of Chapter II. There we had a finite dimensional Hilbert space $\mathcal{H}$ with a basis x_i, orthonormal for the bilinear scalar product $(\ ,\ )$; we built the corresponding basis (x_A) for the (symmetric or) antisymmetric (toy) Fock space, and the Clifford multiplication, which we now denote explicitly by $\diamond$, was defined by the table $x_A \diamond x_B = (-1)^{n(A,B)} x_{A\Delta B}$. Now the antisymmetric Fock space is nothing but the exterior algebra over $\mathcal{H}$, and for $A = \{i_1 < \ldots < i_n\}$ x_A is nothing but the exterior product $x_{i_1} \wedge \ldots \wedge x_{i_n}$. Thus we have two different multiplications on the same space, and we want to put into algebraic language the relation between them, as we did for Wick and Wiener. The main element of structure is the bilinear scalar product, which we denote by $\xi(u,v)$ instead of (u,v).

First of all, $\mathbf{1}$ is the unit for both products. Then for elements of the first chaos, we have

$$(6.1) \qquad u \diamond v = u \wedge v + \xi(u,v)\mathbf{1} \ ,$$

implying the CAR $\{u,v\}_\diamond = 2\xi(u,v)\mathbf{1}$.

We have a "multiplication formula for stochastic integrals" which expresses

$$(u_1 \wedge \ldots \wedge u_m) \diamond (v_1 \wedge \ldots \wedge v_n)$$

as the sum of terms in all the chaos of order $(m-p, n-p)$ involving p contractions. Let us write the simplest non trivial case :

$$(6.2) \qquad u \diamond (v_1 \wedge \ldots \wedge v_n) =$$

$$u \wedge v_1 \wedge \ldots \wedge v_n + \sum_i (-1)^{i+1} \xi(u,v_i) v_1 \wedge \ldots \wedge \widehat{v}_i \wedge \ldots \wedge v_n \ ,$$

omitting the vector with a hat. We leave it to the reader to write down the general formula. These prescriptions can be checked directly on the multiplication table, when each u_i or v_j is a basis element x_k, and then extended to arbitrary elements of the first chaos by linearity. On the other hand, the associativity of the multiplication formula we get is clear, as we are computing within an algebra known to be associative! Since any non–degenerate symmetric bilinear form can be interpreted as a scalar product, we get in this way a general associativity result. Once it is known for non–degenerate bilinear forms it extends to degenerate ones. Finally, the result extends at once to infinite-dimensional spaces.

REMARK. In the symmetric case, it is clear on formula (3.4) that the standard annihilation operators $a^-_{u'}$ act as derivations on all ξ–products. Here we have the same result for fermion operators $b^-_{u'}$, with the only difference that the derivations are graded ones, i.e. $b^-_{u'}(x \diamond y) = (b^-_{u'}x) \diamond y + \sigma(x) \diamond b^-_{u'}(y)$, where σ is the parity automorphism.

Now comes a crucial remark for Le Jan's result : what we have done *does not require the symmetry of ξ* — though of course the proof of associativity by reduction to toy Fock space does. *We postpone the general proof of associativity to the last subsection of this Appendix.* The symmetric form 2ξ must be replaced in the CAR by $\eta(u,v) = \xi(u,v) + \xi(v,u)$: thus all algebras with the same η are isomorphic Clifford algebras, but the way they are concretely realized on the antisymmetric Fock space depends on the choice of ξ. In particular, if ξ is antisymmetric, what we get is a new *Grassmann algebra* structure[1] on Fock space, which we distinguish from the usual one by means of the sign $\diamond$.

We take now a linear space $\mathcal{H}$ with a bilinear form ξ, and denote by $\mathcal{H}'$ a second copy of $\mathcal{H}$ — the $'$ may thus have two meanings, either of suggesting duality when a scalar product is given, or of a mark distinguishing elements of the second summand, and the confusion between these meanings is harmless, or even useful. We define $\mathcal{G} = \mathcal{H} \oplus \mathcal{H}'$, and extend ξ as a bilinear form on $\mathcal{G}$ under which the two summands are totally isotropic spaces :

$$(6.3) \qquad \xi(u + v', w + z') = \xi(u, z) \pm \xi(w, v) .$$

If we choose the $+ (-)$ sign, we get a symmetric (antisymmetric) form. The generators we use for the (incomplete) antisymmetric Fock space over $\mathcal{G}$ are "normally ordered" ones, $u_1 \wedge \ldots \wedge u_m \wedge v'_1 \wedge \ldots \wedge v'_n$. Applying the preceding remark, we define multiplications on this double Fock space, with the following prescriptions : strings of u's or v's are multiplied as in the exterior algebra without correction; the computation of products of mixed strings are expressed as modified exterior products involving contractions. For instance

$$(6.4) \qquad u \diamond (u_1 \wedge \ldots \wedge u_m \wedge v'_1 \wedge \ldots \wedge v'_n) =$$

$$u \wedge u_1 \wedge \ldots \wedge u_m \wedge v'_1 \wedge \ldots \wedge v_n + \sum_i (-1)^{m+i+1} \xi(u, v_i) u_1 \wedge \ldots \wedge u_m \wedge v'_1 \wedge \ldots \widehat{v'_i} \ldots \wedge v'_n$$

$$(6.4') \qquad v' \diamond (u_1 \wedge \ldots \wedge u_m \wedge v'_1 \wedge \ldots \wedge v'_n) =$$

$$(-1)^m u_1 \wedge \ldots \wedge u_m \wedge v' \wedge v'_1 \wedge \ldots \wedge v'_n \mp \sum_i (-1)^{m+i} \xi(u, v_i) u_1 \wedge \ldots \widehat{u_i} \ldots \wedge u_m \wedge v'_1 \wedge \ldots \wedge v'_n$$

---

[1] This was misunderstood in the report [Mey4], which contains a confusion between this "new Grassmann" product and a Clifford product.

(the sign $\mp$ is $-$ for Clifford, $+$ for "new Grassmann").

If $\mathcal{H} = L^2(\mathbb{R}_+)$ and ξ is the standard scalar product, one may also use a probabilistic notation : we take an antisymmetric double Fock space and denote by P and Q the corresponding two complex Brownian motions (which anticommute). Then we read u as $\int u(s)\,dP_s$ and v' as $\int v(s)\,dQ_s$. We also read

$$u_1 \wedge \ldots \wedge u_m \wedge v'_1 \wedge \ldots \wedge v'_n \quad \text{as} \quad P_{u_1} \wedge \ldots \wedge P_{u_m} \wedge Q_{v_1} \wedge \ldots \wedge Q_{v_n}\,.$$

The notation P, Q comes from symplectic geometry : it has nothing to do with the two real Brownian motions on simple Fock space, which satisfied the CCR : we are here in the CAR realm. In both cases we have for all s, t that dP_s anticommutes with dP_t, dQ_s with dQ_t without restriction, and dP_s anticommutes with dQ_t for $s \neq t$. For the Clifford product, we have

$$(6.5) \qquad dP_t \diamond dQ_t = dt = dQ_t \wedge dP_t\,,$$

and therefore $\{dP_t, dQ_t\} = 2\,dt$, while for the "new Grassmann" product, we have

$$(6.6) \qquad dP_t \diamond dQ_t = I\,dt = -dQ_t \diamond dP_t\,,$$

and thus dP_s anticommutes with dQ_t without restriction.

We concentrate our interest on the "new Grassmann" product $\diamond$, and leave aside the Clifford one.

> The ξ–products associated with non–symmetric forms have another interesting application : Le Jan [LeJ3] used them to extend Dynkin's method to non–symmetric Markov processes.

One can define on the incomplete double Fock space a linear functional **Ex** similar to that of symmetric Fock space, and such that

$$\mathbf{Ex}\,[\,u_1 \wedge \ldots \wedge u_m \wedge v'_1 \wedge \ldots \wedge v'_n\,] = \mathbb{E}\,[\,u_1 \diamond \ldots \diamond u_m \diamond v'_1 \diamond \ldots \diamond v'_n\,] =$$
$$\mathbb{E}\,[(u_1 \wedge \ldots \wedge u_m) \diamond (v'_1 \wedge \ldots \wedge v'_n)]\,.$$

Intuitively speaking, **Ex** is the expectation of an antisymmetric Stratonovich integral, which, however, has not been systematically developed in the antisymmetric case : a "trace" operator should be introduced and studied to define **Ex** for classes of coefficients belonging to the completed L^2 spaces.

This linear functional is given by the formula

$$(6.7) \qquad \mathbf{Ex}\,[\,P_{u_1} \wedge \ldots \wedge P_{u_m} \wedge Q_{v_1} \wedge \ldots \wedge Q_{v_n}\,] = \rho(m)\det((u_i, v_j)) \quad \text{if } m = n\,,$$

and 0 if $m \neq n$.

7 We are going now to prove the antisymmetric analogue of (5.4). We consider an element of the second chaos of the form

$$(7.1) \qquad G = P_{u_1} \wedge Q_{v_1} + \ldots + P_{u_n} \wedge Q_{v_n}\,,$$

a finite sum; $u_1 \wedge v'_1 + \cdots + u_n \wedge v'_n$ if you prefer algebraic notation. We associate with it a similar "new Grassman" object

$$(7.2) \qquad \widehat{G} = P_{u_1} \diamond Q_{v_1} + \ldots + P_{u_n} \diamond Q_{v_n}\,.$$

The corresponding exponential series are a finite sums, and no convergence problems arise. The result is the following :

LEMMA 2. *Denoting by A the matrix (v'_j, u_i), we have*

$$(7.3) \qquad \mathbf{Ex}\,[\,e_\wedge^G\,] = \mathbb{E}\,[\,e_\Diamond^{\widehat{G}}\,] = \det(I + A)\,.$$

As in the symmetric case there is a useful extension :

$$(7.4) \qquad \mathbf{Ex}\,[\,P_{a_1} \wedge Q_{b_1} \ldots P_{a_m} \wedge Q_{b_m} e_\wedge^G\,] = \det(b_j, (I+C)^{-1} a_i)\,\det(I+A)\,,$$

where C is the matrix (b_j, a_i) and $I + C$ is assumed to be invertible.

PROOF. We use the standard Grassmann product and **Ex**. We first expand

$$e_\wedge^G = 1 + \sum_k \frac{1}{k!} \sum_{\tau \in I_k} P_{u_{\tau(1)}} \wedge Q_{v_{\tau(1)}} \wedge \ldots \wedge P_{u_{\tau(k)}} \wedge Q_{v_{\tau(k)}}$$

where I_k is the set of all injective mappings from $\{1,\ldots,k\}$ to $\{1,\ldots,n\}$. Moving around a block $P_u Q_v$ produces an even number of inversions and does not change the product; therefore we may reduce I_k to the set J_k of strictly increasing mappings, cancelling the factor $1/k!$. Now we take all the P_u's to the left and get

$$1 + \sum_k \sum_{\tau \in J_k} \rho(k)\, P_{u_{\tau(1)}} \wedge \ldots \wedge P_{u_{\tau(k)}} \wedge Q_{v_{\tau(1)}} \wedge \ldots \wedge Q_{v_{\tau(k)}}\,.$$

Then we apply **Ex** and the factor $\rho(k)$ disappears : we get (putting $a_{ji} = (v_j, u_i)$)

$$1 + \sum_k \sum_{\tau \in J_k} \det((v_{\tau(j)}, u_{\tau(i)})) = 1 + \sum_{B \subset \{1,\ldots,n\}} \det_{i,j \in B} a_{ij}\,.$$

We must show this is the same as

$$\det(I + A) = \sum_\sigma \varepsilon_\sigma \prod_i (\delta_{i\sigma(i)} + a_{i\sigma(i)})\,.$$

We expand the product on the right. Each index i may contribute either a factor $a_{i\sigma(i)}$ or a factor $\delta_{i\sigma(i)}$. Let B the the set of indices of the first kind. If the product is different from 0, σ must leave the indices of the second kind fixed, and therefore arises from a permutation of B, which has the same signature. We then finish the proof by rearranging the sum as a sum over B. The extension is proved as in the symmetric case.

Supersymmetric Fock space

8 We now consider the *supersymmetric* Fock space, *i.e.* the tensor product of the previously considered symmetric and antisymmetric Fock spaces, contructed over $\mathcal{H} \oplus \mathcal{H}'$ — with $\mathcal{H} = L^2(\mathbb{R}_+)$ in the probabilistic interpretation, two complex Brownian motions being thus involved, or four real ones. The classical analogue is the space of all differential forms on a symplectic manifold. We may provide this space with several different products. Those we are interested in are the "Wick" product (Wick $\otimes$ Grassmann) which we denote by $*$, and the "Wiener" product (Wiener $\otimes$ new Grassmann) which we denote by $\diamond$ (sometimes omitted!) as its antisymmetric part.

Consider an operator α of finite rank on $\mathcal{H}$: it can be written (non-uniquely) as

$$(8.1) \qquad \alpha(\cdot) = \sum_i v_i'(\cdot) \, u_i$$

with $v_i' \in \mathcal{H}'$. Then we associate with it an element of the supersymmetric Fock space (which depends only on α, not on the representation)

$$(8.2) \qquad \lambda(\alpha) = \sum_i (Z_{u_i} \overline{Z}_{v_i} - P_{u_i} \diamond Q_{v_i}) = \sum_i (Z_{u_i} : \overline{Z}_{v_i} - P_{u_i} \wedge Q_{v_i}) \ .$$

We have used the probabilistic interpretation, where v_i' becomes v_i. The "supersymmetric miracle" which happens with these elements is the following : if we multiply (5.4) and (7.3), the determinants on the numerator and the denominator cancel, and therefore we have, for ε small enough, the exponential being a Wiener one

$$(8.3) \qquad \mathbb{E}\left[\exp_\diamond(\lambda(\alpha)\right] = 1 \ .$$

Now we insert in front of each term in the sum (8.2) a coefficient ε_i (assumed to be small) and differentiate with respect to all of them. We find that given any finite family of operators of finite rank

$$(8.4) \qquad \mathbb{E}\left[\lambda(\alpha_1)\ldots\lambda(\alpha_n)\right] = 0 \ .$$

If instead of applying (5.4) and (7.3) we apply their extensions, we find that

$$\mathbb{E}\left[Z_a \overline{Z}_b \, e^{\sum_i \varepsilon_i (Z_{u_i} \overline{Z}_{v_i} - P_{u_i} \diamond Q_{v_i})}\right] = (b, (I - (\sum_i t_i v_i' \otimes u_i)^{-1} a)$$

and then if we identify the coefficients of $\varepsilon_1 \ldots \varepsilon_n$ on both sides, we get, putting $\alpha_i = v_i' \otimes u_i$ (an operator of rank 1)

$$(8.5) \qquad \mathbb{E}\left[Z_a \overline{Z}_b \lambda(\alpha_1)\ldots\lambda(\alpha_n)\right] = \sum_\sigma \left(b, \alpha_{\sigma(1)}\ldots\alpha_{\sigma(n)} a\right)$$

By linearity, we may extend this to operators α_i of finite rank. The formula can be extended to Hilbert-Schmidt operators, but still this is a serious restriction.

The right side of (8.5) is the same as (2.1) in section 1, which itself was a rewriting of (1.9), "the n-point function of the occupation field". This has now become the expectation of one single product on supersymmetric Fock space, a simpler result than we could achieve in §1. May be our reader has forgotten the concrete application we had in view, assuming local times do exist : the computation, for F a polynomial in n variables, of the expectation $\mathbb{E}^{k\theta/k}\left[F(L_\infty^{x_1}\ldots L_\infty^{x_n})\right]$!

Properties of the supersymmetric Wiener product

9 We are going to prove a few useful algebraic results concerning the supersymmetric Wiener product. First, let us compute the norm of $\lambda(\alpha)$ in Fock space. Let α and $\tilde{\alpha}$ be two finite rank operators. Then we have

$$(9.1) \qquad\qquad < \lambda(\alpha), \lambda(\tilde{\alpha}) > \; = 2 < \alpha, \tilde{\alpha} >_{HS} \,,$$

where the scalar product on the right is the Hilbert-Schmidt scalar product between operators. To see this, we write

$$\lambda(\alpha) = \sum_i Z_{u_i} : \overline{Z} v_i - P_{u_i} \wedge Q_{v_i}$$

and similarly for $\tilde{\alpha}$ with vectors $\tilde{u}_i, \tilde{v}_i$ (it does not restrict generality to assume the index sets are the same). We understand these elements as random variables in the second chaos of a double complex Brownian motion Fock space. Then the $Z, \overline{Z}$ and P, Q parts decouple, and contribute the same quantity (hence the factor 2), which is

$$< \sum_i \int u_i(s) v_i(t) \, dZ_s d\overline{Z}_t \,, \; \sum_i \int \tilde{u}_i(s) \tilde{v}_i(t) \, dZ_s d\overline{Z}_t >$$

which is easily interpreted as a H–S scalar product.

We now describe a second "supersymmetric miracle". We consider an arbitrary vector u in $\mathcal{H}$ with $(u, u) = q$. We deduce from formula (5.12) that, for s small enough

$$(9.2) \qquad\qquad \exp_{:}(s \, Z_u \overline{Z}_u) = \frac{1}{1+sq} \exp(\frac{sq}{1+sq} Z\overline{Z}) \,.$$

On the other hand, we have

$$(9.3) \quad \exp_\wedge(-s P_u \wedge Q_u) = 1 - s P_u \wedge Q_u = 1 - s(P_u \diamond Q_u - q)$$
$$= (1+sq)(1 - \frac{sq}{1+sq} P \diamond Q) = (1+sq) \exp_\diamond(-\frac{sq}{1+sq} P \diamond Q)$$

Multiplying (9.2) and (9.3) we get

$$(9.4) \qquad\qquad \exp_*(s\lambda(u \otimes u')) = \exp_\diamond(\frac{sq}{1+sq} \lambda(u \otimes u')) \,.$$

We recall the generating function for a family of Laguerre polynomials

$$\exp \frac{tx}{1+t} = \sum_n P_n(x) t^n \quad \text{with} \quad P_n(x) = (-1)^n L_n^{(-1)}(x) \,,$$

and then expanding the exponential gives an expression for supersymmetric "Wick" powers using "Wiener" powers

$$(9.5) \qquad\qquad \lambda(u \otimes u')_*^n = n! \, q^n \, P_n(\lambda(u \otimes u')/q)_\diamond \,.$$

10 After we used the theory of local times as our motivation to get into twenty pages of algebra, it may be disappointing that we devote little space to probabilistic applications.

However, we cannot hide the fact that, for these applications, the best methods are purely probabilistic : a good way to dive into the literature is to look at the volumes XIX to XXI of the *Séminaire de Probabilités*, and to look therein at the papers by Dynkin, Le Gall, Le Jan, Rosen and Yor (and their references). On the other hand, the relations with quantum field theory are described in a series of papers by Dynkin. Here, we merely mention the problems.

Recall that our basic (pre)Hilbert space $\mathcal{H}$ is a space of signed measures with bounded potential, with the energy scalar product. With every measure μ of finite energy, we associated an operator α_μ on the space

$$\alpha_\mu(\nu) = (G\nu)\,\mu \, .$$

If the process has local times, *i.e.* if point measures have finite energy, and if the support of μ is a finite set, then α_μ has finite rank, and we may associate with it an element $\lambda(\mu)$ of the second chaos as explained before, and apply all the previous algebraic computations. In particular, we may associate with every point x an element $Z_x\overline{Z}_x$ of the second symmetric chaos, and an element $\lambda_x = Z_x\overline{Z}_x - P_xQ_x$ of the second supersymmetric chaos. Computations involving the system of local times may be reduced to algebraic computations on these vectors (see Le Jan's paper [LeJ1] for details). Note that λ_x has expectation 0 : its definition implicitly contains a "renormalization".

Assume now that local do not exist. Then we may think of $Z_x\overline{Z}_x$ and λ_x as "generalized random variables" and wonder whether they belong to a true "generalized field", *i.e.* whether we may define something like $\int \lambda_x f(x)\,dx$ for a smooth function with compact support f, and more generally $\int \lambda_x^{*n} f(x)\,dx$ (Wick power). The symmetric part of this integral will then be interpreted as a "renormalized power" of the occupation field. The method to achieve this is the following : we replace the point measure ε_x by a regularization $\varepsilon_x(t)$, λ_x by the corresponding $\lambda_x(t)$, and use the algebra to check whether the corresponding integrals converge in L^2 as $t \to 0$. It turns out that the computation is not too difficult, and that convergence takes place in dimension 2, but not in higher dimensions (the divergence of $g(x,y)$ on the diagonal is logarithmic in dimension 2). For details, we refer to [LeJ2]. Unfortunately, this description of the renormalization problem has reduced it to its abstract core, stripping it of its probabilistic interest : relations with multiple points of Brownian motion, asymptotic behaviour of the volume of the Wiener sausage, etc.

Proof of associativity of ξ-products

11 We are going to prove that the computations involving contractions we performed around (6.2) define an associative product. The following proof is adapted from Le Jan.

We may reduce the infinite dimensional case to a finite dimensional one : consider a space $\mathcal{H}$ of finite dimension N with a scalar product (u,v). We give ourselves a bilinear form $\xi(u,v) = (Au,v)$ on $\mathcal{H}$, and put $\eta(u,v) = \xi(u,v)+\xi(v,u)$. We will assume that ξ and η are non-degenerate. This is not a serious restriction : non-degeneracy holds for $\xi(u,v) + t(u,v)$ except for finitely many values of t, and the associativity result may be extended by continuity to $t=0$.

On the antisymmetric Fock space Φ over $\mathcal{H}$, we denote by b_u^+, b_u^- the standard fermion creation and annihilation operators, satisfying the CAR $\{b_u^+, b_v^-\} = (u,v)$.

Then we put

$$(.1) \qquad\qquad R_u = b_u^+ + b_{Au}^- \quad , \quad S_v = b_{Bv}^+ - b_v^- \ ,$$

where B satisfies $(Au, Bv) = (u, v)$. Knowing how $b_u^\pm$ operates on exterior products, all amounts to proving this : *there exists an associative product $\diamond$ on Φ, with $\mathbf{1}$ as unit, such that for $u \in \mathcal{H}$ and $x \in \Phi$ we have $u \diamond x = R_u x$.*

It is easy to see that

$$\{R_u, R_v\} = (u, Av) + (Au, v) \quad , \quad \{S_u, S_v\} = -(u, Bv) - (Bu, v) \quad , \quad \{Ru, Sv\} = 0 \ .$$

Therefore, the operators R_u, S_v taken together generate a Clifford algebra over the space $\mathcal{H} \oplus \mathcal{H}$ of even dimension $2N$, relative to a non-degenerate quadratic form. Applying the fundamental uniqueness result on Clifford algebras (II.5.5) we see that the operators $R_u S_v$ generate an algebra of dimension 2^{2N}, which is therefore the full algebra $\mathcal{L}(\Phi)$. On the other hand, the two algebras $\mathcal{R}$ and $\mathcal{S}$ generated by the R_u and S_v separately are Clifford algebras of dimension 2^N.

Let Φ^+ (Φ^-) be the even (odd) subspaces of Φ; let $\mathcal{R}^+$ ($\mathcal{S}^+$) and $\mathcal{R}^-$ ($\mathcal{S}^-$) be the even (odd) subspaces of $\mathcal{R}$ and $\mathcal{S}$. Note that R_u and S_u are odd operators in the usual sense, and therefore $\mathcal{R}^+, \mathcal{S}^+$ preserve parity in Φ while $\mathcal{R}^-, \mathcal{S}^-$ change it. We denote below by $R^\pm, S^\pm$ generic elements of $\mathcal{R}^\pm, \mathcal{S}^\pm$.

Let $A \in \mathcal{R}$ such that $A\mathbf{1} = 0$; decompose it into $A^+ + A^-$. Since $A^\pm \mathbf{1}$ belongs to $\Phi^\pm$ we must have $A^\pm \mathbf{1} = 0$. Then applying R^ε for $\varepsilon = \pm$ and using the appropriate commutation relation we find that $A^\pm R^\varepsilon \mathbf{1} = 0$. Then applying S^δ for $\delta = \pm$ and using the appropriate commutation relation with $A^\pm R^\varepsilon$ we find that $A^\pm R^\varepsilon S^\delta \mathbf{1} = 0$. Putting everything together we find that $ARS\mathbf{1} = 0$ for arbitrary $R \in \mathcal{R}$, $S \in \mathcal{S}$. On the other hand, the algebra $\mathcal{RS}$ is the full algebra, and therefore $A = 0$. The mapping $A \longmapsto A\mathbf{1}$ from $\mathcal{R}$ to Φ being injective, a dimension argument shows that it is surjective. Therefore we may use it to identify Φ with $\mathcal{R}$, and define on Φ an associative product $\diamond$, with $\mathbf{1}$ as unit, such that, if x, y are elements of Φ, and R_x, R_y are the unique elements of $\mathcal{R}$ such that $x = R_x \mathbf{1}$, $y = R_y \mathbf{1}$, we have $x \diamond y = R_x R_y \mathbf{1}$.

In particular, if u, v belong to $\mathcal{H}$, we have

$$u \diamond v = R_u R_v \mathbf{1} = (b_u^+ + b_{Au}^-)v = u \wedge v + (Au, v) \ .$$

More generally, we have for $x \in \Phi$ $u \diamond x = R_u R_x \mathbf{1} = R_u x$, the main property we wanted to prove.

REMARK. In Helmstetter's notes [Hel2] (or p. 35 of the more easily accessible paper [Hel1]), one can find an explicit formula for the ξ–product, from which a direct and general proof of associativity can be deduced, including the degenerate cases. We will give now this formula.

Let $\mathcal{E}$ be a finite dimensional space $\mathcal{E}$ with dual $\mathcal{E}'$. We use greek letters for elements of $\mathcal{E}'$ and its exterior algebra. We recall there is a mapping $(\alpha, x) \longmapsto \alpha \lhd x$ from $\bigwedge(\mathcal{E}') \times \bigwedge(\mathcal{E})$ to $\bigwedge(\mathcal{E})$, called the *interior product*, which is a generalized annihilation operator :

1) For $\alpha \in \mathcal{E}'$, $x \in \bigwedge(\mathcal{E})$, we have $\alpha \lhd x = b_\alpha^-(x)$, with the additional convention that $\mathbf{1} \lhd x = x$.

2) $(\alpha \wedge \beta) \triangleleft x = \alpha \triangleleft (\beta \triangleleft x)$.

This product acts as a graded derivation :

3) $\alpha \triangleleft (x \wedge y) = (\alpha \triangleleft x) \wedge y + \sigma(x) \wedge (\alpha \triangleleft y)$, σ denoting the parity automorphism.

> More generally, Helmstetter ([Hel2] p. 25) shows that we may replace $\wedge$ by $\diamond$, an arbitrary ξ-product on E.

We take $\mathcal{E} = \mathcal{H} \oplus \mathcal{H}$, where $\mathcal{H}$ is finite dimensional with dual $\mathcal{H}'$. To distinguish the second component we use a $\tilde{\ }$ since we cannot use the $'$ as before. We identify the exterior algebra $\bigwedge(\mathcal{E})$ with $\bigwedge(\mathcal{H}) \otimes \bigwedge(\mathcal{H})$: this means we use antisymmetry to express elements of $\bigwedge(\mathcal{E})$

$$(u_1 + \tilde{v}_1) \wedge \ldots \wedge (u_p + \tilde{v}_p)$$

as linear combinations of normally ordered products

$$(u_{i_1} \wedge \ldots \wedge u_{i_m}) \otimes (\tilde{v}_{j_1} \wedge \ldots \wedge \tilde{v}_{j_n})$$

with $m+n = p$. The dual $\bigwedge(\mathcal{E}')$ then gets identified with $\bigwedge(\mathcal{H}') \otimes \bigwedge(\mathcal{H}')$. On the other hand, a bilinear form ξ on $\mathcal{H}$ defines an antisymmetric bilinear form on $\mathcal{H} \oplus \mathcal{H}$

$$\xi(u + \tilde{v}, w + \tilde{z}) = \xi(u, z) - \xi(w, v)$$

and therefore $\mathcal{H} \otimes \mathcal{H}$ is imbedded into $\bigwedge_2(\mathcal{E}')$, and has an exterior exponential in $\bigwedge(\mathcal{E}')$. Finally, there is a natural mapping μ from $\mathcal{E}$ to $\mathcal{H}$ given by $\mu(u + \tilde{v}) = u + v$, which extends to the exterior algebras by $\mu(h \wedge k) = \mu(h) \wedge \mu(k)$ for $h, k \in \bigwedge(\mathcal{E})$; for example, given $x, y \in \bigwedge(\mathcal{H})$ we simply have $\mu(x \otimes \tilde{y}) = x \wedge y$.

After all these preliminaries, we can give the closed formula for the ξ-product on $\bigwedge(\mathcal{H})$: given $x, y \in \bigwedge(\mathcal{H})$, $x \otimes y$ belongs to $\bigwedge(\mathcal{E})$, and we take

$$x \diamond y = \mu(\exp_\wedge(\xi) \triangleleft (x \otimes \tilde{y}))$$

For instance, if $x, y \in \mathcal{H}$, the only terms in the exponential that contribute are $1 + \xi$, and we get

$$1 \triangleleft (x \otimes \tilde{y})) = x \otimes \tilde{y} \quad , \quad \xi \triangleleft (x \otimes \tilde{y}) = \xi(x, y) \mathbf{1} ,$$

and applying μ we have $x \diamond y = x \wedge y + \xi(x, y) \mathbf{1}$ as it should be. This is clearly the analogue of the trace formula for Stratonovich integrals.

Appendix 6

More on Stochastic Integration

In this appendix, we prove the representation theorem for operator martingales of Parthasarathy and Sinha, its extension to semimartingales by Attal, and the "true" Ito formula for bounded representable semimartingales, also due to Attal.

§1. EVERYWHERE DEFINED STOCHASTIC INTEGRALS

A strong disadvantage of the H–P stochastic calculus compared to classical stochastic calculus is the fact that operators defined on the exponential domain cannot be composed, and there is no true "Ito formula". Following Attal we are going to define a family of bounded "semimartingales" which is closed under composition, and for which an Ito formula can be proved. We will work directly on a Fock space of arbitrary multiplicity (calling $\mathcal{K}$ the multiplicity space), but for simplicity we take a trivial initial space. Most of the results given here appear in the paper [AtM] by Attal and Meyer.

1 First of all, we are going to prove a formula — not quite explicit outside of the exponential domain — which tells how a "semimartingale of operators" acts on a "semimartingale of vectors". It makes rigorous the heuristic formula (2.6) of Chapter VI, §2 (p. 128) and extends it to arbitrary multiplicity. First of all, let us say what we mean by a semimartingale of vectors : it is an adapted process of vectors

$$(1.1) \qquad f_t = c + \sum_\rho \int_0^t \dot{f}_\rho(s)\, dX^\rho(s)$$

— we recall that $\rho, \sigma...$ may take the value 0, contrary to $\alpha, \beta...$, and that $dX_s^0 = ds$. In this formula c is a constant, $\dot{f}_\rho(\cdot)$ is a complex valued process, square-integrable, measurable and adapted, which satisfies the conditions

$$(1.2) \qquad \int_0^t \| \dot{f}_0(s) \|\, ds < \infty \quad , \quad \sum_\alpha \int_0^t \| \dot{f}_\alpha(s) \|^2\, ds < \infty$$

for t finite. If there is a non-trivial initial space $\mathcal{J}$, c becomes an element of $\mathcal{J}$ and the mappings $\dot{f}_\rho(s)$ take values in $\Psi = \mathcal{J} \otimes \Phi$, but we will not insist on this extension.

We may rewrite (1.1) in a symbolic way as

$$(1.1') \qquad f_t = c + \int_0^t \dot{f}_0(s)\, ds + \int_0^t \dot{\mathbf{f}}(s)\, d\mathbf{X}_s ,$$

where $((\dot{f}_0(s))$ is an adapted process in $L^1_{loc}(\Phi)$, while $(\dot{\mathbf{f}}(s))$ is in $L^2_{loc}(\Phi \otimes \mathcal{K})$ — the component of index 0 is now excluded from it.

We have $\dot{f}_0 = 0$ iff the process (f_t) is a martingale. In all cases, the coefficients $\dot{f}_\rho(s)$ are uniquely determined for a.e. s. In the case of an exponential vector $f_s = \mathcal{E}(uI_{]0,s]})$ we have $\dot{f}_0(s) = 0$, $\dot{f}_\alpha(s) = u_\alpha(s) f_s$. It is always assumed that u is locally bounded.

Next, a "semimartingale of operators" is a family of adapted operators J_t which has a stochastic integral representation

$$(1.3) \qquad J_t = \sum_{\rho\sigma} \int_0^t H_\rho^\sigma(s)\, da_\sigma^\rho(s) .$$

We could add an initial operator, but omit it for simplicity. We assume that all operators involved (J_t and H_ρ^σ) have a common domain $\mathcal{D}$ which contains the exponential domain [1]. Again, we may write (1.3) in compact form

$$(1.3') \qquad J_t = \int_0^t H^+(s)\, da^+(s) + H^\circ(s)\, da^\circ(s) + H^-(s)\, da^-(s) + H^\bullet(s)\, ds ,$$

where $H^\bullet = H_0^0$, H° is the matrix (H_α^β) (an operator on $\Phi \otimes \mathcal{K}$), H^+ is the column (H_0^α), mapping Φ to $\Phi \otimes \mathcal{K}$ and H^- the line (H_α^0) mapping $\Phi \otimes \mathcal{K}$ to Φ. The unusual position of the line or column index is an unfortunate feature of our notation.

On the exponential domain, the meaning of J_t is clear by the H–P theory of stochastic integration, under the standard integrability hypotheses, but on a larger domain it can only be defined by closure (*i.e.* double adjoint). Thus we will also assume the operators J_t and $H_\rho^\sigma(t)$ have adjoints J_t^* and $(H_\rho^\sigma(t))^* = \overset{*}{H}_\sigma^\rho(t)$, defined on the exponential domain, and satisfying on it a similar representation (in the H–P sense)

$$(1.4) \qquad J_t^* = \sum_{\rho\sigma} \int_0^t \overset{*}{H}_\rho^\sigma(s)\, da_\sigma^\rho(s) .$$

Assume now that we have a semimartingale of vectors f_t such that f_t *and all* $\dot{f}_\rho(t)$ *belong to* $\mathcal{D}$ — such a representation will be called *admissible*. Let us define a new set of functions $f_\rho(s)$ by

$$(1.5) \qquad f_\alpha(s) = \dot{f}_\alpha(s) \quad \text{but } f_0(s) = f(s), \text{ not } \dot{f}_0(s).$$

Then we want to show that, under convenient assumptions, we have

$$(1.6) \qquad \boxed{ J_t f_t = \sum_\rho \int_0^t J_s \dot{f}_\rho(s)\, dX_s^\rho + \sum_{\rho\sigma} \int_0^t H_\rho^\sigma(s) f_\sigma(s)\, dX^\rho(s) } \qquad .$$

In the case of martingales ($\dot{f}_0 = 0$) the first sum on the right side bears on α instead of ρ, and in the case of exponential vectors $f_\sigma(s) = u_\sigma(s) f_s$ with $u_0 = 1$, and we get the familiar expressions of Chapter VI.

---

[1] The exponential domain could be replaced for most purposes by the incomplete Fock space (the algebraic sum of the chaos spaces), but we give no details.

In the notation without indexes, the formula is written as follows ($J_s \dot{\mathbf{f}}_s$ means $J_s \otimes I$ acting on $\Phi \otimes \mathcal{K}$)

(1.6′)
$$
\begin{aligned}
J_t f_t = & \int_0^t \left(J_s \dot{\mathbf{f}}(s) + H^+(s)\,f(s) + H^\circ(s)\,\dot{\mathbf{f}}(s) \right) d\mathbf{X}_s \\
& + \int_0^t \left(J_s \dot{f}_0(s) + H^-(s)\,\dot{\mathbf{f}}(s) + H^\bullet(s)\,f(s) \right) ds
\end{aligned}
$$

Formula (1.6) is meaningful under the following conditions, for all $t < \infty$

(1.7) $\displaystyle \int_0^t \| J_s \dot{f}_0(s) \|\, ds < \infty \quad, \quad \sum_\alpha \int_0^t \| J_s \dot{f}_\alpha(s) \|^2\, ds < \infty$,

(1.8) $\displaystyle \int_0^t \| H^\bullet(s)\,f(s) \|\, ds < \infty$,

(1.9) $\displaystyle \int_0^t \| H^-(s)\,\dot{\mathbf{f}}(s) \|^2\, ds < \infty \quad, \quad \int_0^t \| H^+(s)\,f(s) \|^2\, ds < \infty$,

(1.10) $\displaystyle \int_0^t \| H^\circ(s)\,\dot{\mathbf{f}}(s) \|^2\, ds < \infty$.

In each one of these statements involving the norm of a vector, it is assumed this vector exists, which implicitly means some infinite series is strongly convergent. And now we have a satisfactory result, due to Attal :

THEOREM. *Under these conditions which make (1.6) meaningful, it is also true.*

Before we prove this result let us state its main application. Let us assume all operators $H_\rho^\sigma(s)$ and J_t are known to be bounded, and take for $\mathcal{D}$ the whole space. Let us assume that the norm of $H^\bullet(s)$ is locally in L^1, the norms of $H^-(s)$ and $H^+(s)$ are locally in L^2, and the norm of $H^\circ(s)$ locally in L^∞. Then *all the preceding hypotheses are easily seen to be satisfied and (1.6) holds true identically.*

To prove this theorem, we will show that the scalar products of the two sides of (1.6) with an arbitrary exponential martingale $g_t = \mathcal{E}_t(u)$ are equal. To this end, we take adjoints and are reduced to the truth of (1.6) *for the adjoint process, and for exponential vectors* :

$$
J_t^* g_t = \sum_\rho \int_0^t J_s^* \dot{g}_\rho(s)\, dX^\rho(s) + \sum_{\rho\sigma} \int_0^t \overset{*}{H}_\rho^\sigma(s)\, g_\sigma(s)\, dX^\rho(s) .
$$

Note that we wrote as above $\dot{g}_\rho$ in the first integral, and g_σ in the second one, while $\dot{g}_0 = 0$ since (g_t) is an exponential martingale. We want to prove

$$
< g_t, J_t f_t > = < g_t, \sum_\alpha \int_0^t J_s \dot{f}_\alpha(s)\, dX_s^\alpha > + \sum_{\rho\sigma} < g_t, \int_0^t H_\rho^\sigma(s)\, \dot{f}_\sigma(s)\, dX_s^\rho > .
$$

The left hand side is equal to $< J_t^* g_t, f_t >$. We transform the first sum on the right hand side into

$$
\sum_\alpha \int_0^t < \dot{g}_\alpha(s), J_s \dot{f}_\alpha(s) > ds
$$

and replace J_s by J_s^* acting to the left, after which we return to a form

$$< \sum_\alpha \int_0^t J_s^* \dot{g}_\alpha(s) \, dX_s^\alpha, f_t > .$$

In the second sum, we separate the terms $\rho = \alpha > 0$ and $\rho = 0$. In the first case we can rewrite the sum

$$\int_0^t < \dot{g}_\alpha(s), \sum_\sigma H_\alpha^\sigma(s) \dot{f}_\sigma(s) > ds$$

and shift H_α^σ taking an adjoint. Then we again distinguish two cases, $\sigma = \beta > 0$ for which we get

$$\int_0^t < \overset{*}{H_\alpha^\beta}(s) \dot{g}_\alpha(s) \, dX_s^\beta, \dot{f}_t > ,$$

and $\sigma = 0$, for which

$$\int_0^t < \overset{*}{H_\alpha^0} \dot{g}_\alpha(s), f_s > ds = \int_0^t < \overset{*}{H_\alpha^0} \dot{g}_\alpha(s), f_t > ds$$

The second case $\rho = 0$ is similar. Details are left to the reader.

An algebra of operator semimartingales

2   Let us now define the space $\mathcal{S}$ of operator semimartingales J_t of *bounded* operators, admitting a representation

$$(2.1) \qquad J_t = J_0 + \int_0^t H^+(s) \, da^+(s) + H^\circ(s) \, da^\circ(s) + H^-(s) \, da^-(s) + H^\bullet(s) \, ds ,$$

with bounded adapted coefficients $H^\epsilon(s)$ between the appropriate spaces and with

$$(2.2) \qquad \| H^\bullet(s) \| \in L^1_{loc} \quad , \quad \| H^\pm(s) \| \in L^2_{loc} \quad , \quad \| H^\circ(s) \| \in L^\infty_{loc} .$$

The crucial assumption in this statement is that of individual boundedness of the operators J_t ; together with (2.1) it implies the stronger property of *local boundedness* of $\| J_t \|$. Indeed, the operators

$$M_t = J_t - \int_0^t H^\bullet(s) \, ds$$

constitute a martingale of bounded operators, hence $\| M_t \|$ is an increasing function, and $\| J_t \|$ must be locally bounded.

It is clear that $\mathcal{S}$ is stable under passage to adjoints. Let us prove that it is also *closed under products*, and compute the coefficients of the representation of the product[1]. Thus let us consider a second "semimartingale" $(\tilde{J}_t)$ with coefficients $\tilde{H}_\rho^\sigma(t)$ and put

---

[1]   Let us recall incidently that the coefficients of such a representation are unique (Chapter VI, §1, #10), but the proof there is only sketched : a more satisfactory proof is given by Attal [Att1].

$L_t = \tilde{J}_t J_t$ with coefficients K_ρ^σ to be computed. We take a martingale (an exponential one is enough)

$$f_t = c + \int_0^t \hat{\mathbf{f}}_s \, d\mathbf{X}_s$$

and compute first $J_t f_t$ using $(1.6')$, then $\tilde{J}_t(J_t f_t)$ in the same way, and get an expression of the form $(1.6')$, which means that (L_t) is a representable semimartingale of operators. We make its coefficients explicit as follows

$$
(2.3) \quad
\begin{aligned}
K_s^+ &= \tilde{J}_s H_s^+ + \tilde{H}_s^+ J_s + \tilde{H}_s^\circ H_s^+ \quad , \quad K_s^- = \tilde{J}_s H_s^- + \tilde{H}_s^- J_s + \tilde{H}_s^- H_s^\circ \ , \\
K_s^\circ &= \tilde{J}_s H_s^\circ + \tilde{H}_s^\circ J_s + \tilde{H}_s^\circ H_s^\circ \quad , \quad K_s^\bullet = \tilde{J}_s H_s^\bullet + \tilde{H}_s^\bullet J_s + \tilde{H}_s^- H_s^+ \ .
\end{aligned}
$$

It is very easy now to check that these coefficients satisfy again the assumptions (2.1), otherwise stated, (L_t) belongs to $\mathcal{S}$. The computation (2.3) is nothing but the Ito formula, the last term in the expression of each coefficient being the *Ito correction*. It will be convenient later to rewrite this formula as follows

$$(2.4) \qquad d(\tilde{J}_s J_s) = \tilde{J}_s \, dJ_s + d\tilde{J}_s \, J_s + [\![\, d\tilde{J}_s \, , \, dJ_s \,]\!] \ ,$$

where the last term is called the *square bracket* of the two semimartingales, and represents the Ito correction :

$$(2.5) \qquad [\![\, d\tilde{J}_s \, , \, dJ_s \,]\!] = \tilde{H}_s^\circ H_s^+ \, da_s^+ + \tilde{H}_s^\circ H_s^\circ \, da_s^\circ + \tilde{H}_s^- H_s^\circ \, da_s^- + \tilde{H}_s^- H_s^+ \, ds \ .$$

As in the commutative case theory, the process itself is denoted by $[\![\, \tilde{J}, J \,]\!]_t$.

This definition is due to Attal, and he proved that this "square bracket" has many of the properties of the square bracket of classical stochastic calculus. It appears clearly that the "square bracket" has a martingale part consisting of the first three terms, and a "finite variation part" which is the last term : it is naturally called the *angle bracket* of the two semimartingales

$$(2.6) \qquad \langle d\tilde{J}_s \, , \, dJ_s \rangle = \tilde{H}_s^- H_s^+ \, ds \ .$$

We refer to Attal's papers for a detailed study of these "brackets", and limit ourselves to a few remarks.

REMARKS. a) The square bracket of two semimartingales belonging to $\mathcal{S}$ does not belong to $\mathcal{S}$ in general, for the following reason : the representation (2.5) has bounded coefficients with the appropriate norm conditions, *but the operators* $[\![\, \tilde{J}, J \,]\!]_t$ *may be unbounded.*

Thus it seems worthwhile to introduce the following definition : an *adapted process* (J_t) *defined on the exponential domain belongs to the class* $\mathcal{S}'$ *if it has a representation* (1.3') *with bounded coefficients* $H^\varepsilon(s)$, *such that* $\| H^\bullet(s) \|$, $\| H^\pm(s) \|$, $\| H^\circ(s) \|$ *belong locally to* L^1, L^2, L^∞ .

This class is somewhat of a mystery : in the commutative theory it would consist of processes having BMO-like properties, and though it would not act on L^2 by multiplication, it would map into L^2 very large domains. In the non-commutative case nothing is known. But other elements of the class occur naturally. For instance, given two elements J and L of $\mathcal{S}$, it is easy to define the left and right stochastic integrals $\int_0^t L_s \, dJ_s$ and $\int_0^t (dJ_s) \, L_s$, which belong to the class $\mathcal{S}'$, and to write the integration by parts formula

$$(2.7) \qquad d(JL)_t = J_t \, dL_t + (dJ_t) \, L_t + [\![dJ_t, dL_t]\!] ,$$

the left side belonging to $\mathcal{S}$, and the three processes on the right hand side to $\mathcal{S}'$.

It is easy to see that the square bracket is an associative operation on the class $\mathcal{S}'$. On the other hand, it can be approxiated by "quadratic Riemann sums" as in classical probability theory. Namely we have

$$(2.8) \qquad [\![J , L]\!]_t = \lim \sum_i (J_{t_{i+1}} - J_{t_i})(L_{t_{i+1}} - L_{t_i}) ,$$

the limit being taken over dyadic partitions of the interval $[0,t]$, and understood as weak convergence on exponential vectors. This shows that the bracket doesn't depend on the integral representation. There is a similar (and easier) formula for the angle bracket, the quadratic sum being replaced by

$$\sum_i E_{t_i} (J_{t_{i+1}} - J_{t_i})(L_{t_{i+1}} - L_{t_i}) E_{t_i} .$$

§2. REGULAR SEMIMARTINGALES

It is well known that every square integrable martingale on Wiener space has a stochastic integral representation w.r.t. Brownian motion. A natural conjecture is that, similarly, any reasonable martingale of operators on Fock space has a stochastic integral representation w.r.t. the three basic operator martingales. The general validity of this conjecture was disproved by a counterexample of J.-L. Journé (*Sém. Prob. XX*), showing that the martingale of conditional expectations of a fixed bounded operator may not be representable in the usual sense. Hence the main positive result in this direction remains that of Parthasarathy and Sinha, which we present now in Attal's extended version.

Non-commutative "semimartingales"

1 The space $\mathcal{S}$ of "quantum semimartingales" considered in the preceding section may be very convenient, but how can we prove that a process of bounded operators is representable as a stochastic integral, with coefficients satisfying the appropriate norm conditions?

For simplicity, we work on simple Fock space. We consider an operator martingale (1.1) or semimartingale (1.2)

$$(1.1) \qquad M_t = cI + \int_0^t (H_s^+ \, da_s^+ + H_s^\circ \, da_s^\circ + H_s^- \, da_s^-) ,$$

$$(1.2) \qquad M_t = cI + \int_0^t (H_s^+ \, da_s^+ + H_s^\circ \, da_s^\circ + H_s^- \, da_s^- + H_s^\bullet \, ds) .$$

and describe first how one may compute the operators H^ε knowing (M_t) Given a fixed time r and a vector $f_r \in \Phi_r$, the process $Y_t = (M_t - M_r) f_r$ in the case (1.1) is an ordinary martingale with representation

$$(1.3) \qquad\qquad (M_t - M_r) f_r = \int_r^t H_u^+ f_r \, dX_u \ ,$$

and in the case (1.2) it is a semimartingale

$$(1.4) \qquad\qquad (M_t - M_r) f_r = \int_r^t H_u^+ f_r \, dX_u + \int_r^t H_u^\bullet f_r \, du \ .$$

Since the semimartingale Y on $[r, \infty[$ has a unique decomposition, we may recover the process $(H_u^\bullet f_r)$ as the drift of Y (the density of its finite variation part) and the process $(H_u^+ f_r)$ as the density of the angle bracket $< Y, X >$. In principle, knowing $H_u^\bullet f_r$ for $f_r \in \Phi_r$, $r < u$, and using boundedness we may compute $H_u^\bullet f_u$ for $f_u \in \Phi_u$ and then, by adaptation, the operator H_u itself. The same remark applies to H_u^+.

Applying the same procedure to the adjoint process, we could get $(H_t^-)^*$ and then H_t^-. But we can also get H_t^- itself by a similar construction, as follows. Consider the process $Z_t = M_t f_t$. We can see on (1.3) that its drift is equal to $H_t^\bullet f_t + H_t^- \dot{f}_t$. Since we already know $H_t^\bullet$ this allows us to compute $H_t^- \dot{f}_t$, from which the operator H_t^- itself can be computed. Similarly, the density of the angle bracket $< Z, X >$ is $M_t \dot{f}_t + H_t^\circ \dot{f}_t + H_t^+ f_t$, from which (as we already know M_t and H_t^+) H_t° may be computed.

We are going to turn these formal computations into a rigorous reasoning.

DEFINITION. *Let* $M = (M_t)$ *be an adapted process of bounded operators. We say* M *is a regular semimartingale if there exists some increasing absolutely continuous deterministic function* $m(t) = \int_0^t m'(s)\, ds$ *such that, for* $r < s < t$, $f_r \in \Phi_r$ [1]

$$(1.5) \qquad\qquad \| (M_t - M_s) f_r \|^2 \leq \| f_r \|^2 (m(t) - m(s)) \ ,$$
$$(1.6) \qquad\qquad \| (M_t^* - M_s^*) f_r \|^2 \leq \| f_r \|^2 (m(t) - m(s)) \ ,$$
$$(1.7) \qquad\qquad \| (E_s M_t - M_s) f_r \| \leq \| f_r \| (m(t) - m(s)) \ .$$

THEOREM. *A process* $M = (M_t)$ *of bounded operators is a regular semimartingale if and only if it belongs to* S, *i.e. if it has a representation (1.1) with bounded operators* H_t^ε *satisfying the conditions*

> *The mappings* $s \longmapsto \| H_s^\varepsilon \|$ *are locally in* L^1 *for* $\varepsilon = \bullet$, *in* L^2 *for* $\varepsilon = \pm$, *in* L^∞ *for* $\varepsilon = \circ$;

The fact that these assumptions on the coefficients imply the regularity of (M_t) is neither difficult nor very important, and we refer to Attal [Att4] for details. There is one interesting remark, however : the local boundedness of $\| H_t^\circ \|$ is not used in this part of the proof, a local L^2 property is all we need. On the other hand, no known assumption on the coefficients — here as in the classical case — implies the boundedness of M_t

---

[1]   Since the operators are bounded, we may restrict f_r to a dense set.

itself, and this property (together with the regularity of the semimartingale as stated) *implies that* $\| H_t^\circ \|$ *is locally in* L^∞, as we shall see.

To show that regularity implies the representation as stated is the main part of the proof, and for it we will need a lifting theorem. Let us introduce an auxiliary result from measure theory which greatly simplifies the reasoning. People reluctant to use additional axioms for set theory — here, the continuum hypothesis, but Martin's axiom could be used instead — may refer to the original proof of Parthasarathy–Sinha.

2 *A lifting theorem.* This result is due independently to Chatterji and to Mokobodzki (here we follow the latter's version). We add a small remark, which apparently went unnoticed before, but which is very useful in our problem.

THEOREM. *There exists a linear mapping* $f \longmapsto \gamma(f)$ *from the usual Lebesgue space* $\mathcal{L}^\infty(\mathbb{R}_+, dt)$ *to the Banach space* $\mathcal{B}^\infty(\mathbb{R}_+)$ *of bounded everywhere defined Borel functions on* $\mathbb{R}_+$ *with the uniform norm, such that*

   1) γ *is a lifting : for every class* f, $\gamma(f)$ *is a representative of* f.

   2) γ *is order preserving (* $f \leq g$ *a.e. implies* $\gamma(f) \leq \gamma(g)$ *everywhere), norm preserving, and product preserving (* $\gamma(fg) = \gamma(f)\gamma(g)$ *).*

   3) If g *is continuous on* $\mathbb{R}_+$ *and has a limit at infinity, the lifting of the class of* g *is* g *itself (this applies in particular to* $g = 1$ *).*

The additional remark is the *locality* of such a lifting : if a class f vanishes a.e. in some open set U of the line, then $\gamma(f)$ vanishes everywhere in U. Indeed, let h be a continuous function with compact support in U ; then $fh = 0$ a.e., therefore $\gamma(f) h = \gamma(f)\gamma(h) = \gamma(fh)$ vanishes everywhere. In turn, the locality implies that 3) holds for all continuous bounded functions, regardless of their behaviour at infinity.

We will need an obvious extension of this theorem to random variables taking values in a separable Hilbert space $\mathcal{H}$. Let f be such a class. Then for every $u \in \mathcal{H}$ the class $< u, h >$ has a norm in $\mathcal{L}^\infty$ bounded by $\| u \| \, \| h \|_\infty$, and therefore its lifting $\gamma(< u, f >)$ is a Borel function of the same norm, which depends antilinearly on u. Therefore its value at t may be written as $< u, \gamma(f)_t >$ where $\gamma(f)_t$ is an element of $\mathcal{H}$. The mapping $t \longmapsto \gamma_f(t)$ is scalarly Borel, and hence Borel since $\mathcal{H}$ is separable. We denote it by $\gamma(f)$ and we are finished.

3 *The drift term.* We start with the coefficient $H_t^\bullet$. Since we have for $r \leq s < t$

$$\| E_s M_t f_r - M_s f_r \| \leq \| f_r \| (m(t) - m(s))$$

the vector process $M_t f_r$ on $[r, \infty[$ is a *quasimartingale*. Now Hilbert space quasi-martingales were studied by O. Enchev who proved that, as in the scalar case, they can be decomposed into a martingale and a process with finite variation in the norm sense. A report on Enchev's work can be found in *Sém. Prob. XXII*, p. 86–88). Using it leads to considerable simplification of Attal's proof, which in part reproves Enchev's theorem. Its finite variation part is a limit over partitions (t_i) of $[r, t]$

$$A_t = \lim_{(t_i)} \sum_i E_{t_i} M_{t_{i+1}} f_r - M_{t_i} f_r$$

and therefore we have $\| A_t - A_s \| \leq \| f_r \| (m(t) - m(s))$ for $s < t$. Since m is assumed to be absolutely continuous, we may write

$$A_t - A_s = \int_s^t \eta_u(f_r) \, du \quad \text{where} \quad \| \eta_u(f_r) \| \leq \| f_r \| \, m'(u) \text{ a.e..}$$

Let us apply the lifting theorem to the Hilbert space valued mapping $u \longmapsto \eta_u(f_r)/m'(u)$ (defined to be 0 at points where $m'(u)$ does not exist) : we get in this way an everywhere defined mapping, with a norm dominated by $\| f_r \|$, and multiplying by $m'(u)$ again, we get a mapping $u \longmapsto H_u^r(f_r)$ from $]r, \infty[$ into Φ, satisfying an inequality $\| H_u^r f_r \| \leq \| f_r \| \, m'(u)$ (and defined to be 0 whenever $m'(u)$ does not exist).

We may consider H_u^r to be a bounded linear operator from Φ_r to Φ. Let us describe its properties.

Assume $r < s$. Then a vector $f_r \in \Phi_r$ also belongs to Φ_s, and the two functions $\eta_u(f_r)$ corresponding to the two interpretations of f_r agree a.e. on $]s, \infty[$. Therefore, the two functions $H_u^r f_r$ and $H_u^s f_r$ agree everywhere on $]s, \infty[$. Otherwise stated, for fixed u and $r < s < u$ the operator H_u^s restricted to Φ_r is the same as H_u^r. Therefore are restrictions of one single operator H_u defined on $\bigcup_{s<u} \Phi_s$. By continuity, we extend it to Φ_u.

Next, consider again $\eta_u(f_r)$ for $r < u$ and fix $v > r$. Let g be a vector orthogonal to Φ_v. Then $< g, \eta_u f_r > \; = 0$ for a.e. $u \in]r, v[$. Using locality again we have that $< g, H_u^r f_r > \; = 0$ everywhere in the same interval, meaning that $H_u f_r \in \dot{\Phi}_v$ for $r < u < v$. This is extended to $H_u f$ with $f \in \Phi_u$, and then taking an intersection over v we find that H_u applies Φ_u into itself. Then it is trivial to turn it into an adapted operator $H_u^\bullet$ on Φ.

Because of the inequality $\| H_u^\bullet \| \leq m'(u)$, it is clear that the operator $\int_0^t H_u^\bullet \, du$ are bounded — and then, the process $M_t - \int_0^t H_u^\bullet \, du$ is a *martingale of bounded operators*, which satisfies the same regularity conditions (with a different function m). We change notations to denote this martingale by (M_t), and *we are reduced to the case considered by Parthasarathy-Sinha*, discussed in the next subsection.

4 *The case of regular martingales.* We now assume (M_t) is an operator martingale. Given $f \in \Phi_r$, we define an adapted process $h_u(f)$ on $]r, \infty[$ such that, for $r < s < t$

$$M_t f - M_s f = \int_s^t h_u(f) \, dX_u \; .$$

From property (1.5) we know that the (a.e. defined) function $h_\cdot(f)/m'$ belongs to $L^\infty(]r, \infty[, \Phi)$, with norm $\leq \| f \|$. We extend it by 0 on $]0, r[$ and apply the lifting γ. The value of the lifting on $]t, \infty[$ only depends on the restriction of $h_u(f)$ to $]t, \infty[$; therefore, if f happens to belong to some $\Phi_{r'}$ with $r' < r$, its lifting does not change. Multiplying by m', we get a bounded linear operator H_{rt} from Φ_r to Φ_t (zero if $m'(t)$ does not exist), depending measurably on t and such that

$$\text{for all } r < t, \; f \in \Phi_r, \quad H_{rt} f = h_t(f) \text{ a.s. } \; .$$

Besides that, if $f \in \cup_{r<t} \Phi_r$ $H_{rt} f$ depends only on f not on r, and therefore the operator extends by continuity to $\Phi_{t-} = \Phi_t$, as an operator H_t. For $f \subset \Phi_r$ $H_t f$

is strongly Borel, and replacing it by $E_t H_t f$ if necessary we add adaptation to the preceding properties.

To extract H^-, we may either pass to the adjoint, or proceed directly in a similar way, from the definition of H^- as a drift.

5 The number operator coefficient. We recall the notation

$$(5.1) \qquad f_t = c + \int_0^t \dot{f}_s \, dX_s$$

which defines for each t an isometry between the spaces Φ_t and $\mathbb{C} \oplus L_a^2([0,t] \times \Omega)$ (L_a^2 indicates that only adapted processes are considered), and we define for every t an operator N_t on Φ_t as follows

$$(5.2) \qquad N_t f_t = M_t f_t - \int_0^t M_u \dot{f}_u \, dX_u - \int_0^t H_u^+ f_u \, dX_u - \int_0^t H_u^- \dot{f}_u \, du \ .$$

The preceding section implies the following statements :

1) For every t, N_t is a bounded operator on Φ_t.

2) For $s < t$ we have $(N_t - N_s) f_s = 0$.

3) $N_\bullet$ transforms martingales into martingales, otherwise stated $E_s N_t f_t = N_s f_s = N_t f_s$, or finally $E_s N_t = N_t E_s$ for $s \leq t$.

Then we define a constant c' and a process $(\dot{f}'_s)$ by

$$N_t(c + \int_0^t \dot{f}_s \, dX_s) = c' + \int_0^t \dot{f}'_s \, dX_s \ ,$$

(they do not depend on t). Applying E_0 we see that $c' = cN\mathbf{1}$, $N\mathbf{1}$ being a constant. Consider the mapping (between processes) $\dot{f} \longmapsto \dot{f}'$; for every T it is bounded from $L_a^2([0,T], \Phi_T)$ — the space of all square integrable adapted processes on $[0,T]$ — into itself, and commutes with multiplication by $I_{[0,s]}$, $(s < T)$. Then it also commutes with multiplication by Borel bounded functions on $[0,T]$. Composing it with the previsible projection operator, we extend it to a mapping from $L^2([0,T], \Phi)$ into itself, which commutes with multiplication by deterministic Borel bounded functions.

We now use the following classical result, which says that L^∞ is maximal abelian, or that classical probability defines a complete observable in quantum probability : *a bounded operator on $L^2([0,T])$ which commutes with multiplication by bounded functions, it itself a multiplication operator by an essentially bounded function, whose L^∞-norm is the same as its operator norm.* It is not difficult, using the lifting theorem, to extend this result to operators on $L^2([0,T], \mathcal{H})$: *every bounded operator on this space, of norm k, which commutes with multiplication by bounded scalar functions $a(t)$, is of the form $(h_t) \longmapsto (K_t h_t)$, where (K_t) is a strongly measurable family of operators on $\mathcal{H}$ with $\| K_t \| \leq k$ for a.e. t.*

In the case of the mapping $\dot{f} \longmapsto \dot{f}'$ we have $\mathcal{H} = \Phi_T$, and we may replace K_t by $\mathbb{E}_t(K_t)$ without change, i.e. suppose the family is adapted. Also, the local property of the lifting implies that K_t does not depend on the choice of the interval $[0, T]$ (containing t). Then returning to (5.2) we have the formula

$$(5.3) \qquad M_t f_t = \int_0^t (M_s + K_s) \dot{f}_s \, dX_s + \int_0^t H_s^+ f_s \, dX_s + \int_0^t H_s^- \, \dot{f}_s \, ds \ .$$

Then it is clear that K_t is a good version of the number operator coefficient. When f is an exponential vector we get the theorem of Parthasarathy–Sinha.

Let us again emphasize the following result, since it is only implicit in P–S : *If a martingale of bounded operators is representable at all, the number operator coefficients form a family of bounded operators, whose norms not only are locally in L^2, but even locally in L^∞.* This turns out to be important for applications.

REMARK. Let (M_t) be a real valued bounded martingale (M_t) on Wiener space (say)

$$M_t = c + \int_0^t \mu_s \, dX_s \ ,$$

considered as a family of multiplication operators. When is it regular ? Is every bounded martingale regular ?

The regularity condition is $\mathbb{E}\left[(M_t - M_s)^2 f_s^2\right] \leq \|f_s\|^2 (m(t) - m(s))$, or

$$(5.4) \qquad \mathbb{E}\left[(M_t - M_s)^2 \,|\, \mathcal{F}_s\right] = \mathbb{E}\left[\int_s^t \mu_r^2 \, dr \,|\, \mathcal{F}_s\right] \leq m(t) - m(s) \quad a.s. \ .$$

Since m is absolutely continuous, this amounts to saying that μ_u belongs to L^∞ with a locally square integrable norm. It is clear there should exist bounded martingales which do not satisfy this condition, but explicit counterexamples are not easily found. M. Yor indicated us on Wiener space the example of the continuous martingale

$$M_t = L_t X_t = \int_0^t L_s \, dX_s \ ,$$

X denoting Brownian motion and L its local time at 0 ; this martingale is unbounded, so we stop it at the first time T at which $|M_t|$ exceeds C. Then the r.v.'s L_t are not bounded on $\{t \leq T\}$. Indeed, such a property would mean there exists some t and some constants M, c such that

$$(\,|L_s X_s| \leq M \text{ for } s \leq t\,) \Longrightarrow (L_t \leq c) \quad .$$

Choosing $a > c$ and b small enough so that $ab < M$, we would have

$$(L_t \leq a \text{ and } X_t^* \leq b) \Longrightarrow (L_t \leq c) \quad ,$$

and the law of the pair (L_t, X_t^*) would not charge $\,]\,c, a] \times [0, b]$. On the other hand, it can be shown to have a strictly positive density everywhere on $\mathbb{R}_+^2$, and we get a contradiction.

We do not discuss here the very interesting result that the martingale of conditional expectations of a Hilbert-Schmidt operator is always representable, and has in fact a representation with a number operator coefficient of a very special kind. An exposition of the theorem of Parthasarathy and Sinha concerning these "Hilbert-Schmidt martingales" can be found in *Sém. Prob. XXVII*.

The extension of the above result to Fock space with arbitrary multiplicity, or involving a non-trivial initial space, is easy provided one uses the compact notation (1.3') in section 1 for the stochastic integral representation. Details are left to the reader (a complete proof is given in Attal [Att4]).

§3. REPRESENTABLE INTERPRETATIONS OF FOCK SPACE

1 A *probabilistic interpretation* of simple Fock space is a commutative martingale X_t on some probability space, which is *normal* ($X_t^2 - t$ is a martingale), and which has the chaotic representation property : the space L^2 of the σ–field generated by this martingale is isomorphic to Fock space. This definition is not precise enough, in the sense that no relation is given between the standard structure of Fock space (creation and annihilation operators, etc) and the probabilistic structure. We assume for simplicity that $X_0 = 0$

Since X has the chaotic representation property, it has the predictable representation property, and satisfies a *structure equation*

$$(1.1) \qquad d\,[\,X,X\,]_t = dt + \psi_t\,dX_t\;.$$

which simply means that the martingale $[\,X,X\,]_t - t$ has the predictable representation property.

We also denote by X_t and ψ_t the multiplication operators by the corresponding random variables in this probabilistic interpretation. It is reasonable to expect that

$$(1.2) \qquad dX_t = dQ_t + \psi_t\,da_t^\circ\;,$$

where $Q_t = a_t^+ + a_t^-$ as usual. A necessary condition for this equation to be meaningful is that, for an exponential vector $\mathcal{E}(u)$ we should have for t finite

$$(1.3) \qquad \int_0^t \|\,\psi_s\mathcal{E}(u)\,\|^2\,ds < \infty\;.$$

If this property holds, the probabilistic interpretation will be said to be *representable*.

We now prove, following Attal, that (1.2) is true. We call (J_t) the operator process defined by the right hand side. According to the basic formula (1.6) in §1, which holds at least on the exponential domain

$$J_t f_t = \int_0^t J_s \dot{f}_s\,dX_s + \int_0^t (f_s\,dX_s + \dot{f}_s\,ds + \psi_s \dot{f}_s\,dX_s)$$

while using the relation $f_t = c + \int_0^t \dot{f}_s\,dX_s$, the classical Ito (integration by parts) formula and the structure equation, we get

$$X_t f_t = \int_0^t X_s \dot{f}_s\,dX_s + \int_0^t f_s\,dX_s + \int_0^t \dot{f}_s(ds + \psi_s\,dX_s)\;.$$

Taking a difference, we see that the ordinary process $Y_t = J_t f_t - X_t f_t$ is a semimartingale satisfying the ordinary stochastic differential equation $Y_t = \int_0^t Y_s \, dX_s$, whose only solution is 0. In particular, we get that $X_t \mathcal{E}(u)$ belongs to L^2.

An interesting variant of this proof concerns a normal martingale X which has the predictable representation property, but not necessarily the chaotic representation property, so that Fock space may be strictly smaller than the corresponding L^2 space. Let us assume, not only (1.2), but that multiplication by ψ_t maps the exponential domain into Fock space — this is not very natural: some other domain like an incomplete chaos space may be simpler to use here. Then the stochastic integral process (1.2) is well defined, and coincides on the exponential domain with the operator of multiplication by X_t. For instance, these hypotheses seem to be satisfied by the Azéma martingales for all values of the parameter, since $\psi_t = cX_t$ is known to belong to all L^p spaces; it is not known whether this is the case for all exponentials, but the domain of polynomials in the coordinate mappings is likely to be stable under the mappings $f \longmapsto \dot{f}_s$ (this is called "admissible" in Attal's papers).

Attal proves a slightly different result: we assume nothing like (1.3), and consider instead a bounded martingale (f_t) with a predictable representation $c + \int_0^t \dot{f}_s \, dX_s$, such that $\| \dot{f}_t \|_\infty$ and $\| \dot{f}_t \psi_t \|_\infty$ are bounded. There are plenty of such martingales: we start from an arbitrary bounded martingale $g_t = c + \int_0^t \dot{g}_s \, dX_s$, then put

$$ h_t = c + \int_0^t \dot{g}_s \, I_{\{|\dot{g}_s| \le n\}} \, I_{\{|\dot{g}_s \psi_s| \le n\}} \, dX_s $$

to get a martingale which is arbitrarily close to the preceding one for n large, and has a "derivative" $\dot{h}_t$ satisfying the required conditions; then (h_t) itself may not be bounded, but its jumps are smaller than those of (g_t), hence bounded. Then we may construct (f_t) by stopping. Attal then proves that the process of (bounded) multiplication operators by (f_t) belongs to the class $\mathcal{S}$, since it has the everywhere defined stochastic integral representation

$$ f_t = cI + \int_0^t \dot{f}_t \, dQ_t + \dot{f}_t \psi_t \, da_t^\circ . $$

The proof is essentially the same as for X_t above.

2 Let us consider a probabilistic interpretation as above. Denote by $*$ the corresponding multiplication. In all such interpretations, the *curve* $X_t * \mathbf{1} = a_t^+ \mathbf{1}$ is the same. Also in all interpretations, given a vector y, the martingale $y_t = E_t y$ is the same as a curve, and its predictable representation as a vector stochastic integral $c + \int_0^t \eta_s \, dx_s$ depends only on the Fock space structure, not on the product $*$.

On the other hand, there are several interesting operators which depend on the probabilistic interpretation, as explained by Attal. For any vector f, consider the associated martingale $f_t = c + \int_0^t \dot{f}_s \, dX_s$.

1) Given an adapted process (h_t), define

(2.1) $$ J_t f_t = \int_0^t h_s \, df_s = \int_0^t h_s \dot{f}_s \, dX_s . $$

This depends on the interpretation, since the product is involved. It follows from formula (1.6) in §1 that we have

$$(2.2) \qquad J_t = \int_0^t (-J_s + h_s) \, da_s^{\circ} ,$$

where h_s is meant as a multiplication operator in the given interpretation. This formula is rigorous if $\| h_t \|_\infty$ is locally bounded (then J_t is bounded with a locally bounded norm).

2) Let M_t be a martingale with representation $c + \int_0^t m_s \, dx_s$ (as a process of vectors). We consider the operators

$$(2.3) \qquad J_t f_t = \int_0^t f_s \, dM_s = \int_0^t f_s m_s \, dX_s .$$

According to §1, (1.6), we have

$$(2.4) \qquad J_t = -\int_0^t J_s \, da_s^{\circ} + \int_0^t m_s \, da_s^{+} ,$$

again m_s is a multiplication operator.

3) Consider now the angle bracket

$$(2.5) \qquad J_t f_t = \int_0^t \langle df_s, dm_s \rangle = \int_0^t m_s \dot{f}_s \, ds .$$

This time we have

$$(2.6) \qquad J_t = -\int_0^t J_s \, da_s^{\circ} + \int_0^t m_s \, da_s^{-} ,$$

4) Finally, (p_t) being a measurable adapted process with $\| p_\infty \|$ locally bounded, the operator

$$(2.7) \qquad J_t = \int_0^t f_s p_s \, ds$$

has the representation

$$(2.8) \qquad J_t = -\int_0^t J_s \, da_s^{\circ} + \int_0^t p_s \, ds ,$$

In all four cases, there is a contribution of the number operator which prevents the representation from being an explicit one : the left side is involved in its own representation.

References

The references *QP.I–QP.VIII*, concern the volumes on quantum probability and applications, edited by L. Accardi and W. von Waldenfels from vol. II on, whose numbers in the Springer Lecture Notes series are respectively 1055 (1986), 1136 (1987), 1303 (1988), 1396 (1989), 1442 (1990); Vol. VI (1992) Vol. VII (1992) and Vol. VIII (1993) are published by World Scientific (Singapore). The references *Sém. Prob.* concern the *Séminaire de Probabilités*, edited by J. Azéma, P.A. Meyer and M. Yor. The abbreviations LNM, LNP, LNCI refer to the Springer Lecture Notes series in Mathematics, Physics, Control and Information. The order of articles of a given author is arbitrary.

[Acc] ACCARDI (L.). 1. On the quantum Feynman–Kac formula. *Rendiconti Sem. Mat. e Fis. di Milano*, 48, 1978. 2. On square roots of measures, C^*-algebras and their applications, Proc. International School "Enrico Fermi", Varenna 1973, North-Holland. 3. A note on Meyer's note, *QP.III*, p. 1–5. 4. On the quantum Feynman–Kac formula, *Rend. Sem. Mat. Fis. Univ. Polit. Milano*, 48, 1978, p. 135–179.

[AcB] ACCARDI (L.) and BACH (A.). 1. The harmonic oscillator as quantum central limit of Bernoulli processes, 1985, *unpublished preprint*. 2. Central limits of squeezing operators, *QP.IV*, p. 7–19. 3. Quantum central limit theorems for strongly mixing random variables, *Z. für W-theorie Verw. Geb.*, 68, 1985, p. 393–402.

[AcC] ACCARDI (L.) and CECCHINI (C.). 1. Conditional expectations in von Neumann algebras and a theorem of Takesaki, *J. Funct. Anal.*, 45, 1982, p. 245–273. 2. Surjectivity of the conditional expectations on the L^1 spaces, *Harmonic Analysis, Cortona 1982*, LNM 982, 1983, p. 434–442.

[AcF] ACCARDI (L.) and FAGNOLA (F.). Stochastic integration, *QP.III*, p. 6–19.

[AFQ] ACCARDI (L.), FAGNOLA (F.) and QUAEGEBEUR (J.). A representation–free quantum stochastic calculus, *J. Funct. Anal.*, 104, 1992, p. 149–197.

[AFL] ACCARDI (L.), FRIGERIO (A.) and LEWIS (J.). Quantum stochastic processes, *Publ. RIMS Kyoto*, 18, 1982, p. 94–133.

[AJL] ACCARDI (L.), JOURNÉ (J.L.) and LINDSAY (J.M.). On multidimensional markovian cocycles, *QP.IV*, p. 59–66.

[AcM] ACCARDI (L.) and MOHARI (A.). On the structure of classical and quantum flows. *J. Funct. Anal.*, to appear.

[AcP] ACCARDI (L.) and PARTHASARATHY (K.R.). A martingale characterization of the canonical commutation and anticommutation relations. *J. Funct. Anal.*, 77, 1988, p. 211–231.

[AcQ] ACCARDI (L.) and QUAEGEBEUR (J.). A Fermion Lévy theorem. *J. Funct. Anal.*, 110, 1992, p. 131–160.

[AlL] ALICKI (R.) and LENDI (K.). *Quantum Dynamical Semigroups and Applications*, LNP 286, 1987.

[App] APPLEBAUM (D.). 1. Quantum stochastic parallel transport on non-commutative vector bundles, *QP.III*, p. 20–36. 2. Unitary evolutions and horizontal lifts in quantum stochastic calculus, *Comm. Math. Phys.*, 140, 1991, p. 63–80. 3. Towards a quantum theory of classical diffusions on Riemannian manifolds, *QP.VI*, p. 93–111.

[ApH] APPLEBAUM (D.) and HUDSON (R.L.). 1. Fermion Ito's formula and stochastic evolutions, *Comm. Math. Phys.*, 96, 1984, p. 473–496. 2. Fermion diffusions, *J. Math. Phys.*, 25, 1984, p. 858–861.

[Ara] ARAKI (H.). Bogoliubov automorphisms and Fock representations of canonical anticommutation relations, *Operator Algebras and Math. Physics; A.M.S. Contemp. Math.*, 62, 1987, p. 23–141.

[ACGT] ARECCHI (F.T.), COURTENS (E.), GILMORE (R.) and THOMAS (H.). Atomic coherent states in quantum optics, *Phys. Review A*, 6, 1972, p. 2211–2237.

[Att] ATTAL (S.). 1. Problèmes d'unicité dans les représentations d'opérateurs sur l'espace de Fock, *Sém. Prob. XXVI*, LN **1526**, 1992, p. 617–632. 2. Characterizations of some operators on Fock space, *QP. VIII*, 1993, p. 37–47. 3. Characterization of operators commuting with some conditional expectations on multiple Fock spaces, *QP. VIII*, p. 47–70. 4. An algebra of non commutative bounded semimartingales : square and angle quantum brackets, *J. Funct. Anal.*, **124**, 1994, p. 292–331.

[AtM] ATTAL (S.) and MEYER (P.A.). Interprétation probabiliste et extension des intégrales stochastiques non commutatives, *Sém. Prob. XXVII*, LNM **1557**, p. 312–327.

[Azé] AZÉMA (J.). Sur les fermés aléatoires, *Sém. Prob. XIX*, LNM **1123**, p. 397–495.

[Bar] BARGMANN (V.). A Hilbert space of analytic functions and an associated integral transform, *Pure Appl. Math. Sci.*, **14**, 1961, p. 187–214.

[BSW] BARNETT (C.), STREATER (R.F.) and WILDE (I.F.). 1. The Ito–Clifford integral I, *J. Funct. Anal.* 48, 1982, p. 172–212. 2. — II, *J. London Math. Soc.* 27, 1983, p. 373–384. 3. — III, Markov properties of solutions to s.d.e.'s, *Comm. Math. Phys.* 89, 1983, p. 13–17. 4 —IV, a Radon–Nikodym theorem and bracket processes, *J. Oper. Th.* 11, 1984, p. 255–271. 5. Quasi–free quantum stochastic integrals for the CAR and CCR, *J. Funct. Anal.*, **48**, 1982, p. 255–271.

[BaW] BARNETT (C.) and WILDE (I.F.). 1. Quantum stopping times, *QP. VI*, p. 127–136. 2. Natural processes and Doob–Meyer decompositions over a probability gage space, *J. Funct. Anal.*, **58**, 1984, p. 320–334. 3. Quantum Doob–Meyer decompositions, *J. Oper. Th.*, **29**, 1988, p. 133–164.

[Bel] BELAVKIN (V.P.). 1. A quantum non adapted Ito formula and non stationary evolution in Fock scale, *QP. VI*, p. 137–180. 2. A new form and a ⋆–algebraic structure of quantum stochastic integrals in Fock space, *Rend. Sem. Mat. Fis. Milano*, **58**, 1988, p. 177–193. 3. A new wave equation for a continuous nondemolition experiment, and 4. A quantum particle undergoing continuous observation *Phys. Rev. A*, **140**, 1989, p. 355–362. 5. A quantum non adapted Ito formula and stochastic analysis in Fock scale, *J. Funct. Anal.*, **102**, 1991, p. 414–447. 6. Kernel representations of ⋆–semigroups associated with infinitely divisible states, *QP. VII*, 1992, p. 31–50. 7. The quantum functional Ito formula has the pseudo-Poisson structure, *QP. VIII*, 1993, p. 81–86.

[BeL] BELAVKIN (V.P.) and LINDSAY (J.M.). The kernel of a Fock space operator II, *QP. VIII*, 1993, p. 87–94.

[BCR] BERG (C.), CHRISTENSEN (J.P.R.) and RESSEL (P.). *Harmonic Analysis on Semigroups*, Springer, 1984.

[Ber] BEREZIN (F.A.). *The method of second quantization*, Academic Press, New York 1966 (translation).

[Bha] BHAT (B.V. R.). Markov dilations of nonconservative quantum dynamical semigroups and a quantum boundary theory. Dissertation, Indian Statistical Institute, Delhi 1993.

[BhP] BHAT (B.V. R.) and PARTHASARATHY (K.R.). 1. Kolmogorov's existence theorem for Markov processes on C^*–algebras. *Proc. Indian Acad. Sci.*, **104**, 1994, p. 253–262. 2. Markov dilations of nonconservative quantum dynamical semigroups and a quantum boundary theory. To appear.

[Bia] BIANE (Ph.). 1. Marches de Bernoulli quantiques, *Sém. Prob. XXIV*, LNM **1426**, p. 329–344. 2. Chaotic representations for finite Markov chains, *Stochastics*, 30, 1990, p. 61–68. 3. Quantum random walks on the dual of $SU(n)$, *Prob. Th. Rel. Fields*, **89**, 1991, p. 117–129. 4. Some properties of quantum Bernoulli random walks, *QP. VI*, p. 193–204.

[Bra] BRADSHAW (W.S.). Stochastic cocycles as a characterization of quantum flows. *Bull. Sc. Math.*, **116**, 1992, p. 1–34.

[BrR] BRATTELI (O.) and ROBINSON (D.W.). 1.2. *Operator Algebras and Quantum Statistical Mechanics I, II*, Springer 1981.

[Cha] CHAIKEN (J.M.). Number operators for representations of the CCR, *Comm. Math. Phys.* 8, 1968, p. 164–184.

[CaN] CARMONA (R.) and NUALART (D.). Traces on Wiener space and the Onsager–Machlup functional. *J. Funct. Anal.*, 1992.

[Che] CHEBOTAREV (A.M.). 1. Sufficient conditions for conservativity of dynamical semigroups, *Theor. Math. Phys.*, **80**, 1989. 2. The theory of conservative dynamical semigroups and its applications, to appear.

[ChF] CHEBOTAREV (A.M.) and FAGNOLA (F.). Sufficient conditions for conservativity of quantum dynamical semigroups. *J. Funct. Anal.*, **118**, 1993, p. 131–153.

[CFF] CHEBOTAREV (A.M.), FAGNOLA (F.) and FRIGERIO (A.). Towards a stochastic Stone's theorem, *Stochastic Partial Diff. Eq. and Applications III*, Pitman Research Notes in M., vol. 268, 1992, p. 86–97.

[CoH] COCKROFT (A.M.) and HUDSON (R.L.). Quantum mechanical Wiener processes, *J. Multiv. Anal.*, **7**, 1978, p. 107–124.

[CCSS] COHENDET (O.), COMBE (P.), SIRUGUE (M.) and SIRUGUE–COLLIN (M.). 1. A stochastic treatment of the dynamics on an integer spin, *J. Phys. A*, **21**, 1988, p. 2875–2883. See also same J., **23**, 1990, p. 2001–2011. 2. Weyl quantization for $Z^n \times Z^n$ phase space : stochastic aspect. *Stochastic Methods in Math. and Phys., Karpacz 1988*, World Scientific 1989.

[Coo] COOK (J.M.). The mathematics of second quantization, *Trans. Amer. Math. Soc.*, **74**, 1953, p. 222–245.

[CuH] CUSHEN (C.D.) and HUDSON (R.L.). A quantum mechanical central limit theorem, *J. Appl. Prob.*, **8**, 1971, p. 454–469.

[Dav] DAVIES (E.B.). *Quantum Theory of Open Systems*, Academic Press, 1976. 2. On the Borel structure of C^*-algebras, with an Appendix by R.V. Kadison, *Comm. Math. Phys.*, **8**, 1968, p. 147–163.

[D'A] Dell'ANTONIO (G.F.). On the limits of sequences of normal states, *Comm. Pure Appl. Math.*, **20**, 1967, p. 413–429.

[DMM] DELLACHERIE (C.), MAISONNEUVE (B.) and MEYER (P.A.). *Probabilités et Potentiel, Vol. E*, Hermann 1992.

[DeM] DELLACHERIE (C.) and MEYER (P.A.). 1. *Probabilités et Potentiel, Vol. A*, Hermann, Paris, 1975, english translation, North–Holland 1977. 2., Vol. B, 1980, transl. 1982. 3. Vol. C, 1987, transl. 1989.

[Der] DERMOUNE (A.). 1. Formule de composition pour une classe d'opérateurs, *Sém. Prob. XXIV*, LNM **1426**, p 397–401. 2. Opérateurs d'échange et quelques applications des méthodes de noyaux, *Ann. Sci. Univ. Clermont*, **96**, 1991, p. 45–54 (see also p. 55–58).

[Dix] DIXMIER (J.). *Les algèbres d'opérateurs dans l'espace Hilbertien*, Paris, Gauthier–Villars 1969.

[Dyn] DYNKIN (E.B.). 1. Polynomials of the occupation field and related random fields, *J. Funct. Anal.*, **58**, 1984, p. 20–52. 2. Gaussian and non–Gaussian random fields associated with Markov processes, *J. Funct. Anal.*, **55**, 1984, p. 344–376. 3. Functionals associated with self-intersection of the planar Brownian motion, *Sém. Prob. XX*, LNM **1204**, 1986, p. 553–571. 4. Self-intersection local times, occupation fields and stochastic integrals, *Adv. Math.*, **65**, 1987, p. 254–271. 5. Self-intersection gauge for random walks and for Brownian motion, *Ann. Prob.*, **16**, 1988, p. 1–57. 6. Regularized self-intersection local times of planar Brownian motion, *Ann. Prob.*, **16**, 1988, p. 58–74.

[Eme] EMERY (M.). 1. On the Azéma martingales, *Sém. Prob. XXIII*, LNM **1372**, p. 66–87. 2. Sur les martingales d'Azéma (suite), *Sém. Prob. XXIV*, LNM **1426**, p. 442–447. 3. Quelques cas de représentation chaotique, *Sém. Prob. XXV*, LN **1485**, p. 10–23.

[Enc] ENCHEV (O.). Hilbert space valued quasimartingales, *Boll. Unione Mat. Ital.*, **2**, 1988, p. 19–39.

[Eva] EVANS (M.). Existence of quantum diffusions, *Prob. Th. Rel. Fields*, **81**, 1989, p. 473–483.

[EvH] EVANS (M.) and HUDSON (R.L.). 1. Multidimensional quantum diffusions, *QP.III*, p. 69–88. 2. Perturbation of quantum diffusions, *J. London Math. Soc.*, **41**, 1992, p. 373–384.

[EvL] EVANS (D.E.) and LEWIS (J.T.). *Dilations of Irreversible Evolutions in Algebraic Quantum Theory*, Publ. Dublin Institute for Advanced Study, Series A, 1977.

[Fag] FAGNOLA (F.). 1. On quantum stochastic differential equations with unbounded coefficients, *Prob. Theory Rel. Fields*, **86**, 1990, p. 501–516. 2. Pure birth and death processes as quantum flows in Fock space, *Sankhya, ser. A*, **53**, 1991, p. 288–297. 3. Unitarity of solutions to quantum stochastic differential equations and conservativity of the associated quantum dynamical semigroup, *QP. VII*, 1992, p. 139–148. 4. Chebotarev's sufficient conditions for conservativity of quantum dynamical semigroups, *QP. VIII*, 1993, p. 123–142. 5. Diffusion processes in Fock space, preprint. 6. Characterization of isometric and unitary weakly differentiable cocycles in Fock space, *QP. VIII*, 1993, p. 143–164.

[FaS] FAGNOLA (F.) and SINHA (K.B.). Quantum flows with unbounded structure maps and finite degrees of freedom, *J. London M. Soc.*, **48**, 1993, p. 537–551.

[Fan] FANO (U.). Description of states in quantum mechanics by density matrix and operator techniques, *Rev. Mod. Phys.* 29, 1957, p. 74–93.

[Fey] FEYEL (D.). Sur la méthode de Picard (édo et éds). *Sém. Prob. XXI*, LNM **1247**, 1987, p. 515–519.

[Foc] FOCK (V.A.). Zur Quantenelektrodynamik, *Soviet Phys.*, **6**, 1934, p. 425.

[Fol] FOLLAND (G.B.). *Harmonic Analysis in Phase Space*, Ann. Math. Studies, Princeton 1989.

[Fug] FUGLEDE (B.). The multidimensional moment problem, *Exp. Math.*, **1**, 1983, p. 47–65.

[GaW] GARDING (L.) and WIGHTMAN (A.S.). Representations of the commutation relations, *Proc. Natl. Acad. Sci. USA*, **40**, 1954, p. 617–621.

[GvW] GIRI (N.) and Von WALDENFELS (W.). An algebraic version of the central limit theorem, *Z. W-theorie Verw. Geb.*, **42**, 1978, p. 129–134.

[Glo] GLOCKNER (P.). Quantum stochastic differential equations on ∗-bigebras, *Math. Proc. Cambridge Phil. Soc.*, **109**, 1991, p. 571–595.

[GKS] GORINI (V.), KOSSAKOWSKI (A.) and SUDARSHAN (E.). Completely positive dynamical semigroups of n-level systems, *J. Math. Phys.*, **17**, 1976, p. 821–825.

[Gro] GROSS (L.). 1. Existence and uniqueness of physical ground states, *J. Funct. Anal.*, **10**, 1972, p. 52–109.

[Gui] GUICHARDET (A.). *Symmetric Hilbert spaces and related topics*, LNM **261**, 1970.

[Hal] HALMOS (R.). Two subspaces, *Trans. Amer. Math. Soc.*, **144**, 1969, p. 381–389.

[HeW] HE (S.W) and WANG (J.G.). Two results on jump processes, *Sém. Prob. XVIII*, LNM **1059**, 1984, p. 256–267.

[Hel] HELMSTETTER (J.). 1. Monoïdes de Clifford et déformations d'algèbres de Clifford, *J. of Algebra*, **111**, 1987, p. 14–48. 2. Algèbres de Clifford et algèbres de Weyl, *Cahiers Math. Univ. de Montpellier*, **25**, 1982. 3. Algèbres de Weyl et ⋆-produit, *Cahiers Math. Univ. de Montpellier*, **34**, 1985.

[Hid] HIDA (T.). *Brownian motion*, Springer 1980.

[Hol] HOLEVO (A.S.). 1. *Probabilistic and Statistical Aspects of Quantum Theory*, North-Holland, 1982. 2. Time ordered exponentials in quantum stochastic calculus, *QP. VII*, 1992, p. 175–202. 3. Conditionally positive definite functions in quantum probability, *Proc. Intern. Congress Math. Berkeley 1986*, p. 1011–1018. 4. Stochastic representation of quantum dynamical semigroups, *Trudy Mat. Inst. A.N. SSSR*, **191**, 1989, p. 130–139 (Russian).

[HuM] HU (Y.Z.) and MEYER (P.A.). 1. Sur les intégrales multiples de Stratonovitch, *Sém. Prob. XXII*, LNM **1321**, 1987, p. 72-81. 2. On the approximation of Stratonovich multiple integrals, to appear in the volume dedicated to G. Kallianpur, Springer 1992.

[HIP] HUDSON (R.L.), ION (P.D.F.) and PARTHASARATHY (K.R.). Time orthogonal unitary dilations and non-commutative Feynman-Kac formulas, *Comm. Math. Phys.*, **83**, 1982, p. 261–280.

[HKP] HUDSON (R.L.), KARANDIKAR (R.L.) and PARTHASARATHY (K.R.). Towards a theory of non commutative semimartingales adapted to brownian motion and a quantum Ito's formula, *Theory and Applications of Random Fields, Bangalore 1982*, LNCI **49**, 1983, p. 96–110.

[HuL] HUDSON (R.L.) and LINDSAY (J.M.). 1. A noncommutative martingale representation theorem for non-Fock quantum Brownian motion, *J. Funct. Anal.*, **61**, 1985, p. 202-221. 2. Uses of non-Fock quantum Brownian motion and a quantum martingale representation theorem, *QP.II*, p. 276-305.

[HLP] HUDSON (R.L.), LINDSAY (J.M.) and PARTHASARATHY (K.R.). Stochastic integral representation of some quantum martingales in Fock space, *From local times to global geometry, control and physics*, p. 121-131, Pitman Research Notes, 1986.

[HuP] HUDSON (R.L.) and PARTHASARATHY (K.R.). 1. Quantum Ito's formula and stochastic evolutions, *Comm. Math. Phys.*, **93**, 1984, p. 301-303. 2. Stochastic dilations of uniformly continuous completely positive semigroups, *Acta Appl. Math.*, **2**, 1988, p. 353-398. 3. Unification of fermion and boson stochastic calculus, *Comm. Math. Phys.*, **104**, 1986, p. 457-470. 4. Casimir chaos in a Boson Fock space, *J. Funct. Anal.*, **119**, 1994, p. 319-339.

[HuR] HUDSON (R.L.) and ROBINSON (P.). Quantum diffusions and the noncommutative torus, *Lett. Math. Phys.*, **15**, 1988, p. 47-53.

[HuS] HUDSON (R.L.) and STREATER (R.F.). 1. Ito's formula is the chain rule with Wick ordering, *Phys. Letters*, **86**, 1982, p. 277-279. 2. Examples of quantum martingales, *Phys. Letters*, **85**, 1981, p. 64-67.

[Ito] ITO (K.). Multiple Wiener integral. *J. Math. Soc. Japan*, **3**, 1951, p. 157-169.

[JoK] JOHNSON (G.W.) and KALLIANPUR (G.). 1. Some remarks on Hu and Meyer's paper and infinite dimensional calculus on finitely additive canonical Hilbert space, *Teorija Verojat.*, **34**, 1989, p. 742-752. 2. Homogeneous chaos, p-forms, scaling and the Feynman integral, Technical Report n° 274, Center of Stoch. Processes, Univ. of North Carolina, 1989. To appear in *Trans. Amer. Math. Soc.*.

[Jou] JOURNÉ (J.L.). Structure des cocycles markoviens sur l'espace de Fock. *Prob. Theory Rel. Fields*, **75**, 1987, p. 291-316.

[JoM] JOURNÉ (J.L.) and MEYER (P.A.). Une martingale d'opérateurs bornés, non représentable en intégrale stochastique, *Sém. Prob. XX*, LNM **1204**, p. 313-316.

[KaM] KALLIANPUR (G.) and MANDREKAR (V.). Multiplicity and representation theory of purely non-deterministic stochastic processes, *Teor. Ver.* 10, 1965, p. 553-580 (transl.). 2. Semigroups of isometries and the representation and multiplicity of weakly stationary processes *Ark. för Mat.* 6, 1965, p. 319-335.

[Kas] KASTLER (D.). *Introduction à l'électrodynamique quantique*, Dunod, Paris, 1961.

[KlS] KLAUDER (J.R.) and SKAGERSTAM (B.S.) *Coherent States*, World Scientific, 1985.

[Kré] KRÉE (P.). La théorie des distributions en dimension quelconque et l'intégration stochastique, *Stochastic Analysis and Related Topics*, Proc. Silivri 1986, LNM 1316, p. 170-233.

[KrR] KRÉE (P.) and RACZKA (R.). Kernels and symbols in quantum field theory, *Ann. Inst. Henri Poincaré ser. A*, **28**, 1978, p. 41-73.

[Kum] KÜMMERER (B.). 1. Markov dilations on W^*-algebras, *J. Funct. Anal.*, **63**, 1985, p. 139-177. 2. Survey on a theory of noncommutative stationary Markov processes, *QP.III*, p. 154-182.

[Kup] KUPSCH (J.). 1. Functional integration for euclidean Dirac fields, *Ann. Inst. Henri Poincaré, (Phys.)*, **50**, 1989, p. 143-160. 2. Functional integration for euclidean Fermi fields, *Gauge theory and fundamental interactions*, Proc. S. Banach Center 1988, World Scientific, Singapore 1990.

[LaM] de Sam LAZARO (J.) and MEYER (P.A.). Questions de Théorie des Flots V, VI *Sém. Prob. IX*, LNM **465**, p. 52-80.

[LeJ] Le JAN (Y.). 1. Temps local et superchamp, *Sém. Prob. XXI*, LNM **1247**, 1987, p. 176-190. 2. On the Fock space representation of functionals of the occupation field and their renormalization, *J. Funct. Anal.*, **80**, 1988, p. 88-108. 3. On the Fock space representation of occupation times for non reversible Markov processes, *Stochastic Analysis*, LNM **1322**, 1988, p. 134-138.

[Ldb] LINDBLAD (G.). On the generators of quantum dynamical semigroups. *Comm. Math. Phys.*, **48**, 1976, p. 119-130.

[Lin] LINDSAY (J.M.). 1. Independence for quantum stochastic integrators, *QP. VI*, p. 325–332. 2. Quantum and non–causal stochastic calculus, *Prob. Th. Rel. Fields*, to appear.. 3. On set convolution and integral–sum kernel operators, *Prob. Theory and Math. Stat.*, **2**, 1990, p.105–123. 4. The kernel of a Fock space operator I, *QP. VIII*, 1993, p. 271–280.

[LiM] LINDSAY (J.M.) and MAASSEN (H.). 1. An integral kernel approach to noise, *QP. III*, 1988, p. 192–208. 2. Duality transform as *-algebraic isomorphism, *QP. V*, p. 247–250. Stochastic calculus for quantum brownian motion of non–minimal variance.

[LiP] LINDSAY (J.M.) and PARTHASARATHY (K.R.). 1. Cohomology of power sets with applications in quantum probability, *Comm. Math. Phys.*, **124**, 1989, p. 337–364. 2. The passage from random walk to diffusion in quantum probability II, *Sankhya, ser. A*, **50**, 1988, p. 151–170. 3. Rigidity of the Poisson convolution, *QP. V*, 1989, p. 251–262.

[LiW] LINDSAY (J.M.) and WILDE (I.F.). On non–Fock boson stochastic integrals, *J. Funct. Anal.*, **65**, 1986, p. 76–82.

[Lou] LOUISELL (W.H.). *Radiation and Noise in Quantum Electronics*, McGraw Hill, 1964.

[Maj] MAJOR (P.). *Multiple Wiener-Itô integrals*, LNM **849**, 1981.

[Maa] MAASSEN (H.). 1. Quantum Markov processes on Fock space described by integral kernels, *QP. II*, 1985, p. 361–374. 2. Addition of freely independent random variables, *J. Funct. Anal.*, **106**, 1992, p. 409–438.

[Mal] MALLIAVIN (P.). Stochastic calculus of variations and hypoelliptic operators. *Proc. Intern. Conf. on Stochastic Differential Equations*, Kyoto 1976, p. 195-263, Kinokuniya, Wiley 1978.

[MaR] MARCUS (M.B.) and ROSEN (J.). Sample path properties of the local times of strongly symmetric Markov processes via Gaussian processes. To appear in *Ann. Prob.*, 1992.

[MMC] MASSON (D.) and McCLARY (W.K.). Classes of C^∞ vectors and essential self–adjointness, *J. Funct. Anal.*, **10**, 1972, p. 19–32.

[Mey] MEYER (P.A.). 1. Éléments de probabilités quantiques, *Sém. Prob. XX*, LNM **1204**, p. 186–312. 2. *Sém. Prob. XXI*, LNM **1247**, p. 34–80. 3. A note on shifts and cocycles, *QP. III*, p. 209–212. 4. Calculs antisymétriques et supersymétriques en probabilités, *Sém. Prob. XXII*, LNM **1321**, p. 101–123 (correction in *Sém. Prob. XXV*).

[Moh] MOHARI (A.). 1. Quantum stochastic calculus with infinite degrees of freedom and its applications. Dissertation, ISI Delhi 1992. 2. Quantum stochastic differential equations with unbounded coefficients and dilations of Feller's minimal solution. *Sankhya, ser. A*, **53**, 1991, p. 255–287.

[MoP] MOHARI (A.) and PARTHASARATHY (K.R.). 1. On a class of generalized Evans–Hudson flows related to classical Markov processes. *QP. VII*, 1992, p. 221–250. 2. A quantum probabilistic analogue of Feller's condition for the existence of unitary Markovian cocycles in Fock space, preprint.

[MoS] MOHARI (A.) and SINHA (K.B.). 1. Quantum stochastic flows with infinite degrees of freedom and countable state Markov processes. *Sankhya, ser. A*, **52**, 1990, p. 43–57.

[Moy] MOYAL (J.E.). Quantum mechanics as a statistical theory, *Proc. Cambridge Phil. Soc.* 45, 1949, p. 99–124.

[Nel] NELSON (E.). 1. Analytic vectors, *Ann. Math.*, **70**, 1959, p. 572–614. 2. Notes on non commutative integration, *J. Funct. Anal.*, **15**, 1974, p. 103–116.

[Nie] NIELSEN (T.T.). *Bose Algebras : The Complex and Real Wave Representations*, LNM **1472**, 1991.

[Nus] NUSSBAUM (A.). 1. Quasi analytic vectors, *Ark. Mat.*, **6**, 1965, p.179-191. 2. A note on quasi-analytic vectors, *Studia Math.*, **33**, 1969, p. 305-310.

[Par] PARTHASARATHY (K.R.). 1. *An Introduction to Quantum Stochastic Calculus*, Birkhäuser, Basel 1992. 2. What is a Bernoulli trial in Quantum Probability? (unpublished, see *Sém. Prob. XXIII*, LNM **1372**, p. 183–185). 3. Azéma martingales and quantum stochastic calculus, *Proceedings of the R.C. Bose Symp. on Prob. and Stat.*, Wiley Eastern, New Delhi 1990, p. 551–569. 4. Discrete time quantum stochastic flows, Markov chains and chaos expansions, *Proc. 5-th Intern. Vilnius Conf. on Prob. and Stat.*, to appear. 5. The passage from random walk to diffusion in quantum probability, *Celebration Volume in Applied Probability*, 1988, p. 231–245. Applied Prob. Trust, Sheffield. 6. Comparison of completely positive maps on a C^*-algebra and a Lebesgue decomposition theorem, to appear. 7. Maassen kernels and self-similar quantum fields, to appear.

[PSch] PARTHASARATHY (K.R.) and SCHMIDT (K.). *Positive definite kernels, continuous tensor products, and central limit theorems of probability theory*, LNM **272**, 1972.

[PaS] PARTHASARATHY (K.R.) and SINHA (K.B.). 1. Stochastic integral representations of bounded quantum martingales in Fock space. *J. Funct. Anal.*, **67**, 1986, p. 126–151, see also *QP.III*, p. 231–250. 2. Boson–Fermion transformations in several dimensions, *Pramana J. Phys.*, **27**, 1986, p. 105–116. 3. Markov chains as Evans–Hudson diffusions in Fock space. *Sém. Prob. XXIV*, LNM **1426**, 1990, p. 362–369. 4. Representation of a class of quantum martingales, *QP.III*, p. 232–250. 5. Stop times in Fock space stochastic calculus, *Prob. Th. Rel. Fields*, **75**, 1987, p. 431–458. 6. Unification of quantum noise processes in Fock space, *QP.VI*, p. 371–384.

[Per] PERELOMOV (A.). *Generalized Coherent States and their Applications*, Springer 1986.

[Ply] PLYMEN (R.J.). C^*-algebras and Mackey's axioms, *Comm. Math. Phys.*, **8**, 1968, p. 132–146.

[Ped] PEDERSEN (G.K.). C^*-*algebras and their automorphism groups*, Academic Press, 1979.

[Poo] POOL (J.C.T.). Mathematical aspects of the Weyl correspondence, *J. Math. Phys.* **7**, 1966, p. 66–76.

[Pro] PROTTER (P.). *Stochastic Integration and Differential Equations, a New Approach*, Springer 1990.

[ReS] REED (M.) and SIMON (B.). *Methods of Modern Mathematical Physics II, Fourier Analysis, Self-adjointness*, Academic Press, 1970.

[RvD] RIEFFEL (M.A.) and VAN DAELE (A.). A bounded operator approach to Tomita–Takesaki theory, *Pacific J. Math.*, **69**, 1977, p. 187–221. 2. The commutation theorem for tensor products of von Neumann algebras, *Bull. London Math. Soc.*, **7**, 1975, p. 257–260.

[Sch] SCHÜRMANN (M.). 1. Positive and conditionally positive linear functionals on coalgebras. *Quantum Probability II*, LNM **1136**, 1985, p. 475–492. 2. White noise on involutive bialgebras, *QP.VI*, p. 401–420. 3. Non commutative stochastic processes with independent and stationary increments satisfy quantum stochastic differential equations. *Prob. Th. Rel. Fields*, **84**, 1990, p. 473–490. 4. A class of representations of involutive bialgebras, *Math. Proc. Cambridge Phil. Soc.*, **107**, 1990, p. 149–175. 5. The Azéma martingales as components of quantum independent increment processes, *Sem. Prob. XXV*, LNM **1485**, p. 24–30. 6. A central limit theorem for coalgebras, *Probability Measures on Groups VIII*, LNM **1210**, 1986, p. 153–157. 7. Infinitely divisible states on cocommutative bialgebras, *Probability Measures on Groups IX*, LNM **1379**, 1987, p. 310–324. 8. Noncommutative stochastic processes with independent and stationary additive increments, *J. Multiv. Anal.*, **38**, 1990, p. 15–35. 9. Gaussian states on bialgebras, *Quantum Probability V*, LNM **1442**, p. 347–367. 10. Quantum q–white noise and a q–central limit theorem. *Comm. Math. Phys.*, **140**, 1991, p. 589–615. 11. *White Noise on Bialgebras*, LNM, to appear.

[SvW] SCHÜRMANN (M.) and von WALDENFELS (W.). A central limit theorem on the free Lie group, *QP.III*, p. 300–318.

[Seg] SEGAL (I.). 1. Tensor algebras over Hilbert spaces I, *Trans. Amer. Math. Soc.*, 81, 1956, p. 106–134. 2. The complex wave representation of the free Bose field, *Topics on Functional Analysis, Adv. in Math.* Suppl. n° 3, 1978. 3. Mathematical characterization of the physical vacuum, *Ill. J. Math.*, **6**, 1962, p. 500–523. 4. A non commutative extension of abstract integration, *Ann. Math.* **57**, 1953, p. 401–457.

[Sim] SIMON (B.). *The $P(\varphi)_2$ Euclidean Quantum Field Theory*, Princeton Univ. Press, 1974.

[Slo] SLOWIKOWSKI (W.). 1. Ultracoherence in Bose algebras, *Adv. Appl. M.*, **8**, 1988, p. 377–427. 2. Commutative Wick algebras I, *Vector Measures and Applications*, LNM **644**, 1978.

[Spe] SPEICHER (R.). 1. A new example of "independence" and "white noise", *Prob. Th. Rel. Fields*, **84**, 1990, p. 141–159. 2. Survey of stochastic integration on the full Fock space, *QP. VI*, p. 421–437.

[Str] STROOCK (D.W.). 1. The Malliavin Calculus and its applications to second order partial differential equations, *Math. Systems Th.*, **14**, 1981, p. 25-65 et 141-171. 2. The Malliavin Calculus : a functional analytic approach, *J. Funct. Anal.*, **44**, 1981, p. 212-257.

[Sur] SURGAILIS (D.). 1. On Poisson multiple stochastic integrals and associated equilibrium Markov semi-groups, *Theory and Application of Random Fields*, LNCI **49**, 1983, p. 233-248. 2. Same title, *Prob. and Math. Stat.* **3**, 1984, p. 227-231.

[ViS] VINCENT-SMITH (G.F.). 1. Unitary quantum stochastic evolutions, *Proc. London Math. Soc.*, **63**, 1991, p. 1-25 2. Quantum stochastic evolutions with unbounded generators, *QP. VI*, p. 465–472.

[Voi] VOICULESCU (D.). 1. Addition of certain non–commuting random variables, *J. Funct. Anal.*, **66**, 1986, p. 323–346. 2. Multiplication of certain non commuting random variables, *J. Operator Th.*, **18**, 1987, p. 223–235.

[vN] VON NEUMANN (J.). 1. *Mathematische Grundlagen der Quantenmechanik*, Springer, 1932. 2. On infinite direct products, *Compos. Math.*, **6**, 1938, p. 1–77.

[vW] VON WÂLDENFELS (W.). 1. An algebraic central limit theorem in the anticommuting case, *Z. W-theorie Verw. Geb.*, **42**, 1978, p. 135–140. 2. The Markov process of total spins, *Sém. Prob. XXIV*, LNM **1426**, p 357–361. 3. Illustration of the central limit theorem by independent addition of spins, *Sém. Prob. XXIV*, LNM **1426**, p. 349–356. 4. Ito solution of the linear quantum stochastic differential equation describing light emission and absorption, *QP.I*, p. 384–411. 5. Positive and conditionally positive sesquilinear forms on anticocommutative coalgebras, *Probability Measures on Groups VII*, LNM **1064**, 1984.

[Wie] WIENER (N.). 1 The homogeneous chaos, *Amer. J. Math.*, **55**, 1938, p. 897–936.

[Wor] WORONOWICZ (S.L.). The quantum problem of moments I, II, *Reports Math. Phys.*, **1**, 1970, p. 135–145 and p. 175–183.

Index

Adapted operators, processes, II.2.4, VI.1.1, curve, IV.2.6. Additive functional, A5.1.1. Affiliated, operator - to a vNa, A4.3.3. Analytic vectors, I.1.6. Angular momentum commutation relations, II.3.1. Annihilation operator, canonical pair —, III.2.1, Fock space —, IV.1.4. Annihilation operator, elementary —, II.1.3, discrete, II.2.2, spin —, II.3.2, fermion discrete —, II.5.1. Antiparticle, V.1.3. Antipode, VII.1.1. Antisymmetric product, norm, IV.1.1. Approximate unit, A4.2.5. Automorphisms of the Clifford structure, II.5.9. Azéma martingale, IV.2.2, VI.3.14, VII.2.3.

Belavkin's formulas for stochastic integrals, VI.3.3. Belavkin's notation, V.3.2. Bialgebra VII.1.1. Bicommutant, A4.3.2. Bimodule, VI.3.7. Bloch coherent vectors, II.3.3. Bosons, IV.1.1. Bracket, square, angle, A6.1.2.

Cameron–Martin functions, IV.2.3. Canonical, commutation relation, III.1.1, — pair, III.1.1, — transformation, III.1.3. CAR, II.5.1, IV.1.5. CCR algebra, IV.1.9 remark. CCR, canonical commutation relation, III.1.1, IV.1.5. Chaos spaces, IV.1.1. Chaotic representation property, IV.2.2. Characteristic function, III.1.3. Clebsch–Gordan coefficients, II.3.4. Clifford algebra, II.5.5. Clifford product (continuous), IV.3.8. C-M, abbr. for *Cameron–Martin*, IV.2.3. Coalgebra, VII.1.1. Cochain, VI.3.7. Co–commutative, VII.1.1. Cocycle, (multiplicative functional) VI.1.12, (cohomology) VI.3.7. Coherent vector (normalized exponential vector), IV.1.3, V.1.4. Coherent vector, elementary —, II.1.4, discrete —, II.2.5, spin —, II.3.3, harmonic oscillator —, III.2.4. Cohomology, VI.3.7. Combinatorial lemma, IV.3.3. Completely positive mapping, A2.5. Complex gaussian model, III.2.5. Composition formula for kernels, IV.4.3 (3.1), (3.2), V.2.3, V.3.3. Conditionally positive, VII.1.6. Conditioned semigroup or process, A5.1.1. Continuous Markov chain, VI.3.11. Continuous spin field, V.2.1, example. Continuous sum of Hilbert spaces, A2, 4. Contraction, A5.2.2, — of order p, IV.3.1. Contractivity condition, VI.4.4. Convolution, VII.1.3. Coproduct, VII.1.1. Counit, VII.1.1. Creation operator, canonical pair, III.2.1, Fock space, IV.1.4. Creation operator, elementary —, II.1.3, discrete —, II.2.2, spin —, II.3.2, fermion discrete —, II.5.1. Cuntz algebra, IV.1.8. Cyclic vector, V.1.4, A4.2.2.

Density operator, I.1.4 Diagonals, IV.3.1. Differential second quantization, IV.1.6. Dirac's notation, I.1.9. Discrete flows, VI.3.6. Discrete kernels, IV.4.4. Discrete Markov chain, VI.3.8-10. Discrete multiple integrals, VI.3.6. Dominated, operator kernel — by a scalar kernel, V.2.5. Dynamical semigroup, A2.5. Dynkin's formula, A5.1.2.

Energy, A5.1.1. Enlarged exponential domain $\Gamma_1(\mathcal{H})$, IV.1.4 remark. Evans Notation, V.2.1, — delta, V.2.1. Evans–Hudson flow, VI.2.5, VI.4.11. Exchange operators, V.2.1. Exclusion principle, IV.1.1. Exponential domain $\mathcal{E}$, IV.1.3. Exponential s.d.e., right and left, VI.3.1, VI.4.1. Exponential vectors $\mathcal{E}(h)$, IV.1.3, V.1.4 (double Fock space). VI.1.8. Exponential vectors, discrete —, II.2.5, spin —, II.3.5.

Faithful state, A4.2.3. Fermion brownian motion, IV.3.9, kernels, IV.4.6. Fermion creation/annihilation operators II.5.1, IV.1.4. Feynman integral, IV.3.3.Field operators, IV.1.5. Flow, VI.2.5. Fock space, IV.1.2. Fourier–Wiener transform, IV.2.3. Full Fock space, IV.1.2 (remark), IV.1.8 (remark).

Gauge process, IV.2.4. Gaussian states of the CCR, III.1.4. General position (subspaces), A3.1. GNS construction and theorem, A4.2.2-3. Grassmann product, IV.3.8. Guichardet's notation, IV.2.2. Gupta's negative energy oscillator, IV.3.3.

Half–density, VI.3.13. Hamiltonian, III.1.1. Harmonic oscillator, III.2.2. Hellinger–Hahn theorem, A2.4. Hermite polynomials, III.2.3. Hermitian element in an algebra, I.1.1. Hilbert–Schmidt operators, A1.1. Hochschild cohomology, VI.3.7. Hopf algebra, VII.1.1. HS, abbr. for *Hilbert–Schmidt*, A1.1.

Incomplete n–th chaos, IV.1.1, Fock space, IV.1.2. Increasing simplex, IV.1.2. Initial space $\mathcal{J}$, V.2.5. Interpretation of Fock space, A6.3.1. Intrinsic Hilbert space of a manifold, VI.3.13. Isobe-Sato formula, VI.3.2. Isometry, coisometry and unitarity conditions, VI.2.3. Isotropic subspace, A5.2.1. Ito formula, VI.1. (9.3) Ito table, IV.4.3, V.1.5, V.2.1, V.3.1.

Jordan–Wigner transformation, II.5.2, IV.4.2 (2.8). Journé's time reversal principle, VI.4.9-10.

Kaplansky's theorem, A4.3.4. Kernels, IV.4.1, IV.1.5, V.2.2, V.3.3. KMS condition, A4.3.10.

Left cocycle, VI.1.12. Left exponential, VI.3.1. Lévy process, V.2.6. Lifting theorem, A6.2.2. Lindblad generator, A2.8. Local times, A5.1.3. Lusin space, IV.2.2.

Index of Notation

Vol. 1526: J. Azéma, P. A. Meyer, M. Yor (Eds.), Séminaire de Probabilités XXVI. X, 633 pages. 1992.

Vol. 1527: M. I. Freidlin, J.-F. Le Gall, Ecole d'Eté de Probabilités de Saint-Flour XX – 1990. Editor: P. L. Hennequin. VIII, 244 pages. 1992.

Vol. 1528: G. Isac, Complementarity Problems. VI, 297 pages. 1992.

Vol. 1529: J. van Neerven, The Adjoint of a Semigroup of Linear Operators. X, 195 pages. 1992.

Vol. 1530: J. G. Heywood, K. Masuda, R. Rautmann, S. A. Solonnikov (Eds.), The Navier-Stokes Equations II – Theory and Numerical Methods. IX, 322 pages. 1992.

Vol. 1531: M. Stoer, Design of Survivable Networks. IV, 206 pages. 1992.

Vol. 1532: J. F. Colombeau, Multiplication of Distributions. X, 184 pages. 1992.

Vol. 1533: P. Jipsen, H. Rose, Varieties of Lattices. X, 162 pages. 1992.

Vol. 1534: C. Greither, Cyclic Galois Extensions of Commutative Rings. X, 145 pages. 1992.

Vol. 1535: A. B. Evans, Orthomorphism Graphs of Groups. VIII, 114 pages. 1992.

Vol. 1536: M. K. Kwong, A. Zettl, Norm Inequalities for Derivatives and Differences. VII, 150 pages. 1992.

Vol. 1537: P. Fitzpatrick, M. Martelli, J. Mawhin, R. Nussbaum, Topological Methods for Ordinary Differential Equations. Montecatini Terme, 1991. Editors: M. Furi, P. Zecca. VII, 218 pages. 1993.

Vol. 1538: P.-A. Meyer, Quantum Probability for Probabilists. X, 287 pages. 1993.

Vol. 1539: M. Coornaert, A. Papadopoulos, Symbolic Dynamics and Hyperbolic Groups. VIII, 138 pages. 1993.

Vol. 1540: H. Komatsu (Ed.), Functional Analysis and Related Topics, 1991. Proceedings. XXI, 413 pages. 1993.

Vol. 1541: D. A. Dawson, B. Maisonneuve, J. Spencer, Ecole d' Eté de Probabilités de Saint-Flour XXI - 1991. Editor: P. L. Hennequin. VIII, 356 pages. 1993.

Vol. 1542: J.Fröhlich, Th.Kerler, Quantum Groups, Quantum Categories and Quantum Field Theory. VII, 431 pages. 1993.

Vol. 1543: A. L. Dontchev, T. Zolezzi, Well-Posed Optimization Problems. XII, 421 pages. 1993.

Vol. 1544: M.Schürmann, White Noise on Bialgebras. VII, 146 pages. 1993.

Vol. 1545: J. Morgan, K. O'Grady, Differential Topology of Complex Surfaces. VIII, 224 pages. 1993.

Vol. 1546: V. V. Kalashnikov, V. M. Zolotarev (Eds.), Stability Problems for Stochastic Models. Proceedings, 1991. VIII, 229 pages. 1993.

Vol. 1547: P. Harmand, D. Werner, W. Werner, M-ideals in Banach Spaces and Banach Algebras. VIII, 387 pages. 1993.

Vol. 1548: T. Urabe, Dynkin Graphs and Quadrilateral Singularities. VI, 233 pages. 1993.

Vol. 1549: G. Vainikko, Multidimensional Weakly Singular Integral Equations. XI, 159 pages. 1993.

Vol. 1550: A. A. Gonchar, E. B. Saff (Eds.), Methods of Approximation Theory in Complex Analysis and Mathematical Physics IV, 222 pages. 1993.

Vol. 1551: L. Arkeryd, P. L. Lions, P.A. Markowich, S.R. S. Varadhan. Nonequilibrium Problems in Many-Particle Systems. Montecatini, 1992. Editors: C. Cercignani, M. Pulvirenti. VII, 158 pages 1993.

Vol. 1552: J. Hilgert, K.-H. Neeb, Lie Semigroups and their Applications. XII, 315 pages. 1993.

Vol. 1553: J.-L- Colliot-Thélène, J. Kato, P. Vojta. Arithmetic Algebraic Geometry. Trento, 1991. Editor: E. Ballico. VII, 223 pages. 1993.

Vol. 1554: A. K. Lenstra, H. W. Lenstra, Jr. (Eds.), The Development of the Number Field Sieve. VIII, 131 pages. 1993.

Vol. 1555: O. Liess, Conical Refraction and Higher Microlocalization. X, 389 pages. 1993.

Vol. 1556: S. B. Kuksin, Nearly Integrable Infinite-Dimensional Hamiltonian Systems. XXVII, 101 pages. 1993.

Vol. 1557: J. Azéma, P. A. Meyer, M. Yor (Eds.), Séminaire de Probabilités XXVII. VI, 327 pages. 1993.

Vol. 1558: T. J. Bridges, J. E. Furter, Singularity Theory and Equivariant Symplectic Maps. VI, 226 pages. 1993.

Vol. 1559: V. G. Sprindžuk, Classical Diophantine Equations. XII, 228 pages. 1993.

Vol. 1560: T. Bartsch, Topological Methods for Variational Problems with Symmetries. X, 152 pages. 1993.

Vol. 1561: I. S. Molchanov, Limit Theorems for Unions of Random Closed Sets. X, 157 pages. 1993.

Vol. 1562: G. Harder, Eisensteinkohomologie und die Konstruktion gemischter Motive. XX, 184 pages. 1993.

Vol. 1563: E. Fabes, M. Fukushima, L. Gross, C. Kenig, M. Röckner, D. W. Stroock, Dirichlet Forms. Varenna, 1992. Editors: G. Dell'Antonio, U. Mosco. VII, 245 pages. 1993.

Vol. 1564: J. Jorgenson, S. Lang, Basic Analysis of Regularized Series and Products. IX, 122 pages. 1993.

Vol. 1565: L. Boutet de Monvel, C. De Concini, C. Procesi, P. Schapira, M. Vergne. D-modules, Representation Theory, and Quantum Groups. Venezia, 1992. Editors: G. Zampieri, A. D'Agnolo. VII, 217 pages. 1993.

Vol. 1566: B. Edixhoven, J.-H. Evertse (Eds.), Diophantine Approximation and Abelian Varieties. XIII, 127 pages. 1993.

Vol. 1567: R. L. Dobrushin, S. Kusuoka, Statistical Mechanics and Fractals. VII, 98 pages. 1993.

Vol. 1568: F. Weisz, Martingale Hardy Spaces and their Application in Fourier Analysis. VIII, 217 pages. 1994.

Vol. 1569: V. Totik, Weighted Approximation with Varying Weight. VI, 117 pages. 1994.

Vol. 1570: R. deLaubenfels, Existence Families, Functional Calculi and Evolution Equations. XV, 234 pages. 1994.

Vol. 1571: S. Yu. Pilyugin, The Space of Dynamical Systems with the C^0-Topology. X, 188 pages. 1994.

Vol. 1572: L. Göttsche, Hilbert Schemes of Zero-Dimensional Subschemes of Smooth Varieties. IX, 196 pages. 1994.

Vol. 1573: V. P. Havin, N. K. Nikolski (Eds.), Linear and Complex Analysis – Problem Book 3 – Part I. XXII, 489 pages. 1994.

General Remarks

Lecture Notes are printed by photo-offset from the master-copy delivered in camera-ready form by the authors. For this purpose Springer-Verlag provides technical instructions for the preparation of manuscripts.

Careful preparation of manuscripts will help keep production time short and ensure a satisfactory appearance of the finished book. The actual production of a Lecture Notes volume normally takes approximately 8 weeks.

Authors receive 50 free copies of their book. No royalty is paid on Lecture Notes volumes.

Authors are entitled to purchase further copies of their book and other Springer mathematics books for their personal use, at a discount of 33,3 % directly from Springer-Verlag.

Commitment to publish is made by letter of intent rather than by signing a formal contract. Springer-Verlag secures the copyright for each volume.

Addresses:

Professor A. Dold
Mathematisches Institut
Universität Heidelberg
Im Neuenheimer Feld 288
D-69120 Heidelberg
Federal Republic of Germany

Professor F. Takens
Mathematisch Instituut
Rijksuniversiteit Groningen
Postbus 800
NL-9700 AV Groningen
The Netherlands

Springer-Verlag, Mathematics Editorial
Tiergartenstr. 17
D-69121 Heidelberg
Federal Republic of Germany
Tel.: *49 (6221) 487-410